Amorphous and Liquid Materials

NATO ASI Series

Advanced Science Institutes Series

A Series presenting the results of activities sponsored by the NATO Science Committee, which aims at the dissemination of advanced scientific and technological knowledge, with a view to strengthening links between scientific communities.

The Series is published by an international board of publishers in conjunction with the NATO Scientific Affairs Division

A	**Life Sciences**	Plenum Publishing Corporation
B	**Physics**	London and New York
C	**Mathematical and Physical Sciences**	D. Reidel Publishing Company Dordrecht, Boston, Lancaster and Tokyo
D	**Behavioural and Social Sciences**	Martinus Nijhoff Publishers Boston, Dordrecht and Lancaster
E	**Applied Sciences**	
F	**Computer and Systems Sciences**	Springer-Verlag Berlin, Heidelberg, New York
G	**Ecological Sciences**	London, Paris, Tokyo
H	**Cell Biology**	

Series E: Applied Sciences – No. 118

Amorphous and Liquid Materials

edited by

E. Lüscher

Physik-Department
Technische Universität München
D-8046 Garching
FRG

G. Fritsch

Physikalisches Institut FB BAUV/II
Universität der Bundeswehr München
D-8014 Neubiberg
FRG

G. Jacucci

Dipartimento di Matematica e Fisica
Universita degli Studi di Trento
I-38050 Povo
Italy

1987 **Martinus Nijhoff Publishers**
Dordrecht / Boston / Lancaster
Published in cooperation with NATO Scientific Affairs Division

Proceedings of the NATO Advanced Study Institute on "Amorphous and Liquid Materials", Passo della Mendola (Trentino), Italy, August 26-September 7, 1985

ISBN 90-247-3411-8 (this volume)
ISBN 90-247-2689-1 (series)

Distributors for the United States and Canada: Kluwer Academic Publishers, P.O. Box 358, Accord-Station, Hingham, MA 02018-0358, USA

Distributors for the UK and Ireland: Kluwer Academic Publishers, MTP Press Ltd, Falcon House, Queen Square, Lancaster LA1 1RN, UK

Distributors for all other countries: Kluwer Academic Publishers Group, Distribution Center, P.O. Box 322, 3300 AH Dordrecht, The Netherlands

Printed in The Netherlands

PREFACE

Six years passed by since the NATO ASI on "Liquid and Amorphous Metals" was held in Zwiesel, Germany, in September 1979. The present one is the second NATO School devoted to research on disordered condensed matter, mainly liquid and amorphous metals. This time the title contains the word "materials" to explicitely include those aspects of the glassy state of insulators either shared with metallic glasses - e.g. the glass transition - or on the border line with metallic systems - e.g. the metal-non-metal transition.

The long period which purposely elapsed between the two Institutes indicates the intention not to have "just another conference", but to review the state of affairs in the field with a somewhat more durable scope. This is especially important to help basic research to bridge towards applications and to introduce young researchers in this field. In fact, while the understanding of these materials and their properties is a tremendous challenge for experimental and theoretical physicists, glassy substances offer an enormous potential in the development of new materials for technical applications.

To this end, the Institute has brought together insiders and peers from all over the world to discuss basic principles and latest results and to help correlate future research effort. Another important aim was to introduce newcomers to the field.

The Institute consisted of two different parts which were nevertheless strongly related. The first one (first week) was devoted to macroscopic behaviour and microscopic structure of liquids whereas the second considered the same properties of amorphous materials. The connection between the two topics was shown in various talks dealing with both states of matter on the same footing. In particular the reports on amorphous systems in biophysics and localization in fractal systems revealed the common nature of these states of condensed matter.

Thirty-four posters and four contributed lectures supplemented the plenary talks by giving additional experimental and theoretical results. The following is an attempt to describe the essential achievements in the fields presented at the institute.

The static structure of simple liquids, averaged over the angles, can now be calculated starting from two different model systems: the one-component plasma- and the hard sphere model. It was shown that a generalization to alloys is possible, explaining trends in liquid alloy properties. Another way to the understanding of these trends uses certain assumptions on "building blocks" for the short range order which, however, are not easy to prove experimentally. The dynamics of many particle systems are available from neutron scattering. The self-part and the total dynamic structure factor can be determined from the data and analyzed in detail to yield information on single particle and collective motions. The understanding of their behaviour has improved enormously by considering the coupling of this motion to other degrees of freedom. The change of metallic systems from the liquid to the gaseous state has to proceed via a metal-non metal transition. Experiments on liquid Hg, Cs and Rb in the neighbourhood of the critical point have shown, that there may be an additional transition line in the pressure-temperature diagram describing just such a transition.

Amorphous systems can be produced not only by quenching from the melt, but also by sputtering, ion-beam-mixing, laser quenching as well as by solid state reactions. Moreover, it is now possible to obtain bulk metallic amorphous alloys with a volume of several cm^3. In addition new structures have been observed which turned out to be in between amorphous and crystalline. They possess five-fold symmetry but no long range order. Theoretical evidence was presented at the ASI, that the glass transition is not simply a freezing of the subcooled liquid but a new kind of phase transition. Several contributions discussed electronic transport properties of metallic amorphous systems. Trends in superconductivity can now be explained from simple assumptions on "band" structure. Corrections to the weak scattering treatment of the electrical resistivity stemming from multiple scattering and quantum corrections seem to be necessary. They have been studied experimentally in 3-dimensional alloys. The pressure dependence of the resistivity indicates that most of the data may be understood within a modified Ziman model. The electronic density of states are now known for a multitude of systems allowing to estimate the important role of d-electrons in transition metals alloys. Results from positron annihilation (lifetime measurements) exhibit profound differences between the structures of metal-metalloid and metal-metal amorphous alloys. A stability-analysis of those systems as derived from phase diagrams including metastable compounds as well as from the "building block" point of view has been presented in several talks. Finally, it was shown in detail, how computer simulations permit to understand "theoretically" the dynamics of liquid and amorphous alloys. Especially, "optical modes" in liquid metallic alloys are an exciting new possibility.

The book contains texts corresponding to the invited and contributed lectures, and a good many written versions of posters. It features a chapter on rather general topics exhibiting connections with the main field - fractals, biophysics, glass transition, computer simulation - followed by a number of chapters on atomic structure and dynamics of liquid and amorphous systems, on electronic transport properties, on the preparation of glassy substances of interest. The material is mainly in the form of lucid monographic presentations of most relevant fore-front research. During the next few years this book will be an essential tool to researchers both in industry and research institutions needing introduction, information, or an update in the field or its subsections.

Writing the contributions to this book involved much work for which the authors are to be thanked and complimented. We also wish to thank all participants and contributors to the Institute for their invaluable efforts towards its success. Special thanks go to Mrs. M. Gerl as the good spirit and the heart of the meeting, and to the Director of the "Centro de Cultura dell' Università Cattolica di Milano" at La Mendola, Mrs. M. Zanini who hosted the Institute. Its location in the middle of a mountain resort area proved to be both convenient and delightful.

Last but not least we are deeply indebted to the NATO Advanced Study Institute Program for its financial support.

Garching/München in December 1985

E. Lüscher
München

G. Fritsch
München

G. Jacucci
Trento

TABLE OF CONTENTS

1. GENERAL TOPICS

PROTEINS AND GLASSES

H. FRAUENFELDER

Department of Physics, University of Illinois at Urbana-Champaign,
1110 West Green Street, Urbana, IL 61801 USA

The study of biomolecules links biology, biochemistry, chemistry, and physics. For physicists, biomolecules offer challenging and exciting problems. On the one hand, physical investigations of biomolecules promise a deeper understanding of biological medical, and pharmacological questions. On the other hand, biomolecules are complex many-body systems that pose unique problems for physicists and can be considered a "state of matter" different from solids or liquids. In the present lectures I will look at biomolecules, particularly proteins, from the second point of view and will try to indicate where biomolecules can serve as testing grounds for physical theories.

After a brief introduction to biomolecules, I will describe one particular type, hemeproteins, in more detail. Investigations of one simple protein function, binding of small molecules such as dioxygen, leads to the realization that the internal motions of proteins are important for their function and that proteins possess many properties that are glass-like.

1. BIOMOLECULES

1.1 Nucleic Acids and Proteins.

Two types of biomolecules are responsible for most biological activity, nucleic acids and proteins. Nucleic acids store and transmit the biological information, proteins are the machines that perform the actual biological functions (1). Vastly simplified, the connection between nucleic acids and proteins is shown in Fig. 1. Information about proteins is stored in the form of "three-letter words" on a linear deoxiribonucleic acid (DNA) molecule. The DNA is wound around globular proteins called histones. The information stored on the DNA is read and transcribed onto a ribonucleic acid (RNA) molecule. Transported to the protein factory (ribosome) a particular RNA specifies a corresponding amino acid, a building block of the protein to be constructed. A given sequence of words on the DNA translates into a unique sequence of amino acids of the protein. In the ribosome, the amino acids are covalently connected into a linear chain. When the primary polypeptide chain emerges from the ribosome, it folds into the functionally active three-dimensional protein (Fig. 2). The final three-dimensional structure is determined by the order of the amino acids in the primary sequence.

1.2 The Building Blocks.

Nucleic acids, the information carriers, are built from four different building blocks, the nucleotides, which form the four letters of the genetic alphabet. A nucleotide consists of a base, a sugar, and one or more phosphate groups, as indicated schematically in Fig. 3a. Sugar and

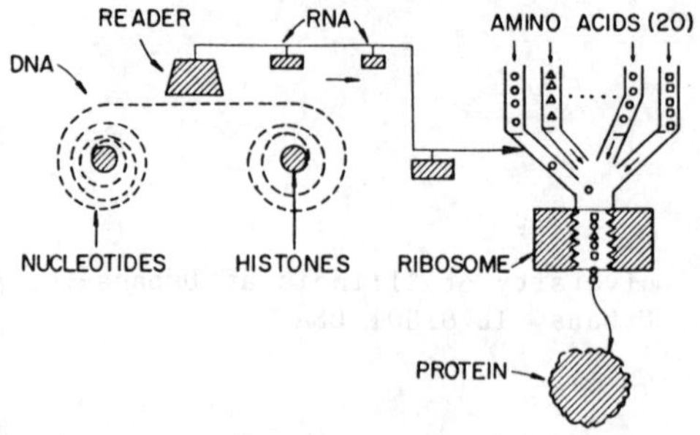

FIGURE 1. The connection between nucleic acids and proteins. The information for the construction of proteins is stored on DNA. The information is read and transcribed onto RNA molecules. On the ribosome, the RNA molecules call for particular amino acids to be used in the proper place in the assembly of the protein.

phosphate group are the same for all nucleotides and the four letters are distinguished by four different bases. In DNA, which stores the information, the bases are adenine (A), cytosine (C), guanine (G), and thymine (T). In RNA, which transfers information, thymine is replaced by uracil (U). Here we are not interested in the molecular structure of the four bases, but simply represent them by the four letters A, C, G, and U.

Proteins are constructed from 20 building blocks, the amino acids. Each amino acid consists of a backbone and a side chain. The backbone is the same in all amino acids, but the side chains are different. One particular amino acid is shown in Fig. 3b. The twenty different amino acids are listed in Table I, together with the standard abbreviation and some of their properties.

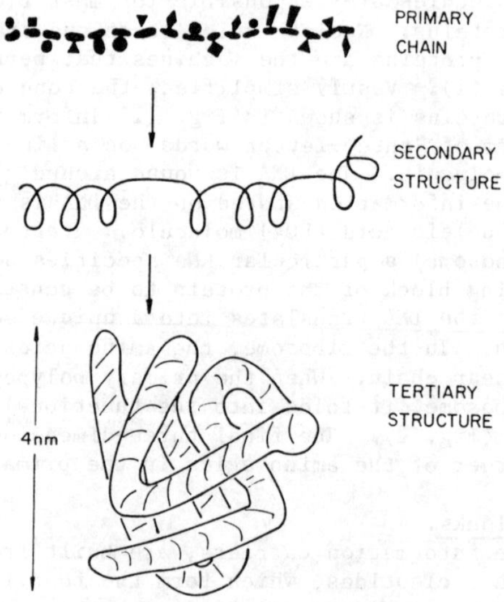

FIGURE 2. The linear polypeptide chain (primary sequence) folds into the final tertiary structure.

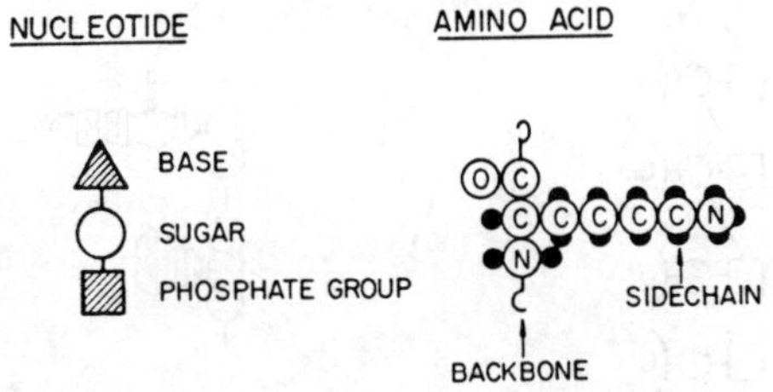

FIGURE 3. The building blocks of nucleic acids and proteins. The nucleic acids have four different bases, the amino acids 20 different side chains.

Table I

Properties of protein building blocks. The length (L) is for the side chain only. The molecular weight is for the entire amino acid – subtract 17.9 to obtain molecular weight of residue. The polarity indicates whether the amino acid is nonpolar (NP) or polar with a net positive, negative, or neutral charge at pH = 6.

Amino Acid	Symbol	Molecular weight (amu)	L (nm)	Polarity	Side chain (X = benzene)
Alanine	ALA	89	0.28	NP	—C
Arginine	ARG	174	0.88	+	—C—C—C—N—C=N N
Asparagine	ASN	132	0.51	0	—C—C=O N
Aspartic Acid	ASP	133	0.50	—	—C—C=O O
Cysteine	CYS	121	0.43	0	—C—S
Glutamine	GLN	146	0.64	0	—C—C—C=O N
Glutamic Acid	GLU	147	0.63	—	—C—C—C=O O
Glycine	GLY	75	0.15	0	—H
Histidine	HIS	155	0.65	+	—C—C=C N C N
Isoleucine	ILE	131	0.53	NP	—C—C—C C
Leucine	LEU	131	0.53	NP	—C—C—C C
Lysine	LYS	146	0.77	+	—C—C—C—C—N
Methionine	MET	149	0.69	NP	—C—C—S—C
Phenylalanine	PHE	165	0.69	NP	—C—X
Proline	PRO	115		NP	C—C C N—C
Serine	SER	105	0.38	0	—C—O
Threonine	THR	119	0.40	0	—C—C O
Tryptophan	TRP	204	0.81	NP	—C—C—C (ring) N
Tyrosine	TYR	181	0.77	0	—C—X—O
Valine	VAL	117	0.40	NP	—C—C C

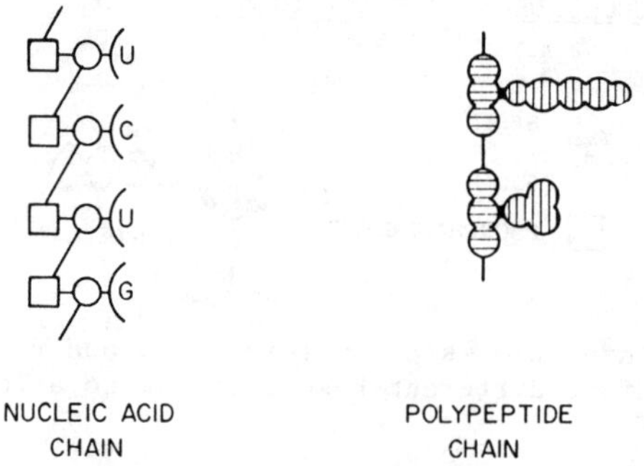

FIGURE 4. Nucleic acids and proteins are both linear chains; the individual building blocks are connected by covalent bonds.

1.3 Language and Translation.

Nucleic acids and proteins are essentially linear systems, as shown in Fig. 4. The building blocks are linked by covalent bonds. The linear arrangement makes the transmission of information and the specification of proteins possible. The "blueprints" for the construction are stored in DNA and transmitted by RNA in the form of three-letter words. Each word specifies a particular amino acid. The translation, called the genetic code, is given in Table II. A particular sequence, for instance ..UCU-UAC-ACG... determines a unique primary sequence of amino acids, in this case ..Ser-His-Ala... . The genetic code is degenerate; most amino acids can be "called" by more than one word.

1.4 Solids, Glasses, and Proteins

Figure 2 gives a somewhat misleading impression of a protein. X-ray diffraction shows that the folded protein is nearly close-packed and looks like an aperiodic crystal. More detailed comparison of structure and function indicates that proteins indeed have some solid-like

Table II

Genetic code

	Second position				
First	U	C	A	G	Third
	PHE	SER	TYR	CYS	U
	PHE	SER	TYR	CYS	C
U	LEU	SER	Stop	Stop	C
	LEU	SER	Stop	TRP	G
	LEU	PRO	HIS	ARG	U
	LEU	PRO	HIS	ARG	C
C	LEU	PRO	GLN	ARG	A
	LEU	PRO	GLN	ARG	G
	ILE	THR	ASN	SER	U
	ILE	THR	ASN	SER	C
A	ILE	THR	LYS	ARG	A
	MET	THR	LYS	ARG	G
	VAL	ALA	ASP	GLY	U
	VAL	ALA	ASP	GLY	C
G	VAL	ALA	GLU	GLY	A
	VAL	ALA	GLU	GLY	G

properties, but that profound differences exist. As will be seen later, in many respects proteins are similar to glasses. A comparison among solids, glasses, and proteins indicates the following features:
(i) Solids possess a periodic, glasses and proteins a nonperiodic spatial structure. While the disorder in glasses is random, it has been carefully selected by evolution in proteins. The "disorder" in proteins is closer to the "disorder" in a chip or a Picasso painting both of which are not periodic. Nonperiodicity consequently described the situation in a protein better than disorder.
(ii) In solids and glasses, the "strong" forces that hold the atoms together are essentially equally strong in all three directions. In proteins, however, the bonds are "strong" (covalent) along the backbone, but the cross connections are "weak" (hydrogen bonds, disulfide bridges, Van der Waals forces). A solid is "dead" and an individual atom can, as a rule, only vibrate around its equilibrium position. In contrast, the weak bonds in a biomolecule can be broken by thermal fluctuations. A biomolecule can therefore execute large motions, it can breathe and can act as a miniature machine.
(iii) Solids are spatially homogeneous, apart from surface effects and from defects. In glasses, inhomogeneities are random and minor. Biomolecules, in contrast, are spatially inhomogeneous; properties such as density, charge, and dipole moment change from region to region within a protein.
(iv) Solids or glasses cannot be modified on an atomic or molecular scale at a particular point; modifications are either periodic or random. In contrast, a protein can be changed at any desired place a the molecular level: Through genetic engineering, the primary sequence is modified at the desired location and this modification leads to the corresponding change in the protein. We have stated in Section 1.3 that the partial sequence .. UCU-UAC-ACG... corresponds to the amino acid sequence ..Ser-His-Ala... . If the letter A in the second codon (word) is changed to G, His is replaced by Arg, as can be understood with Table II. This change actually occurs in hemoglobin Zürich.

The four differences between solids, glasses, and biomolecules already indicate that the physics of biomolecules possesses new, exciting (and difficult) features. Two more aspects make it even clearer that that the field is rich:
(v) The number of "possible" biomolecules is incredibly large. Consider a medium-sized protein, constructed from 150 amino acids. Since there exist 20 amino acids, the number of possible combinations is $(20)^{150} \approx 10^{200}$. If we produce one copy of each combination and fill the entire universe, we need 10^{100} universes to store all combinations. This example demonstrates that biomolecular problems cannot be solved by random experiments.
(vi) A protein with a given primary sequence can fold into a very large number of slightly different conformational substates. Each individual building block can, on the average, assume 2-3 different configurations with approximately equal energy. The entire protein thus possesses about $(2\text{-}3)^{150}$ states of approximately equal energy.

The properties alluded to in (i) to (vi) above imply that biomolecules are complex many-body systems. Their size indicates that they lie at the border between classical and quantum systems. Since motion is essential for their function, collective phenomena play an important role. Moreover, we can expect that many of the features involve nonlinear processes. Function, from storing information, energy, charge, and matter, to transport and catalysis, is an integral characteristic of biomolecules. The physics of biomolecules is a rich field. It stands now where nuclear, particle, and condensed matter physics were around 1930. We can expect exciting progress in the next few decades.

1.5 Heme Proteins.

Nature has evolved far more than 10^7 proteins. Fortunately, many are related; when nature discovers a good trick, it uses it over and over again for many different purposes. This conservative approach suggests that we can find general principles that govern the structure and function of proteins. In order to find such principles, it is best to select a class of proteins that are relatively simple, well studied, and perform many different functions. Heme proteins form such a class (2). They store and transport oxygen, transport electrons, detoxify and oxidize substances, and transform light into chemical energy. Consider myoglobin (Mb) as prototype. Mb stores and transports oxygen in the muscles. It consists of 153 amino acids, has a molecular weight of 17,800 dalton and approximate dimensions $3 \times 4 \times 4$ nm^3. A schematic cross section is shown in Fig. 5. The folded popypeptide chain (globin) encloses the heme group, an organic molecule shown in Fig. 6.

The heme group consists of an organic part and an iron atom. For the chemist, the organic part, protoporphyrin, is made up of four pyrrole groups, linked by methene bridges to form a tetrapyrrole ring. Four methyl, two vinyl, and two propionate side chains are attached to the tetrapyrrole ring. The side chains can be arranged in 15 different ways, but only one arrangement, protoporphyrin IX, is commonly found in biological systems. The iron atom binds covalently to the four nitrogens in the center of the protoporphyrin ring. The iron can form two additional bonds, one on either side of the heme plane.

For physicists, who are usually afraid of any molecular structure with more than two atoms, the heme group can be shown as a disk, about 1 nm in diameter and 0.2 nm thickness (Fig. 6). The disk has an iron atom with two free valences in the center and a one-dimensional electron ring (pi electrons) surrounds the iron.

2. LIGAND BINDING TO HEME PROTEINS

Biomolecules, in contrast to solids or glasses, can be studied in two different ways: Their physical properties can be investigated just as if they were ordinary systems, but their biological activity can also be used as probe. The second approach has yielded surprisingly rich information in the case of heme proteins, where the binding of small molecules such as dioxygen or carbon monoxide can be explored in great

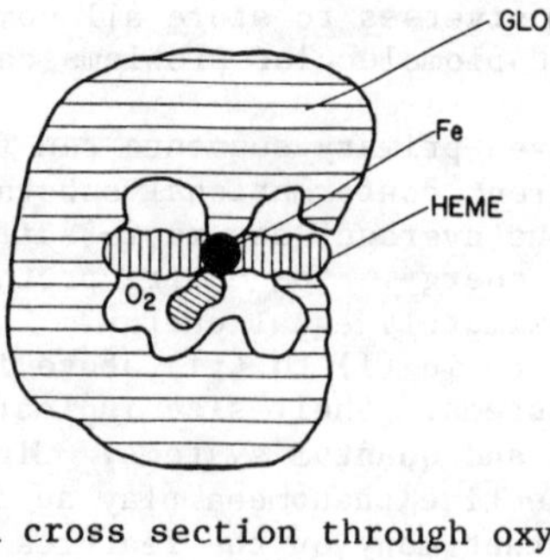

FIGURE 5. A cross section through oxymyoglobin.

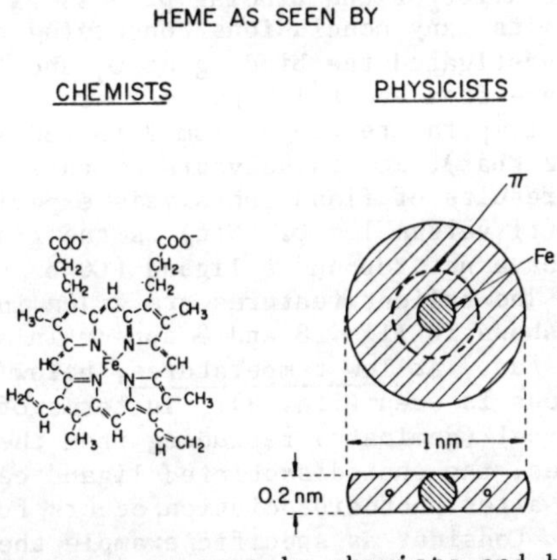

FIGURE 6. The heme group as seen by chemists and by physicists. π indicates the pi-electron ring.

detail (3). Since central concepts and ideas have emerged from such binding studies, we sketch the experimental approach and the most important results here.

The basic idea consists in studying the reaction

$$MbO_2 + \hbar\omega \rightarrow Mb + O_2 \rightarrow MbO_2, \qquad (1)$$

where MbO_2 denotes the state in which each myoglobin molecule has bound an O_2. The method is sketched in Fig. 7. MbO_2 is placed in a cryostat. A laser flash breaks the bond between the iron and the ligand. The ligand molecule moves away from the binding site, but later rebinds. Photodissociation and rebinding can be observed optically because MbCO and Mb, or MbO_2 and Mb, have different optical absorption spectra as

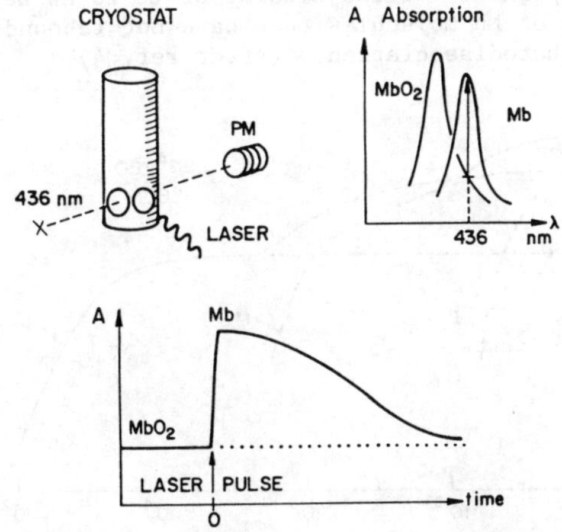

FIGURE 7. Flash photolysis. The protein sample (MbO_2) is placed in a cryostat, the absorbance is monitored at a suitable wavelength, and the MbO_2 is photodissociated by a laser flash. Because MbO_2 and Mb differ in the optical absorbance, photodissociation and rebinding can be monitored.

shown in Fig. 7. The behavior of the binding process as a function of external parameters permits many conclusions concerning the heme protein. In our work, we have investigated the binding of O_2 and CO to many heme proteins in the time range from about 100 ps to 1 ks (13 orders of magnitude in time), the temperature range from 2 to 330 K, the pressure range up to 200 MPa (~ 2 kbar), and in solvents of various viscosities (4). The experimental results of flash photolysis experiments can be given as plots of log N(t) versus log t. N(t) is the fraction of heme protein molecules that have not rebound a ligand (CO or O_2) at the time t after the laser pulse. The salient features are given in Figs. 8 and 9.

The general features shown in Figs. 8 and 9 appear in all heme proteins that we have studied so far. At low temperatures, below about 160 K, only one rebinding process is seen (Fig. 8). In terms of Fig. 5, we interpret I as the internal (geminate) rebinding from the heme pocket. At these low temperatures, the photodissociated ligand cannot leave the heme pocket and binding after photodissociation occurs from a region close to the heme iron. Consider as specific example the binding of CO to Mb. We denote the protein state with CO in the heme pocket with B and the bound state MbCO with A. In state A the heme is planar, the iron has spin 0 and is very close to the mean heme plane. In state B, the heme is

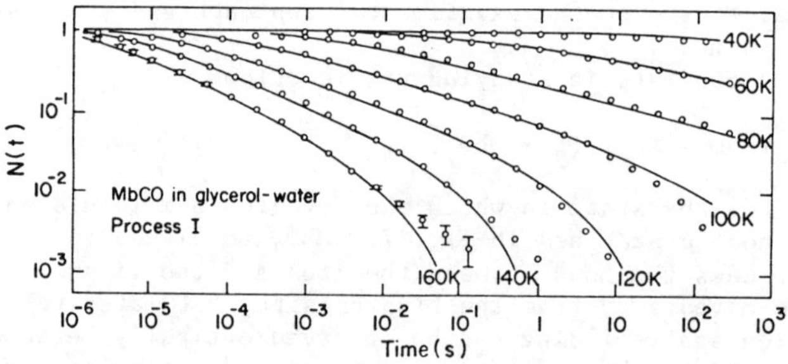

FIGURE 8. Time dependence of the binding of CO to Mb between 40 and 160 K. N(t) is a fraction of Mb molecules that have not rebound a CO molecule at the time t after photodissociation. (After ref. 4)

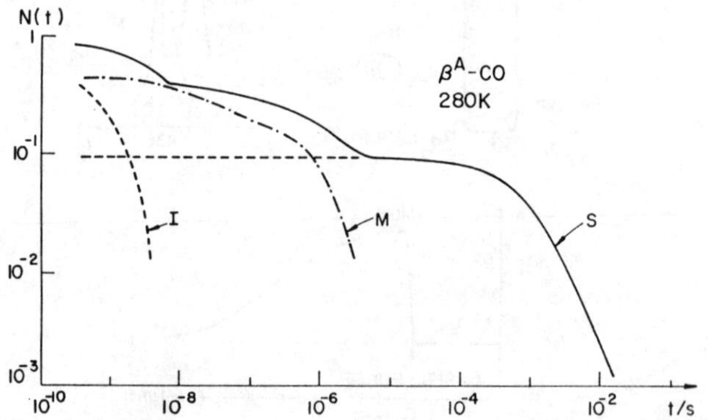

FIGURE 9. Rebinding of CO to the separated beta chain of human hemoglobin at 280 K. The solid line represents the experimental data. I is the internal rebinding from the pocket, M the rebinding from the matrix, and S the binding from the solvent.

domed, the iron has spin 2 and lies about 0.5 Å out of the mean heme plane. The CO molecule is essentially free in the heme pocket. The binding step B → A requires a motion of the iron into the mean heme plane, a spin change 2 → 0 of the iron, a flattening of the heme group, and a simultaneous approach of the CO molecule. The formation of the covalent bond between the ligand and the heme iron thus is a complicated many-body phenomenon. Nevertheless, because the reaction is first order and binding can be followed over a wide range in temperature and time, the heme pocket forms a superb laboratory for studying the physics of the reaction. Three observations, in particular, stand out:

(1) N(t) is not exponential in time, but can be approximated by a power law,

$$N(t) = (1 + t/t_o)^{-n}, \tag{2}$$

where t_o and n are temperature-dependent coefficients.

(2) In a wide range of heme proteins, O_2 and CO bind nearly equally fast.

(3) Above about 40 K, the binding rate coefficient k_{BA} can be described by an Arrhenius relation. Below about 40 K, binding is faster than expected on the basis of classical reaction theory; quantum-mechanical tunneling becomes dominant.

Here we will only consider the first observation and we refer to the literature for the discussion of the reaction mechanism (11) and of tunneling (5,12).

At higher temperatures, above about 200 K, binding after photodissociation becomes more complicated. Fig. 9 displays three different binding processes, denoted by I, M, and S. I and M are independent of the CO concentration in the solvent, S is proportional to it. We interpret I as the direct rebinding from the heme pocket. In the matrix process M, the CO molecule moves into the protein matrix and possibly the hydration shell and rebinds from there. The solvent process S corresponds to the situation where the photodissociated CO leaves the protein and any CO molecule in the solvent can compete for the vacant binding site. In S, the victorious CO moves into the protein matrix, migrates through the matrix to the heme pocket and ultimately binds to the heme iron. The detailed description of the entire binding process is in general difficult. Here we discuss some aspects that are related to protein dynamics.

3. CONFORMATIONAL SUBSTATES

The most striking feature of the low-temperature binding of small ligands to heme proteins is the nonexponential time dependence, exemplified in Fig. 8 for the binding of CO to Mb (4). We have observed such a nonexponential time dependence in all heme-protein-ligand combinations that we have studied. The time dependence observed after repeated flashes ("hole burning") implies that the binding process in each individual protein molecule is, within experimental error, exponential in time, but that different protein molecules possess different rate coefficients (4,13). The simplest explanation for this experimental fact invokes conformational substates: A protein in a given state (say MbCO) can assume a large number of structurally somewhat different substates, each with a different activation energy for ligand binding. Below a critical temperature, transitions among substates are absent and each protein is frozen into a particular substate. Above the critical temperature, transitions among substates occur with an average rate coefficient λ_r. Below the critical temperature the nonexponential rebinding in Fig. 8 can be written as

$$N(t) = \int dH_{BA}\, g(H_{BA})\, \exp(-k_{BA}t), \tag{3}$$

where N(t) is the fraction of myoglobin molecules that have not rebond a CO at the time t after the photodissociation. H_{BA} is the barrier between the pocket (well B) and the covalently bound state (well A) at the heme iron, $g(H_{BA})\, dH_{BA}$ the probability of finding a barrier with height between H_{BA} and $H_{BA} + dH_{BA}$, and k_{BA} is the rate coefficient for the transition $B \rightarrow A$, assumed to be given by an Arrhenius relation

$$k_{BA} = A_{BA} \exp(-H_{BA}/RT). \tag{4}$$

From the measured N(t), the distribution $g(H_{BA})$ is obtained (4,14). A few typical examples are given in Fig. 10. The distributions in Fig. 14 raise the question as to why proteins possess a large number of conformational substates. Fig. 2 suggests an explanation: It is unlikely that folding leads to a unique tertiary structure. A large number of nearly isoenergetic folded states most likely coexist. Sidechains may assume various orientations, hydrogen bonds may differ so that a large number of slightly different spatial structures is plausible. The Gibbs energy of a protein consequently may possess many "energy valleys" or conformational substates - states with essentially the same energy separated by barriers (4,15). Such a structure has been suggested also for spin glasses (16). At temperatures below about 200 K, each protein molecule remains frozen into a particular substate; above 200 K, transitions among substates occur. Below 200 K, rebinding of CO or O_2 after photodissociation consequently is nonexponential in time; above 200 K, it can become exponential (4).

4. EQUILIBRIUM FLUCTUATIONS

The existence of substates leads to two different types of motions in proteins, equilibrium fluctuations (EF) and functionally important motions (fims). EF lead from one substate to another, fims connect different states, for instance MbCO and Mb + CO. At first sight, EF and fims appear to be unrelated. If, however, the states involved in a given fim are structurally similar, EF and fims are connected by fluctuation-dissipation theorems (17).

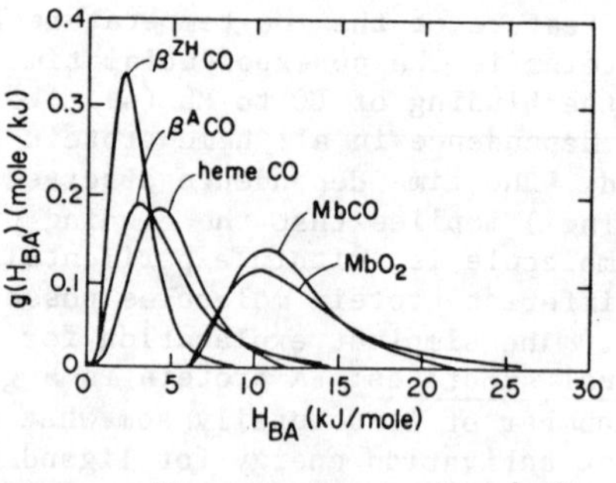

FIGURE 10. Activation energy distributions $g(H_{BA})$ for the low-temperature binding of CO and O_2 to various heme proteins.

Equilibrium fluctuations in proteins have been studied by many different techniques. Some of these, such as nuclear magnetic resonance, fluorescence, and Mössbauer effect, provide information about the time dependence of EF. X-ray diffraction, on the other hand, yields insight into the spatial characteristics of substates: The Debye-Waller factor determines the mean-square deviation, $\langle x^2 \rangle$, of each non-hydrogen atom from its equilibrium position (15,18,19). If proteins were uniform, like small solids, the $\langle x^2 \rangle$ values of all atoms would be nearly the same. The experimental data, for instance the backbone values shown in Fig. 11, suggest that motions vary widely in a protein.

Of course, X-ray diffraction takes a snapshot and cannot see the actual motion. We can only infer from Fig. 11 that the atoms move. The additional information from time-resolved experiments, however, supports the idea that proteins are indeed systems with characteristic fluctuations. Molecular dynamics calculations also provide evidence for flucutating proteins (20-23).

5. FIMS AND PROTEINQUAKES

While studies of equilibrium fluctuations are difficult, but straightforward, exploration of fims involves an additional consideration: The system must be placed into a nonequilibrium state and the relaxation towards equilibrium must be observed. Such studies are routine in, for instance, T-jump experiments. Here, however, a new feature enters: The nonequilibrium state should have functional significance and should be part of a functional pathway. Ligand binding to heme proteins offers the possibility of performing a relevant relaxation experiment (24,25).

Consider again the photodissociation process MbCO + $\hbar\omega \rightarrow$ Mb + CO $\rightarrow$ MbCO. In the bound state MbCO, the heme iron has spin 0, the heme group is planar, with the iron in plane, and the protein is in the liganded structure. In the deoxy state, the iron has spin 2 and occupies a position about 0.5 Å out of the mean heme plane away from the heme pocket, the heme is domed, and the protein is in the deoxy structure. Immediately after photodissociation, the iron has changed spin from 0 to 2 and the iron-CO bond is broken. These changes take place in less than 150 fs (26). The heme and the protein, however, are still in the liganded structure. Two processes now compete. The heme and the protein relax toward the deoxy structure and CO rebinds. We concentrate here on the first process, relaxation.

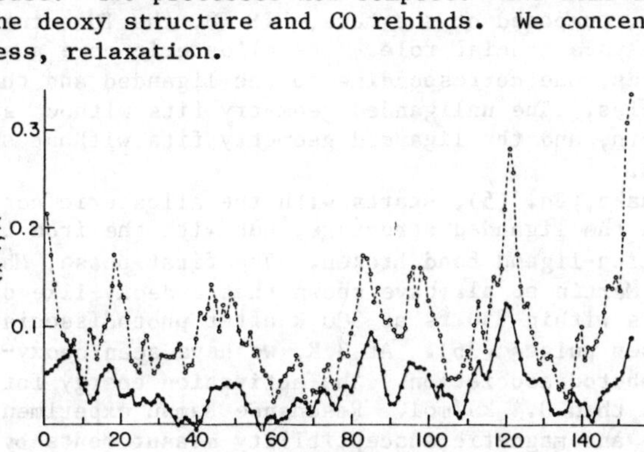

FIGURE 11. Values of the mean-square displacement $\langle x^2 \rangle$ for the backbone of Mb. The points represent averages over the atoms (N, C_α, C) of each backbone, plotted versus residue number. Open circles at 300 K, dots at 80 K. (After ref. 18.)

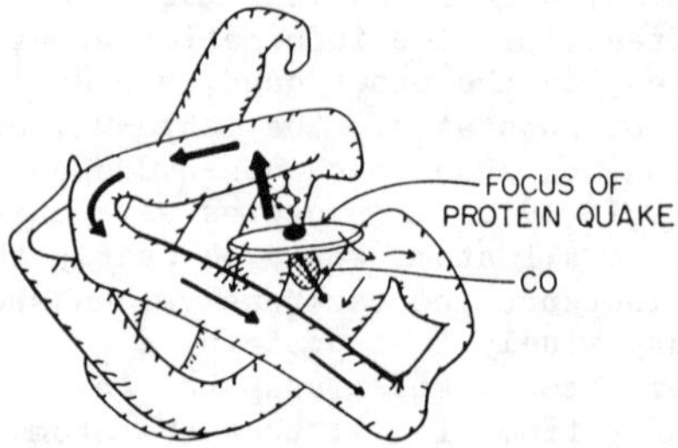

FIGURE 12. Proteinquake. Binding or dissociation of a ligand, for instance CO, causes a proteinquake.

Protein relaxation is similar to an earth or starquake: A stress is relieved at the focus, the heme iron, where the bond is broken. The released strain energy is dissipated in the form of waves (phonons) and through the propagation of a deformation. The situation is shown in Fig. 12. On binding, the protein is stressed, on photodissociation, the stress is relieved. In either case, the protein finds itself in a state far from equilibrium. Return to equilibrium occurs through a protein-quake, a series of motions and rearrangements.

The proteinquake in Mb can be studied by looking at a number of spectroscopic markers as function of temperature and time after photodissociation. Such observations imply that the relaxation of MbCO occurs through at least the following four steps:

$$\text{MbCO} \xrightarrow{h\nu} \text{Mb}_4^* \xrightarrow{\text{fim 4}} \text{Mb}_3^* \xrightarrow{\text{fim 3}} \text{Mb}_2^* \xrightarrow{\text{fim 2}} \text{Mb}_1^* \xrightarrow{\text{fim 1}} \text{deoxyMb}. \qquad (5)$$

Mb_4^* to Mb_1^* are intermediate states. The discussion of the sequence in Eq. (5) requires both experimental data from many different techniques and guidance from theory. In the relevant theoretical work, Karplus and collaborators (27) start from high-resolution X-ray structures for hemoglobin and Mb and minimize the conformational energy. They find that an allosteric core, composed of the heme, HIS F8, the FG corner, and part of the F helix, plays a crucial role. The allosteric core possesses two stable structures, one corresponding to the liganded and the other to the unliganded species. The unliganded geometry fits without strain into the unliganded globin, and the liganded geometry fits without strain into the liganded globin.

The proteinquake, Eq. (5), starts with the allosteric core and the globin still in the liganded structure, but with the iron in a high-spin state and the iron-ligand bond broken. The first phase, $Mb_4^* \xrightarrow{\text{fim 4}} Mb_3^*$, is very fast. Martin et al. have shown that a deoxy-like optical spectrum appears within 350 fs at 300 K after photodissociation with femtosecond laser pulses (26). At 3 K, we have seen deoxy-like features at 1 μs after photodissociation. The activation energy for fim 4 consequently is less than 0.4 kJ/mol. Resonance Raman experiments by Rousseau and Argade (28) and magnetic susceptibility measurements by Roder et al., (29) give more information on fim 4. At 4 K, the Raman spectrum after photodissociation is close to the spectrum for deoxyMb, the iron has spin 2, and the heme core is expanded. These observations suggest that in Mb_3^*, which is metastable at 4 K, the heme is already domed, but the iron is constrained by the unrelaxed globin from moving fully into its deoxy position. The fact that fim 4 occurs within less than one ps is supported by molecular dynamics calculations of Henry, Levitt, and Eaton (30).

In the second phase of the quake, $Mb_3^* \xrightarrow{\text{fim 3}} Mb_2^*$, the iron probably moves completely into the deoxy position and the heme contracts to the deoxy structure. Raman experiments indicate that near 300 K fim 3 starts at about 25 ps and is finished at 10 ns (28,31). In steady-state Raman experiments, Mb_2^* is reached at 50 K. While Mb_2^* has properties close to that of deoxyMb, small differences remain, as is seen particularly clearly in the position of a near-infrared band (32). At about 20 K, the peak of the near-infrared band is at 758 nm in deoxyMb and at 766 nm in Mb_2^*. These observations can be interepreted by assuming that in the intermediate state Mb_2^*, which is metastable at 20 K, the globin has not yet relaxed.

The third place of the proteinquake, $Mb_2^* \xrightarrow{\text{fim 2}} Mb_1^*$, can be studied experimentally by determining the position of the "760 nm" near-infrared band as a function of temperature and time after photodissociation (24,25). Below about 40 K, fim 2 is too slow to be observed. Between 40 and 160 K, fim 2 can be studied in detail and three remarkable features stand out: The band shifts without appreciably broadening; no isosbestic point exists; and the shift is nonexponential in time. It is not yet clear which residues of the protein are involved in figm 2, but the work of Karplus and collaborators (27) suggests that the residues of the allosteric core must move to accommodate the unliganded geometry.

The third phase of the proteinquake, $Mb_1^* \xrightarrow{\text{fim 1}}$ deoxyMb, corresponds to the final rearrangement of the protein structure, including the hydration shell. Details of this motion are still unclear but its existence is assured by the result of a pressure titration experiment (33). The present knowledge of the rates for the four fims are summarized in Fig. 13 (25). Note in particular that fim 2 is nonexponential in time and consequently cannot be characterized by one rate. At each temperature, fim 2 covers a wide range of rates.

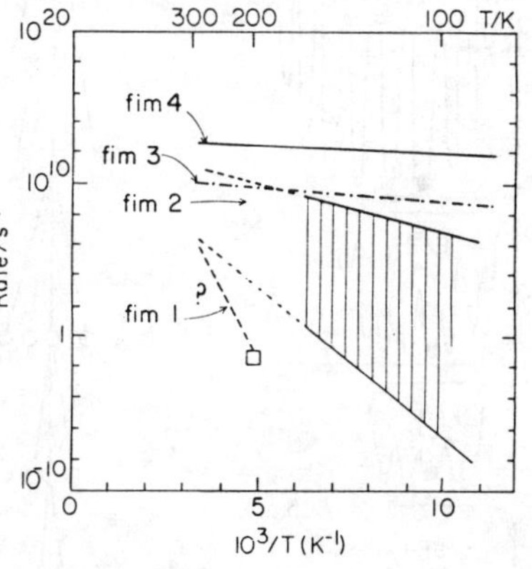

FIGURE 13. Relaxation rates for the four fims (functionally important motions) in myoglobin as function of $10^3/T$. Fim 2 is known to be nonexponential in time and the range of rates is shown. The detailed time dependence of the other motions is not known.

6. THE HIERARCHY OF SUBSTATES

The observation of four different phases of proteinquake or four different fims has consequences for the structure of CS in Mb. Even if the fluctuation-dissipation theorem does not hold rigorously, connections between fluctuations and dissipative motions must exist. Four different phases of the proteinquake call for four different tiers of CS. We thus arrive at the hierarchical model of protein substates given in Fig. 14 (25). A given state, for instance MbCO, shown at the top of Fig. 14 as a single potential well, is subdivided into a large number of conformational substates (CS^1), separated by barriers with heights greater than about 80 kJ/mol. Each of these substates is in turn split into a large number of subsubstates (CS^2) with barriers from 10 to 40 kJ/mol. Each valley in the second tier furcates into a few subsubsubstates (CS^3) with barriers less than a few kJ/mol. Each of these CS^3 again branches into a few CS^4.

The equilibrium fluctuations that can occur in the protein characterized by the substate hierarchy of Fig. 14 depend crucially on temperature. Below about 10 K, the only motions are fluctuations in the CS^4. As the temperature increases, EF3 set in and at about 50 K they are so fast that all CS^3 (and CS^4) are in equilibrium and only the next tier has to be considered. At about 40 K, EF2 among the CS^2 set in. These EF2 also become faster with increasing temperature. Finally, above about 200 K, EF1 among the substates (CS^1) of the first tier begin. At physiological temperatures, the EF in a protein will be complex. In addition to the four tiers of EF with characteristic barrier heights for each

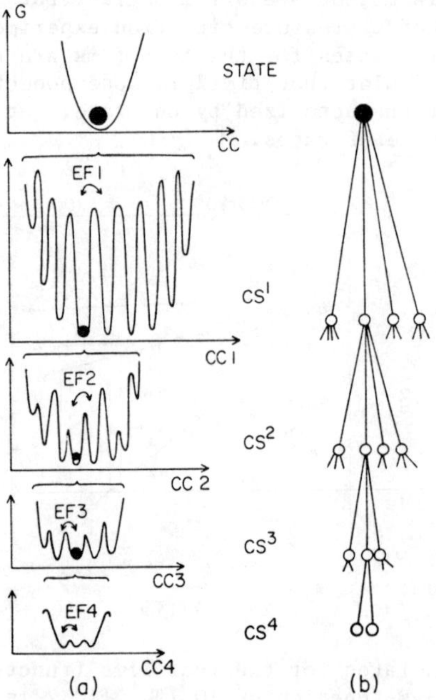

FIGURE 14. Hierarchical arrangement of the CS in myoglobin. (a) Schematic energy surfaces. (b) Tree diagram. G, Gibbs energy of the protein; CC (1-4) a conformation coordinate of tiers 1-4.

tier, EF in one tier do not occur with one rate, but most likely span a range of rates as indicated in Fig. 13.

The results discussed here suggest that proteins have many similarities with glasses and amorphous solids: They possess a hierarchical structure (34), show nonergodic behavior (35), and may be ultrametric (36). The hierarchical arrangement of the CS is shown in Fig. 14. The nonergodic behavior can be stated as follows: Below about 200 K, a protein in a particular substate (CS^1) will remain in that substate for a very long time compared to any measurement time and will therefore not explore all possible substates. The ultrametric property can be explained with Fig. 14. Assume that the protein fluctuates from a CS^3 in one substate (CS^1) to another CS^3 that is a member of a different substate (CS^1) of the first tier. The barriers of the second and third tier then are not important and the fluctuation is dominated by the highest barrier between the initial and final CS^3. The similarity between glasses and proteins may be accidental or it may point to a profound similarity. A deeper understanding of the connection will require more detailed experiments and extension of the investigations to many more proteins and to protein reactions other than simple ligand binding. It may well turn out that glass and spin glass theories will be important for an understanding of protein dynamics. It may even turn out that proteins will be excellent testing grounds for glass and spin glass theories. I believe that we are at the beginning of an exciting period.

This work was supported in part by the U.S. Dept. of Health and Human Services under Grant No. 2GM18051, and the National Science Foundation under Grant No. DMB 82-09616.

REFERENCES

1. Stryer L: Biochemistry: San Francisco: W. H. Freeman, 1981.
2. Lever ABP and Gray HB(eds): Iron Porphyrins: Reading, MA: Addison-Wesley, 1983.
3. Antonini E and Brunori M: Hemoglobin and Myoglobin in Their Reactions with Ligands: Amsterdam: North-Holland, 1971.
4. Austin RH, Beeson, KW, Chan SS, Eisenstein L, Frauenfelder H, and Nordlund TM: Science 192 (1975) 5355.
5. Alberding N, Austin RH, Beeson KW, Chan SS, Eisenstein L, Frauenfelder H, Nordlund TM: Science 192 (1976) 1002.
6. Beece D, Eisenstein L, Frauenfelder H, Good D, Marden MC, Reinisch L, Reynolds AH, Sorensen LB and Yue KT: Biochemistry 19 (1980) 5147.
7. Doster W, Beece D, Bowne SF, DiIorio EE, Eisenstein L, Frauenfelder H, Reinisch L, Shyamsunder E, Winterhalter KH and Yue KT: Biochemistry 21 (1982) 4831.
8. Dlott DD, Frauenfelder H, Langer P, Roder H and DiIorio EE: Proc. Natl. Acad. Sci. USA 80 (1983) 6239.
9. Alben JO, Beece D, Bowne SF, Doster W, Eisenstein L, Frauenfelder H, Good D, McDonald JD, Marden MC, Moh PP, Reinisch L, Reynolds AH, Shyamsunder E and Yue KT: Proc. Natl. Acad. Sci. USA 79 (1982) 3744.
10. Roder H, Berendzen J, Bowne SF, Frauenfelder H, Sauke TB, Shyamsunder E and Weissman MB: Proc. Natl. Acad. Sci. USA (1984).
11. Frauenfelder H and Wolynes PG: Science 229 (1985) 337.
12. Alben JO, Beece D, Bowne SF, Eisenstein L, Frauenfelder H, Good D, Marden MC, Moh PP, Reinisch L, Reynolds AH and Yue KT: Phys. Rev. Letters 44 (1980) 1157.
13. Frauenfelder H: Structure, Dynamics,, Interactions and Dynamics of Biological Macromolecules: C. Helene(ed) Reidel, 1983, p. 227.

14. Young RD and Bowne SF: J. Chem. Phys. 81 (1984) 3730.
15. Frauenfelder H, Petsko GA and Tsernoglou D: Nature 280 (1979) 558.
16. Toulouse G: Helv. Phys. Acta 57 (1984) 459.
17. Kubo R: Rep. Prog. Phys. 29 (1966) 255.
18. Hartmann H, Parak F, Steigemann W, Petsko GA, Ringe Ponzi D and Frauenfelder H: Proc. Natl. Acad. Sci. USA 79 (1982) 4967.
19. Petsko GA and Ringe D: Annu. Rev. Biophys. Bioeng. 13 (1984).
20. Karplus M and McCammon JA: CRC Crit. Rev. Biochem 9 (1981) 293.
21. Karplus M and McCammon JA: Ann. Rev. Biochem. 53 (1983) 263.
22. McCammon JA and Karplus M: Acc. Chem. Res. 16 (1983) 187.
23. McCammon JA: Reports of Progress in Physics (1983).
24. Frauenfelder H: Structure and Motion: Membranes, Nucleic Acids and Proteins: Clementi E, Corongiu G, Sarma MH and Sarma RH(eds): Adenine, Guilderland, NY, (1985) pp. 205-218.
25. Ansari A, Berendzen J, Bowne SF, Frauenfelder H, Iben IET, Sauke TB, Shyamsunder E and Young RD: Proc. Natl. Acad. Sci. USA 82 (1985) 5000.
26. Martin JL, Migus A, Poyart C, Lecarpentier Y, Astier R and Antonetti A: Proc. Natl. Acad. Sci. USA 80 (1983) 173.
27. Gelin BR, Lee AW-M and Karplus M: J. Mol. Biol. 171 (1983) 489.
28. Rousseau DL and Argade PV: Proc. Natl. Acad. Sci. USA, in press.
29. Roder H, Berendzen J, Bowne SF, Frauenfelder H, Sauke TB, Shyamsunder E and Weissman MB: Proc. Natl. Acad. Sci. USA 81 (1984) 2359.
30. Henry ER, Levitt M and Eaton WA: Proc. Natl. Acad. Sci. USA 82 (1985) 2034.
31. Johnston C, Dalickas G, Gupta D, Spiro T and Hochstrasser R: Biochemistry, in press 1985.
32. Iizuka T, Yamamoto H, Kotani M and Yonetani T: Biochim. Biophys. Acta 371 (1974) 126.
33. Eisenstein L and Frauenfelder H: Frontiers of Biological Energetics: Vol. I, pp. 680-688: Academic, New York, 1978.
34. Mézard M, Parisi G, Sourlas N, Toulouse G and Virasoro M: Phys. Rev. Lett. 52 (1984) 1156.
35. Palmer RG: Adv. Phys. 31 (1982) 669.
36. Huberman BA and Kerszberg M: J. Phys. A 18 (1985) 331.

FRACTALS AND LOCALIZATION

B. SOUILLARD
Centre de Physique Théorique, Ecole Polytechnique, F-91128 Palaiseau, France

This talk is concerned with fractals and localization. At first I briefly describe fractal theory with possible relevance to amorphous and glassy materials as a motivation. I then survey localization theory and in particular the scaling theory which allows a simple pedagogical description of it. Finally, localization on fractals is discussed from the theoretical point of view as well as with regard to its possible experimental consequences.

1. FRACTALS

Definition of Fractals

The concept of fractals and its relevance to physics have been introduced by Mandelbrot /1/. In this paper we will say that a set of points (say a set of atoms) is a fractal if it has dilation (scaling) symmetry; this symmetry is understood in an average way. We can then define the fractal dimension D of this set of atoms in the following manner: let d be the embedding dimension that is the (euclidean) dimension of the space to which the atoms belong (d=3, sometimes d=2) and let us draw a sphere B(R) with radius R around some given point of the space; let now N(R) be the number of points of our set inside this sphere, we assume

$$N(R) \sim C\, R^D \,. \qquad (1)$$

For usual (euclidean) sets, e.g. a usual crystal or a disordered crystal, and in fact for all sets which possess some translation invariance at least in the average, equ. (1) hold with D=d. For a fractal set D<d and D is called the fractal dimension.

Do Fractals Exist?

Yes they do! One can certainly construct "artificial" fractals, an example of which is the Sierpinski gasket built by induction as shown in Fig. 1:

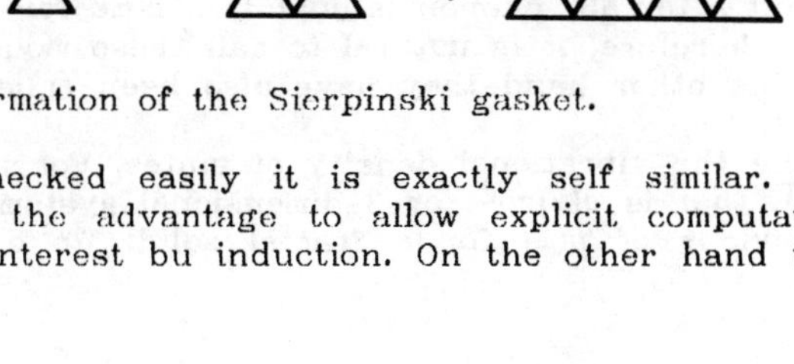

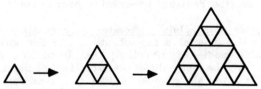

Fig. 1 Formation of the Sierpinski gasket.

As can be checked easily it is exactly self similar. These "regular" fractals offer the advantage to allow explicit computations of various quantities of interest bu induction. On the other hand they may exhibit quite different behaviour compared to the "real" disordered fractals which are more relevant to physics.

The prototype of a "real" disordered fractal is the incipient percolation cluster appearing in the percolation problem when the concentrations p is equal to the critical concentration p_c. In this example and for d=3, the fractal dimension is $D \approx 1.9$.

One of the important things to remember about fractals is that the fractal dimension D is not the only relevant dimension as we will see and that other dimensions and parameters play a crucial role for various physical quantities (see /2/).

"Short Scale Fractals"

In many physical situations, the system that we want to describe is not a pure fractal, but is fractal up to some length scale ξ, that is it exhibits self similarity only up to this scale. Let us take again the example of percolation. At $p=p_c$, the incipient cluster is a fractal, that is a set which has the property that $N(R) \sim R^D$ for all R up to infinity. What now about $p \neq p_c$? It is known then that there exists a unique length ξ (the correlation length) such that

. at scales $L > \xi$ the system looks uniform,

. at scales $L < \xi$ the system looks fractal.

Clearly the existence of this length ξ can be translated into the existence of a frequency ω_0, the frequency of the wave with wavelength ξ, such that for waves travelling

. at frequencies $\omega < \omega_0$, the system looks uniform,

. at frequencies $\omega > \omega_0$, the system looks fractal.

Conjecture

Fractals have shown to be of interest for many physical situations among which are gels and polymers, porous media... Their possible relevance to the topic of this conference stems from the following conjecture by S. Alexander, C. Laermans, R. Orbach and H.M. Rosenberg/3/:"For glasses, amorphous solids, epoxy resine, neutron irradiated quartz, there exists a cross over length ξ such that for length $L < \xi$ the system is fractal". In other words these systems should be "short scale fractals" as defined above.

This conjecture is an attempt to find a universal explanation to the near universal character of the specific heat and thermal conductivity of these very different materials. The existence of this length is conjectured. If it exists its origin is completely open to question.

Anyway it may be of interest to study the physical properties of fractal sets and describe possibly observable properties.

Vibrational Properties of Fractals - Phonons on Fractals - Fractons

Let us consider a fractal array of atoms. As for any solid we can look for its possible modes of vibration; that is to look for phonons of this system. They may be very different from the phonons that we know in usual solids, but after all, phonon is just the name for the vibrational modes of a solid. Therefore, it is natural to call these modes on a fractal phonons too. On the other hand they have also been called fractons for obvious reasons.

Let $\rho(\omega)$ denote the vibrational density of states. For usual euclidean systems $\rho(\omega) \sim \omega^{d-1}$, that is $\rho(\omega) \sim \omega^2$ for 3-dimensional systems. We may ask what is the behaviour of $\rho(\omega)$ for a fractal solid: We expect that the embedding dimension d will not be relevant for its behaviour, and this is indeed the case. Is the fractal dimension D the relevant one? In fact no! It turns out that the relevant dimension is another one, the so called fracton or spectral dimension δ /4,5/ and $\rho(\omega)\sim\omega^{\delta-1}$.

In general for a fractal set we have $\delta<D<d$, whereas these three dimensions become equal for a euclidian set. As an example for percolation we expect $\delta\approx 4/3$ /4/ and thus $\rho(\omega)\sim\omega^{0.33}$, values which are very different from the ω^2 of Euclidan systems.

If we come now to "short scale fractals" we see that we must have

. $\rho(\omega)\sim\omega^{d-1}$ for $\omega\ll\omega_o$, (2)

. $\rho(\omega)\sim\omega^{\delta-1}$ for $\omega\gg\omega_o$. (3)

Thus, there is the question of the connection between these two regimes. The solution to this problem was found /6,7/ to be the continuous line shown in Fig. 2. The dotted lines represent the two regimes where ρ is proportional to ω^2 and $\omega^{0.33}$.

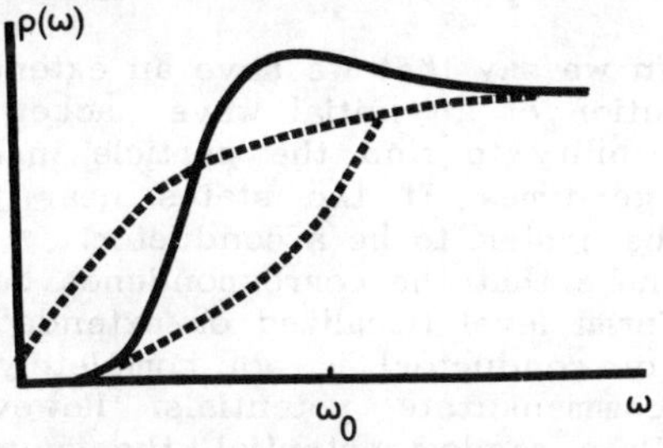

Fig. 2 Density of states for a fractal system. Dotted lines: the regimes ~ ω^2 and ~ $\omega^{0.33}$. Full line: interpolation between the two regimes.

This type of behaviour is in resasonably good agreement with the experimental results of /8/, though the cross-over edge is not very sharp. In the interpretation presented here, this could come from the fact that the characteristic length below which the system would be fractal should be of the order $\xi\approx 20$ Å. This value is very short and thus the influence of the short scale fractal structure should not be very strong. It would be very interesting to find systems with larger ξ for which the influence of a possible short scale fractal structure could be more apparent if the theory is at all correct.

2. LOCALIZATION THEORY

In order to go beyond the facts stated above we now forget completely about fractals and turn to the description of the theory for a usual "euclidean" disordered solid. We will first stick to the problem of propagation of electrons in a disordered system. This problem has attracted most of the attention in localization theory for a long time, but as we will see most of the results do apply also to phonons and to the problems of propagation of many other linear waves in a disordered medium. (For more information we refer to /9,10,11,12/ and to the references stated therein).

We study a system at zero temperature and with no electron-electron interaction. One of the most popular model for such a problem is the

Anderson model which is a tight-binding Hamiltonian

$$(H\Psi)(x) = \Sigma_{|y-x|=1} J(x,y)\Psi(y)+V(x)\Psi(x), \quad (4)$$

where the off-diagonal terms $J(x,y)$ for nearest neighbours are constant. The diagonal terms $V(x)$ are chosen at random between $-W/2$ and $W/2$ for every site. Other models can be thought of; they can differ in some respects but give the same qualitative results.

Nature of the States, Time Evolution, Transport

The basic question dealt with by localization theory is the nature of the states of the Hamiltonian H for a typical sample that is for a typical potential $(V(x))$. We have to find the nature of the physical solutions of the stationary equation $H\Psi=E\Psi$. By physical we mean the solutions of the stationary equation which are bounded.

. if $\int|\Psi|^2dx<\infty$, and in general Ψ will be exponentially decaying, we say that we have a localized state. If we look now to the time evolution of an inital wave packet built with localized states, then the probability to find the particle outside a large box will be small and uniformly in time. If states near the Fermi level are localized, we expect the system to be an insulator.

. if $\int|\Psi|^2dx=\infty$, then we say that we have an extended state. If we look now to the time evolution of an initial wave packet built with extended states, then the probability to find the particle inside a large box will tend to zero for large times. If the states near the Fermi level are extended, we expect the system to be a conductor.

Let us just note here that the correspondence between the nature of the states near the Fermi level (localized or extended) and the transport properties (insulator or conductor) is not completely clear in particular for sytems with incommensurate potentials. However, in case of a disordered system with a random potential, the connection between both properties is certainly correct.

Some History

People who studied disordered systems a long time ago used to believe that disorder introduces only some quantitive corrections. There was no idea that really qualitatively new and important phenomena different from the pure crystal case must be taken into account. This is the so-called weak scattering theory in which one gets always extended states. However in 1958, Anderson argued that for $d=3$ all the states should be exponentially localized if the disorder is large enough.

The qualitative picture for $d=3$ was completed by Mott in 1968: at weak disorder, we should have exponentially localized states near the band edges and extended ones in the middle. The energies corresponding to localized and extended states whould be sharply separated by a mobility edge.

In the preceding, "strong" and "weak" disorder characterize the typical fluctuations of the potential as seen by an electron moving from one site to the other. In the Anderson model as defined above, this statement refers to large or small W.

For the Anderson model, we would have a phase diagram of the type (Fig.3):

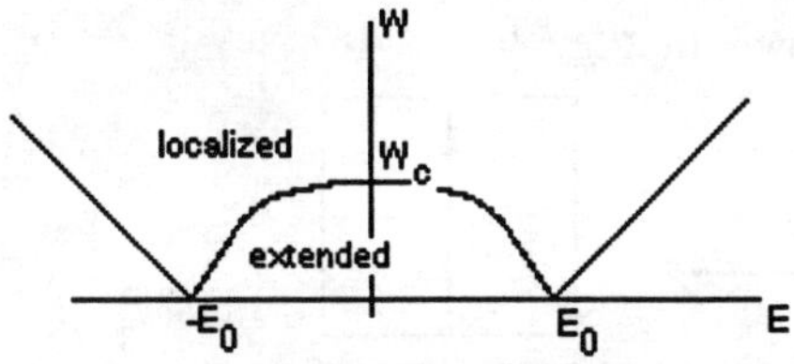

Fig. 3 Phase diagram for the Anderson model.

For zero disorder (W=0), the energy band of the pure crystal is just the interval ($-E_0$, E_0). For disorder W, the band enlarges due to disorder and becomes the interval between the two 45° lines. We should have

localized states outside the mobility edge curve and extended ones inside of it. Thus for large enough W all states in the band are exponentially localized, whereas for smaller W they are localized or extended according to whether their energy is near the edges or in the middle of the band.

A given physical system corresponds to a given disorder and a given Fermi level. If we can move the Fermi level across the mobility edge or change the disorder in the system, we can induce a metal-insulator transition and indeed many metal-insulatortransitions do correspond to this Anderson-Mott transition.

We have presented the physical picture for 3d systems. But it was soon recognized that there are also dimensional effects:

. for d=1, it was argued as early as 1961 by Mott and Twose /13/ that all states should be exponentially localized for any (possibly small) disorder.

. for d=2, the traditional belief had been the same as in the d=3 case until the "gang of four" paper /14/ showed that again exponential localization of all states should hold for any (possibly small) disorder.

In the following we will describe the scaling theory of reference /14/ which is presently the simplest and most intuitive approach to the localization problem.

Localization for Very Large Disorder

In the scaling theory outlined below we will introduce the fact that for very large disorder and in any dimension all states are exponentially localized. This fact is very intuitive although not easy to show analytically. For very large disorder we are in the case of dilute impurites. Coherent tunnelling effects between the local impurity states in order to create extended states are not likely because the phases coming from the contributions of the various states will be random due to the random position in space of these states and thus add destructively.

Scaling Theory

The scaling theory developed in /14/ (a scaling theory starting from microscopic models had been introduced before in /15/) is the simplest way to understand the predictions given above about the phase diagrams (see Fig. 3).

Let us consider the conductance G(L), or better the dimensionless conductance $g(L)=(\hbar e^{-2})G(L)$ for a system of linear size L. This conductance depends on many parameters of the system (e.g. on the size of the system, on the disorder...). Now let us construct a system of size 2L using samples of size L, as shown in Fig. 4:

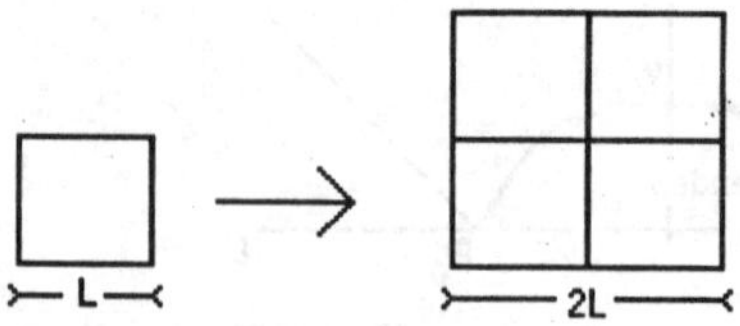

Fig. 4 Construction principle of large systems.

The scaling hypothesis is the assumption that the conductance g(2L) of the system of size 2L can be computed only from the conductance of the system of size L, that is

$g(2L) = f\{g(L)\}$, (5)

where the function f is unknown. In other words the conductance for a system of size 2L is assumed to depend on all the parameters of the system of size L only through its conductance g(L). This hypotheses seems very natural and intuitive, but may deserve some thoughts in view of its consequences!

For theoretical purposes, it is easier to work with a continuous version of (5) which means then that now we take a system of size L and increase its size a little bit by dL. Therefore, we get:

$d \ln g(L)/d \ln L = \beta\{g(L)\}$, (6)

where β is again an unkown function. In this case the scaling hypothesis states that the derivative of the conductance with respect to the logarithm of the size of the system depends on all the parameters of the system only through the conductance itself.

In all the preceding we assumed that the length L is larger than some suitable semi-microscopic length, so that it makes sense to speak of the conductance of a sample of length L.

Now we are going to make use of the fact that we know two asymptotic regimes of the function β(g):

a) when g is large, we have a good conductor and we thus must have Ohm's law in the limit $g \to \infty$ that is $g \sim \sigma L^{d-2}$, which corresponds to $\beta(g) \to d-2$ as $g \to \infty$.

b) when g is small, we have a poor conductor, that is a highly disordered one. Therefore, we must be in the localized regime mentioned above in which all states decay exponentially. In this case it is natural to expect that the conductance will decay exponentially with the size of the system, too. Thus $g \approx g_a \exp(-L/\xi)$ which corresponds to $\beta(g) \approx -L/\xi$ as $g \to 0$.

If we assume that we can interpolate smoothly between the two regimes, we get for the β-function the graph, indicated in Fig. 5:

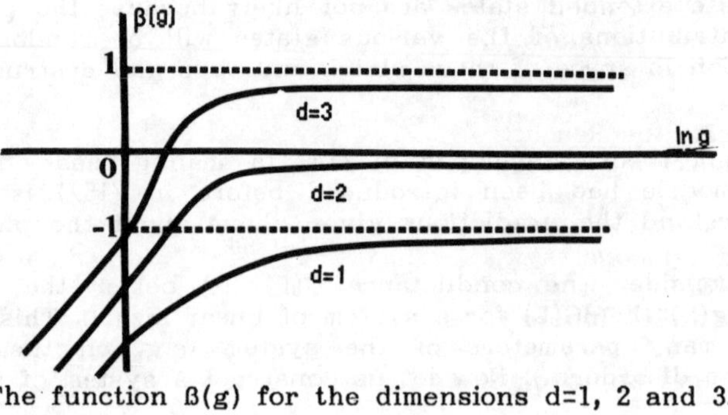

Fig. 5 The function β(g) for the dimensions d=1, 2 and 3.

Hence, for the dimensions 1, 2 and 3 we have the following situation:

. For d=1 we see that ß is always negative. This fact implies, in view of its definition, that if we enlarge the system, the conductance will decrease. For sufficiently large systems we will be in the exponentially localized regime. Thus any one-dimensional system with any degree of disorder exhibits exponential localization of all states if the system is large enough.

· For d=3 in contrast the ß-function changes sign for some value g_c of the conductance. The situation depends on the conductance g of the system. We are starting with: if $g>g_c$ the conductance will increase with L and will tend to the Boltzmann value for very large systems, whereas for $g<g_c$ the conductance decreases with L and sufficiently large systems have their states exponentially localized. We thus have a metal-insulator transition and a mobility edge.

· For d=2 the situation is more delicate since $\beta(g) \to 0$ as $g \to \infty$ and it is thus crucial to know in this limit whether the ß-function starts above or below the axis. Using perturbation theory in g^{-1} it was shown /14/ that the curve starts under the axis, as shown in Fig. 5. Thus for any disorder all states are exponentially localized for a sufficiently large two dimensional system. There is no mobility edge and therefore real two dimensional metals do not exist.

d=3 – The Mobility Edge

It is tempting to study the mobility edge with the methods used in the study of critical points in phase transition theory. In this spirit we introduce critical exponents:

· If the states at the Fermi level E are localized, the corresponding wave functions will behave at infinity modulo oscillations as $\sim \exp\{-|x|/\xi(E)\}$. The characteristic length $\xi(E)$ is called the localization length. We assume that near to the mobility edge E_c where the states cease to be localized,

$$\xi(E) \sim |E-E_c|^{-\nu}, \tag{7}$$

· if the states at the Fermi level E are extended, the conductivity $\sigma(E)$ at zero temperature is non-zero. We assume that it vanishes at the mobility edge, where localization appears, as

$$\sigma(E) \sim |E-E_c|^s . \tag{8}$$

It follows from the scaling theory /14,15/ that $s=(d-2)\nu$. For this reason, s as defined in equ. (8) cannot be zero (for d=3). Therefore σ has to vanish continuously at the mobility edge.

For the Anderson model $s \approx 1$. As in phase transition theory, the exponents are in fact independent of the model - at least in large universality classes. They depend only on the space dimension and some general facts of the problem. For example in the Anderson model we have not considered spin orbit scattering. If we introduce it we will enter another universality class. In fact there are three such universality classes with different critical indices.

Localization from the Resolution of Models

The above scaling theory gives a beautiful insight into the physics of localization. One may however be interested in deriving localization analytically within a given model, say for example the Anderson model.

For one-dimensional systems the most simple approach is the one due to Borland /16/ which has become a classical one. However, this work seemed to contain inherent difficulties in view of counter examples discovered in the theory of incommensurate systems. The conditions under which the Borland approach is valid have been clarified recently /17/ (see also /18,19/). This work leads also to an approach "à la Borland" for localization in three dimensional systems at large disorder.

On the other hand a mobility edge and a Mott-Anderson transition have been obtained on the Bethe lattice /20/. The exponent ν could be computed exactly in this model. The value of this exponent has implications on the possible upper critical dimension for the localization problem /21/ suggesting $d_c=4$.

Localization Lengths and Temperature

The localization length is a characteristic feature of a system at zero temperature. It is important to understand that the localization length is not independent of disorder and energy. According to the value of these parameters it may be short in some cases, but it may be large in some others. Localization will be relevant only for those systems for which the localization length is smaller than the other characteristic lengths involved in the systems. In particular the localization length has of course to be smaller than the size of the sample in order to observe localization! But more importantly it has to be smaller than the inelastic mean free path. In short /22/, if $\xi \ll l_{inelastic}$ then we will be in a position to observe effects of localization since the inelastic processes will not spoil too often the phase coherence of wave functions and thus the interference mechanism leading to localization. In this case the conductivity will be due to activated hopping according to Mott's law. On the other hand if $\xi \gg l_{inelastic}$ the process will be dominated by inelastic scattering. These facts may become clearer later on when we will discuss interference phenomena.

Coulomb Interactions

In all the preceding we have neglected the Coulomb interactions between the electrons. In some cases they can lead to important effects. Some aspects of the theory in presence of interactions have been understood recently, but some are still open to question. For more details see /12/.

About Experimental Results

A description of the experimental results is much outside the scope of this paper. Let us just say that many experiments have confirmed that localization phenomena are indeed relevant in many low temperature experimental situations, and for 1-, 2- and 3-dimensional systems. However, one has to go beyond the brief outline sketched above into more detailed theoretical treatments in order to compare precisely theory with experiment.

Localization of Phonons and Other Waves

It must be understood that localization is fundamentally an interference phenomenon. An incident electron gets scattered by various impurities. If the scattering is elastic the various scattered waves will keep their phase coherence and can interfere constructively or destructively yielding localization or delocalization.

If this fundamental fact is understood, we may assume as I have stressed since several years that essentially any kind of linear waves can exhibit localization and will display the same qualitative behaviour. For example, we can study localization of hydrodynamical shallow surface waves due to the scattering of the waves by a random bottom /23,23/. We can also look for the localization of electromagnetic waves in a plasma the density of which is fluctuating /25/, a phenomenon which has been shown to be relevant in some cases to plasma physics. It may yield new mechanisms for instabilities etc. An interesting case is the one of phonons: in the linear regime, for the same reasons as described above they may exhibit localization.

The transposition of the localization theory to other wave equations yields results very similar to the ones for the electronic case. In particular, one has complete exponential localization for one- and two-dimensional systems for any disorder as well as the existence of a mobility edge in dimensions larger or equal to three. A few differences may appear, however, depending on the case. They are minor as far as localization theory is concerned but they may have important practical consequences. Such an example can be seen in the case of phonons and more generally in all problems which deal with acoustic modes. Because the mode at zero frequency is always "extended", we will have extended phonons at a small enough frequency (for d=3) for any arbitrarily large disorder. For d=1 or 2, all phonons will be localized for any non zero frequency but their localization length will diverge at zero frequency. Therefore, for frequencies small enough the phonons can be considered as extended for any practical purpose.

3. LOCALIZATION ON FRACTALS

In Section 1, we have mentioned that fractals have been suggested /3/ to be relevant to a large class of glassy or amorphous materials. On the other hand they are certainly relevant to disordered materials which are for example of a percolation type /26/. A natural question is thus to ask what are the properties of electronic or vibrational states on fractals and in particular how the localization theory can be modified on a fractal. We would also like to know if the possible fractal character of a system can be tested experimentally by analysing its electronic properties.

Relevant Dimension for Localization on Fractals

In the previous Section we have deduced from a scaling theory that euclidean sytems with dimension d smaller or equal to 2 have all their electronic and vibrational states exponentially localized. If we would like to develop the scaling theory for a system on a fractal what will be the relevant dimension which will allow to discuss the analogous problem? Here again, the answer /5/ is that the relevant dimension is the spectral dimension δ and not the fractal dimension D, and we observe that as long as $\delta \leq 2$, all states are exponentially localized. Note in particular that, since for an incipient percolation cluster $\delta \approx 4/3 < 2$, all states are localized in this case.

"Superlocalization"

The next point we want to investigate is the nature of localization on a fractal. This question is answered in /27/ where it is shown that the localized wave functions (and also the impurity states) in fact exhibit what was called "superlocalization",i.e. a faster than exponential decay with distance. More precisely

$$|\Psi(x)| \sim \exp\{-(|x|/L(E))^{d_{min}}\}, \qquad (9)$$

where $d_{min} > 1$ for fractals (e.g. $d_{min} = 1.7$ for three-dimensional percolation) describes the chemical length exponent or the tortuosity exponent. $|x|^{d_{min}}$ is the length of the shortest path on the fractal connecting the point 0 to the point x. We see that we have a very strong localization.

These ideas should have natural applications in the study of porous materials /27/. Now we turn to their applications on the electronic properties of fractal systems.

Electron Relaxation on Fractals

We consider a fractal such as an incipient percolation cluster for which all electron and all phonon states are localized. As we have just seen they will be very strongly localized. It is thus interesting to wonder if this very strong localization will affect their properties and in particular the interactions between electrons and phonons.

Actually the answer is yes /28/. For example for a localized electronic state relaxing by an emission of a single fracton, the time profile of the electronic state population is of the type

$$P(t) \sim t^{-[\ln t]^{D/d_{min}-1}} \qquad (10)$$

(at least for rapidly relaxed fractions e.g. by anharmonic interactions). We are not going to derive this equation here but we can give the following intuition. Because of superlocalization the interaction between electrons and phonons will be strongly reduced. For an interaction you need not only to be at the right time but also at the right place! It is thus very natural that the decay is slowed down with respect to the usual exponential decay relaxation profile. The decay in equ. (10) is faster than any power but slower than any exponential. The precise form of the exponent depends on the form of equ. (9). Note also that the parameters of the fractal enter the behaviour of the relaxation profile in equ. (10). It would be very different for a euclidean system in which D=d and d_{min}=1. Since this profile should be relevant e.g. to non radiative decay for an optical center in a glass or to electron-hole recombination in amorphous semiconductors in the band tail, we have at least in principle the possibility of an experimental test if a system is a fractal or not. In addition we may have access to some of its parameters.

Other cases can be calculated /29/ for example the relaxation by two fractons, which could be relevant to the spin-lattice relaxation of electronic centers, leading also to predictions for fractal lattices very different from the ones for usual lattices.

Thus, in principle one should be able to test experimentally, through these types of properties, if a system is fractal or not. In particular one should be able to test the conjecture of /3/ which was explained in Section I. In any case, whatever will turn out to be the answer in the particular case of glasses and amorphous materials, no doubt, these ideas have proved to be very stimulating.

REFERENCES

1. Mandelbrot B: Les Objets Fractals (Flammarion, Paris 1975); Fractals: Form, Chance and Dimension (Freeman, San Francisco 1977); The Fractal Geometry of Nature (ibid, 1982).
2. Vannimenus J, J. Phys. (Paris) Lett. 45 (1984) 1071.
3. Alexander A, Laermans C, Orbach R and Rosenberg H, Phys. Rev. B28 (1983) 4615.
4. Alexander S. and Orbach R, J. Phys. (Paris) Lett. 43 (1982) 625.
5. Rammal R and Toulouse G, J. Phys. (Paris) Lett. 44 (1983) 13.
6. Derrida B, Orbach R and Yu K-W, Phys. Rev. B29 (1984) 6645.
7. Aharony A, Alexander S, Entin-Wohlman O and Orbach R., Phys. Rev. B31 (1985) 2565.
8. Bucheneau U, Nucher N and Dianoux, Phys. Rev. Lett. 55 (1984) 2316.
9. Thouless DJ, in *Ill-Condensed Matter*, Proceedings Les Houches, North-Holland, Amsterdam (1979).
10. Bergmann G, Phys. Rep. 101 (1984).
11. Souillard B, in *Common Trends in Particle and Condensed Matter Physics*, Proceedings Les Houches, Phys. Rep. 103 (1984) 41.
12. Lee P and Ramakrishnan TV, Rev. Mod. Phys. 57 (1985) 291.
13. Mott N and Twose W, Adv. Phys. 10 (1961)107.
14. Abrahams E, Anderson PW, Licciardello DC and Ramakrishnan TV, Phys. Rev. Lett. 42 (1979) 673.
15. Wegner F, Z. Phys.B25 (1976)327.
16. Borland RE, Proc. R. Soc. London A 274 (1963) 529.
17. Delyon F, Lévy Y and Souillard B, Phys. Rev. Lett. 55 (1985) 618.
18. Kotani S, in Proceedings of the AMS meeting on Random Matrices, Brunswick 1984.
19. Simon B, Taylor M and Wolff T, Phys. Rev. Lett. 54 (1985) 1589.
20. Kunz H and Souillard B, J. Phys. (Paris) Lett. 44 (1983) 411.
21. Kunz H and Souillard B, J. Phys. (Paris) Lett. 44 (1983) 503.
22. Thouless DJ, Phys. Rev. Lett. 39 (1977) 1167.
23. Guazzelli E, Guyon E and Souillard B, J. Phys. (Paris) Lett. 44 (1983) 837.
24. Devillard P and Souillard B, Localization of surface waves on a rough bottom, to be submitted to J. Fluid Mech.
25. Escande D. and Souillard B, Phys. Rev. Lett. 52 (1984) 1296.
26. Deutscher G, in *Percolation, Structures and Processes*, G.Deutscher, R.Zallen and J.Adler Editors, Annals Isr. Phys. Soc. 5 (1983).
27. Lévy Y and Souillard B, Superlocalization on a fractal, preprint Ecole Polytechnique, 1985.
28. Alexander S, Entin-Wohlman O and Orbach R, J. Phys. (Paris) Lett. 46 (1985) 549.
29. Alexander S, Entin-Wohlman O and Orbach R, J. Phys. (Paris) Lett. 46 (1985) 555.

RELATION BETWEEN MICROSTRUCTURE, TRANSPORT AND OPTICAL PROPERTIES IN COMPOSITE MATERIALS.

F.BROUERS

Department of Physics, University of the West Indies, Kingston, Jamaica and Laboratoire d'Optique des Solides, Université Paris VI, Place Jussieu, Paris 75005.

INTRODUCTION

The relation between the microstructure and the optical and transport properties of composite materials is one of the most important and interesting aspects of the physics of composite materials.

I refer to two recent Conference Proceedings for an overall view of this new challenging interdisciplinary subject (1, 2).

As an illustration of the problems met in this field, I want in this lecture to calculate, in section 2, the percolation threshold critical volume filling factor f_c in cermet films as a function of the depolarisation factor of the inclusions.

PERCOLATION IN METAL-INSULATOR COMPOSITE FILM AS A FUNCTION OF GRAIN GEOMETRY.

Cermet thin films used to fabricate selective surfaces for solar photothermal energy conversion are made of a mixture of metal (for instance Cr,Au,Pt,Co,...) and insulator (Cr_2O_3, Al_2O_3, SiO_2).

When the metallic concentration or volume filling factor f is small, the system is made of metallic grain inclusions in an insulating matrix. By contrast when f tends to one, the system appears as a mixture of insulating inclusions in a metallic matrix.

For a critical value of f(f_c), called the percolation threshold, the system undergoes an insulator-metal transition.

In the vicinity of f_c, the dielectric constant and therefore the optical properties exhibit dramatic changes and as a consequence it is possible to prepare films whose reflectivity can vary from zero to one in a narrow range of frequencies.

The condition for an efficient selective surface is that this transition occurs for a frequency lying between the solar visible spectrum and the infra-red blackbody radiation spectrum of the collector.

The grain formation in these films is the result of several processes, essentially surface diffusion and coalescence and therefore the final microgeometry depends of the nature of the cermet constituents and of the film preparation.

As discussed in a number of recent papers (Ref.3 and references herein), granular composite films generally can be described by one of the two distinct microstructures: an aggregate structure where the metal and insulator grains are interdispersed and topologically equivalent (this is similar on the atomic level to a disordered binary alloy model) and a seperated grain structure where the metal and insulator exhibit an asymmetric topology consisting of either metallic inclusions in an insulating matrix or insulating inclusions in a conducting matrix (this is similar on the atomic level to a binary alloy with a strong short range order where each atom is surrounded by atoms of the other species).

This dichotomy finds its counterpart in the two main theoretical approaches which have been used to interpret the optical and transport properties of these materials.

The traditional ideas of the effective medium approximation have been introduced in this problem, namely a random unit is subjected to an, as yet unknown, effective medium which is determined to be such that the resulting extra perturbation vanishes on the average over all possibilities of the random unit. If this perturbation is taken to be the Lorenz-Mie scattering amplitude in the direction of the impinging light beam $S(\theta=0)$, Niklasson and Granqvist (Ref. 3) have shown that one can obtain elegantly the various approximations. The condition

$$\langle S(0)\rangle = 0 \qquad (1)$$

is a generalization of the optical theorem to non homogeneous systems.

If the random unit is a sphere of component A (concentration f, dielectric constant $\varepsilon_A(\omega)$) or component B (concentration 1-f, dielectric constant $\varepsilon_B(\omega)$) embedded in an effective medium (dielectric constant $\varepsilon^*(\omega)$), the condition (1) reads in this case :

$$S_A(0) + (1-f)S_B(0) = 0. \qquad (2)$$

Using the dipolar approximation of $S(0)$ for a sphere (4) Eq.(1) leads to a relation which allows the calculation of the effective medium $\varepsilon^*(\omega)$:

$$f\cdot\frac{\varepsilon_A - \varepsilon^*_{BA}}{\varepsilon_A + 2\varepsilon^*_{BA}} + (1-f)\cdot\frac{\varepsilon_B - \varepsilon^*_{BA}}{\varepsilon_B + 2\varepsilon^*_{BA}} = 0. \qquad (3)$$

This is the approximation known as Bruggeman Approximation (BA) (Ref. 5). The BA is analogous to the CAP of the theory of random alloys and applies to the aggregate structure.

If the random unit is a coated sphere, for instance an internal sphere A of radius Γ_A surrounded by an external sphere B of radius Γ_B with a filling factor f, defined as $f = (\Gamma_A/\Gamma_B)^3$, using the dipolar approximation of $S(0)$ for a coated sphere having this compositional geometry, the condition $S^{CS}_{AB}(0) = 0$ yields the expression

$$\varepsilon^*_{MG} = \varepsilon_B\,\frac{\varepsilon_A + 2\varepsilon_B + 2f(\varepsilon_A-\varepsilon_B)}{\varepsilon_A + 2\varepsilon_B - f(\varepsilon_A-\varepsilon_B)} \qquad (4)$$

which is the Maxwell-Garnett approximation (MG) for an A inclusion in a B host. If instead one considers a B inclusion in a A host, by making the replacement A $\rightleftarrows$ B and f $\to$ 1-f, one gets an analogous relation for the inverted structure. One can calculate the d.c. conductivity threshold in these two approximations by replacing ε_A by i and ε_B by 0.

For the BA, one gets :

$$\frac{\sigma_{BA}}{\sigma_A} = \frac{3f - 1}{2}, \qquad (5)$$

where σ_{BA} is the conductivity in the composite and σ_A the conductivity of pure A. The percolation threshold is $f_c = 1/3$.

In the MG, for an A (metallic) inclusion in B (insulator), the system is always an insulator and $f_c=1$. For a B inclusion in A, in contrast

$$\frac{\sigma_{MG}}{\sigma_A} = \frac{2f}{3 - f} \qquad (6)$$

and the relative conductivity varies from 0 to 1 without percolation threshold ($f_c=0$).

Since the shape of inclusions in cermets are far from being spherical, the theory can be extended to ellipsoidal inclusions. If g is the depolarisation factor in the direction of the field, Eq. (3),(4),(5) and (6) become respectively:

$$f \frac{\varepsilon_A - \varepsilon^*_{BA}}{\varepsilon^*_{BA} + g(\varepsilon_A - \varepsilon^*_{BA})} + (1-f) \frac{\varepsilon_B - \varepsilon^*_{BA}}{\varepsilon_{BA} + g(\varepsilon^*_B - \varepsilon^*_{BA})} = 0, \quad (3')$$

$$\varepsilon^*_{MG} = \varepsilon_B \frac{g \cdot \varepsilon_A + (1-g) \cdot \varepsilon_B + f(1-g) \cdot (\varepsilon_A - \varepsilon_B)}{g \cdot \varepsilon_A + (1-g) \cdot \varepsilon_B - fg \cdot (\varepsilon_A - \varepsilon_B)}, \quad (4')$$

$$\frac{\sigma_{BA}}{\sigma_A} = \frac{f-g}{1-g}, \quad (5')$$

$$\frac{\sigma_{MG}}{\sigma_A} = \frac{f(1-g)}{1-fg}. \quad (6')$$

For small concentrations of inclusions (f close to 1 or to 0), it appears that a number of composite films are better described by the seperated grain structure. Since the MG approximation is unable to describe the metal-insulator transition, symmetrized MG approximations have been proposed by introducing an average over the two following random units : an inclusion of A coated by a shell of B and an inclusion of B coated by a shell of A. Using this picture the film can be modelled as a mixture of two types of coated ellipsoids. Dielectric- coated metal ellipsoids are denoted as type 1 units and metal-coated insulator ellipsoids are denoted as type 2 units. The problem is to evaluate the proportion of these two random units for a given value of the filling factor. Sheng (4) has introduced a probabilistic growth model where the probability of occurence of these two units as a function of f is given by the number of possible configurations of each type of units. This number is given by the volume available to the internal ellipsoid in the external ellipsoid. In the case of coated confocal spheroidal units, these probabilities are (Ref.4):

$$p_1 = \frac{u_1}{u_1 + u_2} \quad \text{and} \quad p_2 = \frac{u_2}{u_1 + u_2}, \qquad (7)$$

where

$$u_1 = (1 - f^{1/3})^3 \text{ and } u_2 = ((1 - (1-f)^{1/3})^3.$$

For spherical inclusions (g=1/3), two schemes have been used to apply the Sheng approximation, Nicklasson and Grandqvist have used the condition (1) i.e.

$$p_1 S_1(0) + p_2 S_2(0) = 0,$$

where $S_1(0)$ and $S_2(0)$ are the scattering amplitudes for type 1 and type 2 respectively. Gibson and Buhrman (5) have calculated the dielectric constant of type 1 and type 2 units in the MG ε_1^* and ε_2^* and then intoduced these two values in the Brüggeman formula (3) with $\varepsilon_A = \varepsilon_1^*$ and $\varepsilon_B = \varepsilon_2^*$ and $f = p_1$. It can be shown that these approximations yield the same second order equation for ε^* (Brouers 1985 Ref.6) and are therefore identical.

In the case of coated confocal spheroidal units when the depolarisation factors in the direction of the field are respectively g_1 and g_2, the same method can be used and again it can be shown that the two schemes give the same results. In particular, the conductivity of the composite material can be written :

$$\frac{\sigma^*}{\sigma_A} = F \cdot \sigma_2, \qquad (8)$$

where σ_2 is the MG conductivity for the conductive coated spheroid 2 :

$$\sigma_2 = \frac{f(1-g_2)}{1-fg_2} \tag{9}$$

and F is a statistical factor depending on p_2 and the two g's:

$$F = \frac{g_2 - p_2 + p_2\cdot(g_1-g_2)}{g_2 - 1 + p_2\cdot(g_1-g_2)} . \tag{10}$$

The percolation threshold f_c is given by :

$$f_c = f(p_2^c) \quad \text{(obtained from 7)}$$

$$\text{and} \qquad p_2^c = \frac{g_2}{1+g_2-g_1} . \tag{11}$$

The shape of the function σ^*/σ_A and the values of the percolation threshold depend decisively on the values of g_1 and g_2 and therefore on the shape of the incusions (6). This is illustrated by considereing the two extreme limiting cases :

$g_1 \to 1$ (oblate spheroids) where $p_2^c = f_c = 1$ for any g_2

and $g_2 \to 0$ (prolate spheroids) where $p_2^c = f_c = 0$ for any g_1.

REFERENCES

1) " Macroscopic properties of disorder media" Ed.Burridge R. et al. in Lectures in Physics Series, Springer Verlag 1982.

2) " Physics and Chemistry of porous media" American Institute of Physics Conference Proceedings (Ed.Johnson D.L. and Sen P.N) 107, 1984.

3) Nicklasson G.A. and Granqvist C.G. J.Appl.Physics 55, 3382, (1984).

4) Sheng P., Phys.Rev. Lett. 45, 60,(1980).

5) Gibson U.J. and Buhrman R.A. Phys.Rev. B27, 5046,(1983).

6) Brouers F. 1985, to be published.

COMMENTS ON THE MODE COUPLING THEORY OF THE LIQUID GLASS TRANSITION

W. GÖTZE

Physik-Department, Technische Universität München, D-8046 Garching, Germany
and
Max-Planck-Institut für Physik und Astrophysik, D-8000 München, Germany

INTRODUCTION

Recently the mode coupling approximation (MCA) for the dynamics of simple dense liquids has been extended to describe undercooled fluids and some aspects of the liquid glass transition [1,2,3,4], and in these notes I will report the present status of this theory. Since important features of the theory are not worked out yet and since a crucial test of the results by experiment has not yet been performed, I will be very reserved to offer much physical interpretation of the results. I expect that further work will revise those heuristic ideas which guided us during our development of the theory and so it does not seem appropriate to bias the reader too much. Rather I will present the machinery and the results in a self-contained fashion with the primary intension to make it easy for newcomers to join the research in this field.

In the first section the mathematical framework will be summarized. The two main points are the exact representation of the relevant correlation functions in terms of relaxation kernels and the discussion of ergodic versus non-ergodic density fluctuation dynamics in terms of simple singularities. Relaxation kernels have been introduced into the theory of fluids by Zwanzig [5] and they proved to be very fruitful to connect phenomenological concepts uniquely with experimental quantities like dynamic structure factors, to systematically extend phenomenological approaches as done for example in the visco-elastic theory of liquids, to represent experiments in terms of meaningfull parameters as well as to provide a proper starting point for microscopic approximations. Previously the ergodicity problem was of concern only for those working on the mathematical foundations of statistical mechanics. It is of course well known that conservation laws and spontaneous symmetry breaking imply the existence of non-ergodic variables, but there is no use to complement the approved technique [6] of treating the mentioned non-ergodic variables by a discussion of ergodic versus non-ergodic dynamics. It was regularly implied in the previous literature that the liquid glass transition can be discussed approximately as a transition from ergodic to non-ergodic dynamics and in these lectures this view will be adopted also. This picture ignores certain very slow relaxation processes. It is necessary to phrase the mathematical theory such, that ergodic and non-ergodic motion, which is neither caused by conservation laws nor by symmetry breaking, can be treated quantitatively within one frame without hand driving the results one hopes to get. The formalism used here was first proposed in connection with a theory for the disorder induced conductor insulator transition [7].

In the second section the MCA will be used to derive those closed equations for the correlation functions which are the basis of the discussion. The MCA has been introduced into statistical mechanics by Kawasaki [8] to study critical dynamics near second order phase transitions. This approach was used then also for many other examples where low lying excitations cause singularities of spectra. The idea of the approximation is to express relaxation kernels by products of those correlation functions one knows or which one wants to evaluate. The problems are connected with the decision which kernels are to be approximated, which correlation functions are the relevant ones, and with the evaluation of the coupling vertices. The solution of the resulting equations is not a trivial problem either. In the theory of liquids MCA was introduced to derive closed non-linear equations for density and current correlation functions using no other input than the known interparticle interaction and the static pair correlation function. It was shown that the equations could be solved by iteration and that the results for the liquid argon and liquid helium spectra accounted semi quantitatively for the experiments [9]. The theory was extended and improved so as to give a coherent description of the data of liquid argon and rubidium, in particular of the dynamical viscosity, with an accuracy of about 20% [10]. In the mentioned theories the MCA was overdone somewhat; a complete theory should also account for renormalized binary collision effects and for the correct almost free particle dynamics on the short wavelength scale. This extension of the MCA is possible [11] and the evaluation of the incoherent scattering function of argon and rubidium by Wahnström and Sjögren [12] demonstrates the quantitative success of this approach. In the case of anharmonic crystal dynamics the MCA is equivalent to the evaluation of the leading non-trivial Feynman diagram for the phonon self-energy and again the approach was successfully used to describe such a strongly coupled system as solid helium [13]. It was also discovered that the MCA for the current relaxation of particles moving in a static random field brings out a phase transition from the ergodic motion of a conductor to the non-ergodic dynamics of an insulator [7]. This self-consistent current relaxation theory leads to rather non-trivial results for the dynamics of strongly disordered semi conductors [14] and to a microscopic theory of the percolation dynamics which could be checked successfully against molecular dynamics data for the Lorentz model [15]. The mentioned results are the reason why we considered it worthwhile to apply a most simple version of the MCA to a study of the liquid glass transition [2]. There is the work of Cohen and Grest [16], who emphasize the role of percolation physics for the glass state, the paper by Sjölander and Turski [17] who stressed that the liquid glass transition as opposed to ordinary phase transitions is of dynamical origin and the calculation of Geszti [18] who derived the Batchinski Hildebrand law for the viscosity via a MCA. These three papers did not yield a microscopic theory for the liquid glass transition with some predictive power but the mentioned physical features are all contained in our approach [2]. I will briefly indicate some aspects of the mode coupling theory, mainly in order to give a hint to the interested reader, what he should look for in the original papers. Continuous spectra can often be described in terms of homogeneous or inhomogeneous line broadening. Conventionally these effects are either added more or less in an ad hoc fashion or they are obtained via coupling of the system under study with a given external field. Neither approach in adequate in the case of interest here. It is also evident, that the role of inhomogeneous broadening is increased in relation to homogeneous line width effects if the system is driven from the fluid into the glass. So one reason for the success of the MCA, more precisely of its self-consistency properties, is indeed that it describes in a unified fashion homogeneous and inhomogeneous life time effects. It constructs in a self-consistent fashion the propagating modes responsible for homogeneous broadening as well as the inhomogeneities yielding to distributions of oscillator frequencies and relaxation times causing non Lorentzian lines. This feature, which has not been emphazised enough in the previous literature, shall be spelled out in detail for a simplified model of the present theory.

The iterative solution of the approximation equations for zero frequency in the glass phase is equivalent to the study of a fixed point problem in the space of non-ergodicity parameters. This problem will be discussed in section three. First the work will be done for a schematic model where all equations can be solved elementarily. Then the general scenario will be considered. The glass instability point will be identified as a bifurcation point where a stable and an unstable fixed point collapse and disappear. It is possible then to evaluate the asymptotic behaviour of the non-ergodicity parameter and thus of all thermodynamic functions.

The dynamics near the liquid glass transition will be analyzed in section four. For special simplified models the solutions of the MCA are known in some detail [1,2,3,31]. The spectrum consists of two parts. One is described by a scaling law of a kind similar to those obtained for critical dynamics near second order phase transitions. The characteristic scaling frequency ω_c approaches zero at the transition point describing critical slowing down phenomena. The critical exponents are non-trivial. On top of this spectrum there is the elastic line in the glass. In the liquid this line broadens to a quasielastic line described by a second scaling law ruled by a second scaling frequency ω_c'. These results can also be derived for the general theory with explicite expressions for the relevant exponent parameters in terms of microscopic quantities.

In detail the notes are organized as follows:

1. CORRELATION FUNCTION DESCRIPTION

1.1 Mathematical Framework

In this subsection the notations for the correlation function description of classical equilibrium systems, to be used throughout the following, shall be established. The results are taken from the work of Kubo [19], Zwanzig [5] and Mori [20]; a full display can be found for example in Forster's [6] book. Canonical averages of dynamical variables A,B shall be denoted by ⟨A⟩, ⟨B⟩ and the corresponding fluctuations by $\delta A=A-\langle A\rangle$, $\delta B=b-\langle B\rangle$. The A-B-correlation function of time t is defined by

$$\Phi(t) = \langle \delta A(t)*\delta A\rangle \quad (1.1)$$

with time evolution explained by the canonical equations of motion. It will be necessary to represent time dependend functions F(t) like Φ(t), by their Laplace transforms:

$$F(z) =: LT[F(t)](z) =: \pm i\int dt\Theta(\pm t)e^{izt}F(t),\ Imz \gtrless 0 \ . \quad (1.2)$$

These functions are holomorphic for all nonreal values of the complex frequency z. For z approaching a real frequency ω function F(z) may be discontinuous. One expresses then F by its average F'(ω) and by its discontinuity F"(ω):

$$F(\omega\pm i0) = F'(\omega) \pm iF''(\omega) \quad . \quad (1.3)$$

The full function F(z), in particular F'(ω), can be expressed in terms of F"(ω) via a spectral representation

$$F(z) = \frac{1}{\pi}\int d\omega\ F''(\omega)/(\omega-z) \quad . \quad (1.4)$$

Function F"(ω) is called the spectral function and it can be calculated as Fourier transform

$$F''(\omega) = \frac{1}{2}\int dt\ e^{i\omega t}\ F(t) \quad . \quad (1.5)$$

If A=B function Φ is called the A-autocorrelation function. In this case Φ'(ω) and Φ"(ω) are real, i.e. the spectral function is the imaginary part of Φ(z) at the real axis z=ω+i0. One finds also $\Phi''(\omega) \geq 0$ then. Functions Φ(t), Φ(z) and Φ"(ω) are equivalent in the sense that two of them can be evaluated of the third one is known. Function Φ(t) describes the time evolution and so it is often used to phrase the physical picture one wants to achieve; sometimes it is measurable directly. Φ(z) will be used as efficient tool to describe crucial singularities for small frequencies by a function having the most regular properties one can imagine: holomorphy. The spectral function Φ"(ω) is so important, since it is usually measurable directly.

The susceptibilities are other quantities to characterize a system. The A-B-susceptibility $\chi_{AB}(\omega)$ is the response of variable A, $\langle\delta A\rangle_B$, created by a perturbation of strength ß, which varies monochromatically with frequency ω and is coupled to variable B:

$$\lim_{\beta\to 0} \langle\delta A\rangle_B/\beta = \chi(\omega) \quad . \tag{1.6}$$

It can be obtained as real frequency limit of a function $\chi_{AB}(z)$, which is holomorphic for complex z: $\chi(\omega)=\chi'(\omega)+i\chi''(\omega)=\chi(\omega+i0)$. The static susceptibility

$$\chi^o = \chi(z\to 0) = \chi'(\omega\to 0) \tag{1.7}$$

describes the linear change of $\langle\delta A\rangle$ due to a perturbation B, which is switched on arbitrary slowly, while the system is completely isolated. The fluctuation dissipation theorem yields an expression of $\chi(z)$ in terms of the corresponding correlation function

$$\chi_{AB}(z) = [z\Phi_{AB}(z) + \langle\delta A|\delta B\rangle]/(k_B T) \quad . \tag{1.8}$$

Here T is the temperature and k_B denotes Boltzmann's constant.

The bilinear form $(\delta A|\delta B)=\langle\delta A^*\delta B\rangle$ can be used to define a scalar product. In this metric space the time evolution is a unitary transformation generated by the hermitian Liouvillian: $A(t)=\exp(iLt)A$, where

$$LA = -i\partial_t A \quad . \tag{1.9}$$

The correlation function can be written as matrix element of the resolvent $R=(L-z)^{-1}$:

$$\Phi(z) = (A|[L-z]^{-1}|B) \quad . \tag{1.10}$$

The spectrum of Φ is a matrix element of the spectrum of L: $\Phi''(\omega)=\pi(A|\delta(\omega-L)|B)$.

Suppose the two orthogonal projectors P and Q form a resolution of the identity $P+Q=1$. Then elementary algebra yield from $(L-z)R=1$ the equation

$$[PLP-PLQ[QLQ-z]^{-1}QLP-zP]PRP = P \quad . \tag{1.11}$$

Choosing P as projection onto δA, $P=|\delta A)(\delta A|/g$, $g=(\delta A|\delta A)$ one finds for the A autocorrelation function the Zwanzig-Mori representation:

$$[z - \Omega + M(z)]\Phi(z) = -g \quad . \tag{1.12}$$

Here the restoring force is given by $\Omega=\omega/g$, $\omega=(\delta A|L|\delta A)$. If δA has a definitive parity under time inversion $\omega=0$. The relaxation kernel is given by the correlations of the fluctuating forces $F_A(t)=\exp(iQLQt)QLA$: $M(z)=m(z)/g$, where

$$m(z) = LT[\langle F_A(t)^*F_A\rangle](z) = (QLA|[QLQ-z]^{-1}|QLA) \quad . \tag{1.13}$$

The Zwanzig-Mori reduction (1.12) can be repeated for m(z), yielding a representation of $\Phi(z)$ as double fraction etc. One can also consider arrays of N linearly independend variables δA_α, $\alpha=1,2,\ldots,N$. Then the preceding formulae hold with g,ω,Ω and the like to be interpreted as N by N matrices; e.g. $g_{\alpha\beta}=(\delta A_\alpha|\delta A_\beta)$, $\omega_{\alpha\beta}=(\delta A_\alpha|L\delta A_\beta)$. In this case $\Phi_{\alpha\beta}(\omega)''$ is a positive definite hermitian matrix and so is $m_{\alpha\beta}(\omega)''$. In general Ω and M are not hermitian.

1.2 Description of the System

The liquid and the glass phase shall be described by those functions which are introduced conventionally in connection with fluid dynamics. For an exhaustive exposition the reader is referred to some of the known monographs [21].

The most important correlations in the following are those for the density fluctuations of wavevector $\vec{q}$ for the N species of atoms our system is composed of: $\rho_\alpha(\vec{q})=\Sigma_j \exp i\vec{q}\vec{r}_\alpha{}^j$. Here $\vec{r}_\alpha{}^j$ denote the position of atom j of species α; mass and velocity of that particle will be denoted by m_α and $\vec{v}_\alpha{}^j$. The matrix of density correlations

$$\Phi_{\alpha\beta}(q,t) = \langle\delta\rho_\alpha(\vec{q},t)^*\delta\rho_\beta(\vec{q})\rangle \tag{1.14}$$

describes the propagation of density fluctuations in space and time. The initial value is given by the matrix of structure factors $S_{\alpha\beta}(q)=\langle\rho_\alpha{}^*(\vec{q})\rho_\beta(\vec{q})\rangle$. Function S can be analyzed by x-ray scattering experiments and $\Phi_{\alpha\beta}(q,\omega)''$ determines van Hove's dynamical structure factor

$$S(q,\omega) = \Sigma_{\alpha\beta}\ \ell_\alpha\ \Phi_{\alpha\beta}(q,\omega)''\ \ell_\beta \quad . \tag{1.15}$$

Here $S(q,\omega)$ denotes, up to some normalization, the cross section for particle scattering with momentum transfer $\hbar q$ and energy loss $\hbar\omega$. The scatterers, like neutrons, are assumed to couple with scattering length ℓ_α to species α.

If $\vec{j}_\alpha(\vec{q})=\Sigma_j\vec{v}_\alpha \exp i\vec{q}\vec{r}_\alpha{}^j$ denotes the current density, whose longitudinal part will be abbreviated by $j_{\alpha L}(\vec{q})=\vec{q}\vec{j}_\alpha(\vec{q})/q$, the continuity equation for species α reads

$$L\rho_\alpha(\vec{q}) = q j_{\alpha L}(\vec{q}) \quad . \tag{1.16}$$

The current normalization is trivial: $\langle j_{\alpha i}(\vec{q})|j_{\beta k}(\vec{q})\rangle=\delta_{\alpha\beta}\delta_{ij}v_\alpha{}^2$; here i,j denote the cartesian components and $v_\alpha=\sqrt{k_B T/m_\alpha}$ is the thermal velocity for species α. Hence the first Zwanzig-Mori reduction reads

$$[z - q^2 K(q,z)/S]\Phi(q,z) = - S \tag{1.17a}$$

with the longitudinal current correlation function

$$K_{\alpha\beta}(q,t) = \langle j_{\alpha L}(\vec{q}t)^*\ j_{\beta L}(\vec{q})\rangle \quad . \tag{1.17b}$$

This function shall be expressed in terms of the fluctuating forces $F_{\alpha L}(\vec{q}t)=\exp(iQLQt)QLj_{\alpha L}(\vec{q})$, Q projecting perpendicular to the ρ_α and $j_{\alpha L}$:

$$M_{\alpha\beta}(q,t) = \langle F_{\alpha L}(qt)^* F_{\beta L}(q)\rangle / v_\beta^2 \quad , \tag{1.17c}$$

$$\Sigma_\beta [z\delta_{\alpha\beta} - M_{\alpha\beta}(q,z)]\ K_{\beta\gamma}(q,z) = -\delta_{\alpha\gamma}\gamma_\alpha^2 \quad . \tag{1.17d}$$

The physical contents of the preceding formulae is rather complex and in this series of lectures I will not spell it out at all.

To explain the liquid glass transition in simple physical terms I will consider the motion of a tagged particle ; its parameters like mass, position or velocity shall be indicated by an index s. It refers to the concept of self-correlation. So the corresponding correlations of the densities $\rho_s(\vec{q})=\exp[i\vec{q}\vec{r}_s]$ are given by

$$\Phi_s(q,t) = (\rho_s(\vec{q}t)|\rho_s(\vec{q})) = \langle \exp[-i\vec{q}(\vec{r}_s(t)-\vec{r}_s(t{=}0))]\rangle \quad . \tag{1.18}$$

A first step Zwanzig-Mori-reduction leads to

$$\Phi_s(q,t) = \frac{-1}{z + q^2 K_s(qz)} \tag{1.19a}$$

and the longitudinal current correlation can be expressed in terms of a current relaxation kernel $M_s(q,z)$

$$K_s(qz) = \frac{-v_s^2}{z + M_s(q,z)} \quad . \tag{1.19b}$$

The fluctuating force correlations are formed in analogy to eq. (1.17c).

Isotropy can be used derive from equ. (1.18) the formula

$$S(t) = -2\partial\Phi_s(q,t)/\partial(q^2), \qquad q=0, \tag{1.20}$$

where S(t) denotes the averaged displacement squared in one fixed space direction which is experienced by the tagged particle within the time intervall t:

$$S(t) = \langle \Delta x(t)^2\rangle = \frac{1}{3}\langle [r_s(t) - r_s(t{=}0)]^2\rangle$$
$$= \frac{1}{3}\int r^2 \Phi_s(rt)d^3r \quad . \tag{1.21}$$

Here $\Delta x(t)=x_s(t)-x_s(t{=}0)$ with x_s denoting the x component of $\vec{r}_s$; $\Phi_s(rt)=\langle\rho_s(\vec{r}t)\rho_s(\vec{r})\rangle$ is the tagged particle density correlation function in ordinary space as opposed to the Fourier space: $\rho_s(\vec{r},t)=\delta(\vec{r}-\vec{r}_s(t))$. Obviously

$$S(t{=}0) = 0, \quad \partial_t S(t{=}0) = 0, \quad \partial_t^2 S(t) = 2K_s(t) \quad , \tag{1.22}$$

with

$$K_s(t) = \langle v_{sx}(t)v_{sx}(t{=}0)\rangle \tag{1.23}$$

denoting the tagged particle velocity correlation function. Laplace transformation implies

$$LT[S(t)](z) = 2\frac{1}{z^2} K_s(z) \quad , \tag{1.24}$$

and therefore one identifies from equs. (1.19a), (1.20) the velocity correlation function as long wave length limit of the longitudinal current correlations

$$K_s(z) = K_s(q=0,z) \quad . \tag{1.25}$$

From equ. (1.19b) one thus gets a generalized Drude formula

$$K_s(z) = \frac{-v_s^2}{z + M_s(z)} \tag{1.26}$$

with the velocity relaxation kernel given as $M_s(z)=M_s(q=0,z)$.

1.3 Ergodic Versus Non-ergodic Density Dynamics

Ergodicity is a conventional explicit or tacit assumption in most theories of many particle systems. However, in these notes the attitude is taken that a glass state as opposed to a liquid state can be characterized by the density fluctuations to exhibit non-ergodic motion. In this subsection the necessary notations shall be introduced. Let me mention, that the reader can find some more discussion of the ergodicity versus non-ergodicity problem in a paper by Palmer [22], whose results are interesting but will not be used in the following.

The long time behaviour of correlation functions is ruled by those singularities of the Laplace transform, better by its analytic continuation, which are closest to the real frequency axis. Oscillatory long time behaviour will not be of interest in this paper. So we assume $\Phi(z)$ to be analytic on a strip around the real axis with the possible exception of the neighbourhood of the frequency origin. If $\Phi''(\omega)$ is integrable also near frequency zero one finds

$$\Phi(t\to\infty) = 0, \tag{1.27a}$$

$$z\Phi(z) \to 0 \quad \text{for} \quad z \to 0, \quad \text{Im } z > 0. \tag{1.27b}$$

The assumption implies the various kernels, introduced in the preceding subsection, to be continuous for small frequencies; e.g. there exists non-degenerate finite matrices

$$K_{\alpha\beta}''(q,\omega\to 0) = D_{\alpha\beta}(q) \quad , \tag{1.28a}$$

$$M_{\alpha\beta}''(q,\omega\to 0) = \Gamma_{\alpha\beta}(q) \quad . \tag{1.28b}$$

According to equ. (1.27a) fluctuations die out if one waits long enough; the system approaches its equilibrium state in the long time limit. Equation (1.27b) implies for the fluctuation dissipation theorem the equality of the static susceptibility $\chi_{AB}(z=0)$ and the corresponding thermodynamical susceptibility $\langle\delta A^*\delta B\rangle/(k_B T)$, which can be obtained as proper second derivation of the free energy:

$$\chi_{AB}{}^{o} = (\delta A|\delta B)/(k_B T) \quad . \qquad (1.29)$$

Equations (1.27a) and (1.29) are usually derived from the ergodicity hypothesis [19]. I will use here equ. (1.27) as definition of ergodic motion and assume that in a liquid all variables exhibit ergodic dynamics.

For the glass it will be assumed that the density fluctuations are non-ergodic and the breaking of ergodicity shall be quantified by the parameters

$$f_{\alpha\beta}(q) = \Phi_{\alpha\beta}(q,t\to\infty) \quad , \qquad (1.30)$$

$$f_s(q) = \Phi_s(q,t\to\infty) \quad . \qquad (1.31)$$

Equivalently one gets

$$z\Phi_{\alpha\beta}(q,z) \to - f_{\alpha\beta}(q) \quad \text{for} \quad z \to 0, \ \text{Im } z > 0 \qquad (1.32)$$

or

$$\Phi_{\alpha\beta}(q,\omega)'' = \pi f_{\alpha\beta}(q)\delta(\omega) + \text{regular terms;} \qquad (1.33)$$

and similar relations hold for the tagged particle. Equations (1.30, 1.31) imply that density fluctuations, once created, do not die out completely even in the limit of infinitely large times. The fluctuations arrest themselves due to strong particle interaction. For large times the system does not approach the thermal equilibrium state. The state which is approached for large times depends on the initial conditions. If one wants to call the state, which is approached for infinite time an equilibrium state one concludes: there is an infinite number of equivalent equilibrium states and the functions $f_{\alpha\beta}(q)$ are to be used to characterize them. From equ. (1.33) it follows, that matrix $f_{\alpha\beta}(q)$ is non negative definite. If it would have an eigenvalue zero, some linear combination of density variables would behave ergodically. Such situation would describe a disordered superionic conductor rather than an ordinary glass. We will not consider this case and assume therefore $f(q)>0$. Similarly, if $f_s(q)=0$ the tagged particle would move ergodically in a glass, a situation of interest in connection with percolation problems. We will not consider this case either and assume $f_s(q)>0$ in the glass phase.

From equs. (1.15, 1.33) one notices the scattering law to exhibit an elastic line

$$S(q,\omega) = \pi\delta(\omega) \sum_{\alpha\beta} \ell_\alpha f_{\alpha\beta}(q) \ell_\beta + \text{regular terms,} \qquad (1.34)$$

a situation familiar from the particle scattering by fixed atoms. So one can interprete the non-ergodicity parameters $f_{\alpha\beta}(q)$ or $f_s(q)$ also as formfactors of the arrested structure. The corresponding quantity for the tagged particle

$$S_s(q,\omega) = \pi\delta(\omega) f_s(q) + \text{regular terms} \qquad (1.35)$$

describes for example the recoil spectrum in nuclear resonance absorption or emission experiments. So the glass, as opposed to the liquid, shows a Mössbauer effect and $f_s(q)$ is the corresponding Lamb-Mössbauer factor.

From equs. (1.8, 1.32) one notices that the matrix of static susceptibilities $X_{\alpha\beta}{}^{o}(q)=\chi_{\rho\alpha,\rho\beta}(z=0)$ differs from the corresponding matrix of thermodynamic susceptibilities $\chi^{th}{}_{\alpha\beta}(q)=\langle\delta\rho_\alpha{}^*(\vec{q})\delta\rho_\beta(\vec{q})\rangle/(k_B T)$ and the difference is quantified by the non-ergodicity parameters

$$\chi^{th}(q) - \chi^{o}(q) = f(q) > 0 \quad . \qquad (1.36)$$

From equs. (1.17) it is evident that non-ergodicity for the density dynamics is equivalent to kernel K to exhibit a leading linear low frequency decrease

$$K_{\alpha\beta}(q,z)/z \to k_{\alpha\beta}(q), \quad \text{for } z \to 0, \quad \text{Im } z > 0, \qquad (1.37)$$

or to kernel M to show a non-ergodicity pole

$$z\ M_{\alpha\beta}(q,z) \to -\ M_{\alpha\beta}(q), \text{ for } z \to 0, \quad \text{Im } z > 0. \qquad (1.38)$$

Similar results hold for the corresponding quantities of the tagged particle. So the assumption of non-ergodic densities implies non-ergodic forces and vice versa. A microscopic understanding is necessary in order to get an overview on the set of relevant non-ergodic variables. It will be the main task of the next section to establish a self-consistent connection between the non-ergodicity parameters $f_{\alpha\beta}(q)$ and $M_{\alpha\beta}(q)$.

1.4 Hydrodynamics

For the many particle system under consideration there are the conservation laws for the species' number, for the energy and for the momentum. The conservation laws determine the independend thermodynamical variables and they imply the low lying long wave length excitations for diffusion of density and heat, for sound propagation and for shear dynamics. A complete discussion of the mutual interplay of hydrodynamic motion with the dynamics relevant at the glass transition is not available at present. Some interesting results in this context can be found in a recent paper by Bengtzelius and Sjögren [23]. I will not enter seriously a discussion of the hydrodynamics near the glass transition. Rather I adopt the view [2], that the glass transition is a phenomenon ruled by short distance physics and the $q \to 0$ implications of the theory are obtained, hopefully, as special limits. In these lectures I will demonstate the simplest aspects of the problem for the tagged particle motion; in this case previous results from the localization theory [7] can be taken over directly.

From equs. (1.19a) and (1.25) one obtains the Green-Kubo identity

$$\Phi_s(q\to 0,z) = \frac{-1}{z + q^2 K_s(z)} \quad ; \tag{1.39}$$

it expresses in leading non-trivial order the long wave length correlation function in terms of the velocity correlations. In the liquid phase the relaxation spectrum is regular and so the hydrodynamic limit is given by a diffusion resonance

$$\Phi_s(q,\omega)'' = Dq^2/[\omega^2 + (Dq^2)^2] \quad . \tag{1.40}$$

The relevant parameter specifying the hydrodynamics is the diffusivity D, given by the zero frequency limit of the velocity or the relaxation spectrum

$$D = K_s''(\omega) = v_s^2/[M_s''(\omega)] \quad , \quad \omega = 0 \quad . \tag{1.41}$$

So the mathematical frame is set up such, that the known results concerning particle diffusion are obtained trivially. In particular one finds from equs. (1.22) and (1.24) the Einstein result for the asymptotic linear displacement increase:

$$\lim_{t\to\infty} S(t)/t = 2D \quad . \tag{1.42}$$

Thus ergodicity is achieved by the particle orbit to extend arbitrarily far out if the time increases. Equilibrium is approached, since the probability of finding the particle in any finite volume tends to zero for time tending to infinity as can be seen by Fourier back transformation of equ. (1.40).

In the glass phase equs. (1.38, 1.39) imply quite a different behaviour. The hydrodynamic regime can be quantified by a length parameter r_s:

$$r_s^2 = K_s'(\omega)/\omega = v_s^2/[-\omega M_s'(\omega)] \quad , \quad \omega = 0 \quad . \tag{1.43}$$

According to equ. (1.33) density fluctuations become static and the asymptotic particle distribution for long wave length is given by

$$f_s(q) = 1/(1 + r_s^2 q^2) \quad . \tag{1.44}$$

From equs. (1.22, 1.24) follow that the mean square displacement approaches a finite constant

$$\lim_{t\to\infty} S(t) = 2r_s^2 \quad . \tag{1.45}$$

The tagged particle is localized and r_s has the physical meaning of a localization length. On hydrodynamic scales the fluid glass transition is a transition from extended to localized motion.

A truly microscopic theory is necessary in order to quantify the range of q and z values, where the preceding formulae are applicable.

2. THE MODE COUPLING APPROXIMATION

2.1 The Derivation of the MCA

In this subsection the general procedure for deriving the mode coupling approximation (MCA) will be discussed. The aim is the evaluation of kernels $M(t)=(F(t)|F')$, as occuring for example in equ. (1.17c), where fluctuating forces $F(t)$ exhibit a time evolution governed by some reduced Liouvillian QLQ.

All dynamical variables, referring to single particle properties, can be expressed in terms of the distribution functions $f_{\alpha,\vec{p}}(\vec{r}) = \Sigma_j \delta(\vec{r}-\vec{r}_\alpha{}^j)\delta(\vec{p}-m_\alpha \vec{v}_\alpha{}^j)$ which refer to species α at position $\vec{r}$ and momentum $\vec{p}$. It is more convenient to introduce fluctuations of wave vector $\vec{q}$ by Fourier transformation with respect to $\vec{r}$ and to introduce a complete set of functions $H_n(\vec{p})$ in $\vec{p}$-space to write $\vec{p}$-dependend functions as series of H_n's. So all single particle variables can be written as linear combinations of

$$A_\lambda = \Sigma_j e^{i\vec{q}\vec{r}_\alpha{}^j} H_n(m^\alpha \vec{v}_\alpha{}^j) \quad , \tag{2.1a}$$

where index λ abbreviates the triple $(\vec{q},\alpha,n)$. Let us assume the λ's to be ordered in some way. Denoting by $N_{\lambda\lambda'}=(A_\lambda|A_{\lambda'})$ the normalization matrix, $P^{(1)}=\Sigma_{\lambda\lambda'}|A_\lambda)N^{-1}{}_{\lambda\lambda'}(A_{\lambda'}|$ is the projector onto the subspace of single particle variables; $Q^{(1)}=1-P^{(1)}$ projects on the space perpendicular to the A_λ's. The subspace of two-particle variables is then spanned by

$$A^{(2)}_{\lambda_1\lambda_2} = Q^{(1)} A_{\lambda_1} A_{\lambda_2} \quad , \quad \lambda_1 < \lambda_2 \quad . \tag{2.1b}$$

The corresponding normalization matrix will be denoted by $N^{(2)}_{\lambda_1\lambda_2,\lambda_1'\lambda_2'} = (A^{(2)}_{\lambda_1\lambda_2}|A^{(2)}_{\lambda_1'\lambda_2'})$ and the projector onto the subspace by $P^{(2)}$. In this way one can proceed to construct three-particle variables etc. It is not necessary to take care particularly for $\lambda_1=\lambda_2$, since in all sums this subset can be ignored in the thermodynamic limit in comparison with the $\lambda_1 \neq \lambda_2$ terms. Thus the function $M(t)$ can be written as

$$\begin{aligned} M(t) = \Sigma_{\kappa,\kappa'} \Sigma_{\lambda_1<\lambda_2\ldots<\lambda_\kappa} \Sigma_{\mu_1<\mu_2<\ldots<\mu_\kappa} \Sigma_{\lambda_1'<\lambda_2'<\ldots<\lambda_{\kappa'}'} \Sigma_{\mu_1'<\mu_2'<\ldots<\mu_{\kappa'}'} \\ (F|A^{(\kappa)}_{\lambda_1\ldots\lambda_\kappa})\, N^{(\kappa)-1}_{\lambda_1\ldots\lambda_\kappa,\mu_1\ldots\mu_\kappa} \\ (A^{(\kappa)}_{\mu_1\ldots\mu_\kappa}(t)|A^{(\kappa')}_{\mu_1'\ldots\mu_{\kappa'}'}) \\ N^{(\kappa')-1}_{\mu_1'\ldots\mu_{\kappa'}',\lambda_1'\ldots\lambda_{\kappa'}'}\, (A^{(\kappa')}_{\lambda_1'\ldots\lambda_{\kappa'}'}|F') \quad . \end{aligned} \tag{2.1c}$$

The essential step of the approximation consists of replacing the correlation of the multiple-particle variables by products of one-particle variables. So only terms with K=K' are kept and for example $(A^{(2)}_{\mu_1\mu_2}(t)|A^{(2)}_{\mu_1'\mu_2'}) \simeq (A_{\mu_1}(t)|A_{\mu_1'})(A_{\mu_2}(t)|A_{\mu_2'}) + (\mu_1'\longleftrightarrow\mu_2')$. Notice in particular, that the single particle variables are assumed to evolve with the full Liouvillian *L*. So one anticipates that the factorization accounts approximately also for the Q-Projector which restricts the time evolution of F(t). At present no theory is available to estimate the range of validity of the factorization approximation.

Contributions to the non-ergodicity singularity of M(t) can be caused only by those terms, where <u>all</u> the correlation functions $(A_{\mu_i}(t)|A_{\mu_j})$ are non-ergodic. So the most singular contribution $\hat{M}(t)$ is due to all the single particle propagators in equ. (2.1c) to be density correlations. The other terms contribute a term $M^0(t)$, which is ergodic in all cases. Single density contributions do not occur, since the Q-projector in F implies $(F|\rho_\alpha(\vec{q}))=0$. So the dominant contributions are due to pairs of density fluctuations and we are going to ignore the rest. Incorporating of triple modes merely increases the formulae without changing the results qualitatively [4]; they have never been considered in any work on liquids either. As a result one gets $M(t)=M^0(t)+\hat{M}(t)$ with the two mode expression

$$\hat{M}(t) = \Sigma_{\lambda_1<\lambda_2}\Sigma_{\mu_1<\mu_2}\Sigma_{\lambda_1'<\lambda_2'}\Sigma_{\mu_1'\mu_2'}$$

$$(F|\rho_{\lambda_1}\rho_{\lambda_2})N^{-1}{}_{\lambda_1\lambda_2,\mu_1\mu_2}$$

$$[(\rho_{\mu_1}(t)|\rho_{\mu_1'})(\rho_{\mu_2}(t)|\rho_{\mu_2'}) + (\rho_{\mu_1}(t)|\rho_{\mu_2'})(\rho_{\mu_2}(t)|\rho_{\mu_1'})]$$

$$N^{-1}{}_{\mu_1'\mu_2',\lambda_1'\lambda_2'}(\rho_{\lambda_1'}\rho_{\lambda_2'}|F') \tag{2.1d}$$

Here index λ or μ abbreviates the pair $(\alpha,\vec{q}):\rho_\lambda=\rho_\alpha(\vec{q})$. Let us introduce the amplitudes

$$v(\mu_1,\mu_2) = \Sigma_{\lambda_1<\lambda_2}(F|\rho_{\lambda_1}\rho_{\lambda_2})N^{-1}{}_{\lambda_1\lambda_2,\mu_1\mu_2} \tag{2.2a}$$

and let us observe obvious symmetries like $v(\mu_1\mu_2)=v(\mu_2\mu_1)$. Then one can rewrite:

$$\hat{M}(t) = \frac{1}{2}\Sigma_{\mu_1\mu_2}\Sigma_{\mu_1'\mu_2'}v(\mu_1,\mu_2)(\rho_{\mu_1}(t)|\rho_{\mu_1'})(\rho_{\mu_2}(t)|\rho_{\mu_2'})v'(\mu_1',\mu_2')^*. \tag{2.2b}$$

The propagator can be expressed in terms of the density correlation functions, equ. (1.14), if one observes translational symmetry: $(\rho_{\alpha_1}(\vec{k}_1t)|\rho_{\alpha_1'}(\vec{k}_1'))=\delta_{\vec{k}_1\vec{k}_1'}\Phi_{\alpha_1\alpha_1'}(k_1,t)$. So one ends up with the general formulae

$$M(t) = M^o(t) + \hat{M}(t) \quad , \tag{2.3a}$$

$$\hat{M}(t) = \int_o^\infty dk \int_o^\infty dp \sum_{\alpha_1\alpha_2\beta_1\beta_2} V_{\alpha_1\alpha_2,\beta_1\beta_2}(k,p)\Phi_{\alpha_1\beta_2}(k,t)\Phi_{\alpha_2\beta_2}(p,t), \tag{2.3b}$$

$$V_{\alpha_1\alpha_2,\beta_1\beta_2}(k,p)=\frac{k^2p^2}{(2\pi)^6} \int d\Omega_K \int d\Omega_p V(\vec{k}\alpha_1,\vec{p}d_2)v'^*(\vec{k}\beta_1,\vec{p}\beta_2), \tag{2.3c}$$

where $d\Omega_k$ and $d\Omega_p$ refer to the angle integration of $\vec{k}$ and $\vec{p}$ respectively.

Various versions of MCAs differ in the following aspects. First, one has to decide which kernels are to be approximated. If the theory should treat the limit of the weakly coupled fluid, one has to introduce kernels in a kinetic equation [7,11]. Since we are interested here in the strong coupling situation only, we simplify the analysis by using as kernels the self-energies of the density fluctuations, equ. (1.17c). Second, one has to derive approximations for the regular part M^o. In simple case M^o may be neglected completely [7]; if strong short ranged interactions are important it has to be approximated so, that binary collision effects are treated properly [11,15]. In these notes it shall be replaced by a constant, describing white noise relaxation processes [2]. Third, the coupling function V is to be evaluated, i.e. approximations for the equilibrium functions $(F|\rho_{\lambda_1}\rho_{\lambda_2})$ and $(\rho_{\lambda_1}\rho_{\lambda_2}|\rho_{\mu_1}\rho_{\mu_1})$ are to be found. In the theory of the Fermi liquid to Fermi glass transition the functions were evaluated in lowest order perturbation theory with respect to the electron impurity interaction [7]. In our case one has to do better because of the strong short range interparticle forces. For monoatomic systems proper factorization approximations have been carried out, so that V is completely expressed in terms of the liquid structure factor [24]. These results will be taken over, using some standard approximation theory for the structure factor as function of density and temperature. Details for the three types of parameters to be used below, can be found in the original work [2].

2.2 The Self-Consistency Equations for the Density Fluctuations

So far numerical work has been done only for simple systems composed of one species of point particles. Generalizations of the theory to several species merely amounts to more involved notations, as was demonstrated above. So at present nothing is gained by sticking to general formulae and from now on I will specialize to one component systems.

The density correlation function, equ. (1.14), shall be normalized by the structure factor S_q: $\Phi_q(z)=\Phi(q,z)/S_q$. With the characteristic fluid frequency

$$\Omega_q^2 = q^2v^2/S_q \tag{2.4a}$$

we can rewrite equs. (1.17) into the form

$$\Phi_q(z) = \frac{-1}{z - \frac{\Omega_q^2}{z+m_q(z)}} \quad . \tag{2.4b}$$

The current relaxation kernel is written according to equs. (2.3) as

$$m_q(z) = [m_q^o(z) + \hat{m}_q(z)]\Omega_q^2 \quad , \tag{2.5a}$$

where $m_q^o(z)$ is assumed not to exhibit a non-ergodicity singularity, while $\hat{m}_q(z)$ is determined by the mode coupling functional $\hat{m}_q(z) = LT[\hat{m}_q(t)](z)$:

$$\hat{m}_q(t) = F_q(\Phi_K(t)) \quad , \tag{2.5b}$$

$$\hat{F}_q(f_K) = \frac{1}{2}\Sigma_{K_1K_2}V(q;K_1K_2)f_{K_1}f_{K_2} \quad . \tag{2.5c}$$

Here and in the following the K-values are assumed to be discretesized. The mode coupling parameters, equ. (2.3c), are symmetric in the K-variables and they are positive. We assume Ω_q and $V(q;K_1K_2)$ as known functions of the controlparameters like temperature T and density n. In these notes the regular part $m_q^o(t)$ is also considered as known and the approximation by a relaxation rate will be considered sufficient for the low frequency regime to be analyzed:

$$m_q^o(z) \approx m_q^o(z \to i0) = i/\tau_q \quad . \tag{2.5d}$$

The formulae can be combined to

$$\frac{\Phi_q(z)}{1 + z\Phi_q(z)} = (z/\Omega_q^2) + m_q^o(z) + LT[\hat{F}_q(\Phi_K(t))](z) \quad . \tag{2.6}$$

This formula is not very transparent, but it will turn out to be convenient for the following analysis.

Equations (2.4,2.5) or, equivalently, equs. (2.5c,2.6) are closed and they are the basis of the mode coupling theory of the liquid glass transition [2]. The non-linearities of the equations are the cause of the subtleties and they occur in two ways. First the current relaxation kernel $m_q(z)$ rules the density fluctuations; the non-linear relation is elementary in frequency space, equ. (2.4b). Second, the current relaxation kernel is given by the density fluctuations; the fluctuating forces come about because interparticle distances may fluctuate. The corresponding non-linear relation between $m_q(t)$ and pairs of density fluctuations is elementary in time space, equs. (2.5b,c). Since time and frequency space are connected by a Laplace transform, one faces non-linear integral equations. Because of the holomorphy requirement of the function $\Phi(z)$, which is equivalent to the spectral relation (1.4), the integral equations are actually singular ones. So one recognizes, that the difficulties of a glass theory, which everybody expects, are not completely eliminated by invention of the approximations (2.4) to (2.6).

2.3 The Self-Consistency Equations for Tagged Particle Motion

It is not essential for the theory of the liquid glass transition to discuss tagged particle motion; but it is very interesting to do so as already mentioned in section 1.2. According to equs. (1.18,1.19) one gets first the Zwanzig-Mori formulae in analogy to equs. (2.4)

$$\Phi_q{}^s(z) = \frac{-1}{z - \dfrac{(\Omega_q{}^s)^2}{z+m_q{}^s(z)}} \quad ; \quad (\Omega_q{}^s)^2 = v_s{}^2q^2 \quad . \tag{2.7}$$

The MCA yields

$$m_q{}^s(z) = [m_q{}^{os}(z)+\hat{m}_q{}^s(z)](\Omega_q{}^s)^2 \quad , \tag{2.8a}$$

where

$$\hat{m}_q{}^s(t) = \hat{F}_q{}^s(\Phi_K(t),\Phi_p{}^s(t)) \quad , \tag{2.8b}$$

$$\hat{F}_q{}^s(f_K,f_p{}^s) = \Sigma_{Kp}\, V^s(q;k;p)f_K f_p{}^s \quad . \tag{2.8c}$$

So the tagged particle dynamics is not a direct implication of the system dynamics; one has to solve again non-linear equations for $\Phi_q{}^s(t)$ even if the $\Phi_K(t)$ are given. This is obvious, since the dynamics of a tagged particle will depend on its interaction V^s with the system, and this interaction information is independend of the information contained in equs. (2.4,2.5). The preceding formulae can be condensed into the equation

$$\frac{\Phi_q{}^s(z)}{1 + z\Phi_q{}^s(z)} = (z/(\Omega_q{}^s)^2) + m_q{}^{os}(z) + LT[\hat{F}_q{}^s(\Phi_K(t),\Phi_p{}^s(t))](z). \tag{2.9}$$

2.4 The MCA for the Shear Correlations

A liquid glass transition will cause anomalies also in such functions which do not become non-ergodic. As a representative interesting example let us consider the correlation function of the transversal currents $\Phi_q{}^T(z)$. The Zwanzig-Mori reduction is carried out in analogy to equs. (1.17):

$$\Phi_q{}^T(z) = \frac{-1}{z + m_q{}^T(z)} \quad , \tag{2.10}$$

$$m_q{}^T(z) = [m_q{}^{oT}(z) + \hat{m}_q{}^T(z)](\Omega_q{}^T)^2 \quad , \tag{2.11a}$$

where the MCA yields

$$\hat{m}_q{}^T(t) = \hat{F}_q{}^T(\Phi_K(t)) \quad , \tag{2.11b}$$

$$\hat{F}_q{}^T(f_K) = \frac{1}{2}\, \Sigma_{K_1K_2} V^T(q;K_1K_2)f_{K_1}f_{K_2} \quad . \tag{2.11c}$$

$\Omega_q{}^T$ is a characteristic fluid frequency [21]. Notice that the singular part of the current relaxation kernel m_q^T is given directly if the density correlations are known. Remember, that the small wave vector limit of kernel $m_q^T(z)$ yields the frequency dependend shear viscosity

$$D_T(z) = \lim_{q\to 0} m_q^T(z)\ n\cdot m/q^2 \quad . \tag{2.12}$$

The zero frequency spectrum is the shear viscosity $\eta = D_T''(\omega=0)$, which enters the Navier Stokes equations.

2.5 Schematic Models for the Self-Consistency Equations

In this subsection three schematic models for mode coupling equations shall be introduced. They are invented to reflect some of the features of the self-consistency equations without exhibiting all of their horrifying complexity.

One might argue, that the true subtleties of the self-consistency equations are due to the mode coupling integral $LT[\Phi_{k_1}(t)\Phi_{k_2}(t)]$ rather than due to the various sums over wave vectors. So it was suggested [1] to ignore the wave vector dependence of the propagator, $\Phi_q(z)\approx\Phi(z)$; as a result one arrives from equs. (2.4,2.5) at the schematic two mode model for some averaged propagator $\Phi(z)$:

$$\Phi(z) = \frac{-1}{z - \dfrac{\Omega^2}{z+[i\nu+m(z)]\Omega^2}} \quad , \tag{2.13a}$$

$$\hat{m}(z) = C\ LT[\Phi(t)^2](z) \quad . \tag{2.13b}$$

The model then is specified by three parameters: a characteristic oscillator frequency Ω, a regular damping term ν and a mode coupling constant C. In liquids near the glass transition point one expects $\Omega\approx 10^{-12}\mathrm{sec}^{-1}\approx 1/\nu$. The preceding reasoning [1] is not acceptable, however, because the liquid propagator $\Phi_q(z)$ is known to vary strongly with q as is already obvious from the strong oscillatory variation of the characteristic frequency Ω_q, equ. (2.4a). Phase space factors and cutoff formfactors imply the integrand in the decay integral, equ. (2.5c), to be strongly weighted for intermediate wave vectors as opposed to small or large k_1, k_2 even if Born approximations are used to evaluate the vertices. Most important is this: the structure factor S_q of dense systems is strongly peaked for $q\approx q_0$ referring roughly to the inverse interparticle distance. Fig. 1 gives a quantitative example. It is very difficult to induce long wave length density fluctuations because of the strong repulsive short ranged interparticle interaction. The strong favouring of density fluctuations for $q\approx q_0$ is reflected by the whole integral to be dominated by this wave vector regime. Thus it is a justified starting approximation to consider mode coupling only on the wave vector shell $q=q_0$. One gets then the schematic model (2.13) with $\Phi(z)=\Phi_{q_0}(z)$; in addition an expression of C in terms of S_q can be obtained. Having solved for $\Phi(z)$ one can substitute the result into the full equations (2.5b,2.8b,2.11b) to find an approximation for $\hat{m}_q(z)$, $\hat{m}_q{}^s(z)$, $m_q{}^T(z)$. The preceding reasoning of Bengtzelius et al. [2] anticipates in particular: the liquid glass transition anomalies are caused by physical phenomena on a short distance scale. For the present theory, as opposed to ordinary second order phase transition theories, small q fluctuations are of no special importance. Kirkpatrick [25] formulated similar ideas but he arrived at a schematic model somewhat more complicated than equs. (2.13).

It was pointed out, that equs. (2.13) exhibit some artificial simplicity [3,4]. To avoid this artifact let me generalize the original procedure by considering two wave vectors as important. So two correlation functions $\Phi_1(z)=\Phi_{q_1}(z)$ and $\Phi_2(z)=\Phi_{q_0}(z)$, shall be considered. Keeping only the most important contributions to the relaxation kernels an extended two mode schematic model is obtained, which shall be written in the compact form of equ. (2.9):

$$\frac{\Phi_1(z)}{1 + z\Phi_1(z)} = (z/\Omega_1{}^2) + i\nu_1 + LT[C\Phi_1(t)^2 + D\Phi_1(t)\Phi_2(t)](z) \quad (2.14a)$$

$$\frac{\Phi_2(z)}{1 + z\Phi_2(z)} = (z/\Omega_2{}^2) + i\nu_2 + LT[E\Phi_1(t)^2](z) \quad . \quad (2.14b)$$

There are then three coupling parameters: C,D,E; for D=0 the original model (2.13) is obtained. In the limit $E\to\infty$ equ. (2.14b) is solved by $\Phi_2(z)=-(1/z)+0(1/E)$, i.e. $\Phi_2(t)=1$. In this limit the second mode behaves statically compared to the first. So the model reduces to equ. (2.13a) with

$$\hat{m}(t) = \lambda_1\Phi(t) + \lambda_2\Phi(t)^2 \quad . \quad (2.14c)$$

Here $\Phi_1(z)=\Phi(z)$, $\Omega_1=\Omega$, $\nu_1=\nu$ and $C=\lambda_2$, $D=\lambda_1$. This schematic model was introduced originally to eliminate some artifacts of the simplified model $\lambda_1=0$ and to study percolation with interaction [3].

2.6 Some Phenomena Described by the Theory

It is useful to understand the features described by the theory in physical terms and in this subsection I will mention some of them briefly.

a. Oscillatory versus relaxation dynamics

The dynamical compressibility $\chi_q(\omega)$, describing the density change of wave vector q and frequency ω as induced by a unit chemical potential $\delta\mu_q$, varying with frequency ω and wave vector q, is obtained from equ. (1.8) and (2.4) as

$$\chi_q(\omega) = \frac{-(q^2/m)}{\omega^2-\Omega_q{}^2+\omega m_q(z)} \quad . \quad (2.15)$$

If one approximates the relaxation kernel by its zero frequency value $m_q(z)\cong i\gamma_q$ function $\chi_q(\omega)$ exhibits the form well known for a harmonic oscillator whose frequency is Ω_q and whose friction constant is γ_q. The contents of formulae (2.15) is then well understood. For $\gamma_q \ll \Omega_q$ the density performs weakly damped oscillations of frequency Ω_q. For $\gamma_q \gg \Omega_q$ the overdamped oscillator relaxes to the equilibrium with a time constant γ_q/Ω_q^2. In liquids one passes from the first case to the second if q increases. The need to describe the mentioned dynamical phenomena for complicated variables, like densities or shear, was the reason to introduce kernels into the theory of correlation functions.

b. Non-exponential decay - one aspect of the problem

Let us consider the transversal current correlation function, equ. (2.10). Restricting ourselves to small wave vectors and replacing the relaxation spectrum by a constant, one gets a Lorentzian spectrum

$$\Phi_q^{T''}(\omega) = \frac{(1/\tau_q)^2}{\omega^2 + (1/\tau_q)^2} \quad ; \quad 1/\tau_q = (q^2\eta/(nm)). \qquad (2.16a)$$

Equivalently one finds exponential decay of correlations

$$\Phi_q^T(t) = \exp(-t/\tau_q) \ , \qquad (2.16b)$$

or the simple equation of motion

$$\partial_t \Phi_q^T(t) = -(1/\tau_q)\Phi_q^T(t) \quad . \qquad (2.16c)$$

The assumption of a white noise spectrum $D_T''(\omega) \approx D_T''(0)$ underlies in one way or the other all phenomenological theories of relaxation processes; Boltzmann's equation can be mentioned as the most prominent example. If approximation $D_T''(\omega) = D_T''(0)$ is too crude, one can extend the theory in the same spirit, by writing $D_T''(\omega)$ as a Lorentzian. In this way a new time scale is introduced. The procedure goes back to Maxwell and it has some merrits as visco elastic theory of liquids [26].

One interesting feature of strongly interacting systems is, that the procedure discussed above, does not work completely. In Fig. 2 I quote $D_T''(\omega)$ for liquid argon as obtained from molecular dynamics work [27]. Obviously the data do not represent a Lorentzian. Detailed anaysis of $D_T''(\omega)$ brought out, that it consists of two contributions. Similarly, $m_q''(\omega)$ in equ. (2.15) reflects two time scales. A broad background peak, as one would have expected, is superimposed by a narrower low frequency peak [27,28]. So the introduction of relaxation spectra like $m_q''(\omega)$ is useful; they can bring out the non-trivial relaxation dynamics more clearly than the spectra like $\Phi_q''(\omega)$.

c. Mode decay kinematics

A relaxation spectrum, expressed in MCA by density fluctuations, can be written in the form

$$\hat{M}_q''(\omega) = \int d(\omega_1/\pi) \sum_{\vec{k}_1} \sum_{\vec{k}_2} w(\vec{q};\vec{k}_1\vec{k}_2) \ \pi\delta(\omega-\omega_1-\omega_2)$$

$$\delta(\vec{q}-\vec{k}_1-\vec{k}_2) \ \Phi_{k_1}''(\omega_1) \ \Phi_{k_2}''(\omega_2) \qquad (2.17)$$

Here equ. (1.5) was employed. The wave vector δ-function, expressing the translational invariance of the system, has been pulled out explicitely of the vertex, so that $w(q,k_1,k_2)$ is a regular function. The angular integrals have not been carried out for the sake of transparency. In the special limit, that the propagator exhibits sharp resonances, one can write $\Phi_q^{(o)''}(\omega)\approx\pi[\delta(\omega-\Omega_q)+\delta(\omega+\Omega_q)]r_q$ and equ. (2.17) reduces to

$$\begin{aligned}\hat{M}_q''(\omega) = \pi \sum_{k_1}\sum_{k_2} w(\vec{q};\vec{k}_1\vec{k}_2)\ \delta(\vec{q}-\vec{k}_1-\vec{k}_2)\ r_{k_1}r_{k_2} \\ \pi\{\delta(\omega-\Omega_{k_1}-\Omega_{k_2}) + \delta(\omega-\Omega_{k_1}+\Omega_{k_2}) \\ + \delta(\omega+\Omega_{k_1}+\Omega_{k_2}) + \delta(\omega+\Omega_{k_1}-\Omega_{k_2})\}\quad . \end{aligned} \tag{2.18}$$

This is a golden rule formula for a decay rate. It consists of four contributions: decay into two modes with wave vectors $(\vec{k}_1,\vec{k}_2)$, absorption of mode $(-\vec{k}_2)$ and decay into $\vec{k}_2$ etc.

I assume that the reader understands the over all properties of formula (2.18) from other areas of physics. If w and r are replaced by their small wave vector asymptotes, $M_q''(\omega)$ will still exhibit a strong ω variation due to the 4-dimensional δ-function. This δ-function expresses the kinematical restrictions for mode decay processes. In liquids Ω_q varies linearly with q for small q and at $q\approx q_0$ it exhibits a roton like minimum, caused by the structure function peak, equ. (2.4a), Fig. 1. The extrema of the Ω_q versus q-curve imply van Hove singularities of $M_q''(\Omega)$ [9]. In particular the large phase space of the decay into two rotons enhances $M_q''(\omega)$ for intermediate ω. On the basis of equ. (2.18) one would not expect a white noise relaxation spectrum at all. Since, however, in classical liquids the peaks of $\Phi_q''(\omega)$ are rather broad, equ. (2.18) is a poor approximation of equ. (2.17). The whole structure of $M_q''(\omega)$ gets washed out. The kinematic restrictions are weakend and merely imply $M_q''(\omega)$ to exhibit a broad bumb. In this way one feature of the non-exponential decay, the large scale ω-variation of the relaxation spectrum, can be understood microscopically [10].

A special case of mode decay kinematics is the following. For $q\to 0$, $\omega\to 0$ propagators usually can be expressed exactly by their hydrodynamic limit. Then it is easy to evaluate $M_q''(\omega)$ in equ. (2.17) for $q\to 0$ and $\omega\to 0$. The low lying hydrodynamic excitations, like e.g. the diffusion resonance, equ. (1.40), then often imply strong non-analytical variations of the spectra: $M_q''(\omega)\propto|\omega|^{\alpha}$. As a result one finds spectacular deviations from exponential decay: $M_q(t)\propto 1/|t|^{\alpha+1}$ [29].

The preceding remarks demonstrate, that the MCA can achieve more physical insight than a mere parametrization of relaxation kernels. Neither the mentioned large scale variation of the kernels nor the long time anomalies due to hydrodynamic excitations seem to be important for the liquid glass transition.

d. Self-consistent relaxation enhancement

Crucial for the dynamics of liquids, as opposed to the one of dense gases, in particular of supercooled liquids and of the glass transition is the self-consistency feature expressed by equ. (2.17). To understand the point let us contemplate an iterative solution of our equations, dropping m^0 in equ. (2.5a) for the sake of simplicity. In zeroth order $\Phi_q^{(0)''}(\omega)$ does not show any spectrum at all for $\omega=0$. These modes then yield $M^{(0)''}(\omega)$ via equ. (2.18). The processes with $\omega = \pm(\Omega_{k_1}-\Omega_{k_2})$ then lead to $M^{(0)''}(\omega=0)>0$. Substitution of $M^{(0)}(z)$ into equ. (2.4b) yields the first approximation for the density spectrum. One finds $\Phi_q^{(1)''}(\omega\approx 0)=M^{(0)''}(\omega\approx 0)/\Omega_q^2>\Phi_q^{(0)''}(\omega\approx 0)=0$. Substitution of $\Phi_q^{(1)''}(\omega)$ into equ. (2.17) yields, in addition to the kinematically allowed transition contributions, contributions due to all $\Phi_q^{(1)''}(\omega\approx 0)$ for $\vec{q}=\vec{k}_1+\vec{k}_2$. As a result $M^{(1)''}(\omega\approx 0)>M^{(0)''}(\omega\approx 0)$. Substitution into equ. (2.4b) yields $\Phi_q^{(2)''}(\omega\approx 0)=M^{(1)''}(\omega\approx 0)/\Omega_q^2>M^{(0)''}(\omega\approx 0)/\Omega_q^2=\Phi_q^{(1)''}(\omega\approx 0)$, etc. So there is a positive feedback between the low frequency enhancement of the density spectrum $\Phi_q''(\omega\approx 0)$ and the relaxation spectrum $m_q''(\omega\approx 0)$. Since $\int\Phi_q''(\omega)d\omega$ and $\int m_q''(\omega)d\omega$ are fixed by sum rules the low frequency enhancements are connected with suppression of the spectra for intermediate and large frequencies.

The preceding mechanism was proposed [10] as explanation of the two time scale feature of the spectra of dense liquids mentioned in section 2.6b. Fig. 2 shows a quantitative result for $D_T''(\omega)$ of liquid argon near the triple point.

The $\omega=0$ enhancement is stabilized because of the regular term m^0 in equ. (2.5a) and also because the $\omega_1,\omega_2\neq(0,0)$ contributions in equ. (2.17) get suppressed. But in connection with a theory for the Fermi liquid-Fermi glass transition it was observed [7], that stabilization can be achieved only for the coupling w to be lower than some critical w_c. If w approaches w_c the $\omega=0$ spectra diverge and this point was indentified as a transition point to non-ergodic density dynamics. The same observation is correct for the present model, as will be discussed below.

So the known non white relaxation spectra of strongly coupled fluids are explained by the MCA as outcome of the self-consistency to be observed for mode decay processes. Anticipating the results of section 4 one can consider e.g. the narrow peak of the dynamical viscosity $D_T''(\omega)$, Fig. 2, as glass formation precursor.

e. Non-exponential decay - another aspect of the problem

The preceding discussion started with the usual picture for homogeneous line broadening. It was studied, how a density fluctuation decays into pairs of excitations. Non white spectra, i.e. non-exponential decay of correlations with increasing time, resulted from the fact, that for some frequencies more decay channels where open, possibly connected with large coupling coefficients, than for others. It is evident, that this picture becomes incomplete and even wrong, if the self-consistency requirement rules the results. So at least at the divergency point $w\approx w_c$ another aspect of the decay problem is to be understood. Let me explain the point for the schematic model; let me also neglect the regular damping term ν in equ. (2.13a). We intend to solve the equations of motion in the large C limit. The solution shall be found with the ansatz $|\chi(z)|\ll 1$, where

$$\chi(z) = -\Omega^2/z(z+\hat{m}(z)\Omega^2) \quad . \tag{2.19a}$$

One gets:

$$\Phi(z) = -\frac{1}{z} + \frac{1}{z}\chi(z)\ (1+O(\chi)) \quad . \tag{2.19b}$$

The MCA (2.13b) then yields

$$\hat{m}(z) = C[-\frac{1}{z} + \frac{2}{z}\chi(z)\ (1+O(\chi))] \quad . \tag{2.19c}$$

Substitution of $\hat{m}$ into equ. (2.19a) and dropping the $O(\chi)$ terms leads to the self-consistency equation

$$\chi(z) = \frac{-\Omega^2}{z^2 - \Omega_0{}^2 + 2\,\Omega_0{}^2\chi(z)} \quad ; \quad \Omega_0{}^2 = \Omega^2 C. \tag{2.20a}$$

The solution is obtained readily:

$$\chi(z) = \{-[z^2-\Omega_0{}^2] + \sqrt{[z^2-\Omega_0{}^2]^2 - 8\Omega^2\Omega_0{}^2}\}/(4\Omega_0{}^2) \quad . \tag{2.20b}$$

The discussion of this result is left to the reader and I want to stress only four features. First, $|\chi(z)|<|\chi(z=\Omega_0)|=1/\sqrt{2C}$, so that indeed the ansatz was justified and yields the solution for $\sqrt{C}\gg 1$. Second, $\chi''(\omega)$ describes resonances of width Ω centered at $\pm\Omega_0$; the line shape deviates strongly from a Lorentzian in the sense that the wings are suppressed. So $\Phi(t)$ exhibits non-exponential decay. The square root singularities of $\chi''(\omega)$ are artifacts and they disappear, if $\nu\neq 0$ is taken into account by replacing z^2 by $z(z+i\gamma)$ in equ. (2.20b). Third, the solution found describes non-ergodic density fluctuations, since one finds from equ. (2.19b):

$$\lim_{\omega\to 0}\Phi''(\omega) = \pi\delta(\omega)\ (1-\chi_0); \quad \chi_0 = \frac{1}{C}\ (1+O(\chi_0)) \quad . \tag{2.20c}$$

Fourth, remembering the formula $-\zeta+\sqrt{\zeta^2-\omega_0{}^2}=\int d\omega\Theta(\omega_0{}^2-\omega^2)\sqrt{\omega_0{}^2-\omega^2}/(\omega-\zeta)\pi$ one can rewrite the result for $\chi(z)$ into the form

$$\chi(z) = \int_0^\infty d\sigma^2\ P(\sigma^2)\ \frac{-1}{z^2-\sigma^2} \quad . \tag{2.20d}$$

$$P(\sigma^2) = \frac{1}{(4\pi\Omega_0{}^2)}\ \Theta[8\Omega^2\Omega_0{}^2-(\sigma^2-\Omega_0{}^2)^2]\sqrt{8\Omega^2\Omega_0{}^2-(\sigma^2-\Omega_0{}^2)^2} \quad . \tag{2.20e}$$

So $\chi(z)$ describes the superposition of harmonic oscillator responses whose frequency squares, σ^2, are distributed continuously around the average value $\Omega_0{}^2$. The continuous non-Lorentzian spectrum $\chi''(\omega)$ thus describes inhomogeneous line broadening. In the non-ergodic phase the frozen density fluctuations provide a distribution of traps. The particles rattle in these traps. Since the characteristic rattling frequencies exhibit a continuous distribution, density fluctuations will decay via dephazing. As known from other fields, this leads to non-exponential decay of correlations.

3. THE GLASS INSTABILITY

The glass form factors f_q are obtained as solution of the mode coupling equations by considering a special limit. Multiplying equ. (2.6) by z and carrying out the static limit $z \to 0$ one finds with equs. (1.30) (1.32) [2]:

$$f_q/(1-f_q) = \hat{F}_q(f_K) \qquad (3.1)$$

Notice, that the regular kernel $m_q^o(z)$ does not enter the implicit equation (3.1). The non-ergodicity parameters are solely determined by the mode coupling coefficients $V(q;k_1k_2)$, equ. (2.5c). In this section some aspects of the solution of equ. (3.1) will be analyzed.

3.1 Notations

In this subsection some notations will be compiled for later use. First, let us introduce a matrix, to be referred to as stability matrix, which is given by the Jacobian of the functional $\hat{F}_q$:

$$C_{qK} = (\partial \hat{F}_q/\partial f_K)(1-f_K)^2$$

$$= \Sigma_p f_p \; V(q;kp)(1-f_K)^2 \quad . \qquad (3.2)$$

All the numbers C_{qK} are non-negative and thus the theorems of Perron and Frobenius [30] can be applied: there is a largest real eigenvalue $E_o>0$ so that the eigenvalues E_α, $\alpha=0,1,2,...$ of C_{qK} obey the inequality

$$|E_\alpha| < E_o \quad , \qquad \alpha = 1,2,... \qquad (3.3a)$$

The largest eigenvalue E_o is non-degenerate, and so the corresponding eigenvectors

$$\Sigma_k C_{qk} \ell_K = E_o \ell_k \; ; \quad \Sigma_q \hat{\ell}_q C_{qk} = E_o \hat{\ell}_k \qquad (3.3b)$$

are determined uniquely up to a factor. The ℓ_K or the $\hat{\ell}_K$ are all of equal sign and nonzero, so that one can impose

$$\ell_K > 0 \quad ; \quad \hat{\ell}_K > 0 \quad . \qquad (3.3c)$$

Actually, the mentioned theorems require the reservation that C_{qK} is not equivalent to a block matrix. If this accident happened, a small correction added to the vertices would restore the regular situation $C_{qK} \neq 0$. So generically the preceding formulae are valid. Equation (3.3c) determines ℓ_K up to the transformation $\ell_K \to \alpha \ell_K$, $\alpha>0$ and $\hat{\ell}_K \to \beta \hat{\ell}_K$, $\beta>0$. Let us impose

$$\Sigma_K \hat{\ell}_K \ell_K = 1 \quad . \qquad (3.3d)$$

This convention fixes β for given α. To simplify later formulae, we fix α by imposing also

$$\Sigma_K \ell_K \ell_K \hat{\ell}_K (1-f_K) = 1 \quad . \qquad (3.3e)$$

Then both eigenvectors are fixed uniquely. Let us mention that for the extended simplified model one finds: $C_{11}=2Cf_1(1-f_1)^2$; $C_{12}=Df_1(1-f_2)^2$,

$C_{21}=2Ef_1(1-f_1)^2$, $C_{22}=0$. The evaluation of E_0 and of ℓ_K, $\hat{\ell}_K$ is elementary but cumbersome in this case. One finds E_0 to be non-degenerate despite $C_{22}=0$.

We will also need the second variations of functional $\hat{F}_q(f_K)$:

$$C_{q,Kp} = (\frac{1}{2}\, \partial^2 F_q/\partial f_K \partial f_p)(1-f_K)^2(1-f_p)^2$$

$$= \frac{1}{2}\, V(q;Kp)(1-f_K)^2(1-f_p)^2 \quad . \tag{3.4}$$

Later only the average of these quantities, formed with the eigenvectors will enter:

$$\lambda = \Sigma_q \, \Sigma_{Kp} \, \hat{\ell}_q \, C_{q,Kp} \, \ell_K \, \ell_p > 0 \quad . \tag{3.5}$$

This number will be refered to as the exponent parameter for reason which become obvious in section 4. For the simplified model one gets $C_{1,11}=C(1-f_1)^4$, $C_{1,12}=D(1-f_1)^2(1-f_2)^2/2$, $C_{1,22}=0$, $C_{2,11}=E(1-f_1)^4$, $C_{2,12}=C_{2,22}=0$.

Finally, variations of $\hat{F}_q(f_K)$ due to changes of controlparameters for fixed f_K are needed. Controlparameters of interest are densities, temperature or pressure. We indicate only one of them, to be denoted by n. Let us write

$$C_q = n(\partial \hat{F}_q/\partial n)$$

$$= \frac{1}{2}\, n \, \Sigma_{Kp} \, f_K f_p \, [\partial V(q{:}kp)/\partial n] \quad , \tag{3.6}$$

and introduce the number

$$\mu = \Sigma_q \, \hat{\ell}_q \, C_q \quad . \tag{3.7}$$

For the simplified model one may identify n=C and consider D and E as constant. Then one gets $C_1=Cf_1{}^2$, $C_2=0$.

In subsection 3.3 an efficient procedure will be discussed for solving equ. (3.1), for evaluating E_0 and $\hat{\ell}_K, \ell_p$. Anticipating these results one should notice that the evaluation of all the preceding matrices C_q, C_{qK}, $C_{q,Kp}$, vectors $\hat{\ell}_q$, ℓ_K, and numbers E_0, λ, μ is straightforward.

3.2 The Instability Point

In this subsection it will be shown, that equ. (3.1) has a solution for large coupling which becomes unstable for small coupling. The instability point will be identified by the eigenvalue E_0, equ. (3.3a), to approach unity.

Let us symbolize an averaged coupling vertex by v. For large v equ. (3.1) can be solved by Taylor expansion in (1/v) [7]. Let us note only the leading term in the expansion

$$f_q = 1 - f_q^{[1]} + f_q^{[2]} - \ldots \quad , \tag{3.8a}$$

where $f_q^{[1]}=0(1/v^1)$. Substitution into equ. (3.1) leads to

$$f_q^{[1]} = 1/A_q \quad , \quad A_q = \hat{F}_q(f_K=1) = \frac{1}{2}\,\Sigma_{Kp}\, V(q;Kp) \quad . \qquad (3.8b)$$

This result can be substituted into equ. (3.2) to yield the leading approximation for the stability matrix:

$$C_{qK} = \Sigma_p[V(q;Kp)/A_p{}^2] + 0(1/v^2) \quad . \qquad (3.9a)$$

In particular one obtains

$$E_0 = 0(1/v) \quad . \qquad (3.9b)$$

So one generates by trivial iteration a large coupling solution $f_q \lessapprox 1$. It has the meaningful property: with increasing coupling f_q increases towards unity.

Using standard theory of implicit functions one can study the variation of the found solution due to changes of the controlparameter. Writing

$$\delta f_K = (1-f_K)^2 g_K \qquad (3.10)$$

one gets from equ. (3.1) with the notations of the preceding subsection

$$\begin{aligned}\Sigma_K(\delta_{qK}-C_{qK}) &= C_q(\delta n/n)\\ &\quad - (1-f_q)g_q{}^2 + \Sigma_{Kp}\, C_{q,Kp}\, g_K\, g_p\\ &\quad + 0(g\delta n, \delta n^2, g^3). \end{aligned} \qquad (3.11)$$

Because of $C_{qK}=0(1/v)$, equ. (3.9a), there exists the resolvent R_{qK}:

$$\Sigma_q R_{\ell q}(\delta_{qK}-C_{qK}) = \delta_{\ell K} \quad . \qquad (3.12a)$$

It can be evaluated by the Neumann series:

$$R_{k_0 k_\ell} = \Sigma_{\ell=0}^{\infty}\, C_{k_0 k_1}\, C_{k_1 k_2} \ldots C_{k_{\ell-1} k_\ell} \quad . \qquad (3.12b)$$

Thus

$$g_q = [\Sigma_K\, R_{qK}\, C_K](\delta n/n) + 0(\delta n^2) \quad . \qquad (3.13)$$

So the solution is a continuous function of n. Result (3.13) can be used to iteratively construct the solution as long as $E_0<1$.

Let us now examine the problem that E_0 tends to unity for the controlparameter n approaching some critical value n_c. For definity we assume $E_0<1$ for $n>n_c$, so that

$$1 - E_0 \to 0+ \quad , \quad \epsilon = (n-n_c)/n_c \to 0+ \quad . \qquad (3.14)$$

We stick to the generic result (3.3a) for all ϵ:

$$|E_\alpha| \leqslant r < 1 \quad , \quad \alpha = 1,2,\ldots \quad . \qquad (3.15)$$

The solution constructed above will tend to some value

$$f_q \to f_q^c \quad , \quad \epsilon \to 0+ \tag{3.16}$$

Similarly, all the matrices entering equ. (3.11) tend to some limiting value like C_q^c, C_{qK}^c, $C_{q,Kp}^c$. Equation (3.11) can be solved with the expansion $g_q = \sqrt{\epsilon}\, g_q^{[1]} + \epsilon g_q^{[2]} + O(\epsilon\sqrt{\epsilon})$. One finds

$$\begin{aligned}\Sigma_K(\delta_{qK}-C_{qK}^c)g_K &= \epsilon\, C_q^c \\ &\quad - (1-f_q^c)g_q^2 + \Sigma_{Kp}\, C_{q,Kp}^c\, g_K\, g_p \\ &\quad + O(\epsilon\sqrt{\epsilon}) \quad .\end{aligned} \tag{3.17a}$$

So in leading order

$$\Sigma_K(\delta_{qK}-C_{qK}^c)g_K^{[1]} = 0 \quad . \tag{3.17b}$$

Thus $g_K^{[1]}$ is eigenvector of the stability matrix at the critical point corresponding to the largest eigenvalue $E_0^c=1$, equ. (3.3b). Since this eigenvalue is non-degenerate one gets

$$g_K^{[1]} = \alpha \ell_K^c \tag{3.17c}$$

with an unspecified factor α. The next to leading order equation reads then:

$$\Sigma_K(\delta_{qK}-C_{qK}^c)g_K^{[2]} = C_q^c + [-(1-f_q^c)\ell_q^c\ell_q^c + \Sigma_{Kp} C_{q,Kp}^c \ell_K^c \ell_p^c]\alpha^2 . \tag{3.17d}$$

For this equation to be solvable, the right hand side has to perpendicular to the left eigenvector of the matrix $(1-C^c)$. The latter is the uniquely determined vector $\hat{\ell}_q^c$, equ. (3.3b). Observing the convention (3.3e) and using the notations (3.5), (3.7) one gets

$$(1-\lambda^c)\alpha^2 = \mu^c \quad . \tag{3.17e}$$

The possibility $1-\lambda^c=0$ can be ignored as non-generic. $\alpha^2<0$ is impossible, since there was a real solution f_q for $\epsilon>0$. So one gets $\mu^c/(1-\lambda^c)>0$ and near the critical point we find two solutions of equ. (3.1); one corresponds to

$$\alpha = \sqrt{\mu^c/(1-\lambda^c)} > 0 \quad , \tag{3.18a}$$

and the other one is obtained by changing $\alpha \to -\alpha$. The critical point is characterized by the constructed solution to merge with some other solution. For $\epsilon \to 0$ the two solutions of equ. (3.1) read

$$f_q^{\pm} = f_q^c \pm \sqrt{\epsilon}\, \alpha\, (1-f_q^c)^2\, \ell_K^c \quad . \tag{3.18b}$$

Let us consider the behaviour of the largest eigenvalue of the stability matrix corresponding to the two solutions found. We write

$$C_{qK}^{\pm} = C_{qK}^c + \delta C_{qK}^{\pm} \tag{3.19a}$$

and find from equs. (3.2, 3.4)

$$\delta C_{qK}^{\pm} = \mp 2\sqrt{\epsilon}\,\alpha\,[C_{qK}{}^{c}(1-f_K{}^{c})\ell_K{}^{c} - \Sigma_p\, C_{q,Kp}{}^{c}\,\ell_p{}^{c}] + O(\epsilon) \quad . \qquad (3.19b)$$

Thus the eigenvalue equ. (3.3b) reads

$$\Sigma_K(C_{qK}{}^{c}+\delta C_{qK}{}^{\pm})(\ell_K{}^{c}+\delta\ell_K{}^{\pm}) = (1+\delta E_0{}^{\pm})(\ell_q{}^{c}+\delta\ell_q{}^{\pm}) \quad . \qquad (3.20a)$$

In leading order $\sqrt{\epsilon}$ one gets

$$\Sigma_K(\delta_{qK}-C_{qK}{}^{c})\delta\ell_K{}^{\pm} = \Sigma_K\delta C_{qK}{}^{\pm}\ell_K{}^{c} - \delta E_0{}^{\pm}\ell_q{}^{c} \quad . \qquad (3.20b)$$

For this equation to be solvable $\hat{\ell}_q$ has to be perpendicular to the right hand side. Observing equs. (3.3d,e) and equ. (3.5) one gets

$$E_0{}^{\pm} = 1 \mp 2\sqrt{\epsilon\alpha}\,(1-\lambda^{c}) \quad . \qquad (3.21)$$

Now we assume $\mu^{c}>0$. If this was not the case, one merely has to redefine the controlparameter by changing $\epsilon\to-\epsilon$. Thus one gets

$$0 < \lambda^{c} < 1 \quad . \qquad (3.22)$$

Then the constructed solution of equ. (3.1), characterized by $E_0<1$ corresponds to the + alternative in the preceding results. The solution f_q^- of equ. (3.1) is not connected continuously with the strong coupling solution.

Let us also contemplate the weak coupling situation $v\to0$. Since the r.h.s. of equ. (3.1) vanishes one finds $f_q\to0$. Since then $f_q/(1-f_q)=O(f)$, while $F_q(f_K)=O(v)O(f^2)$, equ. (3.1) cannot be solved except by the trivial solution $f_q\equiv0$.

Summarizing one concludes that either $E_0<1$ for all values of the controlparameter. Then there is always a glass form factor; the glass is stable in the whole region of n-values. Or there is a critical value n_c so that for $n\to n_c$, $E_0\to1$. Then the glass becomes unstable for $n\to n_c$. This situation occurs in particular, if the controlparameter can drive the mode coupling parameters into the weak coupling limit. In the following we consider the situation, that n_c exists.

For the schematic model, equ. (2.14), the equ. (3.1) can be rewritten in the form $1/(1-f_1)=G(f_1)=Cf_1+D\alpha/(1+\alpha)$; $\alpha=Ef_1{}^2$, $f_2=\alpha/(1+\alpha)$. So the solution can be discussed as intersection of the $1/(1-f_1)$ versus f_1 curve with the graph of $G(f_1)$. Fig. 3 shows an example. For $C=n>C_0=80/63$ there are two solutions. The one for the larger f_1 increases towards the strong coupling solution if C increases. Both solutions collapse for $C\to C_c$ with $f_1{}^{c}=1/4$, $f_2{}^{c}=15/16$. For $C<C_c$ there is no solution for the non-ergodicity parameters; $\lambda^{c}=0.611$.

3.3 The Fixed Point Equation and the Transition Scenario

In this subsection we want to show, that the solution of equ. (3.1) can be found by a simple iteration procedure. So let us consider the evolution equation

$$f_q^{(i)}/(1-f_q^{(i)}) = \hat{F}_q(f_K^{(i-1)}) \quad , \quad i=1,2,... \qquad (3.23)$$

which creates from a given initial value $f_q^{(o)}$ a sequence of functions $0<f_q^{(i)}<1$ by iterated mapping. If $f_q^{(i)}$ converges towards a fixed point f_q, equ. (3.23) offers a convenient numerical tool to find a set of non-ergodicity parameter f_q. Near a possible fixed point one can write $\delta f_q^{(i)}=f_q^{(i)}-f_q=(1-f_q)^2 g_i^{(i)}$ and one finds the linearized evolution equation

$$g_q^{(i)} = \Sigma_q C_{qK} g_K^{(i-1)} \quad . \qquad (3.24)$$

So the stability matrix rules the iteration close to a fixed point.

Because of equ. (3.3a) all eigenvalues of matrix C have a modulus smaller than unity. Hence the iteration (3.24) is contracting: $g_q^{(i)}\to 0$. Hence the fixed point discussed in section 3.2 as solution of equ. (3.1) for $n>n_c$ is attractive. If $f_q^{(o)}$ is in a neighbourhood of f_q, this function can be obtained by the convergent iteration (3.23).

If the critical point is approached, the convergence of iteration (3.23) becomes slowlier because of equ. (3.14). The second fixed point f_q^-, equ. (3.18b), whose existence was demonstrated for $\epsilon\to 0$, is unstable because $E_o^->1$, equ. (3.21). So the critical point is characterized by the stable physical fixed point to collapse with an unstable unphysical fixed point.

If one approaches n_c, iteration (3.23) becomes difficult. But near n_c equ. (3.18b) gives the solution directly, if one can evaluate $\ell_K{}^c$. The evaluation of the eigenvectors corresponding to the largest eigenvalue is easily done with the aid of the Neumann series [30]:

$$(1/N)\Sigma_{\ell=1}^{N}(C/E_o)^{\ell}_{Kp} = \ell_K\hat{\ell}_p + O(1/N) \quad . \qquad (3.25)$$

This formula determines ℓ_K and $\hat{\ell}_p$ by a monotonous iteration up to some constant.

One can see from Fig. 3 that the large f_1 solution for the schematic model is a stable fixed point for the iteration $1/(1-f_1^{(i)})=G(f_1^{(i-1)})$, while the small f_1 solution is unstable. If $f_1^{(o)}$ is located to the right of the unstable fixed point, the iteration converges towards the physical solution. In the other case $f_1^{(i)}\to 0$. If $C<C_c$ one gets $f_1^{(i)}\to 0$ also.

For the hard sphere model the iteration (3.23) was carried out [2]. Fig. 4 shows the formfactor for $n>n_c$ and for $n\approx n_c$; the corresponding structure factors S_q, which determine the vertices, are shown in Fig. 1. Originally [4], the preceding discussion of the critical point was formulated as a scenario explaining the numerical results of Bengtzelius et. al. [2] generically. In these lectures the Perron-Frobenius theorems have been used in order to proof, that the discussed instability is the only generic possibility described by the mode coupling theory of simple one component systems.

It is plausible, that not all of the preceding results hold for multicomponent systems, where the f in equ. (3.1) are to be read as positive definite hermitian matrices. For the general case, the only case of interest in real experiment, our present lack of knowledge forces us to stick to the original idea of considering our discussion as a scenario.

3.4 Static Quantities near the Instability Point

In this subsection the preceding results shall be summarized by formulating the asymptotic behaviour of formfactors and susceptibilities if the glass instability point is approached. The main result is given by equs. (3.18) and the discussion following it:

$$f_q = f_q^c + h_q \sqrt{\epsilon} + O(\epsilon) \quad , \qquad \epsilon \to +0, \tag{3.25a}$$

$$h_q = \alpha(1-f_q^c)\ell_q^c > 0 \quad . \tag{3.25b}$$

So for $n \to n_c$ the glass formfactor, i.e. the intensity of elastic coherent scattering, decreases rapidly before it exhibits a discontinuity.

We anticipate that singular variations of other quantities are governed by the ones found for the densities. An example of this kind was the variation of the two eigenvalues of the stability matrix, equ. (3.21). Let me illustrate the problem for two other cases.

The transversal current relaxation kernel of the glass phase exhibits a non-ergodicity pole, which is obtained easily if one substitutes equ. (3.25a) into equs. (2.11):

$$-zm_q^T(z) \longrightarrow (\Omega_q^T)^2 f_q^T \quad , \qquad z \to 0, \tag{3.26a}$$

$$f_q^T = f_q^{TC} + h_q^T \sqrt{\epsilon} + O(\epsilon) \quad , \tag{3.26b}$$

$$f_q^{TC} = \frac{1}{2} \Sigma_{kp} V^T(q;kp) f_k^c f_p^c \; , \tag{3.26c}$$

$$h_q^T = \Sigma_{kp} V^T(q;kp) f_k^c h_p \quad . \tag{3.26d}$$

The relevance of these results will become obvious in section 4.

The discussion of tagged particle motion is not so straightforward. From equ. (2.9) one finds the equation for the non-ergodicity parameter

$$f_q^s/(1-f_q^s) = \hat{F}_q^s(f_K;f_p^s) \quad . \tag{3.27}$$

This is a result quite similar to equ. (3.1) and the machinery of subsections (3.1, 3.2) can be applied. Obviously there are two alternatives. First, while varying the controlparameter n the maximum eigenvalue E_0^s of the stability matrix constructed for equ. (3.27) approaches unity if $n \to n_c^s > n_c$. This would imply a localization to non-localization phase transition for the tagged particle in the glass phase of the matrix. In the following we will not consider this interesting type of transition. Rather we will focus on the second alternative: $E_0^s < 1$ for all $n > n_c$. Then the solution of equ. (3.27) can be found as attractive fixed point of the evolution equation

$$f_q^{s(i)}/(1-f_q^{s(i)}) = \hat{F}_q^s(f_K, f_p^{s(i-1)}) \quad . \tag{3.28}$$

Fig. 5 shows the result of such iteration obtained for the hard sphere system [2]. The dominant variation of $\hat{F}_q^s$ near n_c is due to the singular variation of f_K, equ. (3.25a). So with

$$C_{qp}^s = (\partial\hat{F}_q^s/\partial f_p^{(s)})(1-f_p^{(s)})^2 \quad , \tag{3.29}$$

$$C_{qK}' = (\partial\hat{F}_q^s/\partial f_K) \quad , \tag{3.30}$$

one gets for $\delta f_K^s=(1-f_K^{sc})^2 g_K^s$ the equation

$$\Sigma_K(\delta_{qK} - C_{qK}^s)g_K^s = \Sigma_K C_{qK}' \; h_K \sqrt{\epsilon} \tag{3.31}$$

The resolvent $R_{\ell q}^s$ can be evaluated by the Neumann series in analogy to equs. (3.12). As a result one finds

$$f_q^s = f_q^{sc} + h_q^s \sqrt{\epsilon} + O(\epsilon) \quad , \quad \epsilon \to 0+, \tag{3.32a}$$

$$h_q^s = (1-f_q^{sc})^2 \, \Sigma_{Kp} \, R_{qK}^s \, C_{Kp}' \, h_p > 0 \quad . \tag{3.32b}$$

Thus the Lamb-Mössbauer factor f_q^s, the intensity factor for the elastic incoherent scattering, exhibits a similar asymptotic form as found for the glass form factor.

A qualitative difference between coherent and incoherent functions should be emphasized. In defining the relevant funtionals $\hat{F}_q$, $\hat{F}_q^s$, $\hat{F}_q^T$ the original mode coupling expressions have been divided by the characteristics frequencies squares Ω_q^2, Ω_q^{s2}, Ω_q^{T2} (equs. 2.5a, 2.8a, 2.11a). These squares vanish like q^2 for small wavevectors. In the case of the density and current correlations the corresponding vertices vanish also for $q\to 0$ because of the momentum conservation law. Hence the functionals $\hat{F}_q$ and $\hat{F}_q^T$ are well behaved for $q\to 0$. The q=0 point is nothing special in those cases as shown for example in Fig. 4. Momentum conservation does not hold for the tagged particle. Consequently, $\hat{F}_q^s \propto 1/q^2$ for small q. As a result equ. (3.23) implies $f_q^s \to 1$ for $q\to 0$ in the glass phase in agreement with equ. (1.44), Fig. 5. Expanding $f_q^s \to 1-r_o^2 q^2 + O(q^4)$ one gets from equ. 3.27

$$1/(r^s)^2 = \hat{F}_q^s(f_K, f_p^s) q^2 \quad , \quad q = 0 \quad . \tag{3.33}$$

Substitution of equs. (3.25, 3.32) yields the expansion

$$r^s = r^{sc} \, (1 - \frac{1}{2} \, a \, \sqrt{\epsilon} + O(\epsilon)) \tag{3.34a}$$

where

$$a = (r^{sc}q)^2 \, \Sigma_K[(\partial\hat{F}_q^s/\partial f_K)h_K + (\partial\hat{F}_q^s/\partial f_K^s)h_K^s] \quad , \quad q \to 0 \quad . \tag{3.34b}$$

is a positive number. So the localization length increases rapidly up to a finite value r^{sc}. The finite length r^{sc}, which cannot be exceeded in the solid phase, is the present theories quantification of the Lindemann melting criterion (in ref. 4 the second contribution on the r.h.s. of equ. (3.34) was forgotten).

4. LOW FREQUENCY EXCITATIONS NEAR THE TRANSITION POINT

4.1. The Critical Density Spectrum

In this subsection the density correlation functions at the transition point $n=n_c$ will be evaluated for low frequencies. Let us start by exactly rewriting the self consistency equations. We introduce a not yet specified function $0<f_q<1$ and define a correlation function $F_q(t)$ by

$$\Phi_q(t) = f_q + (1-f_q)^2 F_q(t) \quad . \tag{4.1}$$

This is equivalent to $\Phi_q(z)=(-f_q/z)+(1-f_q)^2F_q(z)$. With the notations (3.3, 3.4) one finds for the relevant kernel, equ. (2.5b):

$$m_q(z) = -\frac{1}{z}\hat{F}_q(f_K) + \Sigma_K C_{qK}F_K(z) + \Sigma_{Kp}C_{q,Kp}LT[F_K(t)F_p(t)](z) \quad . \tag{4.2}$$

Thus equ. (2.6) is equivalent to

$$-[f_q/(1-f_q) - \hat{F}_q(f_K)]/z$$
$$+ \Sigma_k(\delta_{qk}-C_{qk})F_K(z)$$
$$- zF_q(z)^2(1-f_q) - \Sigma_{kp}\, C_{q,kp}\, LT[F_k(t)F_p(t)](z)$$
$$= r_q(z) \quad , \tag{4.3a}$$

where

$$r_q(z) = (z/\Omega_q^2) + m_q^o(z) + z^2F_q(z)^3(1-f_q)^2/[1+(1-f_q)zF_q(z)]. \tag{4.3b}$$

A side remark might be in order. If functional $\hat{F}_q(f_K)$, equ. (2.5c), is generalized to contain also single mode, triple mode etc. terms or if only triple modes would enter equ. (2.5c), the result (4.3a) would be unchanged. Merely equ. (4.2) and thus equ. (4.3b) has to be complemented by terms of order $LT[F_q(t)F_K(t)F_p(t)](z)$ [4].

We specialize to $n=n_c$ and choose $f_q=f_q^c$. Then the first bracket in equ. (4.3a) drops out and the matrices C_{qk}, $C_{q,Kp}$ reduce to their values at the critical point. Let us take the frequency z as small parameter. We expand $F_q(z)=F_q(z)^{(1)}+F_q(z)^{(2)}+\ldots$. The solution shall be constructed with the ansatz: $F^{(1)}$, $F^{(2)}$ etc. are successively of higher order and the same holds for the three remaining lines of equ. (4.3a). Then one finds in leading order

$$\Sigma_k(\delta_{qk} - C_{qk}^c)F_k(z)^{(1)} = 0 \quad . \tag{4.4a}$$

From section 3.1 it follows $F_k(z)^{(1)}=\ell_k^cF(z)$. So in leading order function $F_k(z)$ factorizes in the known wave vector dependent function ℓ_k^c, equ. (3.3b), and a not yet specified frequency dependent function $F(z)$, which is independent of wave vector. The next to leading equation reads

$$\Sigma_k(\delta_{qk}-C_{qk}^c)F_k(z)^{(2)} = -(1-f_q)\ell_q^c\ell_q^c\, zF(z)^2$$
$$+ \Sigma_{kp}C_{q,kp}^c\ell_k^c\ell_p^c\, LT[F(t)^2](z). \tag{4.4b}$$

For this equation to be solvable, the r.h.s. has to be perpendicular to $\hat{\ell}_q^c$, equ. (3.3b). Because of equs. (3.3e, 3.5) this is equivalent to

$$\frac{1}{\lambda^c} zF(z)^2 + LT[F(t)^2](z) = 0 \quad . \qquad (4.4c)$$

The latter equation can be solved with the ansatz $F(t)=At^{(\frac{\xi-1}{2})}$.

Since $F(z)=Ai\Gamma(\frac{1+\xi}{2})/(-iz)^{\frac{1+\xi}{2}}$ and $LT(F(t)^2)(z)=iA^2\Gamma(\xi)/(-iz)^\xi$ one finds this function to obey equ. (4.4c) provided $\lambda^c=\lambda(\xi)$,

$$\lambda(\xi) = \Gamma(\tfrac{1+\xi}{2})^2/\Gamma(\xi) \quad . \qquad (4.5a)$$

Function $\lambda(\xi)$ is shown in Fig. 6; it exhibits the maximum unity for $\xi=1$. Hence there are two solutions:

$$\lambda^c = \lambda(\xi) \quad ; \quad \xi = x < 1 \quad ; \quad \xi = y > 1 \quad . \qquad (4.5b)$$

The solution with $\xi=y$ produces a singularity for F which is too strong to be compatible with the spectral representation (1.4). The solution with $\xi=x$ is a positive analytic function if and only if $A>0$. As a result one finds $F^{(2)}(z)\approx zF(z)^2\approx 1/z^x\approx LT[F(t)^2]$. Hence

$F^{(2)}(z)/F^{(1)}(z)\approx zF(z)\approx z^{\frac{1-x}{2}} \longrightarrow 0$ for $z \to 0$; the ansatz was justified and the ratio of to successive lines on the l.h.s. of equ. (4.3a) vanishes for $z\to 0$. The ratio of the three contributions to $r_q(z)$, equ. (4.3b), to the last line on the l.h.s of equ. (4.3a) is z^{1+x}, z^x and $z^{(\frac{1-x}{2})}$ respectively; hence $r_q(z)$ can be neglected in equ. (4.4b) for $z\to 0$.

Therefore

$$\Phi_q(z) = -(f_q^c/z)+i(A/\Omega_0)\Gamma(\tfrac{1+x}{2})h_q[\Omega_0/(-iz)]^{\frac{1+x}{2}} [1+0(z^{(\frac{1-x}{2})})] \quad (4.6)$$

solves equ. (2.6) for $n=n_c$. The coefficient $A>0$ cannot be fixed analytically; it is determined by matching the $z\to 0$ solution with the solution for large frequencies. A frequency Ω_0, representing the dynamics on microscopic scales, was introduced in order for A to be dimensionless. Notice, that the q dependent amplitude h_q is the same, which enters equs. (3.25).

Result (4.6) implies for the scattering law, equ. (1.15), at the transition point.

$$S(q,\omega)=\pi\delta(\omega)F_c(q)+H(q)(A/\Omega_0)\Gamma(\tfrac{1+x}{2})\cos(\tfrac{1+x}{4}\pi([\Omega_0/\omega]^{\frac{1+x}{2}} [1+0(\omega^{\frac{1-x}{2}})]. \quad (4.7)$$

Here $F_c>0$, $H(q)>0$ depend on the microscopic details of the system. The spectrum exhibits a power law divergency. The exponent $(1+x)/2$ is obtained as solution of equs. (4.5). All non-trivial microscopic details are combined in the exponent parameter λ_c, equ. (3.5). Measuring this power law spectrum would yield x and hence λ^c.

4.2 Scaling Laws for the Glass Phase Density Fluctuations

The preceding analysis can be extended to evaluate the correlation functions for the glass phase for small frequencies and small values of the separation parameter $\epsilon=(n-n_c)/n_c>0$. Let us substitute result (3.25) for f_q in equs. (4.1 to 4.3). The first bracket in equ. (4.3a) vanishes and with the aid of equ. (3.19b) an expansion in powers of $\sqrt{\epsilon}$ can be performed. Let us also introduce the substitution

$$f_q(\zeta) =: (\omega_c/\sqrt{\epsilon})F_q(z) \ , \quad \zeta=:z/\omega_c \ . \tag{4.8a}$$

This substitution is equivalent to

$$f_q(\tau) =: F_q(t)/\sqrt{\epsilon} \quad , \quad \tau=:t\omega_c \ . \tag{4.8b}$$

The frequency ω_c is to be fixed below. One finds

$$\Sigma_k(\delta_{qk}-C_{qk}{}^c)f_k(\zeta) =$$

$$\sqrt{\epsilon}\{\Sigma_k(-2\alpha)[C_{qk}{}^c(1-f_k{}^c)\ell_k{}^c-\Sigma_p C_{q,kp}{}^c\ell_p{}^c]f_k(\zeta)$$

$$+(1-f_q{}^c)\zeta f_q(\zeta)^2 + \Sigma_{kp}C_{q,kp}{}^c LT[f_k(\tau)f_p(\tau)](\zeta)\}$$

$$+\sqrt{\epsilon}[\omega_c/\epsilon]\{r_q(\zeta\omega_c) + O(\sqrt{\epsilon})\} \tag{4.9}$$

This equation shall be solved in the limit

$$z \to 0, \quad \epsilon \to 0, \quad \zeta = z/\omega_c \ \text{fixed} \ . \tag{4.10}$$

So ϵ shall be used as the small expansion parameter in equ. (4.9). The strategy will be the same as in section 4.1. Successive terms in equ. (4.9) are successively of higher order, if we assume $\omega_c/\epsilon\to0$ for $\epsilon\to0$. We expand in succession $f_k(\zeta)=f_k(\zeta)^{(1)}+f_k(\zeta)^{(2)}+\ldots$. In leading order one gets equ. (4.4a) with $F_k(z)^{(1)}$ replaced by $f_k(\zeta)^{(1)}$. So we find $f_k(\zeta)^{(1)}=\ell_k{}^c f(\zeta)$ with an unspecified function $f(\zeta)$. In next to leading order one finds

$$\Sigma_k(\delta_{qk}-C_{qk}{}^c)f_k(\zeta)^{(2)} = \sqrt{\epsilon}\ \{\ldots\} \quad , \tag{4.11}$$

where the curly bracket is obtained from the corresponding term in equ. (4.9) with $f_k(\zeta)$ replaced by $f_k(\zeta)^{(1)}$. For this equation to be solvable the r.h.s. has to be perpendicular to $\hat{\ell}_q{}^c$. This condition leads to the equation for $f(\zeta)$: $-\mu f(\zeta)+\zeta f(\zeta)^2/\lambda^c+LT[f(\tau)^2](\zeta)=0$. Here equs. (3.3b, 3.3e, 3.5) have been used and $\mu=2\alpha(1-\lambda^c)/\lambda^c>0$. Replacing $f(\zeta)\to\mu f(\zeta)$ leads to

$$- f(\zeta) + \frac{1}{\lambda^c}\ \zeta f(\zeta)^2 + LT[f(\tau)^2](\zeta) = 0 \ . \tag{4.12}$$

The limit z fixed, $\epsilon\to0$ is equivalent to $\zeta\to\infty$; we have to require, that our solution reproduces the result (4.6). Hence we have to impose the condition $f(\zeta\to\infty)=ia\Gamma(\frac{1+x}{2})/(-i\zeta)^{\frac{1+x}{2}}$. Indeed, in this limit equ. (4.12) is obeyed, since the first term in equ. (4.12) can be neglected compared to the others, so that equ. (4.4c) is reproduced. Hence one finds from equs. (4.1, 4.8a): $\Phi_q(z)=-(f^c/z)+i\Gamma(\frac{1+x}{2})ah_q(\sqrt{\epsilon}/\omega_c)/(-i\zeta)^{\frac{1+x}{2}}$. This result has to agree with equ. (4.6). Therefore $(\sqrt{\epsilon}/\omega_c)\omega_c^{\frac{1+x}{2}}$ has to be ϵ independent and this fixes ω_c up to a constant. We choose

$$\omega_c = \Omega_0|\epsilon|^{\frac{1}{1-x}} \quad . \tag{4.13}$$

Furthermore one gets a=A and so equ. (4.12) has to be solved with the condition

$$f(\zeta\to\infty) = iA\Gamma(\tfrac{1+x}{2})/(-i\zeta)^{\frac{1+x}{2}} \quad . \tag{4.14}$$

Since $\omega_c/\epsilon\propto\epsilon^{x/(1-x)}\to 0$ for $\epsilon\to 0$ the assumptions made at the beginning are valid and we arrive at the solution in the limit specified by equ. (4.10):

$$\Phi_q(z) = [f_q{}^c+\sqrt{\epsilon}h_q+0(\epsilon)](\tfrac{-1}{z}) + h_q(\sqrt{\epsilon}/\omega_c)f(\tfrac{z}{\omega_c})[1+0(\epsilon^{1/2},\epsilon^{\frac{x}{1-x}})] \; . \tag{4.15}$$

This result implies for the scattering law, equ. (1.15), the extension of formula (4.7):

$$S(q,\omega) = \pi\delta(\omega)[F_c(q)+\sqrt{\epsilon}H(q)] + H(q)(\sqrt{\epsilon}/\omega_c)f''(\omega/\omega_c) \quad . \tag{4.16}$$

So scaling laws are obtained of a kind known from the dynamics near second order phase transition points: changes of the controlparameter are equivalent to variations of the frequency scale given by ω_c and variations of the scale for the sepctrum given by $\sqrt{\epsilon}/\omega_c$. The scattering law for $\omega\neq 0$ factorizes into the wave vector independend part $f''(\omega/\omega_c)\sqrt{\epsilon}/\omega_c$, which varies strongly with ϵ, and a frequency and ϵ independend amplitude H(q), which governs the ϵ-dependence of the elastic line intensity also.

The scaling function obeys equs. (4.12) and (4.14); these equations are invariant under the substitution

$$\zeta\to\zeta\Omega \; , \quad f\to f/\Omega \; , \quad A\to A/\Omega^{(\frac{1-x}{2})} \quad . \tag{4.17}$$

So choosing different prefactors A merely amounts to a rescaling of f and ζ. This rescaling can be achieved by a redefinition of Ω_0. As a result one concludes: up to the definition of Ω_0 the scaling functions $f(\zeta)$ of various systems are determined by the one parameter λ^c only. This implies in particular that a numerical evaluation of the scaling function can be done for a convenient schematic model, provided it can generate the desired λ^c. Fig. 7 gives a representative example for x=0.25; it was obtained by solving the differential equation for the model, defined by equs. (2.14c) [31].

4.3 Transversal Sound in the Glass

Let us examine the implications of the preceding results on the transversal current correlations. The following analysis will also exemplify how to draw conclusions from the present theory for macroscopic experiments. Result (4.15) can be substituted into equs. (2.11). Using the notations (3.26) one finds in leading order in the scaling regime, defined by equ. (4.10):

$$\Phi_q^T(z) = -z/\{z^2-(\Omega_q^T)^2[f_q^{Tc}+\sqrt{\epsilon}h_q^T[1-(z/\omega_c)f(z/\omega_c)]]\} \quad . \qquad (4.18)$$

Let us restrict ourselves to small wavevectors. Then one can expand in leading order $(\Omega_q^T)^2 f_q^{Tc} =: C_o^2 q^2$, $h_q^T/f_q^{Tc} =: h > 0$. One finds for the long wavelength excitations the propagator

$$\Phi_q^T(z) = -z/\{z^2-C_o^2q^2[1+\sqrt{\epsilon}h[1-(z/\omega_c)f(z/\omega_c)]]\} \quad . \qquad (4.19)$$

Since $\zeta f(\zeta)\to 0$ for $\zeta\to 0$ one concludes: there are transversal sound excitations in the glass phase; this holds even at the transition point $n=n_c$ itself. This means, there are resonances of $\Phi_q^T(\omega)''$ whose position $\omega(q)$ scales linearly with q and whose width $\gamma(q)$ becomes arbitrarily small compared to the frequency if wavevector q tends to zero:

$$\lim_{q\to o} \omega(q)/q = C = C_o(1+\sqrt{\epsilon}h/2+O(\epsilon)) \quad , \qquad (4.20a)$$

$$\lim_{q\to o} \gamma(q)/\omega(q) = 0 \quad . \qquad (4.20b)$$

The transversal sound velocity C decreases to a non-zero value C_o if the transition point is approached; near C_o velocity C varies rapidly with the separative parameter.

One has to distinguish the <u>hydrodynamic regime</u> from the critical regime. The first one is defined by $|z|\approx|Cq|\ll\omega_c$. Then the scaling function can be approximated by its zero frequency value and from equ. (4.19) one gets

$$\omega(q) = Cq \quad ; \quad \gamma(q) = (Cq)^2[hf''(0)/2]\sqrt{\epsilon}/\omega_c \quad . \qquad (4.21)$$

Like in the theory of elastic media one finds a linear dispersion law and a damping function proportional to $\omega(q)^2$. Approaching the critical point, the hydrodynamic regime shrinks to zero, equ. (4.13); simultaneously the damping constant exhibits a power law divergency $\gamma\propto 1/\epsilon^{\left(\frac{1+x}{2(1-x)}\right)}$. The <u>critical regime</u> is defined by $|z|\approx|Cq|\gg\omega_c$. Then the scaling law can be approximated by its large ζ asymptote, equ. (4.14), with the result

$$\omega(q) = Cq[1+\delta\cos(\tfrac{1-x}{4}\pi) \quad (C_oq/\Omega_o)^{\frac{1-x}{2}}] \quad , \qquad (4.22a)$$

$$\gamma(q) = Cq[\delta\sin(\tfrac{1-x}{4}\pi) \quad (C_oq/\Omega_o)^{\frac{1-x}{2}}] \quad ; \qquad (4.22b)$$

here $\delta = hA\Gamma(\frac{1+x}{2})/2 > 0$. So the sound velocity increases with q with an anomalous exponent; the damping exhibits a similar q dependence. Damping and corrections to the sound velocity are asymptotically independent of the controlparameter.

4.4. Density Fluctuations of the Liquid on ω_c-scale

The technique for deriving scaling laws shall be extended to determine the density correlations of the liquid for the frequency regime of order ω_c. A new frequency scale ω_c' for the liquid dynamics will be identified.

Instead of separating the non ergodicity contribution of the glass in equ. (4.1) one can take an alternative route and split off the form factor at the critical point f_q^c. Actually, then the treatment of equ. (4.3a) is easier than the calculation in section 4.1 since neither the ϵ expansions of f_q nor of C_{qk} are needed. The procedure works also on the glass side, $\epsilon>0$, as well as on the liquid side, $\epsilon<0$. So we will proceed and will rederive the results of section 4.2 as a byproduct. Let us write in equ. (4.1) $F_q(z)=\hat{f}_q(\zeta)\sqrt{|\epsilon|}/\omega_c$, $\zeta=z/\omega_c$. Expansion of functional $F_q(f_k)$ in equ. (4.3a) is given in terms of $C_q{}^c$, equ. (3.6):

$$\Sigma_k(\delta_{qk}-C_{qk}{}^c)\hat{f}_k(\zeta) =$$

$$\sqrt{|\epsilon|}\{(\mp C_q{}^c/\zeta) + (1-f_q{}^c)\zeta\hat{f}_q(\zeta)^2 + \Sigma_{kp}C_{q,kp}{}^c LT[f_k(\tau)\hat{f}_p(\tau)](\zeta)\}$$

$$+ \sqrt{|\epsilon|}(\omega_c/\epsilon)[r_q(\omega_c\zeta) + 0(\sqrt{|\epsilon|})] \quad . \tag{4.23}$$

This result corresponds to equ. (4.9); $\mp$ refer to glass and liquid respectively. The further analysis follows the one in section 4.2 so closely, that it will not be repeated. One finds the formula for the density fluctuations

$$\Phi_q(z) = [-f_q{}^c/z]+h_q[\sqrt{|\epsilon|}/\omega_c]\hat{f}(z/\omega_c)[1+0(\epsilon^{1/2},\epsilon^{(\frac{x}{1-x})})] \quad . \tag{4.24}$$

This implies for the scattering law

$$S(q,\omega) = \pi\delta(\omega)F_c(q)+H(q)[\sqrt{|\epsilon|}/\omega_c]\hat{f}''(\omega/\omega_c) \quad . \tag{4.25}$$

The scaling function obeys the equation

$$(\mp\hat{\mu}/\zeta) + \frac{1}{\lambda^c}\zeta\hat{f}(\zeta)^2 + LT[\hat{f}(\tau)^2](\zeta) = 0 \quad , \tag{4.26}$$

where equ. (3.7) yields $\hat{\mu}=\mu^c/\lambda^c>0$. The results hold in the scaling limit, defined by equ. (4.10). The scaling frequency is defined by equ. (4.13) and the asymptote (4.14) is to be required. Notice that all the parameters h_q, $H(q)$, μ, λ^c and A are independent of ϵ; they are the same for the glass and the liquid. The difference between the two phases is reflected only in the sign alternative in equ. (4.26); this alternative will lead to very different scaling functions.

The first term in equ. (4.26) requires $\hat{f}(\zeta)$ to have a singularity for $\zeta\to 0$. Let us make the simplest assumption $\hat{f}(\zeta)=(-f_0/\zeta)+f(\zeta)$, $\hat{f}(\tau\to\infty)=f_0$, $\zeta f(\zeta)\to 0$ for $\zeta\to 0$. One gets $\pm\hat{\mu}=f_0^2(1-\lambda)/\lambda$. The spectral representation, equ. (1.4) requires f_0 to be real. Hence the assumption works for the glass phase only. For $f(\zeta)$ one obtains then equ. (4.12). This analysis makes it somewhat clearer, why the same amplitude h_q enters the glass form factor and the continuous spectrum, equ. (4.15); both terms belong to the same scaling function $\hat{f}(\zeta)$.

From now on I restrict the discussion to the liquid phase. The substitution $\hat{f}=f\sqrt{\hat{\mu}}$ simplifies the scaling equation to

$$1/\zeta + \frac{1}{\lambda^c}\zeta f(\zeta)^2 + LT[f(\tau)^2](\zeta) = 0 \quad . \tag{4.27}$$

The singularity for $\zeta\to 0$ has to be stronger than $1/\zeta$. Hence, for $\zeta\to 0$, the scaling equation reduces to equ. (4.4c). The power law ansatz for the solution works, but as exponent we have to choose $\xi=y$, equ. (4.5b). As a result one finds

$$f(\zeta\to 0) = -iB\Gamma(\tfrac{1+y}{2})/(-i\zeta)^{\frac{1+y}{2}} \quad . \tag{4.28a}$$

Again, all systems exhibit the same scaling function provided the exponent parameter λ^c is the same. Fig. 8 demonstrates $f''(\omega/\omega_c)$ for the same $\lambda^c=0.5676$ used in Fig. 7; $y=2.74$. Notice the strong increase of the spectrum for $\omega<\omega_c$:

$$f''(\omega/\omega_c) = -B\cos(\tfrac{1+y}{4}\pi)\Gamma(\tfrac{1+y}{2})/(\omega_c/\omega)^{\frac{1+y}{2}}, \quad B>0. \tag{4.28b}$$

The result (4.28) for the spectral increase can be rewritten: for $z/\omega_c\to 0$ one gets

$$\Phi_q(z) = \frac{1}{\omega_c'}\{(-f_q^c/\zeta') + h_q i[-B\Gamma(\tfrac{1+y}{2})]/(-i\zeta')^{\frac{1+y}{2}}\} \quad , \tag{4.29a}$$

$$\zeta' = z/\omega_c' \quad , \tag{4.29b}$$

where

$$\omega_c' = \Omega_0|\epsilon|^{(\frac{1}{1-x}+\frac{1}{y-1})} \quad . \tag{4.30}$$

So in the liquid the theory yields a cross over from an ϵ independent power law spectrum, equ. ((4.6), to another power law spectrum which is characterized by the strongly ϵ dependent scale ω_c'. The cross over frequency ω_c becomes smaller if the critical point is approached, but $\omega_c'/\omega_c\to 0$ for $\epsilon\to 0$.

Some remarks concerning the low frequency divergency of the scaling function might be worthwhile. Firstly, a singularity of type (4.28) is not compatible with the spectral representation. From equ. (1.4) one gets $\mathrm{Im}\Phi(z)>0$ for $\mathrm{Im}z>0$ for example. This property follows from equ. (4.29a) if $|\zeta'|\gg 1$ but it is violated for $|\zeta'|\ll 1$. So the results lead to problems if applied on scale ω_c'. The same conclusion is obtained, if one remembers, secondly, that $|zF_q(z)|\ll 1$ was used in equ. (4.3b). This inequality is obeyed by equ. (4.29a) if $|\zeta'|\gg 1$ but it is violated for $|\zeta'|\ll 1$. So the preceding results require

$$|z| \gg \omega_c' \quad . \tag{4.31}$$

The new scale limits the range of validity of the scaling laws of the liquid on the low frequency side. A new mathematical technique is necessary to analyze the spectra on the ω_c' scale. Thirdly, the inequality (4.31) is not in contradiction to the formal derivation of equ. (4.27) in the scaling limit. Since $\omega_c'/\omega_c \to 0$ for $\epsilon \to 0$ the range of scaled frequencies ζ, where there are problems, is asymptotically given by $\zeta=0$ only. On scale ω_c the problems on scale ω_c' are not visible. Finally, if the scaling laws should show up in the spectra, ω_c has to be smaller than the typical microscopic frequency Ω_0. Therefore standard inelastic neutron scattering experiments will have difficulties to map out the scattering law on ω_c scale. Fig. 9 gives an example for the expected result, obtained for $S(q,\omega)$ within the simple schematic model for $\epsilon=0.05$; scales are gauged to meet the hard sphere model [2]. Near the transition the dynamics become so slow, that neutron scattering techniques presumably cannot analyze the anomalies on ω_c' scales and thus the results of sections 4.2, 4.4 [3,4] may be sufficient for scattering data analysis.

4.5 Propagating Shear Excitations of the Liquid

To continue the discussion of transversal sound propagation one has to modify equ. (4.19) in a straightforward way. One gets for the liquid phase in the scaling regime

$$\Phi_q^T(z) = -z/\{z^2-C_0^2q^2[1-h\sqrt{|\epsilon|}\cdot(z/\omega_c)f(z/\omega_c)]\} \ . \tag{4.32}$$

Equivalently, one can introduce the viscosity function, equ. (2.12), to write

$$\Phi_q^T(z) = -1/\{z+q^2D_T(z)/(nm)\} \quad , \tag{4.33a}$$

where

$$D_T(z) = nmC_0^2[-1+h\sqrt{|\epsilon|}(z/\omega_c)f(z/\omega_c)]/z \quad . \tag{4.33b}$$

In the whole scaling regime the preceding formulae imply modes whose frequencies $\omega(q)$ are larger than their width $\gamma(q)$. In the <u>first critical region</u>,

$$\omega_c \ll C_0 q \ll \Omega_0 \quad , \tag{4.34}$$

the same asymptotical expansion (4.14) can be used as in section 4.3. As a result one gets equs. (4.22) with C replaced by C_0. So on this scale the only difference between transversal sound modes in glass and in liquid is that for $\epsilon>0$ the velocity varies rapidly with the separation parameter, while for $\epsilon<0$ there is only a weak dependence of $\omega(q)$ on ϵ. If one is in the <u>second critical region</u>,

$$\omega_c' \ll C_0 q \ll \omega_c \quad , \tag{4.35a}$$

the resonance (4.28) alters the results to

$$\omega(q) = C_0 q[1+\hat{\delta}\,\mathrm{tg}(\tfrac{1+\mu}{4}\pi)(\omega_c'/C_0 q)^{\frac{\mu-1}{2}}] \quad , \tag{4.35b}$$

$$\gamma(q) = C_0 q\,\hat{\delta}\,(\omega_c'/C_0 q)^{(\frac{\mu-1}{2})} \quad , \tag{4.35c}$$

where $\hat{\delta}=h[-B\cos(\frac{1+\mu}{4}\pi)]\Gamma(\frac{1+\mu}{2})/2>0$. In this region the damping constant increases sublinearly with increasing wavevector. More important is that with decreasing q the resonance falls below C_0q. For $\omega_c'/C_0q\to 1$ the mode gets overdamped. So one identifies

$$q_c' = \omega_c'/C_0 \tag{4.35d}$$

as a characteristic small wavevector cut-off for the shear excitations. In the window $\omega<\omega_c'$, $q<q_c'$ one does not observe propagating modes. If the critical point is approached the window for propagating shear excitations becomes larger since $q_c'\to 0$ for $\epsilon\to 0$.

The dynamical shear viscosity reads in the scaling region

$$D_T''(\omega) = D_0[\sqrt{|\epsilon|}/\omega_c]f''(\omega/\omega_c) \quad , \tag{4.36}$$

where $D_0=nmC_0^2h>0$. Near the glass transition the theory predicts a much more subtle frequency dependence of the stress fluctuations than the Lorentzian, which is assumed in the viscoelastic theory [26]. For $\omega_c\ll\omega\ll\Omega_0$ a power law decay with exponent (1+x)/2 is obtained, equ. (4.14), while for $\omega_c'\ll\omega\ll\omega_c$ a power law with exponent (1+y)/2 is predicted, equ. (4.28). Fig. 8 presents an example. Extension of the molecular dynamics work on simple liquids (compare Fig. 2 and reference [27]) into the supercooled region might be a way to test the present theory. For the time dependent functions our results are equivalent to

$$D_T(t) = \begin{cases} D_0A/(t\Omega_0)^{(\frac{1-x}{2})} \quad , & 1 \ll t\omega_c \ll \omega_c/\Omega_0 \\ \text{const}-D_0B(t\omega_c')^{\frac{\mu-1}{2}} \quad , & 1 \ll t\omega_c' \ll (\omega_c'/\omega_c) \;. \end{cases} \tag{4.37}$$

Notice, that the result for the shorter time intervall is independent of ϵ. For the long times the scale ω_c' enters, which depends sensitively on the controlparameter.

4.6 The Velocity Spectra near the Transition Point

In this subsection the velocity correlation function, $K_s(z)$ shall be discussed within the scaling regime. This might be of interest since

$K_s(t)$, equ. (1.23), is well suited for molecular dynamics studies. Let us start by considering the tagged particle density correlation function. We write $\Phi_q{}^s(t)=f_q{}^{sc}+(1-f_q{}^{sc})^2F_q{}^s(t)$, $\Phi_q(t)=f_q{}^c+(1-f_q)^2F_q(t)$. Let us also introduce scaled functions as done before $\hat{f}_q{}^s(\zeta)=(\omega_c/\sqrt{|\epsilon|})F_q{}^s(z)$, $\hat{f}_q(\zeta)=(\omega_c/\sqrt{|\epsilon|})F_q(z)$, $\zeta=z/\omega_c$. Expansion of equ. (2.9) in analogy to the procedure carried out in connection with equ. (4.3), leads to

$$\Sigma_k(\delta_{qk}-C_{qk}{}^{sc})f_k{}^s(\zeta) = \Sigma_k C_{qk}{}'^c(1-f_k{}^c)^2\hat{f}_k(\zeta) + O(\sqrt{|\epsilon|}) \ . \tag{4.38}$$

Notations (3.29, 3.30) were used. The result is so simple compared to the ones derived in section 4.1, 4.2, because we could restrict ourselves to the leading terms. Since the inverse of $[1-C^{sc}]$ exists, next to leading terms do not enter. Using the result (4.24) and the notation (3.32b) one arrives at the formula for the tagged particle correlations in the scaling limit:

$$\Phi_q{}^s(z) = [-f_q{}^{sc}/z] + h_q{}^s[\sqrt{|\epsilon|}/\omega_c]\hat{f}(z/\omega_c) \ . \tag{4.39}$$

$\hat{f}(\zeta)$ is, up to normalizations, the scaling function for the glass or the liquid discussed before in connection with equ. (4.26). So the low frequency spectra for the tagged density are the same as the ones for the coherent density except for a change of the amplitude factor h_q to $h_q{}^s$. The latter was considered in section 3.4.

Let us carry out the long wave length expansion studied at the end of section 3.4: $f_q{}^{sc}=1-(r^{sc}q)^2+O(q^4)$, $h_q=\hat{a}(r^{sc}q)^2+O(q^4)$. Here r^{sc} is the localization length at the critical point and $a\propto\hat{a}>0$ rules the controlparameter dependence of this quantity in the glass. Combining equs. (1.20, 1.24) to $K_s(z)=z^2\partial\Phi_q{}^s(z)/\partial(q^2)$, $q=0$, one arrives at the desired expression

$$K_s(z) = (r^{sc})^2\{z + az^2[\sqrt{|\epsilon|}/\omega_c]\hat{f}(z/\omega_c)\} \ . \tag{4.40}$$

So the scaling functions determine the velocity correlations easily. Again one has to distinguish various cases. In the critical region a rise of the spectrum is obtained from equ. (4.14) according to

$$K_s''(\omega) \propto \omega^{(\frac{3-x}{2})} \quad ; \qquad \omega_c \ll \omega \ll \Omega_0 \ . \tag{4.41a}$$

This formula holds for the glass as well as for the liquid. This critical spectrum does not depend on the controlparameter. The result is equivalent to a power law decay of the velocity correlations:

$$K_s(t) \propto -1/t^{(\frac{5-x}{2})} \quad , \qquad 1/\Omega_0 \ll t \ll 1/\omega_c \ . \tag{4.41b}$$

In the glass there is a finite zero frequency limit of the spectral function $f''(0)>0$ and so the velocity spectrum in this hydrodynamic regime increases proportional to ω^2 like the phonon spectral density in a solid:

$$K_s''(\omega) = (r^{sc})^2af''(0)(\sqrt{\epsilon}/\omega_c)\omega^2 \quad , \quad \omega \ll \omega_c, \ \epsilon > 0 \ . \tag{4.42}$$

The spectrum in this regime depends strongly on the controlparameter.

In the second critical region of the liquid one gets from equ. (4.28) a sublinear rise of the spectrum, which depends strongly on ϵ:

$$K_s''(\omega) \propto \omega_c'(\omega/\omega_c')^{(\frac{3-y}{2})} \quad ; \quad \omega_c' \ll \omega \ll \omega \ , \quad \epsilon < 0 \quad . \qquad (4.43a)$$

The corresponding decay law in time reads

$$K_s(t) \propto -\omega_c'^2/(t\omega_c')^{(\frac{5-y}{2})} \qquad (4.43b)$$

4.7 The Quasi Elastic Spectrum

In this subsection I will contemplate the simplest possible ansatz for the correlation functions of the liquid on scale ω_c'; but I will not be able to demonstrate completely the consistency of the ansatz with the mode coupling equations. Because of these uncertainties I will neither work out all the implications of the results for the various experimental quantities like viscosity, self diffusion coefficient and the like. The alert reader can derive the results by himself easily.

Let us write $\zeta'=z/\omega_0$ and $\varphi_q(\zeta')=\omega_0\Phi_q(z)$, where ω_0 is a not yet specified scaling frequency. Equ. (2.6) then reads

$$\frac{\varphi_q(\zeta')}{1+\zeta'\varphi_q(\zeta')} = \omega_0[(\omega_0\zeta'/\Omega_q^2) + m_q^o(\omega_0\zeta')] + LT[\hat{F}_q(\varphi_k(t'))](\zeta') \quad .$$

Let us assume $\omega_0 \to 0$ for $\epsilon \to 0$. Then one finds in the limit $\epsilon \to 0$, ζ' fixed, a new scaling limit:

$$\frac{\varphi_q(\zeta')}{1+\zeta'\varphi_q(\zeta')} = LT[\hat{F}_q^c(\varphi_k(t'))](\zeta') \quad . \qquad (4.44)$$

The controlparameter in equ. (4.44) is to be taken at the critical point. Hence $\varphi_q(\zeta')$ is independent of ϵ. Notice that the regular term has disappeared again. Equ. (4.44) is invariant under the substitution $\varphi(\zeta') \to \varphi(\zeta'/\Omega)/\Omega$; this substitution is equivalent to a redefinition $\omega_0 \to \Omega\omega_0$. So if there is a solution of the non-linear equations (4.44) one gets whole series' of rescaled functions as solutions, too.

Let us assume $\varphi_q(t' \to 0)=f_q'$ to exist. This implies for the Laplace transforms in the large ζ' limit $\zeta'\varphi_q(\zeta') \to -f_q'$ and similarly $\zeta' LT[\hat{F}_q^c(\varphi_k(t'))](\zeta') \to -\hat{F}_q^c(f_k')$. Hence one finds from equ. (4.44): $f_q'/(1-f_q')=F_q^c(f_q')$. From equ. (3.1) we know this equation to have the solution f_q^c. So we obtain the first result

$$\varphi_q(\zeta') = -(f_q^c/\zeta') + (1-f_q^c)^2 G_q(\zeta') \quad ,$$

$$\zeta' G_q(\zeta') \to 0 \quad \text{for} \quad \zeta' \to \infty \quad . \qquad (4.45a)$$

Because of the spectral representation (1.4) this is equivalent to the area of the shape function $\varphi_q''(\omega')$ to be normalized by f_q^c

$$\frac{1}{\pi}\int d\omega' \ \varphi_q''(\omega') = f_q^c \quad . \tag{4.45b}$$

Equation (4.44) can be rewritten into an equivalent equation for $G_q(\zeta')$; the work has been done before in connection with equs. (4.3):

$$\begin{aligned} &\Sigma_k(\delta_{qk}-C_{qk}^c)G_k(\zeta') \\ &= \zeta' G_q(\zeta')^2(1-f_q^c) + \Sigma_{kp}C_{q,kp}^c LT[G_k(t')G_p(t')](\zeta') \\ &\quad + \zeta'^2 G_q(\zeta')^3(1-f_q^c)^2/[1+(1-f_q^c)\zeta' G_q(\zeta')] \quad . \end{aligned} \tag{4.46}$$

In the limit $\zeta'\to\infty$ successive lines are successively of higher order because of equ. (4.45a). So in leading order one gets from equ. (3.3b): $G_k(\zeta')=\ell_k G(\zeta')$ with an unspecified $G(\zeta')$. For the next to leading order equation to be solvable, the second line in equ. (4.46) is to be perpendicular to $\hat{\ell}_k$, equ. (3.3b). This brings us back to equ. (4.4c). That power law solution discussed there, with $\xi=y$ is compatible with equ. (4.45a). Therefore we arrive at $G(\zeta'\to\infty) = iC\Gamma(\frac{1+y}{2})/(-i\zeta')^{\frac{1+y}{2}}$. Remembering equ. (3.25b) one notices: the large ζ' expansion for $\Phi_q(z)$, equ. (4.45a), is identical in leading and next to leading order with the small $\zeta=z/\omega_c$ expansion, equ. (4.29), if and only if $\omega_0=\omega_c'$. Let us choose $\omega_0=\omega_c'$.

Summarizing one gets: in the new scaling limit

$$\epsilon \to 0 \quad , \quad z \to 0 \quad , \quad \zeta' = z/\omega_c' \quad \text{fixed} \quad , \tag{4.47}$$

the mode coupling equations are solved by

$$\Phi_q(z) = \varphi_q(z/\omega_c')/\omega_c' \quad , \tag{4.48}$$

provided the scaling function obeys the equ. (4.44). The solution obeys the large frequency asymptoties

$$\varphi_q(\zeta') = (-f_q^c/\zeta') + h_q i[-B\Gamma(\tfrac{1+y}{2})]/(-i\zeta')^{(\frac{1+y}{2})}[1+0_q(1/(\zeta')^{\frac{y-1}{2}}]. \tag{4.49}$$

If one assumes equ. (4.44) to have a solution with the necessary causality properties, which obeys the required asymptotics, and exhibits a continuous spectrum $\varphi_q''(\omega)$, the mode coupling equations would be understood near the transition point. The spectrum of the liquid on ω_c' scale would be a narrow line of width ω_c' and area f_q^c. The shape would be given by $\varphi_q''(\omega')$. Approaching the critical point this line would converge in a generalized sense towards the elastic line of the glass:

$$\lim_{\epsilon\to 0} \Phi_q(\omega) \to \pi f_q{}^c \delta(\omega) \quad , \qquad \omega \ll \omega_c \quad . \tag{4.50}$$

Most encouraging is the result: the quasi elastic line is not Lorentzian whenever $\lambda^c \neq 1/2$. The large-on-ω_c'-scale frequency wing is identical with the small-on-ω_c-scale part of the first scaling law, section 4.4, and thus it exhibits a power law decrease with the anomalous exponent $(1+y)/2$, equs. (4.29, 4.49). Relaxation in the liquid near the transition point is not given by a Debye law [3].

For the simple schematic model (2.13a), studied originally [1,2] with the two mode kernel (2.13b), the solution of equ. (4.44) is trivial if $m(t)=\lambda_k \Phi(t)^k$, $k=2,3,\ldots$. One gets a Debye relaxation law $\varphi(t')=f_k \exp(-t')$, where $f_k=(k-1)/k$. Experiments hardly ever exhibit Debye relaxation, however.

For the extended schematic model, equ. (2.14c), the numerical solution yields quite a smooth shape function $\varphi''(\omega)$. In Fig. 10 [31] a set of susceptibility spectra $\omega\Phi''(\omega)$ are shown in the standard semilogarithmic plott, used e.g. by experimentalists to present dielectric loss data. They scale according to equ. (4.48) with equ. (4.30) for ω_c'. The parameters of the model are the same as used in Figs. 7,8; $x=0.25$, $y=2.74$ corresponding to $\lambda=0.5676$. Two points are to be emphasized. First, even the 14% change of λ has brought already a change of the width of the susceptibility curves from the 1.14 decade, characteristic for Debye relaxation, to about 1.4 decade. Second, the asymptotic wing, equs. (4.29, 4.49) is practically unvisible in the Fig. 10. Multiplication of the spectrum $\Phi''(\omega)$ by ω distorts the curves considerably and the figure is dominated by the shape function for $\omega<\omega_c'$, where there is no simple analytic result available.

More work is necessary to understand the shape function for the quasi elastic line. As a first step systematic analysis of the numerical solution for the shape function $\varphi_q(\zeta')$ might be helpful. It can be obtained by first rewriting the scaling equation (4.44) as an integro-differential equation for the Laplace back transform $\varphi_q(\tau)$:

$$\varphi_q(\tau) - \hat{m}_q(\tau)(1-f_q{}^c) = - \int_0^\tau \hat{m}_q(\tau-\tau')\partial_{\tau'}\varphi_q(\tau')d\tau' \quad , \tag{4.51a}$$

$$\hat{m}_q(\tau) = \hat{F}_q{}^c(\varphi_k(\tau)) \quad . \tag{4.51b}$$

The mode coupling functional is to be taken at the critical point. Discretization yields an algorithm to determine $\varphi_q(\tau)$ at an array of discrete points provided one knows the initial behaviour. The latter follows from equ. (4.49) to

$$\varphi_q(\tau) = f_q{}^c - h_q B\, \tau^{\left(\frac{y-1}{2}\right)} + O(\tau^{(y-1)}) \tag{4.51c}$$

FIGURES

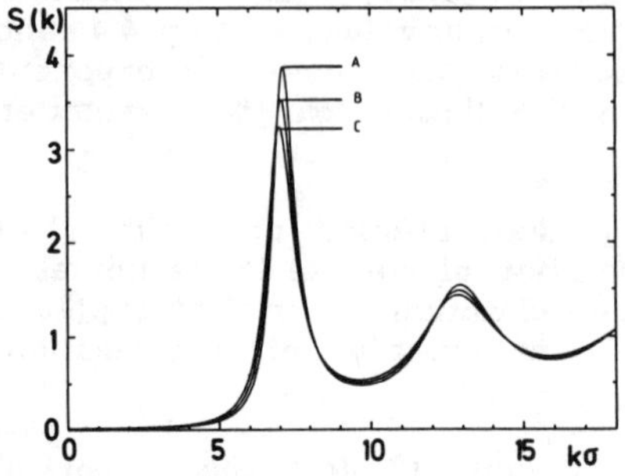

Fig. 1 Wertheim-Thiele approximation for the structure factor S(q) of a hard sphere system at packing fractions η=0.530 (A), 0.515 (B), 0.500 (C), from Bengtzelius et.al. [2].

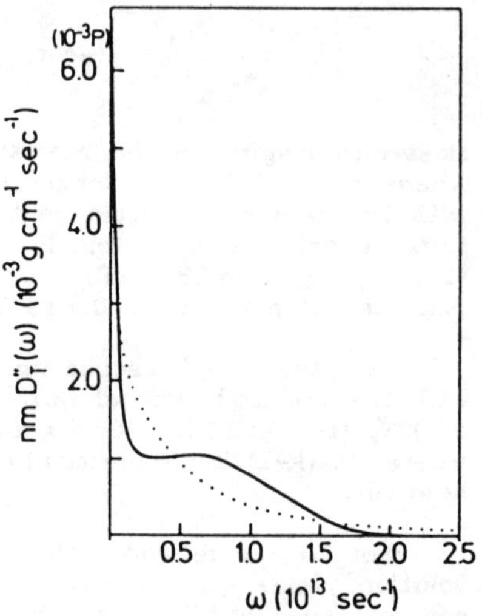

Fig. 2 Transversal viscosity spectrum $D_T''(\omega)$ of liquid argon near the triple point. Dotted curve: molecular dynamics results of Levesque et.al. [27]; full curve mode coupling theory of Bosse et. al. [10].

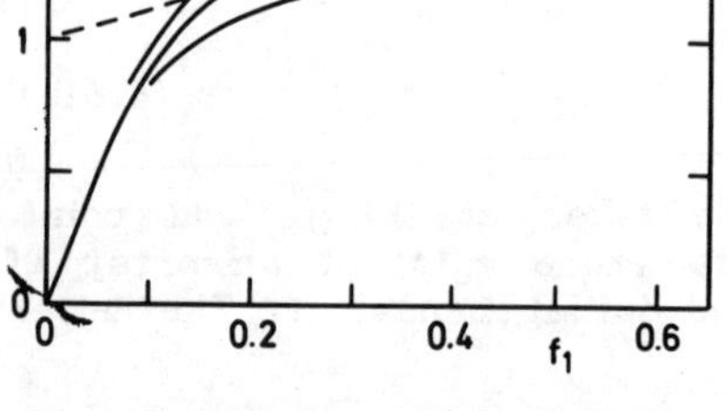

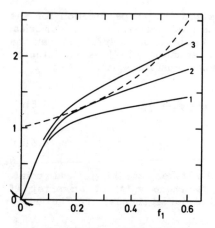

Fig. 3 Solution of the equation $1/(1-f_1)=G(f_1)$ for the model (2.14) with D=1.08, E=240, and C=40/63, 80/63, 120/63 for curve 1,2,3.

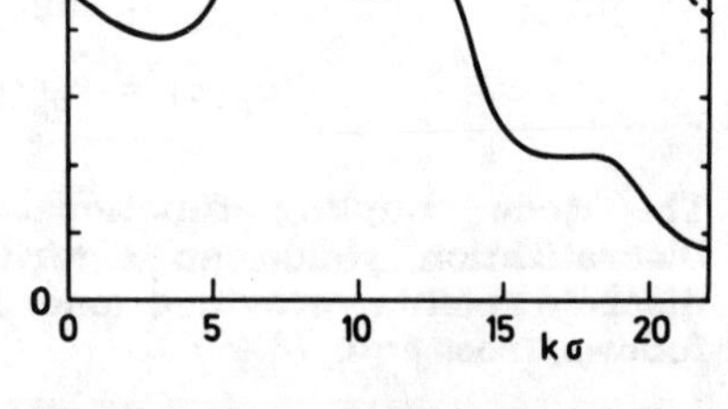

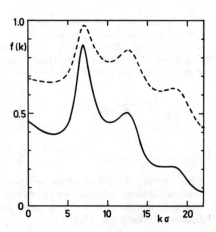

Fig. 4 Glass form factors f_q of the hard sphere system for packing fractions η=0.516 (full curve) and 0.550 (dashed curve), from Bengtzelius et. al. [2].

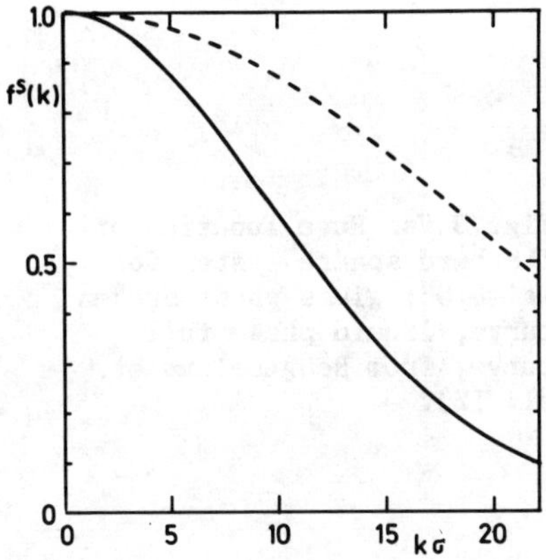

Fig. 5 Mössbauer–Lamb factors $f_q{}^s$ of the hard sphere system; parameters as in Fig. 4 [2].

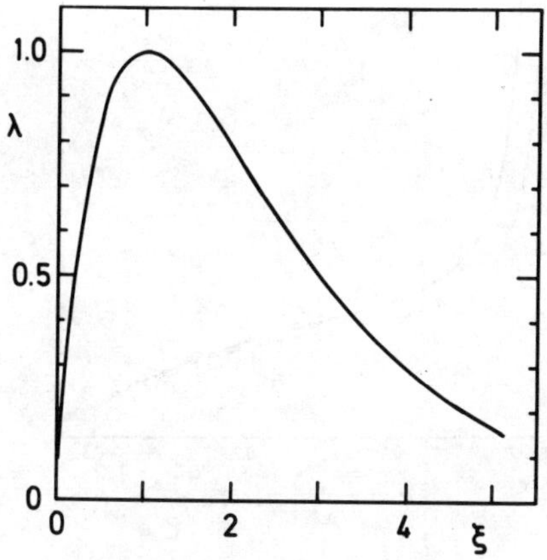

Fig. 6 Exponent parameter λ as function of ξ, equ.(4.5a).

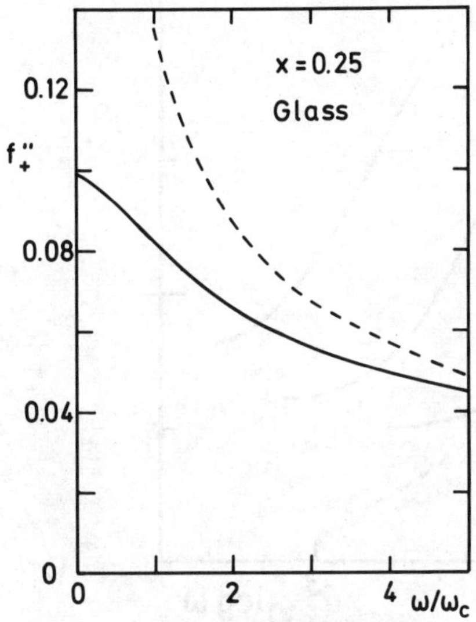

Fig. 7 Scaling function f" of the glass; from DeRaedt and Götze [31].

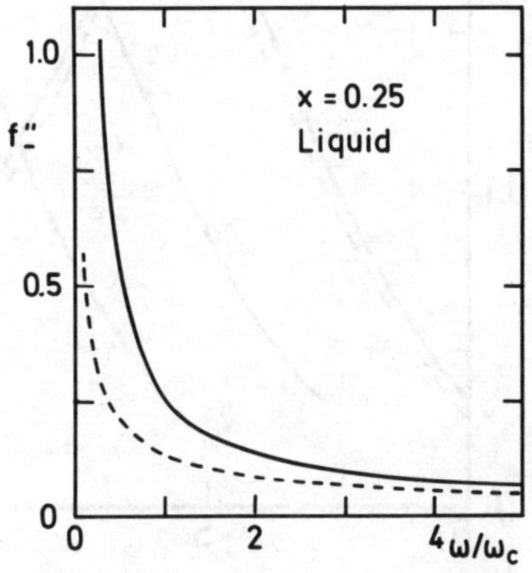

Fig. 8 Scaling function f" of the liquid; from DeRaedt and Götze [31].

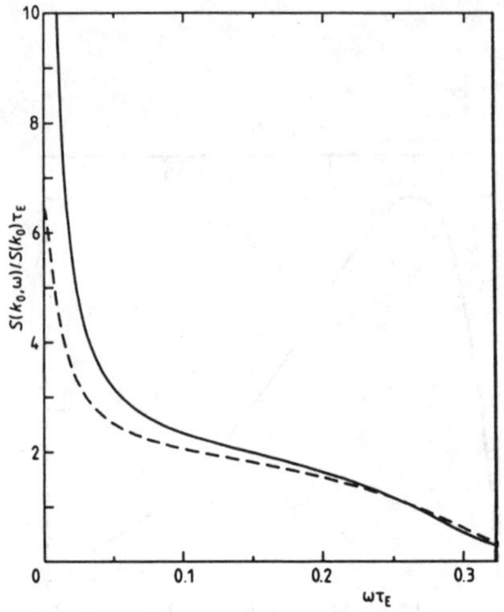

Fig. 9 Van Hove function of the hard sphere system for $|\epsilon|=0.05$; glass phase broken curve, liquid phase full curve, from Bengtzelius et. al. [2].

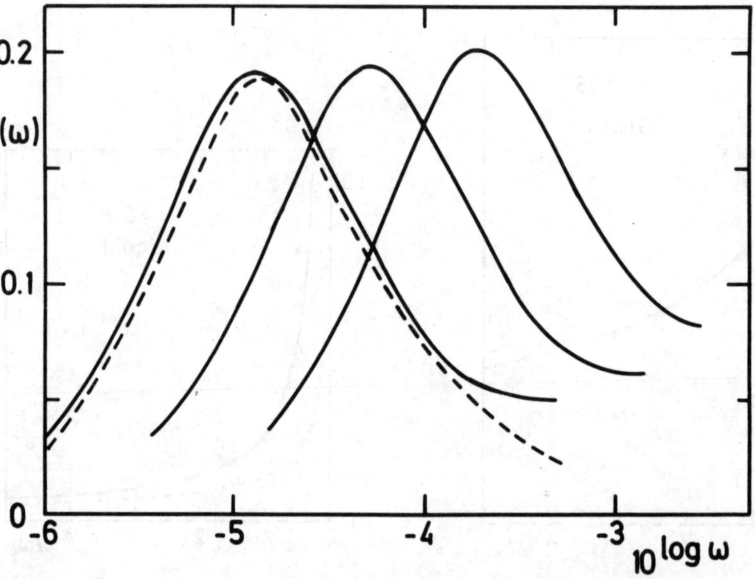

Fig. 10 Susceptibility spectra $\chi''(\omega)=\omega\Phi''(\omega)$ for the schematic model, equ. (2.14c), for x=0.25, y=2.74, λ_c=0.57 for ϵ=0.10, 0.05, 0.025; from DeRaedt and Götze [31].

REFERENCES

1. Leutheusser, E. Phys. Rev. A 29 (1984) 2765.
2. Bengtzelius, U., Götze, W. and A. Sjölander. J. Phys. C 17 (1984) 5915.
3. Götze, W. Z. Phys. B 56 (1984) 139.
4. Götze, W. Z. Phys. B 60 (1985) 195.
5. Zwanzig, R. J. Chem. Phys. 33 (1960) 1338 - Phys. Rev. 124 (1961) 983.
6. Forster, D. "Hydrodynamic Fluctuations, Broken Symmetry, and Correlation Functions", New York, Benjamin Inc. (1975).
7. Götze, W. Solid State Comm. 27 (1978) 1393 - J. Phys. C 12 (1979) 1279 - - Phil.Mag. B 43 (1981) 219 - "Recent Developments in Condensed Matter Physics" T.T. Devreese edit., New York, Plenum Press, Vol.I (1981) 133.
8. Kawasaki, K. Phys.Rev. 150 (1966) 291.
9. Götze, W. and M. Lücke. Phys. Rev. A 11 (1975) 2173 - Phys. Rev. B 13 (1976) 3825.
10. Bosse, J., Götze. W. and M. Lücke. Phys. Rev. A 17 (1978) 434 - Phys. Rev. A 17 (1978) 447 - Phys. Rev. A 18 (1978) 1176.
11. Sjögren, L. and A. Sjölander. J. Phys. C 12 (1979) 4369.
12. Wahnström, G. and L. Sjögren. J. Phys. C 15 (1982) 401
13. Horner, H. J. Low Temp. Phys. 8 (1972) 511
14. for a review see Götze, W. in "Localization, Interaction and Transport Phenomena", Kramer et.al. editors, Springer Verlag (1985) p.62.
15. Götze, W., Leutheusser E. and S. Yip. Phys. Rev. A 23 (1981) 2634 - A 24 (1981) 1008 - A 25 (1982) 533.
16. Cohen, M.H. and G.S. Grest. Phys. Rev. B 20 (1979) 1077.
17. Sjölander, A. and L.A. Turski. J. Phys. C 11 (1978) 1173.
18. Geszti, T. J. Phys. C 16 (1983) 5805.
19. Kubo, R. J. Phys. Soc. Japan 12 (1957) 570 - "Lectures in Theoretical Physics", Vol. 1, W.E. Brittin and L.C. Dunham edit., Interscience, New York 1959.
20. Mori, H. Progr. Theor. Phys., Kyoto 33 (1965) 423.
21. Boon, J.P. and S. Yip. "Molecular Hydrodynamics", McGraw-Hill, 1980.
22. Palmer, R.G. Adv. Phys. 31 (1982) 669.
23. Bengtzelius, U. and L. Sjögren. J. Chem. Phys., in print.
24. Sjögren, L. Phys. Rev. A 22 (1980) 2866.
25. Kirkpatrick, T.R. Phys. Rev. A 31 (1985) 939.
26. Copley, J.R.D. and S.W. Lovesey. Rep. Prog. Phys. 38 (1975) 461.
27. Levesque, D., Verlet, L. and J. Kurkijärvi. Phys. Rev. A 7 (1973) 1690.
28. Kahol, P.K., Bansal, R. and K.N. Pathack. Phys. Rev. A 14 (1976) 408.
29. For a review see: Pomeau Y. and P. Résibois. Phys. Rep. 19 (1975) 63.
30. Bellman, R. "Introduction to matrix analysis", Ch. 16, McGraw-Hill, New York, 1960.
31. DeRaedt, H. and W. Götze. preprint 1985.

COMPUTER SIMULATION STUDIES OF ATOMIC STRUCTURE AND DYNAMICS RELEVANT TO LIQUID AND AMORPHOUS ALLOYS

Marco Ronchetti and Gianni Jacucci

Centro del Consiglio Nazionale delle Ricerche e Dipartimento di Fisica, Universita di Trento, 38050 Povo, Italy
and
Department of Physics and Materials Research Laboratory
University of Illinois at Urbana-Champaign
Urbana, Illinois 61801

ABSTRACT

The recent observation of icosahedral atomic order in diffraction experiments from quasi-crystalline aluminium-manganese alloys had a precursor in the revelation by computer simulation of icosahedral bond orientational order in a supercooled monoatomic liquid.

The revelation by computer simulation of a short wavelength plasma oscillation in the partially ionic lithium lead liquid alloy has prompted an extension in that region of thermal neutron scattering experiments.

Computer simulation, and Molecular Dynamics in particular, is playing a central role in the investigation of the properties of liquid and amorphous metals and alloys. These two examples will be described in some detail.

I - Background for Icosahedral Quasi-Crystals

The fluid phase of a monoatomic system exhibiting a liquid-gas critical point contains both low density quasi-ideal gas states in which the atoms take up completely random positions, and high density states characterized by steric constraints and short range order. A continuum path around the critical point joins these quite different states without crossing phase transformation lines. Upon cooling the fluid in the liquid density side of the liquid-gas coexistence, the system hits the liquid-crystal first order transition line. At lower temperatures the thermodynamically stable phase is characterized by the onset of long range order in atomic positions and interatomic bond orientations. The long range order takes the form of a periodic lattice satisfying the usual symmetry requirements. Each crystalline lattice has a characteristic space group symmetry. A space group includes symmetry operations (i.e. operations that displace the lattice between configuration in which the lattice coincides with itself) consisting of a three dimensional translation τ combined with a three dimensional rotation or reflection R. The so called Bravais lattice is formed by the end points of the translation vector τ as its components take up integer values along three basic vectors $\tau 1$, $\tau 2$, and $\tau 3$. The rotation and reflection operations R define the so called point group, that leaves one point in space invariant The very existence of the Bravais lattice reduces the possible point

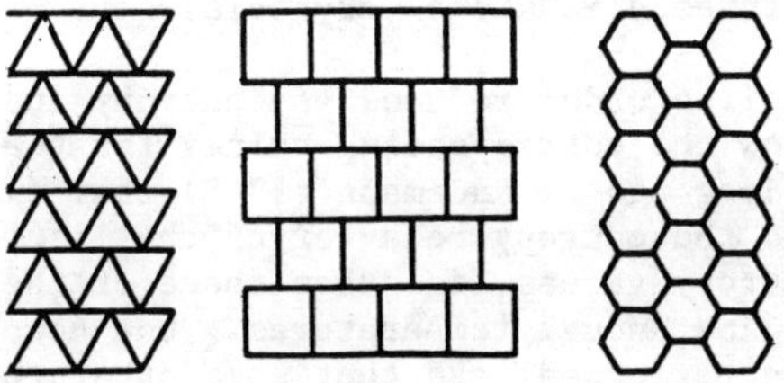

Fig. 1. Equilateral triangles, squares and hexagons are the only regular polygons that tile the plane.

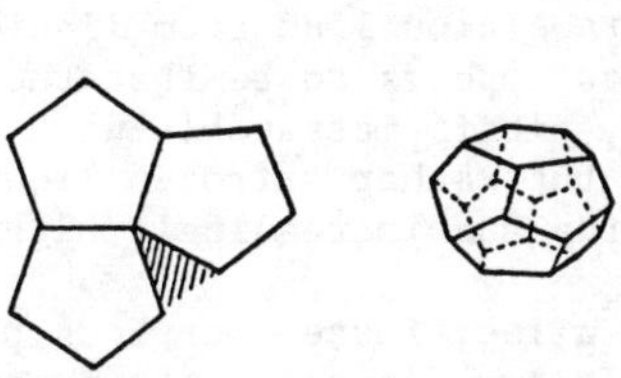

Fig. 2. (left) Regular pentagons do not tile the plane.
(right) The dodecahedron is a sphere "tiled" with pentagons. Similarly, icosahedrons do not fill ordinary space, but they do fill curved space (from Kleman and Sadoc, 1979).

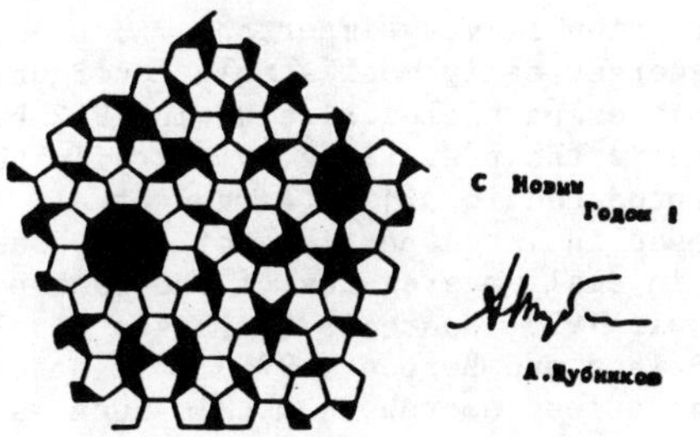

Fig. 3. Attempts to tile the plane with pentagons leave empty spaces (from a new year's card by A. V. Shubnikov).

groups. Only 2-fold, 3-fold, 4-fold and 6-fold rotation symmetries are allowed. Sharp diffraction peaks with one of these symmetryies are characteristic of all three dimensional crystals.

Long range crystalline order reduces the entropy and the free energy of the solid phase below the corresponding values for the liquid phase. However, it was noted long ago by Kauzmann (1948) that extrapolating towards low temperature the entropy behavior of the liquid in the supercooled region, entropy values less then those of the ordered phase would be predicted for low enough temperatures. One concludes that the extrapolation cannot be performed, and that some other thermodynamic accident must occur before reaching the region of the paradox. An accident indeed occurs: quenching of the liquid in the supercooled region produces amorphous or glassy states. These states, however, are not only metastable with respect to the crystalline solid phase, but they are also out of thermodynamic equilibrium in the sense that the kinetics of the system has slowed down so much that only a small portion of the available states are visited in any reasonable amount of time. Quench rate dependence and hysteresis phenomena become apparent. The very definition of the location of the transition line from liquid to glass is quench-rate dependent. The conclusion appears to be that the amorphous state is not a good thermodynamic phase, albeit metastable with respect to the crystalline solid phase, but rather a frozen liquid configuration with vanishingly small diffusion and increasingly high viscosity.

Yet, the search for a legitimate amorphous phase, the fourth state of matter, has continued. In two dimension the existence of a phase intermediate between crystalline and liquid has been postulated.(Halperin and Nelson, 1978)) The so called "exatic phase" has short range translational order, but it exhibits long range nearest-neighbor bond orientational order. In three dimensions the search for analagous bond orientational order in a simple liquid supercooled to 10% below its normal melting point has met with success in the milestone computer simulation work by Steinhard, Nelson and Ronchetti (1981). In that work, indications were found for the existence of a phase transition, upon supercooling the Lennard-Jones liquid, to a new phase exhibiting orientational order, in a range of thermodynamic parameters allowing atomic diffusion and normal values of kinetic relaxation times. The observed order has icosohedral symmetry.

This last observation is very important. Icosahedral symmetry was noted to yield the energetically most stable configuration in small clusters of Lennard-Jones particles already in 1952 by Franck. Furthermore, the icosahedron is the prevailing symmetry in the structure of small aggregates of argon produced in a jet (Farges et al, 1973); it is the symmetry locally obeyed in large portions of amorphous Lennard-Jones packing, as well as in small aggregates of the same particles, in the computer (Barker et al, 1975, Hoare 1976, Farges et al, 1980, Chaudari and Turnbull, 1978). Briant and Burton (1975) have noted that the scattering patterns observed in various amorphous metals such as NiP, $Ni_{32}Pd_{53}P_{15}$, PdSi, and CuHg are quite similar to those from icosahedral clusters and qualitatively dissimilar to those from microcrystals.

The relevant point in all these findings is that the icosahedron packing is locally preferred to a crystallographic one, say fcc, until one gets to clusters of several hundred atoms. The icosahedron appears

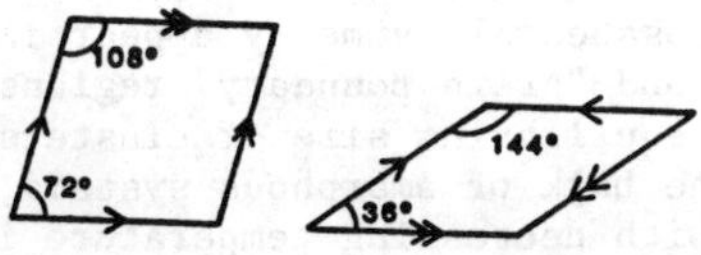

Fig. 4. Arrowed rhombuses, which are the building blocks of the two dimensional Penrose patterns (from Nelson and Halperin, 1985).

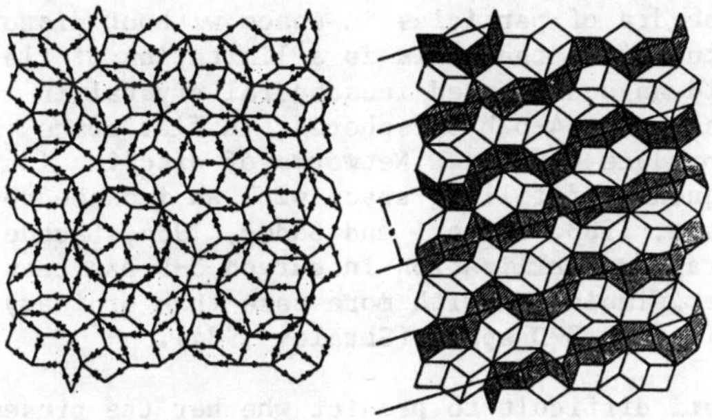

Fig. 5. (left). Two-dimensional Penrose tiling, illustrating how the the arrowed rhombuses (Fig. 4) fit together. Three decagons, all with the same orientation, are shown in boldface. All decagons contain five thick and five thin rhombuses. (right) A regularly spaced diffraction grid is superimposed on the Penrose pattern (right) Each shaded row of tiles may be associated with a particular grid line. Four similar sets of grid lines with the same spacing d may be constructed at angles of 36° and 72° to the set shown. (from Nelson and Halperin, 1985).

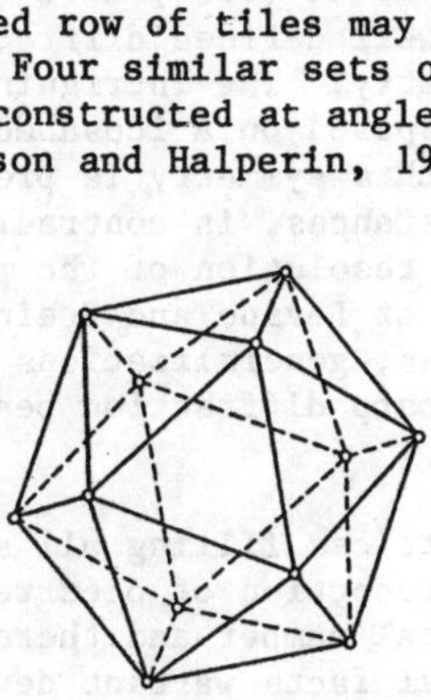

Fig. 6. View of the 12 atoms making up the surface of an icosahedron. Six fivefold symmetry axis pass through the 12 vertices. The three- and two-fold symmetry axis are associated with the 20 faces and 30 edges, respectively.

whenever the system has a special symmetry point that plays the role of the center. This happens in small spherical clusters, and in the case of a charged ion in a simple liquid (Ciccotti and Jacucci, 1975). If on the other hand, nuclei of icosahedral symmetry appear in the bulk phase, they compete with each other and "grain boundary" regions are formed. Hoare has speculated that the equilibrium size of clusters having icosahedral structure embedded in the bulk of amorphous systems, the so-called "amorphons", increases with decreasing temperature in supercooled liquids. Frank already argued that the experimentally observed ability to supercool simple liquid metals well below the equilibrium melting temperature must be due to the prevalence of these icosahedral clusters with respect to germs of crystal with cubic symmetry. It is only when the structure has to ill all space, that the crystalline lattice is thermodynamically preferred.

The tentative to fill all space with one icosahedral structure meets with topological frustration. An icosahedron is made up of tetrahedra. One cannot pack tetrahedra of particles in space without distorting the tetrahedra. It turns out that this is a limitation of flat 3-D space, and that one can build an undistorted icosahedral crystal in curved 3-D space, e.g. on the surface of a 4-D hypersphere. In flat space the distortions will eventually produce defects. Networks of disclination lines will be the inevitable consequence of filling space with an icosahedron (Kleman and Sadoc, 1979, Nelson, 1983, Mosseri and Sadoc, 1984, Sachdev and Nelson, 1984). Icosahedral crystallization in curved 3-D has been observed to happen in computer simulation with more ease than ordinary fcc crystallization in flat 3-D space (Straley 1984).

It is a priori difficult to predict whether the presence of the defects due to topological frustration will inhibit long range order. There has been question of whether even the long range bond orientational order can in fact survive. These questions have recently acquired a central importance in this field, and have received new impetus from experiment and from theory.

First of all, Shechtman et al (1984) have reported the discovery of a phase of $Al_{0.86}$ $Mn_{0.14}$ with well defined diffraction spots obeying an icosahedral point group symmetry. The intriguing fact is that the spots are both well defined and disposed on a icosahedral pattern, implying that order characterized by this symmetry is present in the structure of the system over very long distances, in contradiction with traditional crystallographic rules. The resolution of the paradox came almost simultaneously with the work of Levine and Steinhardt (1984). They showed that "quasi-periodic" lattices, generalizations of Penrose tiles (Penrose 1974, Mackay 1982) produce sharp diffraction peaks and have icosahedral symmetry.

Penrose patterns are lattices filling all space with a mixture of two different unit cells. The proportion of occurrence of the two unit cells in the lattice is an irrational number and there is no periodic repetition in the lattice. These unusual facts warrant deviations from the rules of classical crystallography, which assumes that a single unit cell is repeated cyclically throughout the lattice. The requirements for Penrose quasi crystals are so stringent that only very few geometrical forms are allowed for the unit cells. In two dimensions the most famous Penrose tiling consists of a thick and a thin rhombus with internal angles all multiple of 36°. The ratio of the frequency of occurrence of the fat rhombus to the thin one is the golden mean $\tau = (\sqrt{5} + 1)/2$.

The fact that all the angles are multiples of one tenth of a turn causes the edges of the rhombuses to point toward the vertices of a decagon. As a result there is tenfold symmetry in the orientational long range order in the system. Penrose tailing are also characterized by a special kind of long range translational order. Although the atomic positions do not recur periodically in space, the disposition of the rhombuses along the special directions form lines in which rhombuses of different shapes and orientations repeat aperiodically. These lines repeat in a parallel array, again without periodic matching, but with a wavelength related to the side of the rhombuses. The fivefold symmetry exhibited by this Penrose tiling imposes the existence of two irrationally related length scales, that are not simply periodic. The presence of long range orientational and translational order has suggested that these lattices be called quasi-crystals.

In summary, the fivefold symmetry has finally shown up in condensed matter physics, in its own right and full glory. It has done so in a rather complicated way that requries a bit more mathematics than traditional crystallographic symmetries. Rapidly quenched alloys have been described as exhibiting local icosahedral ordering, broken up at longer range by disclination lines. Long range icosahedral order, like the one observed in Al-Mn alloys, can be produced by a quasi-crystal consisting in a 3-D generalization of Penrose tiling. Finally, the existence of an icosahedral liquid crystal phase supporting long range bond-orientational order but only short range translational order, more akin to the findings of the simulation study of Steinhardt et al (1981) has been recently predicted (Jaric, 1985). It has also been connected to the interpretation of experiments on quenched Al-Mn quasi-crystals (Bancel et al, 1985). In short, the relation between icosahedral packing, the existence of intermediate phase between liquid and cystalline solids, and the amorphous structure of rapidly quenched alloys is still much an open question.

It could not go without saying that the stability of icosahedral patterns in density fluctuations of condensed matter, with respect to cubic patterns (fcc or bcc) was predicted in a lucid paper by Alexander and McTague (1978) on the Landau theory of solidification and crystal nucleation (Landau 1937). But the prediction was dismissed on the basis that the regular icosahedron, with its fivefold axis, cannot form a periodic structure. Now their approach is extensively used to demonstrate the stability of structures exhibiting long range icosahedral order, whether translational or bond orientational.(Bak, 1985, Nelson and Sachdev, 1985, Levine et al, 1985, Mermin and Troian, 1985).

II - Computer Simulation and Bond-Orientational Order in Liquids and Glasses

In certain anisotropic fluids, e.g. nematic liquid crystals, orientational order refers to alignment of the anisotropy axis of the constituent molecules. In fluids made up of spherical molecules the orientational anisotropy refers to the "bonds" linking the center of neighboring molecules. It is possible to disrupt the translational order of a periodic crystal lattice maintaining at the same time a great deal of long range orientational correlation. For example, (Nelson and Toner, 1981) simple crystalline solids disordered by large equilibrium concentration of unbound

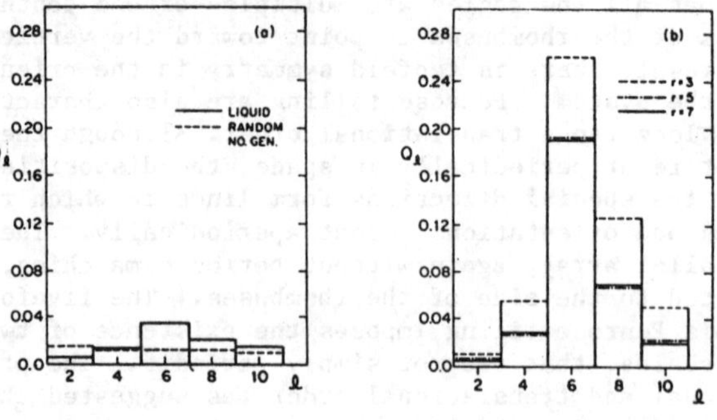

Fig. 7. (a) Quadratic invariants Q_ℓ for a high-temperature Lennard-Jones liquid (solid lines). The dashed lines indicate averages obtained using a random-number generator to produce the orientations of an equivalent number of bonds. (b) Q_ℓ histograms for a supercooled Lennard-Jones liquid. The dependence on the radius r of the averaging volume is shown: dashed line, r = 3; dotted-dashed line, r = 5; solid line, r = 7. (from Steinhardt et al., 1983).

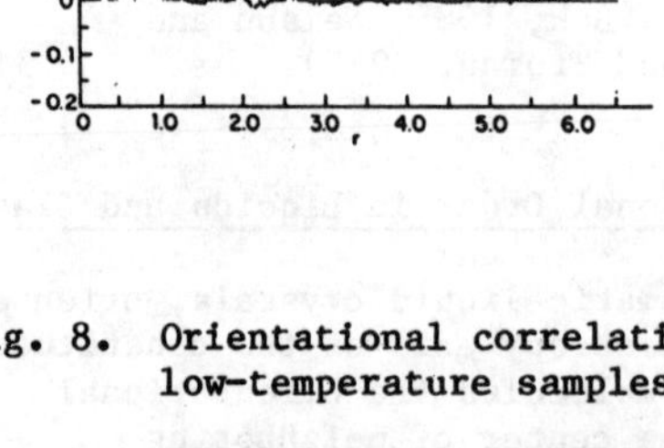

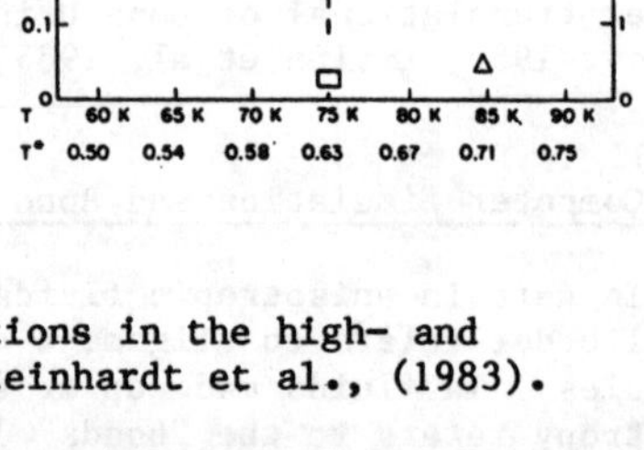

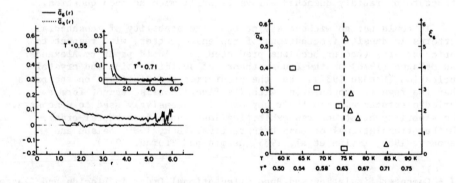

Fig. 8. Orientational correlation fucnctions in the high- and low-temperature samples (from Steinhardt et al., (1983).

Fig. 9. Orientational correlation length ξ_6 (triangles) and "order parameter" Q_6 (squares) as a function of temperature. The temperature is given both in reduced units and in units appropriate to liquid argon (from Steinhardt et al., 1983).

dislocation loops retain long range cubic orientational order. However, the structure factor of the system prepared in this way may not look very different from that of a rapidly quenched liquid.

The degree and character of the bond orientational order present in a given structure must be investigated introducing proper tools. Steinhardt et al (1981, 1983) have developed ways of measuring both local and extended orientational symmetries in computer generated models of dense liquids and glasses. To every bond (line) joining near neighbor atoms they associate a spherical harmonic:

$$Q_{\ell m}(\vec{r}) \equiv Y_{\ell m}\left(\theta(\vec{r}), \phi(\vec{r})\right) \qquad (1\text{-}1)$$

where the $\{Y_{\ell m}(\theta(\vec{r}), \phi(\vec{r}))\}$ are the usual symbols for spherical harmonics, and $\theta(\vec{r})$ and $\phi(\vec{r})$ are the polar angles of the bond measured with respect to some reference coordinate system. $\vec{r}$ is the midpoint of the bond. Averages over suitable sets of bonds will be indicated by

$$\overline{Q}_{\ell m} \equiv \langle Q_{\ell m}(\vec{r})\rangle. \qquad (1\text{-}2)$$

It is important to consider rotationally invariant combinations of $Q_{\ell m}$ for given values of ℓ:

$$Q_\ell \equiv \left\{ \frac{4\pi}{2\ell + 1} \sum_{m=-\ell}^{\ell} \left| \overline{Q}_{\ell m} \right|^2 \right\}^{1/2} \qquad (1\text{-}3)$$

this is the second order invariant. The third order invariant W_ℓ can be similarly defined. Q_ℓ and W_ℓ are found by Steinhardt et al to be the key to a kind of cluster "shape spectroscopy" in liquids and glasses. Furthermore, the orientational order parameters $Q_{\ell m}(\vec{r})$ allow the determination of the range of orientational order. Their results mainly pertain to a 864 atoms system included in periodic boundaries and interacting with the Lennard-Jones potential. In an isotropic system one would expect that all $Q_{\ell m}(\vec{r})$ except Q_{oo} vanish when averaged over sample volume. And this they found for temperatures above the melting point. However, upon supercooling the liquid some 10% below T_m, an extended orientational order develops of predominantly icosahedral character without apparent increase in translational correlation. They interpret their result in terms of an orientational phase transition. What they see is that the orientational correlation length $\xi_6(T)$ of the $\ell = 6$ second order invariant increases with decreasing temperature until it greatly exceeds the translational correlation length ξ and it finally reaches the size of the sample.

Finite size effects are of course substantial in a 864 atoms system, and the determination of a phase transition is not easy. Furthermore, the cubic symmetry imposed by periodic boundary conditions is certainly contributing to the orientational order parameter with cubic character, i.e. $\ell = 4$.

The quench procedure employed to produce equilibrated LJ models is a delicate point. Even the slowest possible quenches on computers avaialble today are about five order of magnitude faster than those of the laboratory (10^{12} degrees per second versus 10^7). However, one can verify that the properties of the state reached do not depend on the history of preparation. Another problem is avoiding crystallization. One can monitor pressure and energy of the system during the quench to test against the happening of this transition. Obviously nucleation of a cubic phase would strongly influence the behavior of certain orientational order parameters.

More information on the bond orientational order can be obtained from the bond-angle correlation function

$$G_\ell(\vec{r}) = \frac{4\pi}{2\ell+1} \sum_{m=-\ell}^{\ell} \langle Q_{\ell m}(\vec{r})\, Q_{\ell m}(\vec{o})\rangle / G_o(\vec{r}) \qquad (1\text{-}4)$$

where $G_o(\vec{r}) = 4\pi \langle Q_{oo}(\vec{r})\, Q_{oo}(\vec{o})\rangle$, and the angular brackets indicate an average over all atoms separated by $\vec{r}$. $G_o(\vec{r})$ is the bond-density autocorrelation function and provides a useful normalization.

The results show that in the higher temperature sample, i.e. the triple point liquid, the correlation functions for ℓ=4 and ℓ=6 fall off rapidly, with a correlation length of the order of the interparticle distance or less. Similarly, the value of the quadratic invariants Q_4 and Q_6 are very small, almost indistinguishable from values obtained with a random number generator. In contrast, the hystogram of Q_ℓ vs ℓ for the supercooled liquid suggests extended correlations in the orientations of bonds possibly of icosahedral symmetry. At the same time, $G_6(\vec{r})$ appears to approach a non zero value for large $\vec{r}$. There are however, no persistent correlations in $G_4(\vec{r})$, suggesting no cubic bond-orientational order. The radial distribution function does not change significantly upon the onset of the long range orientational correlation. A temperature plot of $Q_6(T)$ shows that its value is non zero below $T\sim.63\varepsilon$ while $\xi_6(T)$ increases strongly and appears to diverge approaching $T\sim.63\varepsilon$ from above. This behavior suggests a transition to an icosahedral phase at $T_c{}^* = .63\varepsilon$, indicated by an increase of orientational order.

Behavior very similar to that found for $G_6(\vec{r})$ in supercooled 3-D LJ liquid has been found for an hexatic orientational correlation function in 2-D LJ by Frenkel and McTague (1979).

It would be very interesting to extend Molecular Dynamics investigations along these lines to include recent theoretical insight on icosahedral quasi-crystals. It is clear that periodic boundaries will force the quasi crystal within a periodic array of supercells. This can be accommodated only introducing suitable defects in the structure. However, it would be very interesting to study the dynamics of such a structure, looking in particular for the three extra vibrational modes of acoustic character called phasons (Bak, 1985), connected with the 6-D character of the 3-D icosahedral quasi-crystal. It is possible that the imposition of periodic boundaries would amount to a finite gap at zero k-vector in their dispersion relation.

III - Dynamics of Atoms in Fluids

As an example of computer simulation of atomic dynamics in fluid systems we shall now describe a recent Molecular Dynamics investigation of dynamical structure factors in a liquid alloy (Jacucci, Ronchetti and Shirmacher, 1984). We shall start by briefly recalling the dynamical properties of classical liquids and liquid mixtures.

In a single-component simple classical liquid a central role is played by density fluctuations. These are related to differential cross sections for inelastic scattering of neutrons by the scattering laws:

$$\frac{d^2\sigma}{d\Omega dE} = \frac{k}{ko}\left\{b^2_{coh} S(Q,\omega) + b^2_{incoh} S_s(Q,\omega)\right\} =$$

$$\left\{\frac{d^2\sigma}{d\Omega dE}\right\}_{coh} + \left\{\frac{d^2\sigma}{d\Omega dE}\right\}_{incoh} \tag{2-1}$$

Here the momentum transfer is $\hbar\vec{Q} = \hbar(\vec{k} - \vec{k}_o)$ and the energy transfer is $\hbar\omega = (\hbar^2/2M)(k^2 - k_o^2)$. In isotropic liquids $d^2\sigma/d\Omega dE$ depends only on $|\vec{Q}|$. S and S_s are the dynamical structure factors and are related by Fourier transformation to the Van Hove correlation functions:

$$G_s(\vec{r},t) = \left(\frac{1}{2\pi}\right)^3 \int d^3\vec{Q} \int d\omega\, S_s(Q,\omega) \exp\{i\omega t - i\vec{Q}\cdot\vec{r}\} =$$

$$= \left(\frac{1}{2\pi}\right)^3 \int d^3\vec{Q}\, F_s(Q,t) \exp\{\, i\vec{Q}\cdot\vec{r}\,\} \tag{2-2}$$

and

$$G(\vec{r},t) = \left(\frac{1}{2\pi}\right)^3 \int d^3\vec{Q} \int d\omega\, S(Q,\omega) \exp\{\, i\omega t - i\vec{Q}\cdot\vec{r}\}$$

$$\left(\frac{1}{2\pi}\right)^3 \int d^3\vec{Q}\, F\,(Q,t) \exp\{i\vec{Q}\cdot\vec{r}\} \tag{2-3}$$

G_s describes individual motions of the particles: It gives the probability for a specific particle appearing at r at time t if it started at the origin at t = 0. G describes the collective motion of the particles: it gives the probability for any particle being at time t in r given that the same or another one was at the origin at t = 0. Their connection to the probability density of a tagged particle $N_o(\vec{r},t) \equiv \delta(\vec{r} - \vec{r}_o(t))$ and to the number density $N(\vec{r},t) \equiv \Sigma_{i-1}^{N} \delta(\vec{r}-\vec{r}_i(t))$ is simple:

$$G_s(\vec{r},t) = \langle N_o(\vec{r},t)\, N_o\,(\vec{0},0)\rangle$$

$$G\,(\vec{r},t) = \frac{1}{N} \langle N(\vec{r},t)\, N(\vec{0},0)\rangle \tag{2-4}$$

The intermediate scattering functions F obey the following sum rules at t=0:

$$Fs(Q, t = 0) = 1$$

$$F(Q, t = 0) = S(Q) = 1 + \rho_o \int d^3r \left(g(r)-1\right) \exp (iQr) \qquad (2\text{-}5)$$

$$\lim_{Q \to 0} S(Q) = \rho_o k_B T \chi_T$$

where S(Q) is the static structure factor, g(r) is the pair-correlation function, and χ_T is the isothermal compressibility.

S_s (Q,ω) is dominated, in the limiting case of small wavelength, by free particle motion, and in the opposite limit of large wavelength by diffusion. For small wavelength we have the so called collisionless regime: Q<<2π/a, where a is the mean interparticle distance:

$$S_s(Q,\omega) \cong S_{\text{ideal}}(Q,\omega) = \left(\frac{M}{2\pi k_B TQ^2}\right)^{1/2} \exp \left\{ \frac{M\omega^2}{2k_B TQ^2} \right\} \qquad (2\text{-}6)$$

For large wavelengths we have the hydrodynamic regime:

$$S_s(Q,\omega) = \frac{1}{\pi} \quad \frac{DQ^2}{\omega^2 + (DQ^2)^2} \qquad (2\text{-}7)$$

with $D = (K_B T/M)\int_0^\infty Z(t)dt$ being the diffusion coefficient, and

$Z(t) = (M/3K_B t) \langle v(t')\ v(t' + t)\rangle$ the velocity auto correlation function.

Collective motion is seen in S(Q,ω), yet only in the hydrodynamic and intermediate regimes. Again for Q>>(2π/a) the collisionless regime of individual, uncorrelated motion is recovered. In the hydrodynamic regime S(Q,ω) as a function of ω at constant Q appears as a three peaked function centered at the origin. The two side peaks are called Brillouin peaks and represent non-overdamped propagating density waves positioned at $\omega_b(Q) = \pm uQ$, with u the sound velocity. The width of these peaks increases with Q^2 and it is related to the system viscosity. The central peak is called Rayleigh peak and represents entropy fluctuations. These do not propagate and decay in time with a relaxation time related to the peak's width $\Delta_q = D_q Q^2$, with D_q the thermal diffusivity. Entropy fluctuations couple to density fluctuation through a non zero thermal expansion coefficient.

An interesting question is: up to which values of Q non-overdamped collective excitations exist? In the intermediate region of Q values, the dispersion curve ω = ω(Q) levels out, after the low Q linear increase, while the width Δ becomes comparable to the distance of the peak position from the origin. This and other interesting questions have been addressed with success using Molecular Dynamics (MD).

Computer simulations using the MD technique consist in the integration of equations of motion for $10^2 \div 10^3$ particles with periodic boundary conditions, interacting via a pair-wise additive potential, using a constant force approximation for a time intervals $\Delta t \sim 10^{-2} \div 10^{-3}$ psec. The basic loop is an iteration to find x^+ from x^o and x^- and knowledge of $\ddot{x}^o$. Accurate trajectories in phase space are obtained for Δt small enough. About 10^5 time steps can be integrated yielding histories corresponding to $10^2 \div 10^3$ psec. With a good choice of initial conditions, and a few corrections of the total kinetic energy of the system, desired equilibrium thermodynamic states can be reached and thermal energies of equilibrium and time dependent properties can be evaluated making use of the ergotic theorem.

$S(Q,\omega)$ can be evaluated in this way. An important limitation, however comes from the periodicity of the system that selects discrete allowed $\vec{Q}$ values:

$$\vec{Q} \equiv \frac{2\pi}{L}(n_x, n_y, n_z) \qquad \text{with L the basic box edge.}$$

To compute $S(\vec{Q},\omega)$, the row data in the form of time dependent particle positions at successive discrete values of the time $\{r_i(t_s)\}$, usually recorded on a magnetic tape, are employed to form the spatial Fourier transform of the particle number density at each time t_s:

$$N(\vec{Q},t_s) = FT_Q \left\{\Sigma_{i=1}^{N} \delta\left(\vec{r} - \vec{r}_i(t_s)\right)\right\}$$

$$= \Sigma_{i=1}^{N} \exp\left(i\vec{Q}\cdot\vec{r}_i(t_s)\right) . \tag{2-8}$$

Then two routes are available to get $S(\vec{Q},\omega)$:

a) $$F(\vec{Q},t) = \langle N(\vec{Q},t)\, N(-\vec{Q},0)\rangle \frac{1}{N}, \qquad S(\vec{Q},\omega) = FT_\omega \{F(\vec{Q},t)\} \tag{2-9}$$

b) $$N(\vec{Q},\omega) = FT_\omega\{N(\vec{Q},t)\}, \quad S(\vec{Q},\omega) = |N(\vec{Q},\omega)|^2 \tag{2-10}$$

Route b) is often convenient for data handling. An important issue in the (discrete) Fourier transform to frequency is the determination of the correct resolution $\Delta\omega$ to be employed. $\Delta\omega$ is related to the maximum correlation in time τ allowed: $\Delta\omega = (2\pi/2\tau) = (1/m)(\pi/\Delta t)$, where $\tau = m\Delta t$. The maximum frequency component that can be obtained from a discrete time function is $\omega_{max} = (2\pi/2\Delta t) = (\pi/\Delta t)$.

It is apparent that to properly conserve the information contained in a discrete time function, known at m time steps one has to use m frequency values. Of course the free spectral range of the transform ends ar ω_{max}. For larger frequencies the spectrum repeats itself. Care must be used to minimize statistical noise and avoid missing sharp features in the frequency data. To do this one has to cut off the phase of sin's and cos's after m integration time steps, and restart from zero the process for subsequent segments, in evaluating the frequency Fourier transform. Segments so treated must be squared separately and averaged in eqn. 10. Cutting off the correlation in the harmonic phases in the Fourier transform calculation enforces the chosen value of the resolution $\Delta\omega$. Longer time correlations result in possibly missing sharp features of the spectrum between values of the discrete ω axis, or reduces averaging over the frequency interval $\Delta\omega$ of the computed values of the Fourier amplitudes at ω_i.

IV - Collective Modes in Li_4-Pb Liquid Alloys

Let us now extend the description of the dynamical structure factors to two components. C_1 and C_2 will denote the concentrations: $N_1 = C_1N$, $N_2 = C_2N$. The partial densities are

$$N_\nu(\vec{r},t) = \Sigma_{i=1}^{N_\nu} \delta(\vec{r} - \vec{r}_i(t)) \quad \text{with } \nu = 1,2$$

and the partial dynamical structure factors:

$$S_{\mu\nu}(Q,\omega) = \frac{1}{4\pi N_\mu^{1/2} N_\nu^{1/2}} \int d^3\vec{r} \int d\omega \;\cdot$$

$$\cdot \; \langle N_\mu(\vec{r},t)\, N_\nu(\vec{0},0) + N_\nu(\vec{r},t)\, N_\mu(\vec{0},0)\rangle \exp\{i\vec{Q}\cdot\vec{r} - i\omega t\} \qquad (2\text{-}11)$$

$\frac{d^2\sigma}{d\Omega dE}$ is a weighted sum of these three functions with scattering lengths squared as weights.

An average physical picture arises if one considers overall number fluctuations;

$$N(\vec{r},t) = N_1(\vec{r},t) + N_2(\vec{r},t), \qquad (2\text{-}12)$$

and deviations from random mixing, i.e. concentration fluctuations, are described by

$$C(\vec{r},t) = \frac{1}{N}\{N_1(\vec{r},t) - C_1N(\vec{r},t)\} = \frac{1}{N}\{C_2N_1(\vec{r},t) - C_1N_2(\vec{r},t)\} \qquad (2\text{-}13)$$

New $S(Q,\omega)$'s are defined with N and C: $S_{NN}(Q,\omega)$ is the double Fourier transform of the time correlation function $\langle N(\vec{r},t)\, N(0,0)\rangle$, $S_{CC}(Q,\omega)$ of $\langle C(\vec{r},t)\, C(0,0)\rangle$, and the cross function $S_{NC}(Q,\omega)$ of $\langle N(\vec{r},t)C(0,0) + C(\vec{r},t)\, N(0,0)\rangle$.

A sum rule similar to eqn. 5 relates $F_{NN}(Q,t = 0)$ to $\rho_o k_B T \chi_T$, with a small correction from the cross correlation F_{NC}. Again we see that in the average number density picture, density fluctuations are related to the compressibility of the system. An important new sum rule appears for the concentration fluctuations:

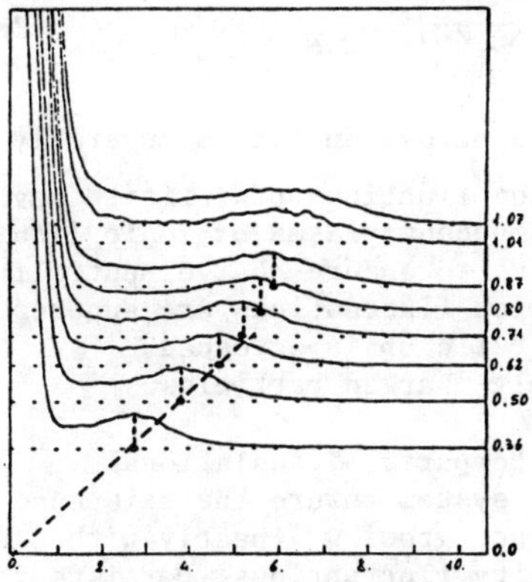

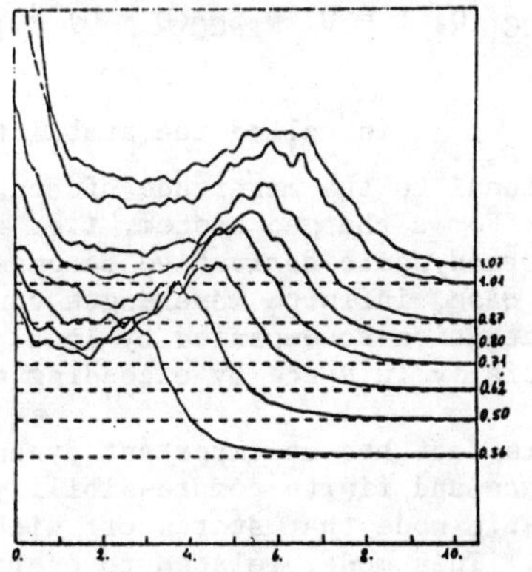

Fig. 10. Concentration-concentration dynamic structure factor S_{CC} (Q, ω) for Li_4Pb liquid alloy. The high frequency propagating concentration wave has an acoustic type linear dispersion relation at small Q (from Jacucci et al, 1984).

Fig. 11. Li-Li partial dynamic structure factor. The high frequency concentration wave seen in S_{CC} (Q, ω) is sustained by Li motion (from Jacucci et al, 1984).

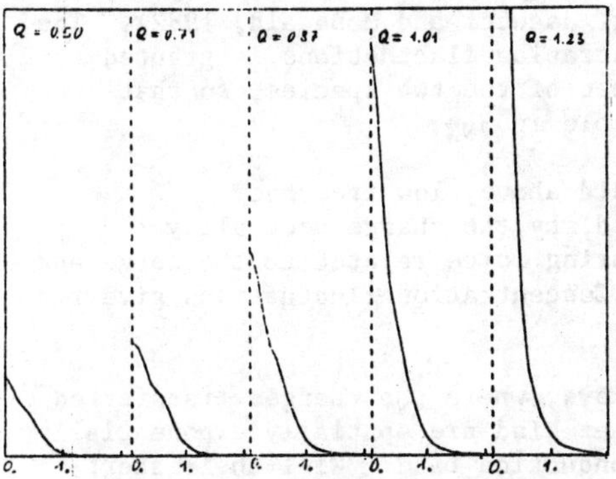

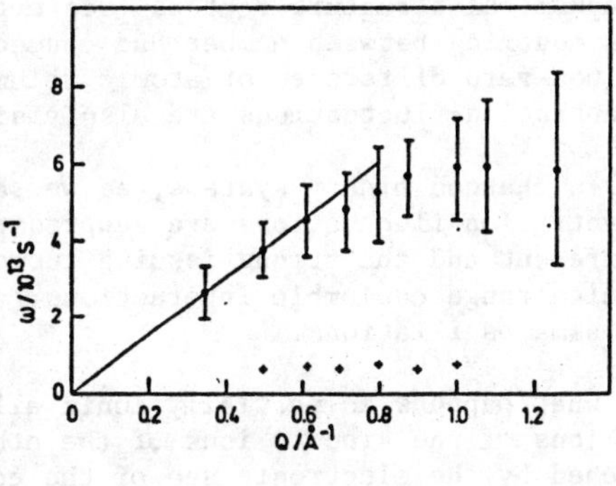

Fig. 12. Pb-Pb partial dynamic structure factor. The high frequency con-centration wave seen in S_{cc} (Q, ω) has no participation from Pb atoms. The ordinary sound acoustic mode is unresolved because overdamped, and produces the shoulder in the central component of the spectrum (from Jacucci et al, 1984).

Fig. 13. Dispersion relation for the two propagating modes present in Li_4Pb liquid alloy. The width of the higher frequency lithium sustained concentration fluctuations is also shown. The portion of the Brillouin peak of ordinary sound is indicated by crosses showing the approximate location of the shoulder in the central component of the overall density dynamic structure factor (from Jacucci et al, 1984).

$$\lim_{Q\to 0} F_{CC}\ (Q,\ t = 0) = S_{CC}(Q = 0) = \rho_o k_B T / \left(\frac{\partial^2 G}{\partial C_1^{\ 2}}\right)_{\rho,T,N} \qquad (2\text{-}14)$$

$\left(\frac{\partial^2 G}{\partial C_1^2}\right)_{\rho,T,N}$ is called the stability of the alloy, and it is inversely proportional to the magnitude of concentration fluctions of infinite wavelength. For a charged system, i.e. a two component plasma or ionic liquid, this thermodynamic derivative becomes infinite to ensure charge neutrality. In this case, infinite wavelength concentration fluctuations are suppressed. Any fluctuation accompanied by local charge build up is screened exponentially in space by exceeding oppositely charged particles.

This fact has an important dynamic counterpart. Translational invariance and finite compressibility of the system ensure the existence of an acoustic mode that starts off with frequency growing linearly with Q at small Q. This mode, related to overall density fluctuations, persists for larger Q albeit somewhat dispersed (i.e. with deviations from $\omega_0 = \pm uQ$), until it becomes overdamped. Charge neutrality and vanishing electrical susceptibility introduce a frequency gap at zero Q in the concentration fluctuation mode. This is the well known optical mode or plasma oscillation that starts off with a constant value at Q = 0.

For uncharged binary mixtures and alloys, there is no charge neutrality requirement, so that the stability of the alloy is finite, the new sum rule eqn. 14 gives a finite value, and there is no optical mode. Usually, however, it is also not possible to see any propagating concentration fluctuations. These fluctuations contribute a zero frequency diffusive peak, instead, to the dynamical structure factors (see e.g. Jacucci and McDonald, 1982). The cross coupling between number and concentration fluctuations is granted by a non zero difference of atomic volumes of the two species, so that concentration-fluctuations are also visible in S_{NN}.

In charged binary systems, as we said above, low frequency concentration fluctuations are suppressed, by the charge neutrality requirement and the strong issuing restoring force related to the large and infinite range coulombic interactions. Concentration fluctuations give rise to plasma oscillations.

What happens in partially ionic alloys, where the charges transferred from ions of one kind to ions of the other kind are spatially exponentially screened by the electronic see of the conduction band? Will their short distance Coulomb interaction prevail in determining the character of concentration fluctuations, producing a plasma oscillation, or will their long range neutral interaction prevail, producing an acoustic type mode for the dispersion of these fluctuations?

Recently, a computer simulation study of the Li_4Pb alloy (Jacucci, Ronchetti and Shirmacher, 1984) has accompanied anelastic thermal neutron scattering experiments on the same system (Soltwish, Quitman, Ruppersberg and Suck, 1983). This is a system made up of metallic constituents, showing strong deviations from ideal mixing at compositions determined by chemical stoichiometry, and a tendency towards charge transfer and ionic bonding. Its interaction pair potential is believed to contain a screened Coulomb term (Copestake and Evans, 1982; Ruppersberg and Reiter, 1982).

The dynamical structure factors computed in the simulation are very remarkable. The concentration one, S_{CC}, is dominated by propagating concentration fluctuations, in contrast to the S_{CC} in Na-K liquid alloys, (Jacucci and McDonald, 1982) where it is dominated by a non propagating interdiffusion mode. The velocity connected with the acoustic behavior of this mode is very high indeed, about five times higher than the sound velocity in the system. As a consequence, as Q reaches the intermediate region, $\omega_c = \pm u_c Q$ has attained large values typical of optical plasma oscillations.

The answer to our previous question is as follows. The screening of the Coulomb charge divides up Q-space in two regions. For small Q -values, the charge is screened, and the concentration mode is acoustic in character. For large Q-values the charges are not screened, and the concentration mode appears as an optical plasma oscillation. The crossover value of Q_c is dictated by the screening length λ and is also related to the value of the sum rule eqn. 14. Larger screening length means higher stability of the alloy and a smaller value of $S_{CC}(Q = 0)$. In fact, if the crossover value of Q is smaller, the value of ω_c related to the plasma frequency has to be attained faster (i.e. for lower Q_c, and with a higher u_c). An interesting feature, albeit somewhat secondary to the present discussion, is that in the system in question the mass of the two particles are very different. (About a factor of 30!) In the concentration fluctuation it therefore applies a kind of Born-Oppenheimer separation in which the lithium ions oscillate around essentially still lead ions. While the value of the cross-over wave vector Q_c does not depend on the masses, thee terminal plasma oscillation frequency, and hence u_c, do.

This is we believe the first observation of finite Q plasma oscillations starting off as an acoustic made in a screened Coulomb system. (Bosse, Jacucci, Ronchetti and Shirmacher, 1985). This mode was not open to observation in the experiment (Soltwish et al, 1983) because it would appear at much higher exchange energies than the ones monitored. A new experiment is being undertaken, including the interesting region, as a result of the computer simulation findings.

Acknowledgments

We are indebted to Franco Nori for discussions of recent theories of icosahedral quasi-crystals, and to Jürgen Bosse for discussion of the behavior of charged density waves in screened Coulomb systems. Discussions with Walter Schirmacher are also acknowledged. We acknowledge support from the National Science Foundation through the University of Illinois Materials Research Laboratory, Grant No. NSF-DMR-83-16981.

References

Alexander, S. and McTague, J. P. (1978), Phys. Rev. Letters 40, 702.
Bak, P. (1985) Phys. Rev. Letters 54, 1517; Phys. Rev B, in press.
Bancel, P.A., Heiney, P.A., Stephens, P. W., Goldman, A. I., and Horn, P. H. (1985), Phys. Rev. Letters 54, 2422.
Barker, J. A., Hoare, M. R., and Finney, J. L. (1975) Nature (London) 257, 120.
Bosse, J., Jacucci, G., Ronchetti, M., and Shirmacher, W. (1985) to be published.
Briant, C. L., and Burton, J. J. (1975) J. Chem. Phys. 63, 2045.
Chaudari, P., and Turnbull, D. (1978), Science 199, 11.
Ciccotti, G., and Jacucci, G. (1975) Phys. Rev. Letters 35, 789.
Copestake, A. P., Evans, R. (1982) J. Phys. C.: Solid State Phys. 15, 4961.
Farges, J., de Feraudy, M. F., Raoult, B., and Torchet, G. (1980), J. Phys. Colloq. (France) 41, C3/263; (1983) J. Chem. Phys. 78, 5067.
Farges, J., Raoult, B. and Torchet, G. (1973), J. Chem. Phys. 59, 3454.
Franck, F. C. (1952), Proc. R. Soc. London, Ser A 215, 43.
Frenkel, D., and McTague, J. P. (1979), Phys. Rev. Letters 42, 1632.
Halperin, B. I., and Nelson, D. R. (1978), Phys. Rev. Letters 41, 121; 41, 519(E); (1979) Phys. Rev. B 19, 2457.
Hoare, M. R. (1976), Ann. NY Acad. Sci. 279, 186; (1978) J. Non Cryst. Solids 31, 157; (1979) Adv. Chem Phys., 40, 49.
Jacucci, G., and McDonald, I. R. (1980) in Liquid and Amorphous Metals, E. Luscher and H. Coufal, eds. (Sijthoff and Noordhoff).
Jacucci, G., Ronchetti, M., and Shirmacher, W. (1984), in Condensed Matter using Neutrons, S. W. Lovesey and R. Scherm Eds., Plenum Press.
Jarie, M.V. (1985), Phys. Rev. Letters 55, 607.
Kauzmann, W. (1948), Chem. Rev. 43, 219.
Kleman, M. and Sadoc, J. F. (1979), J. Phys. (Paris) Lett. 40, L569.
Landau, L. D. (1937), Phys. Z. Soviet 11, 26, 545.
Levine, D., Lubensky, T. C., Ostlund, S., Ramaswamy, S., Steinhardt, P. J. and Toner, J. (1985), Phys. Rev. Letters, 54, 1520.
Levine, D. and Steinhardt, P. J. (1984), Phys. Rev. Letters 53, 2477.
McKay, A. L. (1982), Physica 114A, 609.
Mermim, N. D., and Troian, S. M. (1985), Phys. Rev. Letters 54, 1524.
Mosseri, R. and Sadoc, J. F. (1983) in Structure of Non Crystalline Materials, R. H. Gaskell, J. M. Parker, and E. A. Davis edts. (Taylor and Francis, NY).
Nelson, D. R. (1983). Phys. Rev. B 28, 5515.
Nelson, D. R., and Halperin, B. I. (1985), Science 229, 233.
Nelson, D. R. and Sachdev, S. (1985), Phys. Rev. B 32, 689, 1480.
Nelson, D. R., and Toner, J. (1981), Phys. Rev. B 24, 363.
Penrose, R. (1974), Bull. Inst. Math. Appl. 10, 266.
Ruppersberg, H. and Reiter, H. (1982), J. Phys. F.: Met. Phys. 12, 1311.
Sachdev, S., and Nelson, D. R. (1984), Phys. Rev. Letters 53, 1947.
Soltwish, M., Quitmann, D., Ruppersberg, H., and Suck, J. B. (1983). Phys. Rev. B 28, 5583.
Steinhardt, P. J., Nelson, D. R., and Ronchetti, M. (1981) Phys. Rev. Letters 47, 1297; (1983) Phys. Rev. B 28, 784.
Shechtman, D. S., Blech, I., Gratias, D., and Cahn, J. W. (1984), Phys. Rev. Letters 53, 1951.
Straley, J. P. (1984), Phys. Rev. B 30, 6592.

ON SURFACE PROPERTIES OF THE CLASSICAL ONE-COMPONENT PLASMA

M. HASEGAWA

Faculty of Integrated Arts and Sciences, Hiroshima University, Hiroshima 730, Japan

1. INTRODUCTION

The classical one-component plasma (OCP) is a system of charged particles (which we call ions) in a rigid compensating charge background. It is probably the simplest model of Coulombic systems but describes quite well some realistic systems under special circumstances. The OCP also provides a useful starting point for understanding real systems such as liquid metals. Bulk properties of the OCP have been investigated by a number of theoretical and computer-simulation studies and are now relatively well known [1].

This contribution is concerned with the OCP surface properties, which have also received increasing theoretical attention [2-5]. The density functional formalism provides a useful means of studying inhomogeneous systems [6] and the density-gradient expansion, one of the approximation schemes in the formalism, has been developed for the OCP[2,3,5,7].Recently,Rosinberg et al [4] have used this method for calculating surface properties of the OCP and found serious discrepancies between their calculations and Monte Carlo (MC) simulations [8]. They argued, based on these calculations, that the density-gradient expansion is inadequate for studying the OCP surface even in the weak coupling regime. However, we have found that two important points are overlooked in their argument. The first point is that the validity of the coefficient of the square-gradient term has not yet been fully investigated. In fact, we have found that the coefficient used by them is quite different from that obtained by using an improved approximation and a much better agreement between theory and MC results is achieved by incorporating such an improved coefficient [5]. The second point overlooked by Rosinberg et al is the influence of the finite-size which could be substantial in the MC simulations for a finite spherical system [8]. We have estimated such an effect on surface properties and found that theoretical calculations for a planar surface cannot be compared with the MC results in the weak coupling regime [5,9,10]. The presence of a hard wall, which must be introduced for a finite system, also must be taken into account in interpreting MC data. In the following sections we briefly summarize these results of theoretical investigations.

2. THE DENSITY-GRADIENT EXPANSION

We consider a system of ions with charge Ze embedded in a non-uniform charge background of density $-en(\mathbf{r})$. The relevant quantity in the density functional theory is the functional which is identified with the thermodynamic potential for the equilibrium ion density $\rho(\mathbf{r})$:

$$\Omega[\rho] = G[\rho] + \frac{e}{2}\int d\mathbf{r}\phi(\mathbf{r})\,[Z\rho(\mathbf{r}) - n(\mathbf{r})] - \mu\int d\mathbf{r}\rho(\mathbf{r}), \tag{1}$$

where $G[\rho]$ is the non-Coulombic part of the free energy functional, $\phi(\mathbf{r})$ is the electrostatic potential and μ is the chemical potential. The equilibrium ion density can be determined by the stationary condition of $\Omega[\rho]$ or the Euler-Lagrange equation [6].

One of approximation schemes to manipulate the unknown functional $G[\rho]$ is the density-gradient expansion truncated at second order

$$G[\rho] = \int d\mathbf{r}[g_0(\rho(\mathbf{r})) + g_2(\rho(\mathbf{r}))|\nabla\rho(\mathbf{r})|^2]. \quad (2)$$

For the free energy density $g_0(\rho)$ a very accurate expression is known which has been fitted to the MC data [11]. On the other hand, the coefficient of the square-gradient term is given by [6,7]

$$g_2(\rho) = \frac{1}{12\beta}\int d\mathbf{r}\,r^2\tilde{c}(r), \quad (3)$$

where $\beta = (k_BT)^{-1}$ and $\tilde{c}(r)$ is the non-Coulombic part of the direct correlation function $c(r)$, i.e. $\tilde{c}(r) = c(r) + \beta(Ze)^2/r$. Evans and Sluckin [7] used the approximation scheme proposed by Baus and Hansen (BH)[12] to calculate $g_2(\rho)$ and this result has been used for surface calculations [3,4]. On the other hand, Ballone et al [2] used a mean-spherical approximation (MSA) in their attempt to calculate $g_2(\rho)$. Recently, we have investigated these results and found that the intermediate-range correlation is not properly taken into account in these calculations [5]. We have used a modified HNC (MHNC) approximation proposed by Rosenfeld and Ashcroft [13], a much better approximation than the BH and MSA, and obtained a result which is quite different from the previous ones.

Figure 1 shows the comparisons of these results in terms of the dimensionless quantity $Y(\Gamma)$ defined by [3]

$$g_2(\rho) = [(Ze)^2/528]\cdot(3/4\pi)^{1/3}\rho^{-4/3}Y(\Gamma), \quad (4)$$

where Γ is the plasma parameter defined by $\Gamma = \beta(Ze)^2/a$, a being the ion-sphere radius $4\pi a^3/3 = 1/\rho$. We find that the MHNC result for $Y(\Gamma)$ is much smaller than the BH and MSA results in the strong coupling regime. Using the MHNC result for $g_2(\rho)$ we have performed variational calculations for a planar surface with step background density. For $\bar{\Gamma} = 10$, where $\bar{\Gamma}$ is the bulk plasma parameter, our result for the surface energy U_s is 0.553 in units of $\bar{a}\bar{\rho}k_BT$ ($\bar{a}$ and $\bar{\rho}$ are bulk quantities). This value is much smaller than the value 0.857 obtained by Rosinberg et al [4] using the BH result for $g_2(\rho)$ and in much better agreement with the MC result (see Fig. 4). For $\bar{\Gamma} = 1$ the gradient term plays a minor role and both approximation schemes produce much the same value for U_s and these are much smaller than the MC data. This discrepancy will be discussed in the next section.

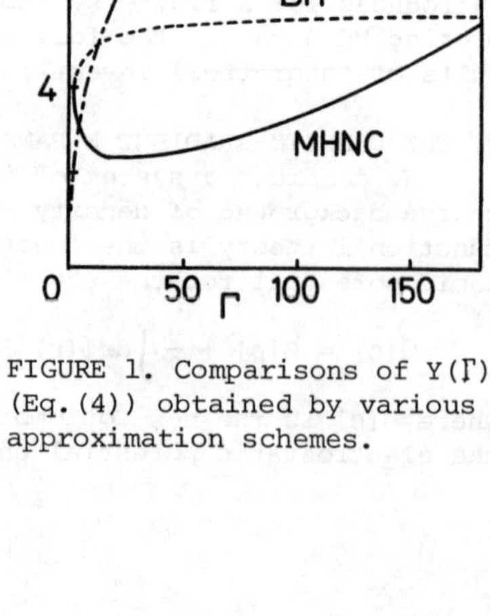

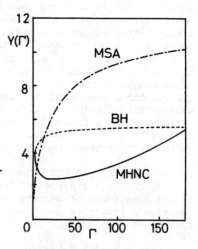

FIGURE 1. Comparisons of $Y(\Gamma)$ (Eq.(4)) obtained by various approximation schemes.

3. THE FINITE-SIZE EFFECT

We have performed two types of approximate calculations to estimate the finite-size effect which could be substantial in the MC simulations for a finite spherical system [8]. We consider a system in which the charge background density $n(\mathbf{r})$ is uniform in a spherical volume of radius R_0, i.e. $n(r) = \bar{n}$ for $r < R_0$ and $n(r) = 0$ for $r > R_0$. We also introduce a hard wall at $r = R_H$ ($R_H > R_0$), which is essential for a finite system [8].

3.1. The Poisson-Boltzmann approximation

We first consider the Poisson-Boltzmann (PB) approximation, which consists of taking $G[\rho]$ to be the one for an ideal gas. It is valid in the weak coupling limit. In this approximation we can derive a differential equation for $\rho(r)$ from the Euler-Lagrange and Poisson equations. We have solved this differential equation numerically and calculated the surface energy U_s [10].

Figure 2 and 3 show the results of these calculations for $\bar{\Gamma} = 1$ and their comparisons with the MC data. The calculated ion density $\tilde{\rho}(r) = \rho(r)/\bar{\rho}$ is in reasonable agreement with the corresponding MC results. We find that $\tilde{\rho}(r)$ strongly depends on the size of a sphere and $\tilde{\rho}(r)$ for a planar surface is quite different from that for a finite system. The surface energy U_s also shows a strong size-dependence and the large MC values of U_s can be explained by this size effect (curve A in Fig.3). The calculated results in Fig.3 indicate that quite different results would have been obtained if different values of R_H/R_0 had been used in the MC simulations.

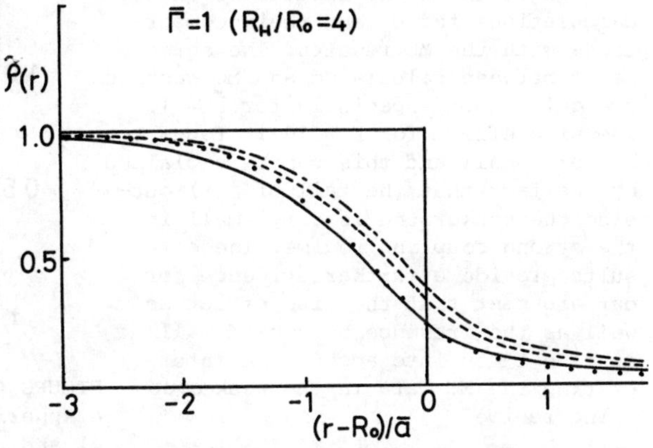

FIGURE 2. Density profiles calculated in the PB approximation for $R_0/\bar{a} = 6.903$ (full curve), 20 (broken curve) and ∞ (chain curve) and their comparisons with the MC results for $R_0/\bar{a}= 6.903$.

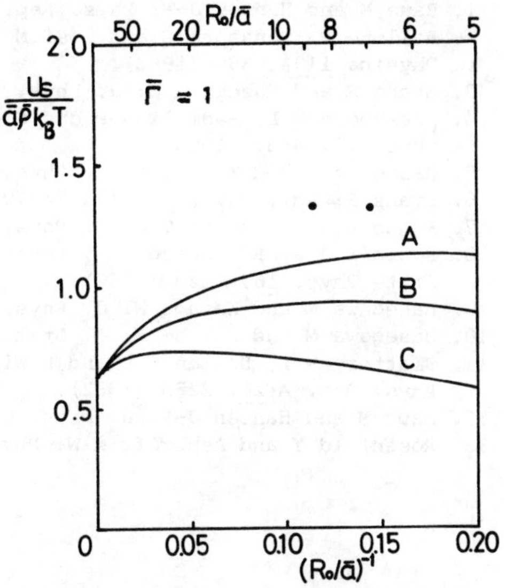

FIGURE 3. Size-dependences of U_s. Curves are results in the PB approximation for R_H/R_0 = 4 (curve A), 3 (curve B) and 2 (curve C). Circles are MC results for R_H/R_0 = 4 [8].

3.2. Variational calculations

We have also performed variational calculations to give a further evidence for the size effect predicted by the calculations in the PB approximation [9]. In these calculations we have used a three-parameter trial ion density

$$\rho(r) = \begin{cases} 1 - Ae^{\alpha(r-r_0)}\cos\gamma(r-r_0) & \text{for } 0 < r < r_0 \\ B/(r-r_1)^2 & \text{for } r_0 < r < R_H \end{cases}$$

The parameters r_0, r_1 and B can be eliminated by using the continuity conditions of $\tilde{\rho}(r)$ and $\tilde{\rho}'(r)$ at r =

r_0 and the overall charge neutrality condition. The remaining parameters A, α and γ are determined variationally by the minimum condition of the surface free energy. We have used an accurate expression for $g_0(\rho)$ and the MHNC result for $g_2(\rho)$ (see Fig.1).

Fig.4 shows the results of these calculations for U_s and their comparisons with the MC results. The agreements between calculated and MC results are quite good especially for $\bar{\Gamma} = 1$. The size effect for $\bar{\Gamma} = 10$ is found to be very small and this can be explained by the fact that the tail of $\rho(r)$ outside the background is very small in the strong coupling regime. These results provide a further evidence for our argument that the size effect as well as the presence of a hard wall must be taken into account in interpreting the MC data in the weak coupling regime.

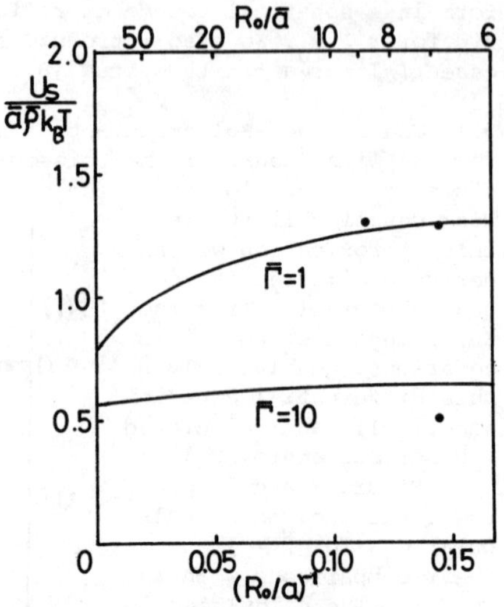

FIGURE 4. Size-dependences of U_s for a spherical system at $\bar{\Gamma} = 1$ ($R_H/R_0 = 4$) and $\bar{\Gamma} = 10$ ($R_H/R_0 = 2.5$): Curves are results of variational calculations and circles are MC results [8].

REFERENCES

1. Baus M and Hansen J-P: Phys. Rep. 59, 1 (1980).
2. Ballone P, Senatore G and Tosi M P: Nuovo Cimento 65B, 293 (1981); Physica 119A, 356 (1983).
3. Evans R and Hasegawa M: J. Phys. C: Solid State Phys. 14, 5225 (1981).
4. Rosinberg M-L, Badiali J-P and Goodisman J: J. Phys. C: Solid State Phys. 16, 4487 (1983).
5. Hasegawa M and Watabe M: J. Phys. C: Solid State Phys. 18, 2081 (1985).
6. Evans R: Adv. Phys. 28, 143 (1979).
7. Evans R and Sluckin T J: J. Phys. C: Solid State Phys. 14, 3137 (1981).
8. Badiali J-P, Rosinberg M-L, Levesque D and Weis J J: J. Phys. C: Solid State Phys. 16, 2123 (1983).
9. Hasegawa M and Watabe M: J. Phys. C: Solid State Phys. 18, L573 (1985).
10. Hasegawa M and Watabe M: J. Stat. Phys. to be published (1985).
11. Slattery W L, Doolen G D and DeWitt H E: Phys. Rev. A 21, 2087 (1980); Phys. Rev. A 26, 2255 (1982).
12. Baus M and Hansen J-P: J. Phys. C: Solid State Phys. 12, L55 (1979).
13. Rosenfeld Y and Ashcroft N W: Phys. Rev. A 20, 1208 (1979).

2. STRUCTURE OF LIQUIDS

RELATIONS BETWEEN ATOMIC ARRANGEMENT AND ELECTRONIC STRUCTURE IN LIQUID METALS AND ALLOYS

W. van der LUGT and J.A. MEIJER

Solid State Physics Laboratory, Material Science Centre,
University of Groningen, Melkweg 1, 9718 EP Groningen
The Netherlands

1. DIFFRACTION MODEL

In the beginning was the diffraction model. It establishes a clearcut relation between structure and electronic properties, and it works well for N.F.E. metals. For elements reliable experimental structure factors are known. In monovalent metals inaccuracies arise from the near-coincidence of the steep slope of the structure factor and the node in the pseudopotential in the important region of q just below $2k_F$. Yet accuracies within 10%, typically, can be obtained in calculations of the resistivity, ρ, and the thermopower, Q. In alloys difficulties arise because partial structure factors are usually not known from experiment and one has then to rely upon HSPY (hard sphere Percus-Yevick) structure factors, which obviously involve two serious approximations: that of the Percus-Yevick approximation and that of hard-sphere potentials. But by reasonable choice of structural parameters in the HSPY expressions agreement within, say, 20% can be obtained.

By the famous factorization of the resistivity integrand, structure factors and electron-ion potentials enter the Ziman formula as if they were independent properties. Of course, they are not. Both depend on the behaviour of the valence electrons. Indeed, the possibility of calculating pair potentials from pseudopotentials is one of the important achievements of the diffraction model (1) and it has been shown (2) that a unified approach to calculating a variety of physical properties of N.F.E. metals is quite feasible within the diffraction model.

In this paper we want to emphasize the intimate relationship between transport properties and structure. This relationship becomes even stronger and more difficult to disentangle, in systems with strong interactions, which are not described by the N.F.E. model. In that case more or less saturated chemical bonds may replace the metallic cohesion and the particle-particle interactions are not adequately described by pair potentials. Evidently, the diffraction model is then no longer applicable. In the last part of this paper we will focus attention to such systems. First we deal with two cases which are usually treated within the diffraction model, but still offer some unsolved problems: Na-Cs and the amalgams.

2. SODIUM-CAESIUM

The liquid system Na-Cs exhibits the following anomalies.

Around 20% Cs: a peak in $S_{cc}(0)$ indicating a tendency to phase separation (3,4,5).

Around 25% Cs: strong anomalies in $d^2\rho/dT^2$ as well as a sharp maximum in the sound absorption (6,7) (see figs. 1 and 2).

Around 60% Cs: a sharp minimum in $d\rho/dT$ (6).

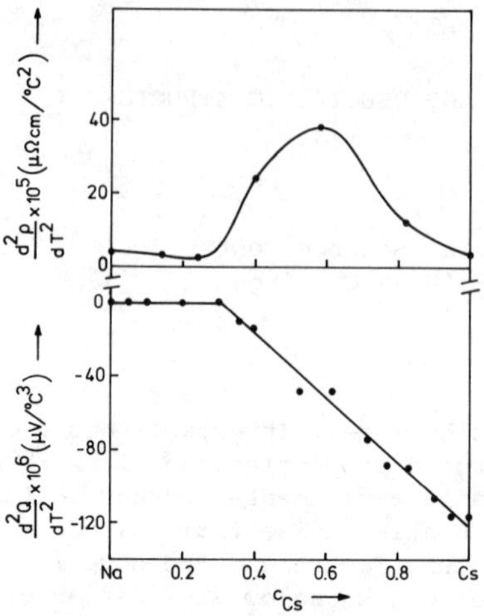

FIGURE 1. Plots of $(d^2\rho/dT^2)$ and (d^2Q/dT^2) as a function of composition in liquid Na-Cs alloys (6).

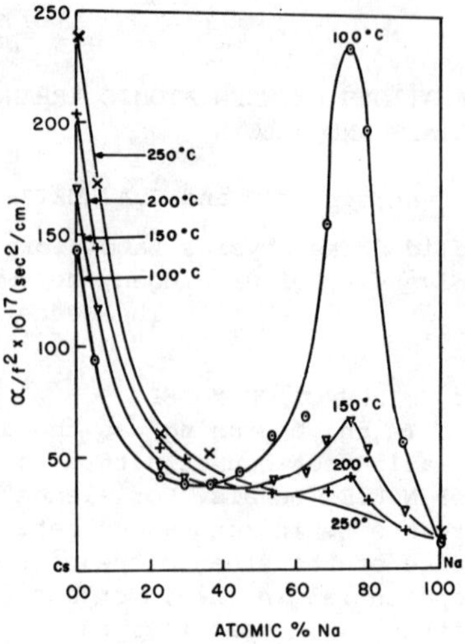

FIGURE 2. Sound absorption in liquid Na-Cs alloys (7).

As far as the authors know, no comprehensive model has been proposed to explain these anomalies, although the phase separation appears to be reproduced by a simulation based on Heine-Abarenkov pair potentials (3). A suggestion may be that, though the alkalis are all relatively electropositive, the electronegativities vary considerably among them: for Na-Cs the electronegativity difference on the Miedema scale, $\Delta\phi^*$, is 0.75, still 60% of that for Li-Pb. The importance of charge transfer has been noted explicitly by Lai and Wang (8). Indeed, they find a minimum in $d\rho/dT$, but they arrive at this result in a rather roundabout way.

3. RELATIVISTIC EFFECTS

In the heavy metals, the electron states of the free atoms and ions are subject to strong relativistic effects. As a consequence, the electron-ion potential and the chemical properties deviate from what one would expect from a mere extrapolation down the corresponding column of the Periodic Table. These anomalies are, in turn, the cause of the interesting properties exhibited by some of the liquid heavy elements and their alloys.

A discussion of the relativistic effects is facilitated considerably by the appearance of a short review paper by Pyykkö and Desclaux (9), which we will follow closely. The following effects are discerned by these authors.

1) A contraction of the s- and p-shells due to the relativistically enhanced effective mass. This effect is accompanied by a lowering of the energy.

2) The d-electrons are not susceptible to this effect, because they have a small density near the nucleus but they are, to the contrary, subject to an indirect effect: due to the enhanced screening by s- and p-electrons, their energies rise by approximately the same amount as the s-electron energies are lowered . As a consequence of 1) and 2) the d- and s-levels

approach each other closely (fig 3).

3) There is a large spin-orbit coupling.

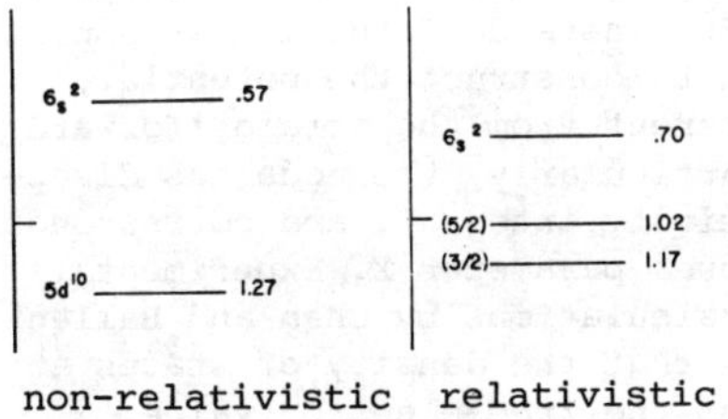

FIGURE 3. Non-relativistic and relativistic energy levels of mercury (10).

The relativistic effects become significant when going from the fifth to the sixth row of the periodic system and have the following interesting consequences for the sixth-row elements.

GOLD has one 6s valence electron. The 6s level is deep and the Au^- ion very stable. We all know, that AuCs is an ionic semiconductor. Also, the diameter of gold is relatively small and the (covalent) bonds relatively strong: Au_2 and the iso-electronic $(Hg_2)^{2+}$ are relatively stable, the Au-Au bond being shortened by no less than 0.35 Å due to relativistic effects.

CAESIUM. Calculations show, that in Cs the relativistic contraction is still compensated by the normal shell-structure expansion. Accordingly, Cs is the largest and most easily ionized stable atom, providing another reason for the ionicity of AuCs. But the proximity of the d-band to the Fermi level might be a cause of the anomalies found in the Na-Cs system (see section 2) (62).

MERCURY. We have seen that the 6s-shell has relatively low energy. Consequently, the mercury $(6s)^2$ closed shell configuration behaves rare-gas like. The interatomic Hg-Hg potential is relatively shallow and mercury is a liquid at room temperature. Keeton and Loucks (10), among others, showed that the reduced separation of the 5d- and 6s-states leads to intersection of the conduction band by 5d-bands.

LEAD and TIN, spin-orbit coupling. There is a tendency to jj-coupling and we have to distinguish between $p_{1/2}$ and $p_{3/2}$ states, widely separated by the spin-orbit interaction. Accordingly, the lead configuration $(6s)^2(6p_{1/2})^2$ resembles a rare-gas configuration and bismuth can occur as a monovalent element. Also the 6s-6p splitting is rather large due to relativistic effects and hybridization does not take place easily: lead is predominantly divalent due to the "inert pair" $(6s)^2$.

4. MERCURY

The properties of liquid mercury have been reviewed by Mott (11). Its resistivity and thermopower are considerably higher than for most metallic elements and, most remarkably, the resistivity drops sharply when it is alloyed with most other metals (although the resistivity is raised by addition of, e.g., caesium). Finally, Q, and the temperature derivatives of Q and ρ exhibit extrema upon admixture of approximately 5% of certain metals (see next section: mercury alloys). Mott, employing some of the concepts developed by him for explaining localized states in disordered metals, invoked the existence of a deep minimum in the density of states, associated with the proximity of a Brillouin-zone boundary in the solid state, as an explanation for this behaviour.

Although some of the anomalies can be explained in this way, Evans et al. (12) proposed a different, more quantitative approach in terms of the diffraction model and Heine-Animalu model potentials. They noted that the usual, simple way of finding the well-dephts by linear extrapolation of term values is invalidated by the proximity of the d-bands, resulting in a non-negligible effect of the d-phase shift. Instead of the linear approximation the quantum-defect method was used to construct the potential, which, indeed, turns out to be quite different from the straightforward Heine-Animalu potential (fig. 4). More particularly, the node has disappeared, resulting in a much larger resistivity integral, and correspondingly increased values of ρ and the thermopower parameter X. Experimental structure factors were used. Subsequent calculations by Chan and Ballentine (13) and by Itami and Shimoji (14) showed that the density of states at the Fermi level is not deviating strongly from the free-electron value.

5. MERCURY ALLOYS

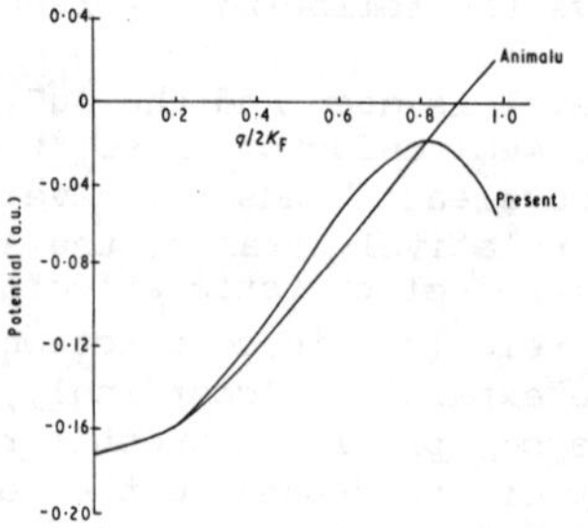

FIGURE 4. Heine-Animalu- and Evans' ("present") potentials for Hg (12).

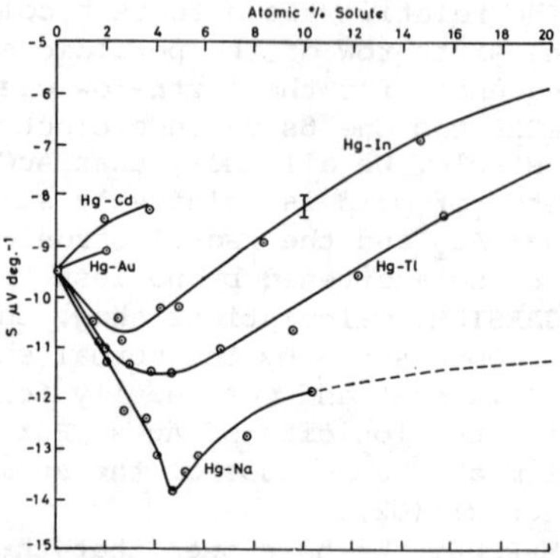

FIGURE 5. Examples of minima of the thermopower in mercury alloys (Fielder (16)).

The steep drop in the resistivity upon alloying was easily explained by Evans (15): the Faber-Ziman formula contains a cross term involving the product w_1w_2, which in the upper limit of the resistivity integrand is negative for the corresponding mercury alloys (because $w_{Hg} < 0$ and $w_{impurity} > 0$ (see fig. 4), whereas it is positive in most other alloys. He could also nicely explain the overall behaviour of the thermopower in Hg-Tl and Hg-In alloys, but not the remarkable minimum in Q at approximately 5% In or Tl admixture (16) (fig. 5).

The latter category of phenomena formed the subject of extensive investigations, involving many amalgams, by Itami and coworkers (17,18,19). They showed that not only Q, $d\rho/dT$ and dQ/dT may exhibit extrema but also the expansion coefficient α_p(20). By a rather subtle analysis of the transport data they arrive at the conclusion that an anomaly in the $dS_{\alpha\beta}/dT$ is essential. This quantity is closely related to higher-order structural correlations; the same applies to α_p.

Furthermore, they established a distinct relation between the sign of the (Miedema) electronegativity difference and the occurence of the anomalies around 5% impurity. Probably, the effect is caused by electron transfer to Hg. Also, it correlates with a negative heat of mixing. This substantiates the remark of Itami et al. (17) that ordering plays an essential role.

Another Japanese group, at Niigata university, has paid attention to the thermodynamic properties of Hg-Na, in which the electronegativity difference is large (1.50). EMF measurements (21) demonstrate that $S_{CC}(0)$ is distinctly below its ideal value, indicating compound formation. Volume contraction in liquid Hg_2Na is 18%, another indication of (ionic?) ordering.

An X-ray diffraction study (22) shows, that no particular effects, like a superstructure prepeak, are observable around 5% Na.

We may make the following concluding remarks. The effects around 5% admixture of a more electropositive element are beyond any doubt, and they are accompanied by structural effects. But this raises new questions: what is the cause of the structural effects, why is there no prepeak, why are the effects for the transport properties so much smaller than in, say, Li-Pb, with approximately the same electronegativity difference (1.25), and why are they happening just at 5%, far from any stoechiometric composition? The present authors may conjecture that the cause of all the anomalies, including the structural ones, is to be found in the peculiar quantum states of the mercury atom. These may also constitute the origin of the strange chemical behaviour of mercury. Mercury often assumes valency 1 and in crystals it may occur in dumb-bells and squares (23). The electronic structure of liquid mercury and particularly the 6s-6p hybridization are sensitive to changes in the coordination (24). It is still an open question how it is influenced by the combined action of charge transfer and reduction of coordination number by the introduction of foreign atoms. Mercury still appears to offer an interesting, though subtle field of investigation.

6. COMPOUND FORMING ALLOYS: SIMPLE IONIC SYSYEMS

In the following two chapters we are primarily concerned with alkali-non-alkali alloys. We will particularly emphasize possible correspondences between solid and liquid compounds. As a consequence of their low electronegativity, alkali atoms tend to donate an electron to their partners in an alloy, thus forming a more or less ionic compound.

Van der Marel (25,26) carried out a systematic investigation of lithium and sodium alloys with increasing ionicity. Although in Li-Mg (27,28) and Li-Cd (29) alloys the electronegativity difference $\Delta\phi^*$ is small (0.60 and 1.20, respectively), the charge transfer is clearly reflected in a decrease of the alkali Knight shift K on alloying. This can be interpreted as being essentially due to a decrease of the Fermi-surface s-electron contact density. The electrical resistivity is somewhat higher than in alkali-alkali mixtures. Although Ruppersberg et al. (30) discovered distinct indications for short-range order in Li-Mg its resistivity can still be understood largely within the diffraction model. In liquid Li-Cd the Knight shift exhibits a distinct bend at the equiatomic composition, perhaps a reminiscence of the solid compound LiCd.

The electronegativity difference in the Li-In system is 1.05. It is tempting to associate the rather sharp maximum in the resistivity and the bends in both the In and the Li Knight shifts, occuring at the composition Li_3In (31) with the existence of a solid compound of the same composition, but that should be too rash a conclusion. First, Li_3In does not obey a simple octet rule and second there are besides Li_3In many other lithium-rich compounds in the phase diagram.

An indication of what happens is provided by considering the solid compound LiIn. According to the old "Zintl model" (32,33) an electron is transferred from Li to In, making the latter isoelectronic with, e.g. Si. Indeed, the crystal structure is that of two interpenetrating diamond lattices (B32). Quite recently, an account of the electronic structure of this kind of compounds has been given by Christensen (34). The term charge transfer needs some qualification. This concept is difficult to define. It is better to say: the original s-electron is governed by the deep anion potentials. The remaining charge distribution around the Li atom may have lost its s-character, and thus the Knight shift is diminished. The resistivity, including the strongly negative $d\rho/dT$, can be explained remarkably well within the diffraction model, the maximum in ρ being a consequence of the passage of the Fermi level through the main liquid diffraction peak. It is remarkable that also the atomic volume has a distinct minimum (18%) at Li_3In, so one must believe, that indeed this is a stable liquid compound.

The electrical resistivities of some alkali-group III alloys are shown in figs. 6 and 7. With increasing atomic weight of the alkali component, the resistivity maximum shifts to the equiatomic composition. There is a remarkable similarity in the behaviour of corresponding Ga-, In- and Tl-systems (see fig. 8).

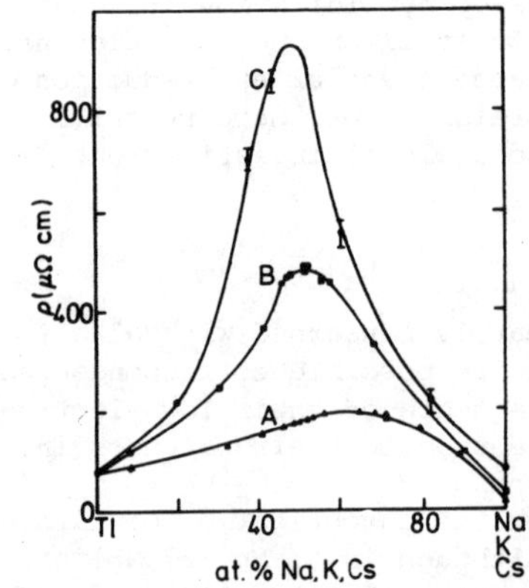

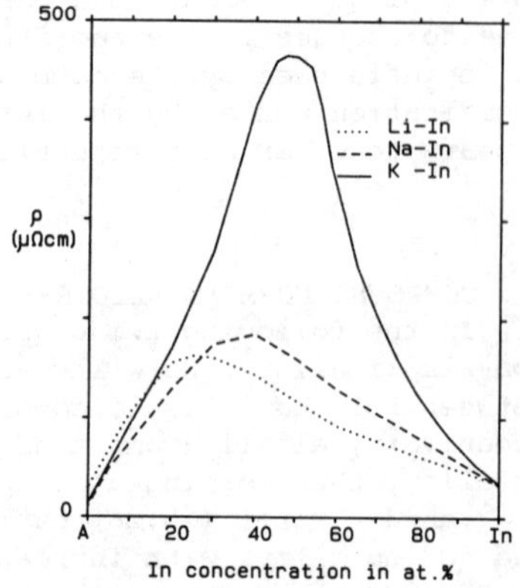

FIGURES 6 (left) and 7 (right). Resistivities of alkali-Tl (35) and alkali-In (36) alloys. Fig. 6. A: Na-Tl; B: K-Tl; C: Cs-Tl.

Much more distinct are the effects in the alkali-group IVB alloys, Li-Pb being the classical example $(\Delta\phi^* = 1.25)$. In these alloys, physical effects due to alloying (like extrema in ρ and $d\rho/dT$, in K, in $S_{CC}(0)$, in the density etc.) are so strong, and the compositions at which they occur so sharply defined that the term stoechiometric compound is fully justified. Such a compound is Li_4Pb. The measured structure factor has a well-developed prepeak and the degree of ordering can be fairly accurately estimated (38,39). It is a "simple ionic compound", with octet configuration, in between a normal alloy and a molten salt. The resistivity of Li-Pb is outside the scope of the diffraction model, which yields far too small values, though, interestingly, the maximum occurs at the correct composition. It should be emphasized, that the most typical feature of the structural ordering, the prepeak, does not contribute significantly to the resistivity integral. The real effect is the gradual formation of saturated chemical bonds. For the same reason, tight-binding methods replace the NFE model in

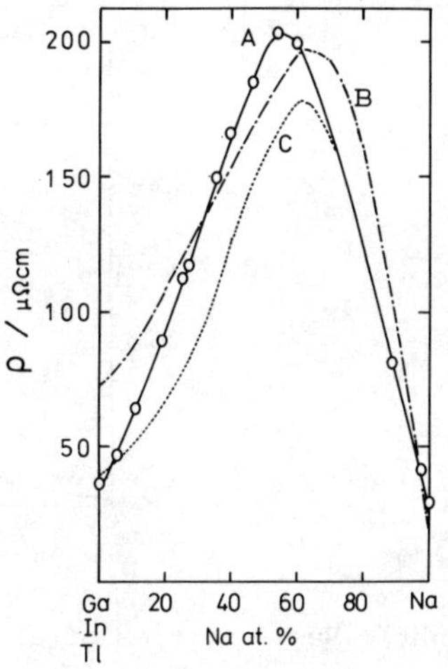

FIGURE 8. Resistivities of Na-Ga(A), Na-Tl(B) and Na-In(C) (35,36,37).

the calculations. There exists a strong correlation between structural properties and electronic properties, described in an elegant model by Franz, Brouers and Holzhey (40).

In a tight-binding model (input parameters: ΔE between the alkali-s and Pb-p levels, transfer integrals t_{ij}) they perform a self-consistent calculation, in which the charge transfer and ionic ordering are determined by minimizing the free energy. The width of the bands is roughly proportional to the average transfer integral $\langle t \rangle$. For small $\Delta E/\langle t \rangle$ the bands are wide and overlapping; we have a metallic system and small charge transfer. When $\Delta E/\langle t \rangle$ increases, the bands get narrower and charge transfer is increased. Charge transfer and ionic ordering mutually reinforce each other: charge transfer favours ionic ordering via the Madelung energy and, vice versa, ionic ordering (preferential coordination by atoms of different kind) pushes the electron bands apart, thus causing a larger charge transfer. Furthermore, the Madelung term is enhanced, because, at decreasing $N(E_F)$, screening becomes less effective. Consequently, there is a rapid transition from a metal to an ionic semiconductor as $\Delta E/\langle t \rangle$ increases.

This model was applied to Li-Pb (40) and to the alkali-gold alloys (41,42,43). In the latter, there is a marked increase in $\Delta E/\langle t \rangle$, going from Li to Cs, the most important effect deriving from the decrease of the t_{ij} due to the rapidly increasing size of the alkali ions. For equiatomic alkali-gold alloys the transition from metal to semiconductor is expected somewhere between K and Rb; this is borne out by experiment (44,45). Actually, the band gap is probably not between Au-6s and Cs-6s states, but between the Au-6s and the Au-6p bands (Robertson (46)), but this does not detract from the essential merits of the model.

The Holzhey-Franz-Brouers picture also bears relevance to the alkali-group IIIB, IVB and VB systems, though, of course, the actual electronic structure might be more complicated than is suggested by their model. Generally we observe an increase of the maximum resistivity from the Li- to the Cs-alloys. It is remarkable, that, in the same order, there is also a general tendency for the resistivity maximum to shift to the equiatomic

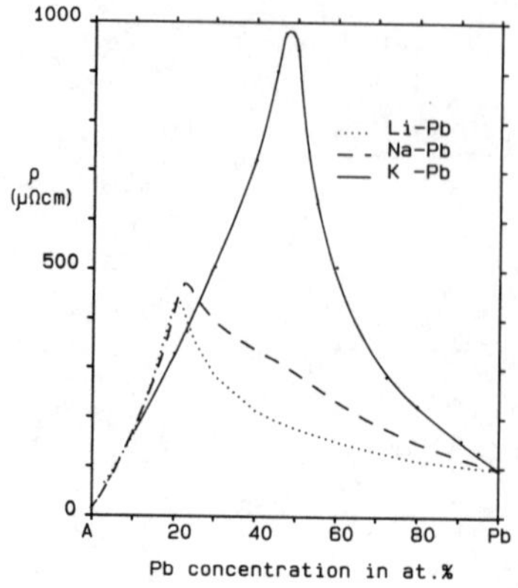

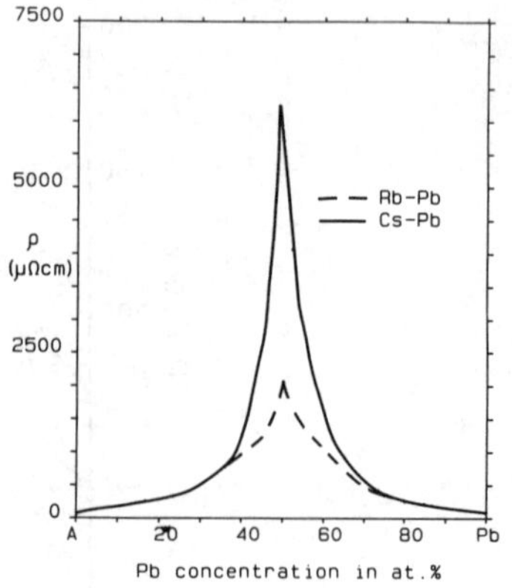

FIGURES 9 (left) and 10 (right). Resistivities of all alkali-lead systems at $T_{liq}+$ 10 K.

composition (see figs. 6,7,9 and 10). This effect will be discussed in the next section.

7. COMPOUND-FORMING ALLOYS: CLUSTERING.

Meijer made a study of all the alkali-lead alloys (34,47). A sudden change in the stoechiometric composition takes place between Na-Pb and K-Pb: from 20% Pb in Na-Pb (octet compound) to 50% Pb in K-Pb (figs. 9 and 10). The same phenomenon had already been observed in Na-Sn, in which both compounds, their compositions sharply defined, occur. Geertsma has made a link to the crystal structure of the solid compound NaSn. In almost all alkali-group IVB equiatomic compounds tetrahedral clusters occur, the origin of which can be explained in terms of the Zintl model mentioned above. The alkali donates an electron to, say, Pb and the latter becomes isoelectronic with P, which is known to form tetrahedral molecules P_4 in the gas phase. Apparently, these tetrahedra form an important structural element. Materials like NaSn are known to be semiconducting. The energy levels and electron distribution of P_4 are well known. The tetrahedra, with angles of 60° (and not of 108°), are strained and electron clouds protrude from them at the side faces, accommodating the stripped Na atoms (48) (see fig. 11).

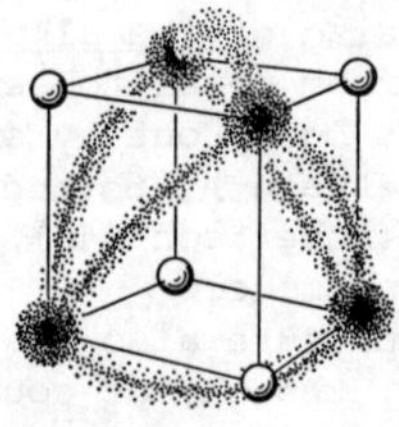

FIGURE 11. Sketch of electron distribution in P_4 (48).

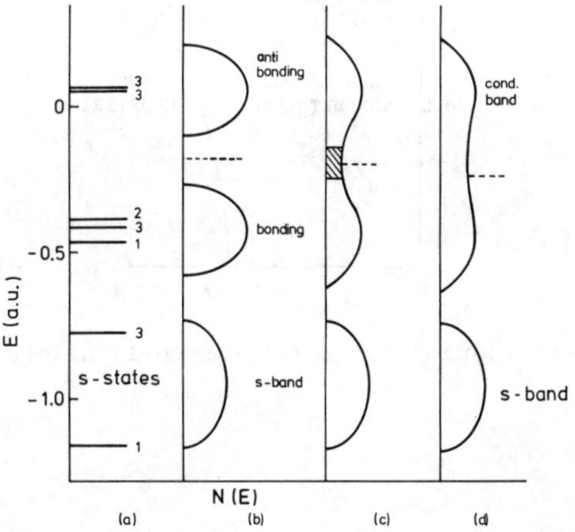

FIGURE 12. Formation of bands of, e.g. $(Sn_4)^{4-}$. Left: the level structure of P_4. From (a) through (d) the overlap is assumed to increase.

The energy levels form the basis for a more intuitive understanding, as well as theoretical calculations of the band structure of NaSn.

Fig. (12) shows how one may imagine the development of bands out of the P_4 term scheme. From left to right, increasing interaction between the clusters is assumed.

Geertsma, in a series of papers (49,50,51) developed a tight-binding method, based on a "Modified Bethe Lattice", in which two interaction parameters are introduced: U decribes the interaction with Z_U partners within one cluster, and V that with Z_V partners in different clusters. By suitable choice of Z_U and Z_V this model can be adapted to describe many types of cluster configurations. Finally, a third interaction parameter, T, representing metallic interaction is introduced. A discussion of clustering phenomena is given in terms of the reduced quantities V/U and T/U, which leads to the "phase diagram" of fig. 13, in which three "phases" can be discerned: non-clustering, clustering metallic and clustering-nonmetallic. By applying common rules for scaling of interaction parameters with interatomic distance, Geertsma was able to determine the position in the phase diagram of each of the equiatomic alkali-group-IVB alloys. The result is in almost perfect agreement with experiment.

Finally, Springelkamp et al. (52) carried out an ASW band structure calculation of solid NaSn. The results perfectly confirm our intuitive picture (fig. 14). From left to right we see first the isolated pair of bonding and antibonding s-states, bonding p-states, a gap containing the Fermi energy and the antibonding p-states.

We have postulated that the band structure of the liquid resembles that of the solid, with some allowance for the disorder in the liquid. Liquid KPb is then ideally an ionic solution of K^+-ions and $(Pb_4)^{4-}$ complex ions. Thus, the structural similarity between solid and liquid is deduced indirectly from the similarity of their electronic properties. It may be noted that in simple metallic systems making a comparison between liquid and solid properties is rarely fruitful. So our reasoning strongly needs further

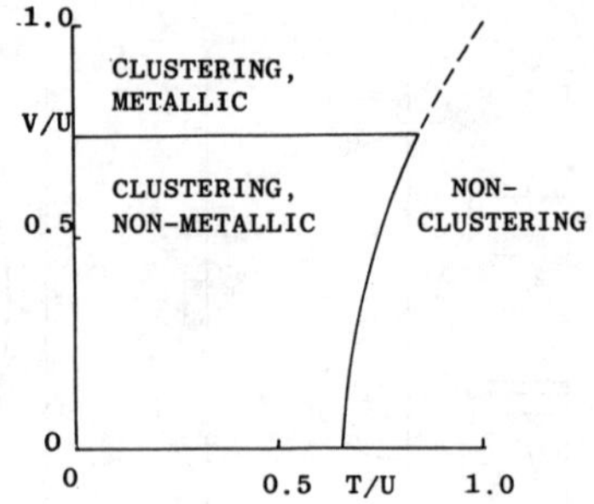

FIGURE 13. "Phase Diagram" for alkali-group-IV alloys (52).

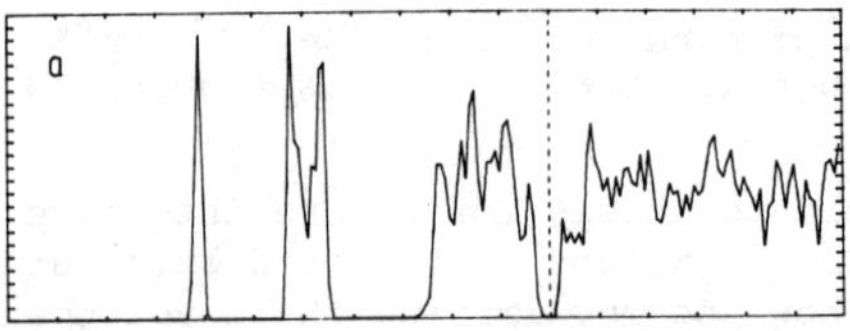

FIGURE 14. Density of states of solid NaSn (52).

confirmation. For this reason, Alblas et al. (53) performed neutron diffraction measurements in liquid Na-Sn alloys. Some evidence for the proposed liquid structure is provided by the existence of a shoulder on the main peak and a small shift of the prepeak with composition. However, the measurements were carried out under adverse conditions and will be repeated on a friendlier material, KPb.

In an earlier paper (54) we have pointed out that essentially the same reasonings can be applied to understand the alkali pnictides, which show comparable effects of compound formation. Groups in Marburg, Warwick and Stuttgart have carried out extensive research on these materials; we refer to their papers for further study (55,56,57).

8. CONCLUSION: BEYOND THE DIFFRACTION MODEL.

In the two preceding sections, we discussed liquid alloys systems, which are definitely non-metallic in part of the composition range. Close to compound-forming compositions, the electron states do not resemble free electron states; k is no longer a good quantum number and k_F and v_F cease to have a clear meaning. From the chemical point of view, we have left the realm of metallic, isotropic, non-saturated bonds and have entered that of covalent/ionic, possibly anisotropic, saturated bonds. In many cases, the assumption of pair-wise additive interactions is no longer valid and higher order correlations have to be included. For the description of the structure of isotropic systems the MSA provides relief (58,59,60,61).

Having reluctantly taken our leave from the diffraction model, we look for a worthy replacement. The straightforward approach would be to include triplet and higher order correlation functions in our description of the liquid structure, and to extend the calculations of the electronic properties to all orders of perturbation theory. But such calculations would prove to be of an unmanageable complexity; the inclusion of higher order correlations in the description of the structure is still a challenge to the modern theory of liquids.

At present, the best one may do is to construct a plausible model for the liquid structure, and to calculate the electronic structure in a tight-binding approximation. We have shown in the preceding two sections, that much insight can be gained in this way. As a more fundamental approach one might try to perform an exact calculation of the electronic structure of a model system, large enough to simulate a disordered system.

ACKNOWLEDGEMENTS.

The authors acknowledge with thanks discussions with and contributions from Dr. W. Geertsma and Dr. C. van der Marel. This work forms part of the research program of the "Stichting voor Fundamenteel Onderzoek der Materie" (Foundation for Fundamental Research on Matter, FOM) and was made possible by financial support from the "Nederlandse Organisatie voor Zuiver Wetenschappelijk Onderzoek" (Netherlands Organization for the Advancement of Pure Research, ZWO).

REFERENCES.

(1) Hafner J. and Kahl G., J.Phys.F: Met.Phys.14 (1984) 2259-2278.
(2) Hafner J. Z.Phys.B22 (1975) 351-357; ibid. B24 (1976) 41-52.
(3) Huijben M.J., Van der Lugt W., Reimert W.A.M., de Hosson J.Th.M., and Van Dijk C., Physica 97B (1979) 338-364.
(4) Ichikawa K., Granstaff S.M.Jr. and Thompson J.C., J.Chem.Phys.61 (1974) 4059.
(5) Neale F.E. and Cusack N.E., J.Phys.F: Metal Phys.12 (1982) 2839-50.
(6) Feitsma P.D., Hennephof J. and Van der Lugt W., Phys.Rev.32 (1974) 295-297.
(7) Kim M.G. and Letcher S.V., Chem.Phys.55 (1971) 1164-1170.
(8) Lai S.K. and Wang S., Phys.Lett.85A (1981) 239.
(9) Pyykkö P. and Desclaux J.-P., Acc.Chem.Res.12 (1979) 276.
(10) Keeton S.C. and Loucks T.L., Phys.Rev.152 (1966) 548-555.
(11) Mott N.F., Phil.Mag.13 (1966) 989-1014.
(12) Evans R., Greenwood D.A., Lloyd P. and Ziman J.M., Phys.Lett.30A (1969) 313.
(13) Chan T. and Ballentine L.E., Can.J.Phys.50 (1972) 814.
(14) Itami T. and Shimoji M., Phil.Mag.25 (1972) 229.
(15) Evans R., J.Phys.C: Metal Phys.Suppl.No.2 (1970) 137.
(16) Fielder R.L., Adv.Phys.16 (1966) 681.
(17) Itami T., Wada T. and Shimoji M., J.Phys.F: Met.Phys.12 (1982) 1959-1970.
(18) Itami T., Takahashi N. and Shimoji M., J.Phys.F: Met.Phys.13 (1983) 1225.
(19) Itami T., Takahashi S. and Shimoji M., J.Phys.F: Met.Phys.14 (1984) 427.
(20) Sato T., Itami T. and Shimoji M, J.Phys.Soc.Japan 51 (1982) 2493.

(21) Ishiguro T., Takeda S. and Tamaki S., J.Phys.F: Met.Phys.12 (1982) 845-856.
(22) Tamaki S., Waseda Y. and Tsuchiya Y., J.Phys.F: Met.Phys.12 (1982) 1101-1109.
(23) Wells A.F., Structural Inorganic Chemistry, Oxford, at the Clarendon Press, 1962.
(24) Mattheiss L.F. and Warren W.W.Jr., Phys.Rev.B 16 (1977) 624.
(25) Van der Marel C., Thesis, Groningen 1981.
(26) Van der Marel C. in: Liquid and Amorphous Metals, ed. E. Lüscher and H. Coufal, Sijthoff & Noordhoff, 1980, p. 627.
(27) Feitsma P.D., Lee T. and Van der Lugt W., Physica 93B (1978) 52-58.
(28) Van der Marel C. and Van der Lugt W., Physica 112B (1982) 365-368.
(29) Van der Marel C. and Van der Lugt W., J.Phys.F: Met.Phys.10 (1980) 1177.
(30) Ruppersberg H., Saar J., Speicher W. and Heitjans P., J.Phys.41 (1980) C8-595.
(31) Van der Marel C., Brandenburg E.P. and Van der Lugt W., J.Phys.F: Met.Phys.8 (1978) L273.
(32) Zintl E. and Woltersdorf G., Z.Elektrochem.41 (1935) 876.
(33) Zintl E. and Brauer G., Z.Phys.Chem.Abt.B 20 (1933) 245.
(34) Christensen N.E., Phys.Rev.B 32 (1985) 207-228.
(35) Kitajima M. and Shimoji M., in: Liquid Metals, Inst.Phys.Conf.Ser. no.30 (1977) 226.
(36) Meijer J.A., Geertsma W. and Van der Lugt W., J.Phys.F 15 (1985) 899-910.
(37) Itami T. and Meijer J.A., unpublished result.
(38) Ruppersberg H. and Egger H., J.Chem.Phys.63 (1975) 4095.
(39) Ruppersberg H. and Reiter H., J.Phys.F 12 (1982) 1311.
(40) Franz J.R., Brouers F. and Holzhey C., J.Phys.F: Met.Phys.12 (1982) 2611.
(41) Franz J.R., Brouers F. and Holzhey C., J.Phys.F: Met.Phys.10 (1980) 235.
(42) Holzhey C., Brouers F. and Franz J.R., J.Phys.F: Met.Phys.11 (1981) 1047.
(43) Holzhey C., Brouers F. Franz J.R. and Schirmacher W., J.Phys.F: Met.Phys.12 (1982) 2601.
(44) Schmutzler R.W., Hoshino H., Fischer R. and Hensel F., Ber.Bunsenges. Phys.Chem.80 (1976) 107.
(45) Nicoloso N., Schmutzler R.W. and Hensel F., Ber.Bunsenges.82 (1978) 621.
(46) Robertson J., Phys.Rev. B 27 (1983) 6322.
(47) Meijer J.A. and Van der Lugt W., to be published in J.Phys.F: Met.Phys.
(48) Hart R.M., Robin M.B. and Kuebler N.A., J.Chem.Phys.42 (1965) 3631-3638.
(49) Geertsma W., Dijkstra J. and Van der Lugt W., J.Phys.F 14 (1984) 1833-1845.
(50) Dijkstra J. and Geertsma W., J. Non-Cryst. Solids 69 (1985) 317-324.
(51) Geertsma W., J.Phys.C 18 (1985) 2461.
(52) Springelkamp F., De Groot R.A., Geertsma W., Van der Lugt W. and Mueller F.M., Phys.Rev. B 32 (1985) 15 August issue.
(53) Alblas B.P., Van der Lugt W., Dijkstra J. and Van Dijk C., J.Phys.F 14 (1984) 1995-2006.
(54) Van der Lugt W. and Geertsma W., J Non-Cryst. Solids 61&62 (1984) 187-200.

(55) Redslob H., Steinleitner G. and Freyland W., Z.Naturforsch. 27a (1982) 587.
(56) Lamparter P., Martin W., Steeb S. and Freyland W. Z.Naturforsch. 38a (1982) 329.
(57) Dupree R., Kirby D.J. and Freyland W., Phil.Mag. B 46 (1982) 595.
(58) Copestake A.P. and Evans R., J.Phys.C 15 (1982) 4561.
(59) Copestake A.P., Evans R., Ruppersberg H. and Schirmacher W., J.Phys.F: Met.Phys. 13 (1983) 1993.
(60) Hafner J., Pasturel A. and Hicter P., J.Phys.F: Met.Phys.14 (1984) 1137-1156.
(61) Hafner J., Pasturel A. and Hicter P. J.Phys.F:Met.Phys.14 (1984) 2279-2295.
(62) Tamaki S., Ross R.G., Cusack N.E. and Endo H., The Properties of Liquid Metals, ed. S. Takeuchi, Taylor and Francis Ltd. (1973) 289.

LIQUID STRUCTURE AND FREEZING OF METALS AND MOLTEN SALTS

M. Rovere and M.P. Tosi

Dipartimento di Fisica Teorica dell'Università, and
International Centre for Theoretical Physics, Trieste, Italy

ABSTRACT

The density-wave theory of freezing is briefly reviewed, with main emphasis on (i) phenomenological insight gained on the liquid-solid transition in simple metals and salts, and (ii) quantitative limitations emerging from detailed calculations on liquid-solid coexistence in model systems.

1 INTRODUCTION

The coexistence between solid and liquid phase presents both thermodynamic and structural aspects. Though speculations on the nature of this phase transition go back to classical antiquity, experimental determinations of its thermodynamic properties (entropy and density change, melting curve) started only in the 1700's and were well developed in the course of the past century. The subsequent development of diffraction techniques for structural studies has placed emphasis on the change in the state of order across the transition. Under both aspects similarly bonded materials show similar behaviours, with little sensitivity to the details of the interatomic forces.

Thus, a number of empirical melting relations are known and find a justification in the melting properties of a model with inverse-power repulsions (1). This becomes directly relevant to experiment in the limit of high temperatures (2). Notable instances are the hard sphere system, with a melting curve $P \propto T$, and the classical plasma in a rigid neutralizing background, with a melting curve $n \propto T^d$ in d dimensions (this is perhaps better known for the 3D plasma as $e^2/(ak_BT) = \Gamma_c$, a fixed critical value for the coupling strength at melting, with $4\pi a^3/3 = 1/n$). Although the rigorous validity of such relations is upset by addition of attractive forces, corresponding-states scaling between similar materials remains useful. For instance, a simple model for alkali halides (Coulombic attractions and repulsions plus inverse-power overlap repulsions between first neighbours) leads to $e^2/(ak_BT)$ = constant for low-pressure melting, in reasonable accord with observation through this family of salts excepting the Li halides (3). High-pressure work (2) has also motivated a simple empirical relation between entropy and density changes on melting, which holds approximately for both metals and ionic materials (4,5).

Broad structural regularities are also well known. Under this aspect, the criterion for melting proposed a long time ago by Lindemann (6) has been complemented with a criterion for freezing by Hansen and Verlet (7). This draws attention to the height of the main peak in the liquid structure factor $S(k)$, which is empirically found to reach a value of about 3 in simple monatomic liquids near freezing. According to the classical fluctuation-dissipation theorem, generalizing the relation of Ornstein and Zernike between $S(k \to 0)$ and the liquid compressibility, $S(k)$ gives the compliance of the liquid against a weak external modulation of its density at wave vector $\underline{k}$. Thus the liquid on cooling is becoming "soft" (though not unstable) against the appearance of density waves with wave vectors in the region of the main peak in $S(k)$. It is reasonable to associate these with the first star of reciprocal lattice vectors (RLV) of the corresponding crystal.

Recent theoretical developments have concerned, on the one hand the evaluation of melting curves for specific systems from interatomic force laws, and on the other progress in the statistical mechanics of freezing motivated by the Hansen-Verlet criterion. The latter line of work is of main interest here. We refer to the calculations of Moriarty et al. (8) on the melting curve of Aℓ for an example of the former and references to similar earlier work.

2. DENSITY WAVE THEORY OF FREEZING

The single-particle density in a monatomic lattice is a periodic function represented by the Fourier series

$$n_s(\underline{r}) = n_s + \sum_{\underset{\sim}{G}}{}' n_{\underset{\sim}{G}} \exp(i\underset{\sim}{G}\cdot\underset{\sim}{r}), \tag{2.1}$$

where $\underset{\sim}{G}$ denotes the RLV and the primed sum means $\underset{\sim}{G} \neq 0$. The density profile (2.1) would have to be generated by a periodic external potential $U(\underset{\sim}{r})$ in a thermodynamic state in which the liquid (at constant density n_ℓ) is stable. At coexistence the order parameters of the phase transition, which are the density change $n_s - n_\ell$ and the Fourier components $n_{\underset{\sim}{G}}$, take spontaneously finite values. The quantities $|n_{\underset{\sim}{G}}|^2$ measure the Debye-Waller factor at the Bragg reflections and as such are related, at least in the harmonic approximation, to the mean-square atomic displacement entering Lindemann's criterion for melting.

The possibility of relating the order parameters directly to the observed liquid structure, without appealing to interatomic force models, was proposed in a seminal paper of Ramakrishnan and Yussouff (9). It is useful to start by recalling some general results of the theory of an inhomogeneous monatomic classical system, with density $n(\mathbf{r})$ and in a potential $U(\mathbf{r})$. According to the theorem of Hohenberg, Kohn and Mermin (10), the Helmholtz free energy can be written as

$$F = k_BT \int d\mathbf{r}\; n(\mathbf{r}) \left[\ell n(\lambda^3 n(\mathbf{r})) - 1\right] + \int d\mathbf{r}\; n(\mathbf{r})U(\mathbf{r}) + F_e[n(\mathbf{r})], \tag{2.2}$$

where λ is the thermal de Broglie wavelength and the excess free energy F_e is a functional of $n(\mathbf{r})$. The variational principle yields the equilibrium density as

$$n(\mathbf{r}) = \lambda^{-3} \exp[C(\mathbf{r}) - \beta U(\mathbf{r}) + \beta\mu], \tag{2.3}$$

where $\beta = 1/k_BT$, μ is the chemical potential and

$$C(\mathbf{r}) = -\beta\delta F_e/\delta n(\mathbf{r}) . \tag{2.4}$$

Alternatively, invariance under rigid translation yields the equilibrium condition as

$$\beta\nabla[\mu - U(\mathbf{r})] = \frac{\nabla n(\mathbf{r})}{n(\mathbf{r})} - \int d\mathbf{r}'\; c(\mathbf{r},\mathbf{r}')\; \nabla n(\mathbf{r}'), \tag{2.5}$$

where $c(\mathbf{r},\mathbf{r}')$ is the direct correlation matrix of the inhomogeneous system at density $n(\mathbf{r})$. The exact free-energy functional involves integration of this matrix in the space of density profiles (11).

The results of Ramakrishnan and Yussouff (9) can be derived by suitable approximations on (2.3) (12) or on (2.5) (13). In the former approach, one first notices that at coexistence, letting $U(\mathbf{r})$ vanish and using the conditions for thermal and chemical equilibrium of the two phases ($T_s = T_\ell$ and $\mu_s = \mu_\ell$), the equilibrium profiles are related from (2.3) by

$$n_s(\mathbf{r})/n_\ell = \exp[C(\mathbf{r}) - C_\ell] . \tag{2.6}$$

An expansion of the exponent in (2.6) around the liquid phase,

$$C(\mathbf{r}) - C_\ell = \int d\mathbf{r}'\; c(|\mathbf{r} - \mathbf{r}'|)[n_s(\mathbf{r}') - n_\ell] + \cdots \tag{2.7}$$

relates next the order parameters to the direct correlation function $c(r)$ of the liquid and to higher-order direct correlation functions (not shown in (2.7)). Explicitly,

$$1 + \eta = \exp[\hat{c}(0)\eta] \int d\mathbf{r}\; \exp\left[\sum_{\mathbf{G}}{}' \hat{c}(G)\rho_{\mathbf{G}} \exp(i\mathbf{G}\cdot\mathbf{r}) + \cdots\right] \tag{2.8}$$

and

$$\rho_{\mathbf{G}} = \exp[\hat{c}(0)\eta] \int d\mathbf{r}\; \exp\left[i\mathbf{G}\cdot\mathbf{r} + \sum_{\mathbf{G}'}{}' \hat{c}(G')\rho_{\mathbf{G}'} \exp(i\mathbf{G}'\cdot\mathbf{r}) + \cdots\right], \tag{2.9}$$

where $\eta = (n_s - n_\ell)/n_\ell$, $\rho_{\underset{\sim}{G}} = n_{\underset{\sim}{G}}/n_\ell$ and $\hat{c}(G)$ is the Fourier transform of c(r) , related to the liquid structure factor by

$$\hat{c}(G) = 1 - 1/S(G) \quad . \tag{2.10}$$

Finally, the condition for mechanical equilibrium, $P_s = P_\ell$ with $P = (N\mu - F)/V$, yields the further equation

$$[-1 + \hat{c}(0)]\eta + \tfrac{1}{2}\sum_{\underset{\sim}{G}}{}' \hat{c}(G)|\rho_{\underset{\sim}{G}}|^2 + \cdots = 0 \quad . \tag{2.11}$$

We also remark at this point that the appropriate macroscopic conditions for a plasma freezing at constant density on a rigid neutralizing background are $n_s = n_\ell$ and $F_s = F_\ell$, with $\mu_s - \mu_\ell = (P_s - P_\ell)/n \neq 0$ and balanced by an interfacial potential drop. The above microscopic equations still apply, with $\eta = 0$ and $\hat{c}(0)\eta$ replaced by $\beta(\mu_s - \mu_\ell)$ (14).

Clearly, the microscopic equations relate the order parameters to liquid pair structure if the expansion (2.7) is truncated at the terms explicitly shown. Assuming that the first-order transition is sufficiently weak for the expansion to apply, the leading corrective terms involve the triplet correlation functions $c^{(3)}(\underset{\sim}{G},\underset{\sim}{G}')$ of the liquid. In particular, the functions $c^{(3)}(G,0)$ and $c^{(3)}(0,0)$ can be related to the density dependence of the structure factor S(k) and of the compressibility K , respectively. The latter turns out to be quantitatively quite important in the evaluation of the density change η (15), while the former could also play a role in the selfconsistent determination of η and of the scale of length for RLV from (2.8). The function $c^{(3)}(\underset{\sim}{G},\underset{\sim}{G}')$ for finite $\underset{\sim}{G}$ and $\underset{\sim}{G}'$, on the other hand, describes couplings between microscopic order parameters which could play a role in determining the stable crystal structure. Angular correlations are clearly crucial in cases where $\eta < 0$ (cf the signs of the two terms in (2.11), where $1 - \hat{c}(0) \propto 1/K$ is strongly positive).

In practice, the low-order theory has proved to be very useful in correlating observed liquid structure with hot-solid properties, while its limitations are emerging from detailed calculations on model systems. Several illustrations will be given in the next section.

3 APPLICATIONS

3.1 The Hansen-Verlet Criterion

In the simplest form of the theory, one might look upon crystallization as a cooperative process involving the density change η and microscopic order parameters at RLV underlying the main peak in S(k) , with a multiplicity depending on crystal structure. The early calculations of Ramakrishnan and Yussouff (15) considered freezing of sodium into the b.c.c. lattice and of argon and hard spheres into the f.c.c. lattice. A good fit of structural data near freezing was achieved by including two sets of microscopic order parameters, corresponding to stars of RLV chosen to fall at the main peak and near the second peak of $S(k)$. The detailed results show that inclusion of the first set only yields already a good result for the peak value of $\hat{c}(k)$ for freezing into b.c.c. ($\hat{c}_{peak} = 0.69$, against $\hat{c}_{peak} = 0.66$ from experiment), but an excessive one ($\hat{c}_{peak} \simeq 0.95$) for freezing into f.c.c. As already noted, an account of $c^{(3)}(0,0)$ was necessary in all cases to obtain a reasonable value for the density change η.

The reciprocal of the f.c.c. lattice is the b.c.c. lattice and the lengths of the RLV of its first two stars differ only by a factor $2/\sqrt{3}$. The main peak of $S(k)$ in an f.c.c.-freezing liquid thus overlies two sets of microscopic order parameters. Using this argument and simplifying further the microscopic equations by taking the limit of an incompressible liquid ($\eta \to 0$ and $K \to 0$ with η/K finite), D'Aguanno (16) estimates $\hat{c}_{peak} \approx 0.56 \div 0.70$ at freezing into b.c.c., f.c.c. and h.c.p. structures, against measured values $\hat{c}_{peak} = 0.58 \div 0.66$ for 24 metals at temperatures as close as possible to freezing.

These simple estimates agree with those of a much more detailed discussion given earlier by Baus (17) for freezing of an incompressible liquid into the cubic lattices. He reported critical values of $\hat{c}_{peak}$ equal to 0.69 and 0.59 for freezing into the f.c.c. and the b.c.c. lattice, respectively.

We may add explicitly at this point that the use of a restricted set of order parameters in the treatment of the phase transition does not imply that the remaining ones are zero. As is clear from equations (2.8)-(2.11) the order parameters become essentially irrelevant in determining the phase transition once the corresponding values of $\hat{c}(G)$ are close to zero (i.e. $S(G)$ is close to unity). Such "irrelevant" order parameters have nevertheless finite values, determined by the relevant ones from the infinite set of equations (2.9). Not surprisingly, if this fact is overlooked, one finds from (2.1) a density profile $n(\underset{\sim}{r})$ which is negative in the interstitial regions. Equations (2.6) and (2.7), of course, describe a positiv definite particle density.

In conclusion, though several microscopic order parameters are clearly playing a role in the phase transition, the results already justify a viewpoint which focusses attention on the first few peaks of the liquid structure factor near freezing. A critical height of the main peak can therefore be a good gauge for freezing of simple monatomic liquids, as expressed in the criterion of Hansen and Verlet.

3.2 Freezing of the 3D Classical Plasma

The 3D classical plasma is especially well suited for a test of the microscopic theory of freezing. The thermodynamic properties and the liquid structure of this model are known in great detail from computer simulation work (18) and freezing into the b.c.c. lattice at constant density is known to occur at a coupling strength $\Gamma_c \simeq 178$ (19). Accurate theories of its liquid structure, which embody consistency with thermodynamic properties, are available (20). Since freezing is at constant density, the scale of length for RLV is known a priori. As already noted, the density change on freezing is replaced by the potential drop at the solid-liquid interface, which can be eliminated from the calculation of the microscopic order parameters through the use of (2.8).

The direct correlation function $\hat{c}(k)$ for the plasma at $\Gamma = 160$, as calculated by two alternative theories of liquid structure, is compared in Figure 1 with computer simulation results of Hansen (18) in the region of wave number of main interest for freezing. We note immediately that the first star of RLV for constant-density-freezing into the b.c.c. lattice (the (110) star) lies very close to the position of the main peak in $\hat{c}(k)$, while the third and fourth star underlie the second peak. The (110) and (211) stars were those included in the calculations on sodium by Ramakrishnan and Yussouff (15) that we referred to in section 3.1. On the other hand, the (200) star lies in the region of the main minimum of $\hat{c}(k)$, indicating that the liquid is rather rigid against a modulation by density waves at these wave vectors. One expects, therefore, that the corresponding order parameter should be driven by the others rather than generated spontaneously.

This is confirmed by detailed calculations in the framework of the low-order theory of freezing (14). A solution of the equilibrium equations could be found only on assuming that ρ_{200} is small, and by setting it equal to zero one predicts a lower limit for the critical coupling strength at freezing which is in the range $140 \div 160$, depending on liquid structure theory. A rather large number of order parameters (up to 19 stars of RLV) had to be included in order to achieve convergence in the calculated value of Γ_c, and the results are clearly very sensitive to the theory of liquid structure which is being used. The calculations yield also a reasonable value for the entropy change on freezing ($\Delta S = 0.8 \div 1.0\ k_B$, against a value of $0.8\ k_B$ from simulation), whereas the calculated interfacial potential drop is wrong by a factor of two. A quantitatively important role of $c^{(3)}(110,200)$ (i.e. of couplings between order parameters at the first two stars of RLV) was suggested from considerations on the local structure in a hot b.c.c. solid and its liquid.

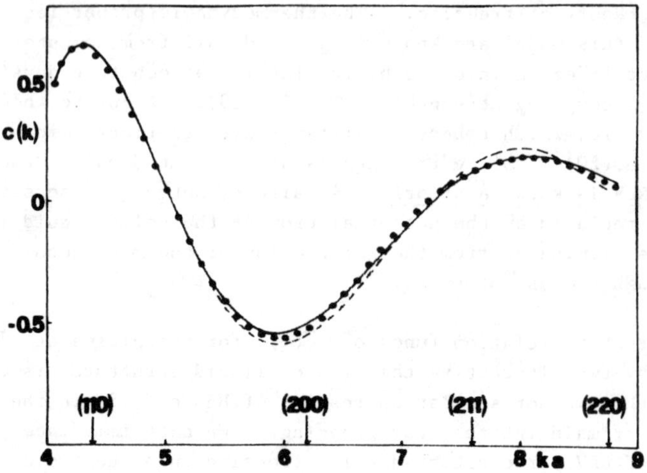

Figure 1. Direct correlation function $\hat{c}(k)$ of the classical 3D plasma at $\Gamma = 160$. Dots are simulation data of Hansen (18) and full and dashed lines are from GMSA and MHNC calculations, respectively.

Similarities between the freezing of alkali metals and that of the plasma have often been noticed in the past (21, 22). For instance, the ionic coupling strength at freezing is in the range $181 \div 211$ for the alkalis from Cs to Li, against $\Gamma_c \simeq 178$ for the plasma; both the Lindemann criterion and the Hansen-Verlet criterion correlate phenomenologically all these systems. It was first noticed empirically by Minoo et al. (23) and later justified by electron screening theory (24) that the liquid structure factor of the freezing plasma resembles closely the observed structure factors of freezing alkalis in the region of wave number shown in Figure 1. In this viewpoint the main role of the conduction electrons in liquid structure and freezing is to screen long-wavelength fluctuations of the ionic density as well as the interfacial dipole layer between the two coexisting phases. A relatively small density change ($\eta \simeq 2.5\%$) then ensues.

3.3 Freezing of Hard Spheres

The other important model for a test of the theory, in relation to freezing into the f.c.c. lattice, is the system of hard spheres. Computer simulation work has made it clear that the main features of freezing of simple liquids are already present in hard-sphere crystallization (25). The Wertheim-Thiele analytic solution of the Percus-Yevick approximation (26) offers easy access to the description of liquid structure, although its thermodynamic inconsistencies (27) may be quite relevant in the problem at hand. An accurate parametrization of simulation data for liquid structure, which embodies the Carnahan-Starling equation of state, is also available from work of Verlet and Weis (28).

A detailed discussion of freezing of the hard sphere liquid in the Percus-Yevick approximation has been given by Haymet (29). Relative to the case of the plasma that we have discussed above, the scale of length of RLV is not known a priori and, in principle, ought to be determined self-consistently with the density change η. Haymet makes the reasonable assumption that the first star of RLV lies at the main peak of the liquid structure factor. Preliminary calculations by D'Aguanno (16) indicate that a requirement of selfconsistency could shift appreciably the solution of the microscopic equations. Other points to be remarked are the need to include up to 15 stars of RLV for a reliable assessment of the freezing density and the inclusion of three-body correlations of the types $c^{(3)}(0,0)$ and $c^{(3)}(G,0)$. The main results can be summarized as follows: (i) the phase change is found to take place at a reduced liquid density $n_\ell\sigma^3 = 0.976$, against Monte Carlo data (30) in the range $0.939 \div 0.948$, with a fractional volume change $\eta = 0.06$, against simulation data in the range $0.09 \div 0.11$; (ii) the order parameters after the first few ones satisfy quite well the harmonic-like relation

$$\rho_{\underset{\sim}{G}} = (1 + \eta) \exp[-\alpha|\underset{\sim}{G}|^2/|\underset{\sim}{G}_1|^2], \tag{3.1}$$

where $\underset{\sim}{G}_1$ denotes the first star and α is a constant. This corresponds to a density profile in the crystal which is approximately a sum of Gaussian centred at the lattice sites. Better agreement with the simulation data on the location of the freezing point and on the fractional volume change has been reported (31) on the basis of the Verlet-Weis empirical structure factor.

In subsequent work, Haymet (32) has reported that, if the calculation is continued to still higher densities, the solid eventually becomes again less stable than the liquid and finally unstable. Possible reasons for this artifact are the inapplicability of the low-order perturbation expansion far away from equilibrium freezing and/or that of wide extrapolations of structural and thermodynamic properties into a highly compressed liquid region. Baus and Colot (33) have approximately summed over higher-order terms by making use, in the formal approach of Saam and Ebner (11), of the Percus-Yevick $\hat{c}(k)$ at a liquid density such that the peak of $S(k)$ lies at the first star of RLV. The order parameters are also included to infinite order through the assumption of Gaussian profiles. They find again quite reasonable results for equilibrium freezing and remove the artificial "remelting" pointed out by Haymet, finding a stable hard-sphere f.c.c. solid up to the density of crystal close packing. Work on the freezing of the hard sphere liquid, which is based directly on (2.6)-(2.7) and requires therefore knowledge of $c(r)$, has also been presented by Tarazona (34).

A final remark to be made here regards consistency in the evaluation of the thermodynamic parameters of the phase transition with finite density change (35). The entropy and density change can be obtained from the theory in two alternative ways, namely (i) from the function $\Delta P(T,n_\ell)$ at $\Delta\mu = 0$, or (ii) from the function $\Delta\mu(T,n_\ell)$ at $\Delta P = 0$. Although these routes ought to lead to identical results, route (ii) may turn out to be more useful in practice. A spurious dependence of the entropy change on atomic mass could arise in route (i) as a result of an approximate treatment of the phase transition.

3.4 Freezing of Ionic Materials

As we have seen in the two preceding sections, detailed studies on the models of the 3D plasma and of the hard sphere system have tested the limitations of the theory and indicated directions for its improvement. At the phenomenological level illustrated in section 3.1, on the other hand, the theory has found a variety of further useful applications. We focus here on the work on freezing of some molten salts (36, 37).

The native point defects in simple ionic crystals can be Schottky or Frenkel defects, the former being the equilibrium defects in alkali halides such as $NaC\ell$ and the latter (of anionic type) in fluorite-type alkaline-earth dihalides such as $SrC\ell_2$. Though point defects cannot be defined outside the crystalline phase, the observed liquid structures of these salts reflect the nature of the dominant point defect in the hot solid.

For molten $NaC\ell$ we report in Figure 2 the Bhatia-Thornton structure factors $S_{QQ}(k)$ and $S_{NN}(k)$, as constructed from neutron-diffraction data (38) and from calculations in a pair-potentials model (39). These express the correlations of the charge density fluctuations and of the particle number density fluctuations, respectively, the cross correlations between these fluctuations (the charge-number structure factor $S_{QN}(k)$) being very small. This feature and the shapes of S_{QQ} and S_{NN} in Figure 2 are characteristic of a binary liquid in which the two components have similar internal orders and strong relative order. We expect on freezing to find a solid with similar magnitudes of the order parameters for the two components (i.e. containing Schottky defects), the freezing transition being driven by the relative order of cations and anions (measured by the peak in S_{QQ}) and assisted by the density change which is very large (25%). The compressibility of the melt is measured by $S_{NN}(k \to 0)$.

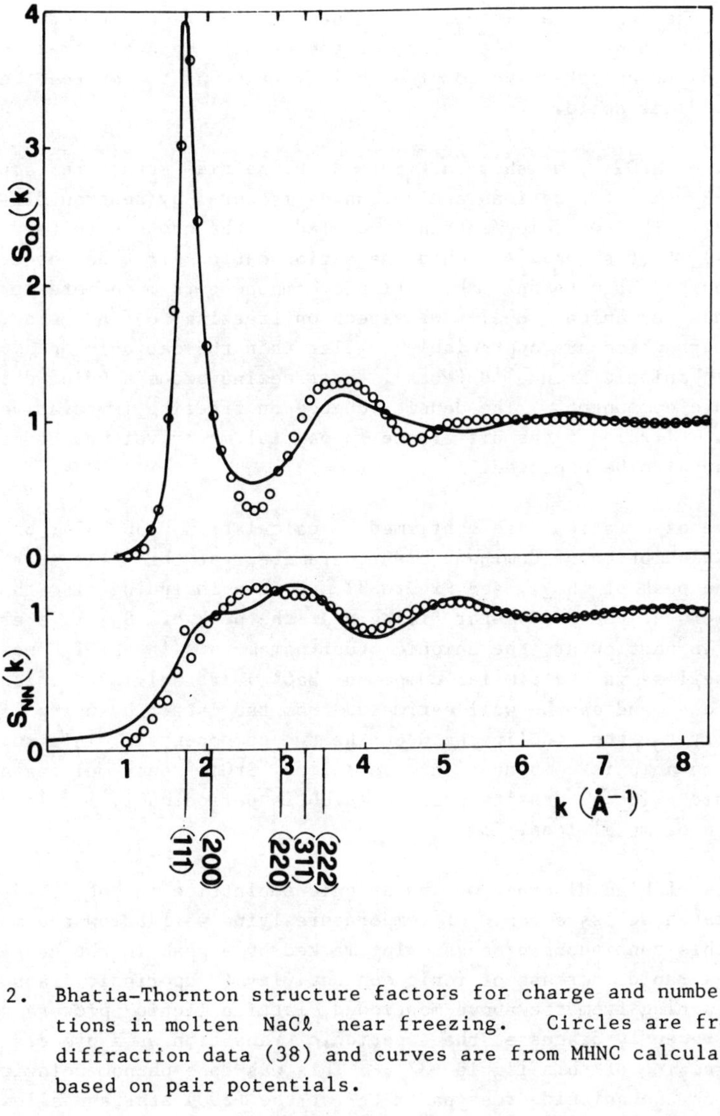

Figure 2. Bhatia-Thornton structure factors for charge and number fluctuations in molten NaCℓ near freezing. Circles are from neutron diffraction data (38) and curves are from MHNC calculations based on pair potentials.

The even Bragg reflections of the crystalline phase (f.c.c. Bravais lattice with a two-ions basis) are weighted by the average form factor, as is $S_{NN}(k)$ in the diffraction pattern of the liquid: they describe the s.c. lattice to which the crystal structure reduces if the differences between the two ionic species are neglected. The odd Bragg reflections, on the other hand, are weighted by the difference of the form factors as is $S_{QQ}(k)$. The stars of RLV are shown at the bottom of Figure 2: we have placed the (111) star at the peak of S_{QQ} and the (200), (220) and (222) stars are seen to underlie the broad peak in S_{NN}. Detailed numerical calculations (36) indicate that these are indeed the main order parameters of the phase transition. Nonlinear terms are again very important to obtain a reasonable value for the density change: empirically, it suffices to use an effective compressibility taken as the average between liquid and hot solid.

Turning to $SrCl_2$, we show in Figure 3 the partial structure factors $S_{++}(k)$ and $S_{--}(k)$ for cations and anions as measured by neutron diffraction (40). The two components are coupled by the cross structure factor $S_{+-}(k)$ (not shown). Though the cation-cation first neighbour distance is quite large (~ 5Å), the cationic component is much more strongly structured than the anionic one. We expect on freezing to find anionic order parameters which are appreciably smaller than the cationic ones (i.e. a crystal with anionic Frenkel defects), the freezing process being driven by the cationic component. The density change on freezing is relatively small (4.2%). A role of the difference in partial molar volumes of cations and anions may also be expected.

The above expectations are confirmed by calculations (36, 37), based first on the assumption of dominant order parameters at the (111) star (placed at the peak of S_{++}, see Figure 3) and then including also the order parameters at the (200) star (lying near the peak of S_{--}, see Figure 3). In particular, the anionic order parameters in $SrCl_2$ near melting (as well as in the similar compound $BaCl_2$) are only a fraction of the cationic ones and can be well estimated from the latter by using linear response theory for the coupling between the two components. This supports a picture of the anionic component in crystalline $SrCl_2$ near melting as a "lattice liquid" with a density profile which is periodically modulated by the sublattice of metal ions.

The state of high disorder on the anionic sublattice in hot $SrCl_2$ is in fact generated across a range of temperature lying well below the melting point T_m, this continuous process being marked by a peak in the heat capacity and a rapid increase of ionic conductivity ("superionic transition"). Starting from the above mentioned "lattice liquid" picture near T_m, we have recently discussed the superionic transition as a process of continuous freezing of this liquid (37). This was done phenomenologically by focussing on the anionic order parameter at the (200) star and allowing the corresponding $\hat{c}_-(G)$ to grow continuously on cooling. The order parameter grows continuously from its finite value at T_m and saturates at low temperatures; its growth is therefore accompanied by a peak in the heat capacity at a temperature T_c, identified with the superionic transition temperature. Semiquantitative agreement with structural and thermodynamic data on $SrCl_2$, as well as with the Hansen-Verlet criterion ($\hat{c}(200) \simeq 0.76$ at T_c) is found.

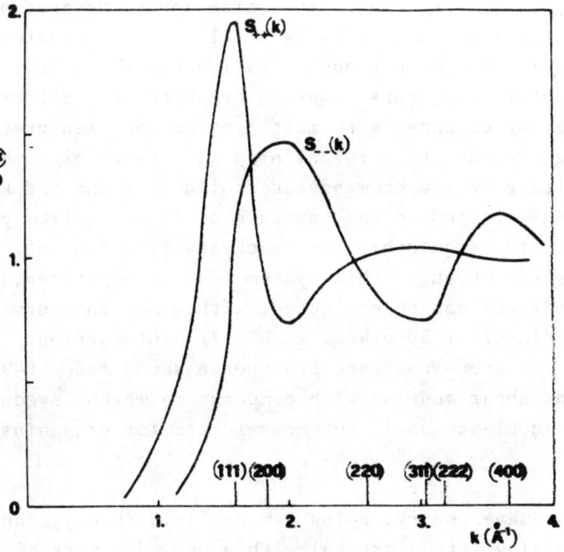

Figure 3. Sr-Sr and Cℓ-Cℓ structure factors in molten $SrC\ell_2$ near freezing, from neutron diffraction experiments of McGreevy and Mitchell (40).

3.5 First-order Freezing of 2D Liquids

Experimental and computer simulation studies of melting in 2D have been greatly stimulated both by interest in physical adsorption phenomena (41) and by the dislocation-unbinding theory due to Nelson and Halperin (42), based on the ideas of Kosterlitz and Thouless (43). They predicted for a triangular lattice on a smooth substrate the existence of a hexatic phase characterized by a power-law decay of orientational order, between the solid (with algebraic decay of translational order and long range orientational order) and the isotropic liquid phase. Thermodynamically continuous transitions lead from solid to hexatic (dislocation unbinding) and from hexatic to liquid phase (disclination unbinding). However, a first-order melting transition driven by premature disclination unbinding cannot be ruled out. In the presence of a periodic substrate, dislocation unbinding should still be applicable to melting of an incommensurate adsorbate solid into a fluid phase.

The available evidence has been discussed recently by Ramakrishnan (44) and Morph (45). Most recent computer simulations on r^{-n} and Lennard-Jones systems show a first-order transition and no hexatic phase. However, the interpretation of these results is still somewhat controversial, the main question concerning the defect equilibration times near the phase transition. On the experimental side, rare gases on graphite show different melting behaviours: e.g. argon appears to melt from an incommensurate solid to liquid with a higher-order transition for a wide range of coverage (46), while first-order melting from a commensurate solid is found for krypton (47). A film of electrons trapped on the surface of liquid helium provides a case where the substrate is smooth: Wigner crystallization into a triangular lattice has been observed in this system (48) in experimental conditions where quantal effects can be neglected, with a melting curve $n \propto T^2$ as expected for a classical 2D plasma with $1/r$ interactions. Measurements of the coupled electron-substrate transverse sound mode (49) yield a behaviour of the shear modulus with temperature which is consistent with the Kosterlitz-Thouless model, but no evidence for or against a hexatic phase.

The density-wave theory, being a mean field theory, can only lead to first-order freezing into a crystal with a periodic particle density. As discussed in more detail by Ramakrishnan (44) for the case of an ideally smooth substrate, it is nevertheless worthwhile to examine its predictions for 2D systems, since the free energy balance may not be very much affected by 2D peculiarities. Dislocation unbinding could be preempted by first-order melting or, as may be happening in simulations, the system could still be perceived as a solid at temperatures above dislocation unbinding if defect relaxation times are long. His results, obtained for r^{-n} systems by using two microscopic order parameters, are in reasonable agreement with simulation data, but correspond to a noticeably weaker first-order transition

A more detailed study on the 2D plasma with $1/r$ interactions has been presented in the same spirit by Ballone et al. (50), on the basis of an accurate MHNC theory of liquid structure. Figure 4 shows the calculated $\hat{c}(k)$ of this liquid at a coupling strength $\Gamma = 90$ (with the definition $\Gamma = e^2(\pi n)^{\frac{1}{2}}/(k_B T)$), in comparison with simulation data (51) and in juxtaposition with the RLV of the triangular lattice at the liquid area density n. Freezing is taken to occur at constant density but with a finite interfacial potential drop, as already discussed for the 3D plasma. It is evident from Figure 4 that the first three stars of RLV lie in close correspondence with the two main peaks of $\hat{c}(k)$ and that the remaining stars lie near nodes of $\hat{c}(k)$. The liquid is therefore structurally predisposed to freeze into the triangular lattice and the first three order parameters are clearly dominant. The numerical results for first-order freezing underestimate somewhat the range of stability of the solid phase (the theory yields $\Gamma_c = 149$ against $\Gamma_c = 137 \pm 15$ from experiment (48) and $\Gamma_c = 118 \div 130$ from simulation (52)) and grossly overestimate the entropy change on freezing and the interfacial potential drop.

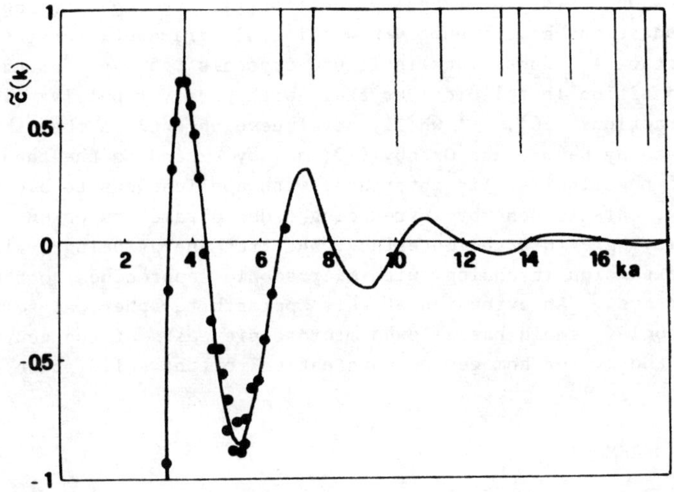

Figure 4. Direct correlation function $\hat{c}(k)$ of the classical 2D plasma with $1/r$ interactions at $\Gamma = 90$. Dots are simulation data of Gann et al. (51) and the curve is from MHNC calculations.

Freezing of 2D liquids on a periodic substrate has been examined by Sokolowski and Steele (53) for models appropriate to rare gases adsorbed on graphite. The structure of the modulated 2D liquid is first calculated in an approach using eqn. (2.5) and an expansion of $c(\underset{\sim}{r},\underset{\sim}{r}')$ around a homogeneous fluid, with results which compare rather well with simulation data (54). Results for freezing of the substrate-modulated liquid into a $\sqrt{3} \times \sqrt{3}$ triangular lattice are next obtained for hard discs modelling argon and xenon and for a Lennard-Jones system modelling krypton. A general result is that liquid nonuniformity causes a decrease of the fluid density at which freezing occurs. From calculations using a single microscopic order parameter, the argon-like system is found to freeze into an incommensurate triangular lattice while the xenon-like system, with a higher translational barrier along the surface, freezes at densities higher than commensurate. In the more detailed study of the krypton/graphite system, a multiplicity of order parameters is considered and the stability of different orientations of the 2D solid relative to the substrate is investigated The results predict freezing at a reduced solid density $n^* = 0.82$ into a registered $\sqrt{3} \times \sqrt{3}$ structure over a wide temperature range, as indicated by experiment and simulation.

3.6 Further Applications

We should mention here for completeness various other applications of the density-wave theory of freezing. Firstly, with regard to complex liquids, discussions have been given within this framework for the isotropic-nematic-smectic A transitions in liquid crystals (55) and for the order-disorder transition in colloidal crystals with purely repulsive electrostatic interactions (56). A wholly novel development, on the other hand, has been given by Haymet and Oxtoby (12, 57) by extending the theory to the treatment of the liquid-solid interface, with applications to b.c.c. metals. Very briefly, this is done by introducing order parameters which are spatially varying across the interface, the free energy being evaluated by a gradient expansion in analogy with microscopic approaches to the liquid-vapour interface. An extension of this approach to spherical solid nuclei in an undercooled liquid has allowed microscopic tests of the conventional capillarity theory for homogeneous nucleation of the solid phase (58).

4 CONCLUDING REMARK

A mean-field theory of the solid-liquid transition ought to be simple in order to be useful. The idea of describing the phase transition merely in terms of the pair structure of the liquid, which is originally due to Kirkwood, is implemented in the work of Ramakrishnan and Yussouff through a direct appeal to the observed liquid structure, thereby bypassing the question of interatomic potentials. A number of useful insights and novel developments have already ensued from this approach.

REFERENCES

1. W.G. Hoover and M. Ross, Contemp. Phys. 12, 339 (1971).
2. S.M. Stishov, Sov. Phys. Uspekhi 17, 625 (1975).
3. H. Reiss, S.W. Mayer andJ.L. Katz, J. Chem Phys. 35, 820 (1961).
4. M. Lasocka, Phys. Lett. 51A, 137 (1975).
5. J.L. Tallon, Phys. Lett. 87A, 362 (1982) and Solid State Commun. 42, 243 (1982).
6. F.A. Lindemann, Zs. Phys. 11, 609 (1910).
7. J.P. Hansen and L. Verlet, Phys. Rev. 184, 151 (1969).
8. J.A. Moriarty, D.A. Young and M. Ross, Phys. Rev. B30, 578 (1984).
9. T.V. Ramakrishnan and M. Yussouff, Solid State Commun. 21, 389 (1977). For a discussion of the approach to freezing based on the first equation of the BBGKY hierarchy, see e.g. H.J. Raveché and R.F. Kayser Jr., J. Chem. Phys. 68, 3632 (1978) and Phys. Rev. B22, 424 (1980); L. Feijoo and A. Rahman, J. Chem. Phys. 77, 5687 (1982).

10. P. Hohenberg and W. Kohn, Phys. Rev. 136, B864 (1964); N.D. Mermin, Phys. Rev. 137, 1441 (1965).

11. W.F. Saam and C. Ebner, Phys. Rev. A15, 2566 (1977).

12. A.D.J. Haymet and D.W. Oxtoby, J. Chem. Phys. 74, 2559 (1981).

13. N.H. March and M.P. Tosi, Phys. Chem. Liquids 11, 129 (1981).

14. M. Rovere and M.P. Tosi, J. Phys. C18, 3445 (1985).

15. T.V. Ramakrishnan and M. Yussouff, Phys. Rev. B19, 2775 (1979).

16. B. D'Aguanno, Ph. D. Thesis, ISAS-SISSA, Trieste 1984.

17. M. Baus, Molec. Phys. 50, 543 (1983).

18. S.G. Brush, H.L. Sahlin and E. Teller, J. Chem. Phys. 45, 2102 (1966); J.P. Hansen, Phys. Rev. A8, 3096 (1973); M. Baus and J.P. Hansen, Phys. Repts. 59, 1 (1980).

19. E.L. Pollock and J.P. Hansen, Phys. Rev. A8, 3110 (1973); W.L. Slattery, G.D. Doolen and H.E. DeWitt, Phys. Rev. A21, 2087 (1980) and A26, 2255 (1982).

20. Y. Rosenfeld and N.W. Ashcroft, Phys. Rev. A20, 1208 (1979); D.K. Chaturvedi, G. Senatore and M.P. Tosi, N. Cimento B62, 375 (1981); H. Iyetomi and S. Ichimaru, Phys. Rev. A25, 2434 (1982); F.J. Rogers, D.A. Young, H.E. DeWitt and M. Ross, Phys. Rev. A28, 2990 (1983).

21. H.M. van Horn, Phys. Lett. 28A, 706 (1969).

22. V.A. Ivanov, I.N. Makarenko, A.M. Nikolaenko and S.M. Stishov, Phys. Lett. 47A, 75 (1974).

23. H. Minoo, C. Deutsch and J.P.Hansen, J. Physique Lettres 38, L191 (1977).

24. D.K. Chaturvedi, G. Senatore and M.P. Tosi, Physica 111B, 11 (1981); G. Pastore and M.P. Tosi, Physica 124B, 383 (1984).

25. See e.g. J.P. Hansen and I.R. McDonald, "The Theory of Simple Liquids" (Academic Press, New York 1976).

26. M.S. Wertheim, Phys. Rev. Lett. 10, 321 (1963); E. Thiele, J. Chem. Phys. 39, 474 (1963).

27. J.A. Barker and D. Henderson, Rev. Mod. Phys. 48, 587 (1966).

28. L. Verlet and J.J. Weis, Phys. Rev. 45, 939 (1972); D. Henderson and E.W. Grundke, J. Chem. Phys. 63, 601 (1975).

29. A.D.J.Haymet, J. Chem. Phys. 78, 4641 (1983).

30. W.G. Hoover and F.H. Ree, J. Chem. Phys. 49, 3609 (1968).

31. A.D.J. Haymet and D.W. Oxtoby, in the course of publication.

32. A.D.J. Haymet, J. Phys. Chem. 89, 887 (1985).

33. M. Baus and J.L. Colot, J. Phys. C18, L365 (1985).

34. P. Tarazona, Molec. Phys. 52, 81 (1984) and Phys. Rev. A31, 2672 (1985).

35. B. D'Aguanno, M. Rovere and G. Senatore, in the course of publication.

36. B. D'Aguanno, M. Rovere, M.P. Tosi and N.H. March, Phys. Chem. Liquids 13, 113 (1983).

37. M. Rovere and M.P. Tosi, Solid State Commun., in the press.

38. S. Biggin and J.E. Enderby, J. Phys. C15, L305 (1982).

39. P. Ballone, G. Pastore and M.P. Tosi, J. Chem. Phys. 81, 3174 (1984).

40. R.L. McGreevy and E.W.J. Mitchell, J. Phys. C15, 5537 (1982).

41. See e.g. "Ordering in Two Dimensions" (ed. S.K. Sinha; North Holland, New York 1980).

42. D.R. Nelson and B.I. Halperin, Phys. Rev. B19, 2457 (1979).

43. M. Kosterlitz and D.J. Thouless, J. Phys. C6, 1181 (1973).

44. T.V. Ramakrishnan, Phys. Rev. Lett. 48, 541 (1982).

45. R.H. Morf, Helv. Phys. Acta 56, 743 (1983).

46. J.P. McTague, J. Als-Nielsen, J. Bohr and M. Nielsen, Phys. Rev. B25, 7765 (1982).

47. R.J. Birgenau, E.M. Hammons, P. Heiney and P.W. Stephens, in ref. (41), p. 29.

48. C.C. Grimes and G. Adams, Phys. Rev. Lett. 42, 795 (1979); D.S. Fisher, B.I. Halperin and P.M. Platzman, Phys. Rev. Lett. 42, 798 (1979).

49. F. Gallet, G. Deville, A. Valdès and F.I.B. Williams, Phys. Rev. Lett. 49, 212 (1982).

50. P. Ballone, G. Pastore, M. Rovere and M.P. Tosi, J. Phys. C (in press).

51. R.C. Gann, S. Chakravarty and G.V. Chester, Phys. Rev. B20, 326 (1979).

52. R.K. Kalia, P. Vashishta and S.W. de Leeuw, Phys. Rev. B23, 4794 (1981).

53. S. Sokolowski and W. Steele, J. Chem. Phys. 82, 2499 and 3413 (1985).

54. S. D. Prasad and S. Toxvaerd, J. Chem. Phys. 72, 1689 (1980).

55. T.J. Sluckin and P. Shukla, J. Phys. A16, 1539 (1983); M.D. Lipkin and D.W. Oxtoby, J. Chem. Phys. 79, 1939 (1983).

56. S. Alexander, P.M. Chaikin, D. Hone, P.A. Pincus and D.W. Schaefer, in the course of publication.

57. D.W. Oxtoby and A.D.J. Haymet, J. Chem. Phys. 76, 6262 (1982).

58. P. Harrowell and D.W. Oxtoby, J. Chem. Phys. 80, 1639 (1984).

THE STRUCTURE OF LIQUID BINARY ALLOYS

G. Kahl and J. Hafner

Institut für Theoretische Physik, Technische Universität Wien,
A-1040 Karlsplatz 13, Wien, Austria

1. INTRODUCTION

The aim of this paper is to present a method to calculate the partial structure factors and pair correlation functions for a binary system. Our starting point is a set of pair interactions $\Phi_{ij}(r)$, $i,j=1,2$, $\Phi_{ij}=\Phi_{ji}$, based on pseudopotential theory. We then split each of these potentials into a repulsive part $\Phi^0_{ij}(r)$ and an attractive part $\Phi^1_{ij}(r)$ at or near the first minima of the Φ_{ij}:

$$\Phi_{ij}(r)=\Phi^0_{ij}(r)+\Phi^1_{ij}(r) \qquad (1)$$

$$\Phi^0_{ij}(r)=\begin{cases}\Phi_{ij}(r)-\Phi_{ij}(r_{ij}) & r<r_{ij}\\ 0 & r>r_{ij}\end{cases} \qquad (2)$$

$$\Phi^1_{ij}(r)=\begin{cases}\Phi_{ij}(r_{ij}) & r<r_{ij}\\ \Phi_{ij}(r) & r>r_{ij}\end{cases} \qquad (3)$$

We treat each part of the interaction by suitable perturbation theories: the Weeks-Chandler-Andersen (WCA) approximation for the repulsive part and the Optimised Random Phase Approximation (ORPA) for the attractive part /1,2/. For monoatomic systems this is a very effective way to derive the liquid structure from the interatomic forces /3-9/. Both these perturbative methods are introduced in the subsequent sections, the final section contains a discussion of results.

2. THE WEEKS-CHANDLER-ANDERSEN (WCA) APPROXIMATION

The WCA approximation establishes a relationship between the properties of a model alloy, whose atoms interact via the purely repulsive pair potentials $\Phi^0_{ij}(r)$ and a mixture of hard spheres (HS) with the same density and concentration with the set of diameters $\sigma=\{\sigma_{11},\sigma_{22},\sigma_{12}\}$. With the definition of the Mayer function f_{ij}

$$f_{ij}(r)=\exp(-\beta\,\Phi_{ij}(r))-1 \qquad (4)$$

we may express the free energy F of the model alloy as a functional Taylor series in terms of the difference between the Mayer functions of the repulsive potentials and of the reference potentials

$$\Delta f_{ij}(r)=\exp(-\beta\Phi^0_{ij}(r))-\exp(-\beta\Phi_{\sigma_{ij}}(r)) \qquad (5)$$

Denoting the pair correlation function of the HS-system by $g_{\sigma_{ij}}$ and defining

$$y_{\sigma_{ij}}(r)=\exp(\beta\Phi_{ij}(r))g_{\sigma_{ij}}(r) \qquad (6)$$

it may be shown that (c_i stands for the concentration of species i)

$$F=F_\sigma+\frac{1}{2}\sum_{i,j} c_i c_j \int B_{\sigma_{ij}}(r)d\vec{r} + O(\Delta f^2) \qquad (7)$$

with the "blip-functions"

$$B_{\sigma_{ij}}(r)=y_{\sigma_{ij}}(r)\Delta f_{ij}(r) \qquad i,j=1,2 \quad . \tag{8}$$

For a given set of σ the pair correlation functions for the repulsive system are expressed as

$$g_{ij}(r)=g_{\sigma_{ij}}(r)+B_{\sigma_{ij}}(r) \quad , \tag{9}$$

a similar relation holds in q-space for the S_{ij}.

In the one component case the HS diameter σ is chosen to make the integral over the blip-function vanish. Andersen et al. /10/ have shown that in this case convergence of this expansion is optimised, the first non-vanishing corrections being of order $O(\Delta f^4)$.

In the two component case optimised convergence in (7) is achieved for

$$I_{ij}= \int B_{\sigma_{ij}}(r)d\vec{r} = 0 \, . \tag{10}$$

Simultaneously these three conditions can only be satisfied for a non-additive HS-mixture, i.e. if we allow $\sigma_{12}\neq(\sigma_{11}+\sigma_{22})$. Unfortunately, a solution of the Percus-Yevick equations for non-additive HS-diameters is available only for a restricted range of parameters. Therefore, Sandler /11/ proposed to minimise one of the functions I_1 or I_2:

$$I_1= \sum_{ij} c_i c_j |I_{ij}| \, , \tag{11}$$

$$I_2= \sum_{ij} |I_{ij}| \, , \tag{12}$$

whereas Lee and Levesque /12/ used

$$\sum_{ij} c_i c_j I_{ij}=0 \tag{13}$$

Note that the minimisation conditions (11) and (12) determine σ_{11},σ_{22} uniquely, whereas (13) is satisfied for many pairs of $(\sigma_{11},\sigma_{22})$. In any case the individual I_{ij} do not vanish, and as a consequence the partial static structure factors are unphysical in the long-wavelength limit (see Fig.2). Note that this also yields unphysical results for the thermodynamic quantities such as the isothermal compressibility.

So we proposed /13/ to proceed as follows: in default of solutions of non-additive HS-systems we had to use the solution of the Percus-Yevick equations for the additive case. We determined in the first step σ_{11} and σ_{22} by requiring $I_{11}=I_{22}=0$, keeping $\sigma_{12}=(\sigma_{11}+\sigma_{22})/2$. Then we varied the separation point r_{12} (see (2)) in such a way to make $I_{12}=0$. The positions of the r_{ij} are depicted in Fig. 1 for the system KCs. Indeed it turns out that r_{12} has to be shifted only about 3%. That the separation of Φ_{12} is now different from the original one causes no problem: after adding the long-range interactions $\Phi^1_{ij}(r)$ in the ORPA, the final separation of Φ_{ij} into Φ^0_{ij} and Φ^1_{ij} has to be determined self-consistently (see below) - so that the initial ansatz for Φ_{12} is ultimately irrelevant.

3. THE OPTIMISED RANDOM PHASE APPROXIMATION (ORPA)

The optimised random phase approximation is defined by the coupled Ornstein-Zernike equations for the partial direct and total correlation functions $c_{ij}(r)$ and $h_{ij}(r)$ in conjunction with the closure relations

$$c_{ij}(r)=c^0_{ij}(r)-\beta\Phi^1_{ij}(r) \qquad r>\sigma_{ij} \, , \tag{14}$$

$$g_{ij}(r)=h_{ij}(r)+\delta_{ij}=0 \qquad r<\sigma_{ij} \tag{15}$$

Here the c^0_{ij} stand for the direct correlation functions of a HS-reference system (determined in the WCA-approach). (14) expresses the random-phase approximation for the long-range interactions, (15) states the impenetrability of the ionic cores. (14) and (15) are equivalent to the mean spherical approximation for a HS-reference system and to the variational condition

$$\frac{\delta F[\Phi_{ij}(r)]}{\delta\Phi_{ij}(r)}=0 \qquad r<\sigma_{ij} \qquad (16)$$

(16) determines an "optimised attractive" interaction Φ^{1*}_{ij} for $r<\sigma_{ij}$. As of course the total interaction remains invariant, this also sets a new repulsive interaction via

$$\Phi_{ij}(r)=\Phi^{0*}_{ij}(r)+\Phi^{1*}_{ij}(r) \qquad (17)$$

a self-consistent separation has to be found by iterating the combined WCA+ORPA procedure.

4. RESULTS

We now present results of our calculations for liquid alkali alloys. Fig. 1 demonstrates the positions of the r_{ij} and the shifted r_{12} according to our method. The resulting change in Φ^0_{12} is very small and is easily absorbed in the iterative algorithm of the WCA- and ORPA-steps. Fig. 2 shows how drastic the consequences of a non-zero I_{ij} are for the long-range behaviour of the respective $S_{ij}(q)$. It clearly may be seen that our solution yields a correct small q-behaviour of the $S_{ij}(q)$ for <u>all</u> three pairs of i and j. Comparison of WCA-results (repulsive parts of the potentials only!) with experiment /14/ show remarkable improvement over the

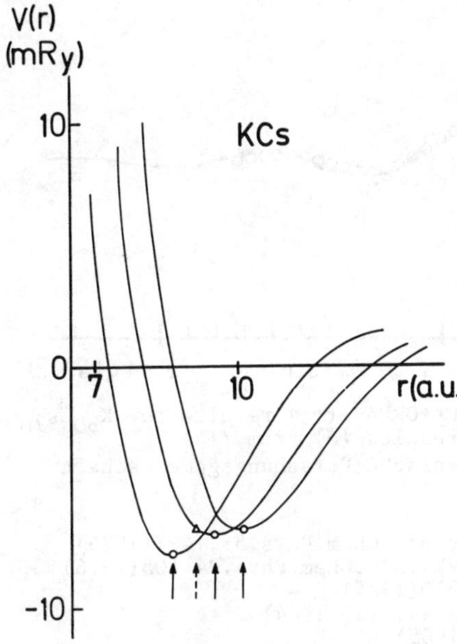

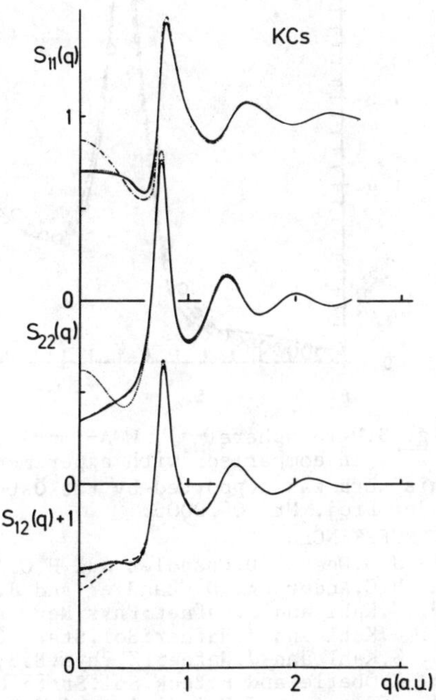

<u>Fig. 1</u> (left): Interatomic potential for KCs, 1-1, 1-2 and 2-2 interactions (from left to right). Full arrows and circles denote separation points at the first minima, broken arrow and triangle indicate redefined r_{12}.
<u>Fig. 2</u> (right): Partial structure factors for KCs (WCA only). Different symbols denote different methods of determining HS-diameters: (i) $I_{11}=I_{22}=0$, $I_{12}\neq 0$: – –; (ii) $I_{11}=I_{12}=0$, $I_{22}\neq 0$: ·····; (iii) $I_{12}=I_{22}=0$, $I_{11}\neq 0$: –·–; (iv) $I_{11}=I_{22}=I_{12}=0$, with modified r_{12}: ——.

HS approximation for the whole concentration range (see /13/).

Fig. 3 shows results of a full WCA+ORPA calculation for a $K_{30}Cs_{70}$-alloy in comparison with X-ray scattering data /14/. HS- and WCA-curves are also shown besides the final results. The strong higher order oscillations of the HS-approximation are dampened in the WCA, the positions of the minima and maxima remain unchanged. The height of the first maximum is still too low, this is corrected by the ORPA, which increases the peak and leaves the heights of the other maxima more or less unchanged. The other important influence of the attractive interactions is the shift of the oscillations to smaller q-values, thus improving agreement with experiment. The hump at about .3 a.u. is a well known problem caused by the WCA and may be removed by the Jacobs-Andersen modification of the WCA /15/ As in the one component case we may find (not depicted here) that in the attractive regions of the interaction Φ_{ij} the corresponding g_{ij} is enhanced whereas in the repulsive regions the pair correlation function is dampened.

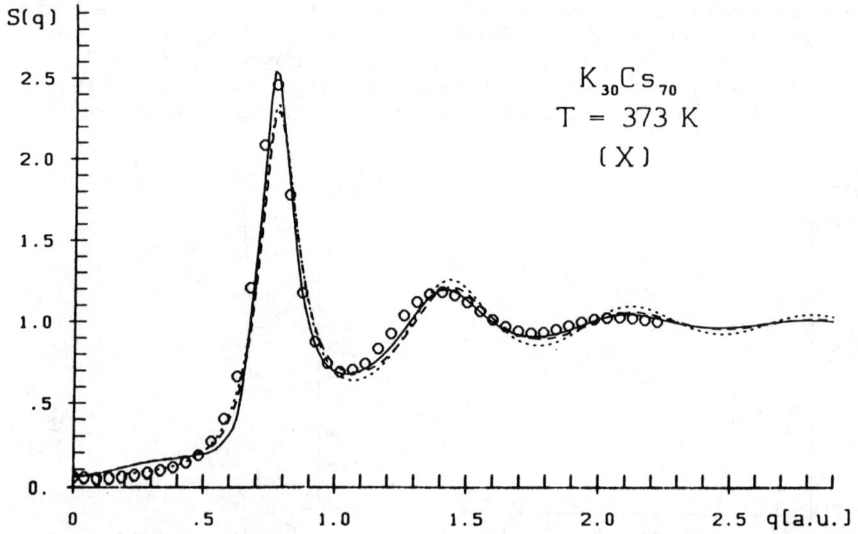

Fig. 3:Hard-sphere(••••), WCA- (--) and WCA+ORPA- (—) results for $K_{30}Cs_{70}$ in comparison with experimental results (o) from /14/.

This work was supported by the Österreichische Forschungsgemeinschaft under Proj. Nr. 09/0005.

5. REFERENCES

1. J.D.Weeks, D.Chandler and H.C.Andersen:J.Chem.Phys.55,1497(1976)
2. H.C.Andersen, D.Chandler and J.D.Weeks:Adv.Chem.Phys.34,105(1976)
3. G.Kahl and J.Hafner:Phys.Rev. A29,3310(1984)
4. G.Kahl and J.Hafner:Sol.State Commun.49,1125(1984)
5. G.Kahl and J.Hafner:Z.Phys.B58,283(1985)
6. R.Oberle and H.Beck:Sol.State Commun.32,959(1979)
7. C.Regnaut, J.P.Badiali and M.Dupont :Phys.Lett.74A,245(1979)
8. C.Regnaut and J.L.Bretonnet:J.Phys.F14,L85(1984)
9. J.L.Bretonnet and C.Regnaut:Phys.Rev.B31,5071(1985)
10. H.C.Andersen, D.Chandler and J.D.Weeks:Phys.Rev.A4,1597(1971)
11. S.I.Sandler:Chem.Phys.Lett.33,351(1975)
12. L.L.Lee and D.Levesque:Mol.Phys.26,1351(1973)
13. G.Kahl and J.Hafner:J.Phys.F15,1627(1985)
14. B.P.Alblas,W.van der Lugt,O.Mensies and C.van Dijk:Physica B106,22 (1981)
15. R.E.Jacobs and H.C.Andersen:Chem.Phys.10,73(1975)

GLASS TRANSITION IN SIMPLE MOLTEN SALTS

UDO KRIEGER AND JÜRGEN BOSSE
FREIE UNIVERSITÄT BERLIN, D-1000 BERLIN 33, GERMANY

The following discussion of the liquid-glass transition in the symmetric molten salts (SMS) presents the first application of the theory of Leutheußer [1] and Bengtzelius, Götze and Sjölander [2] to a two-component fluid; thus a step towards more realistic glass formers.

In the SMS, due to charge inversion symmetry, mass-density (μ=+1) and charge-density (μ=-1) relaxation functions do not couple directly. However, friction kernels $K_\mu(z)$ are determined by mode-decay processes thus resulting in a coupling of $\phi_\pm(z)$

$$\phi_\mu(z) = -1/(z-\Omega_\mu^2/(z+K_\mu(z))) \tag{1}$$

(we refer the reader to Ref. [3] for details). Although to our knowledge it has not been possible to produce a glass by quenching a simple molten salt, computer simulations [4] of the SMS with plasma parameter Γ=64 as well as mode-coupling theory [5] for the same system show typical precursor effects of the glass transition (Fig. 1).

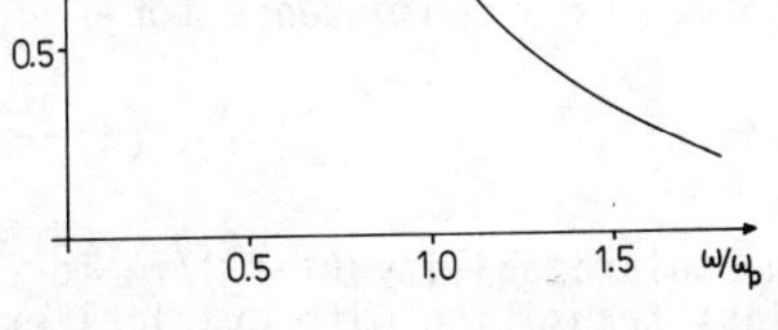

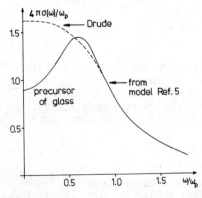

Fig. 1: Ac-conductivity of the SMS in the liquid phase (Γ=64). Note the strong decrease of $\sigma(\omega)$ at small frequencies.

Motivated through mode-coupling theory we studied the following simplified model for the friction-kernels:

$$\begin{aligned} K_+(z) &= \lambda\Omega_+^2 \; i \int_0^\infty dt \; e^{itz} \; (\alpha\,\phi_+^2(t) + (1-\alpha)\,\phi_-^2(t)) + K_+^{rest}(z) \;, \\ K_-(z) &= \lambda r \Omega_-^2 \; i \int_0^\infty dt \; e^{itz} \; \phi_+(t)\,\phi_-(t) + K_-^{rest}(z) \;. \end{aligned} \tag{2}$$

Here $K_\mu^{rest}(z)$ denotes a regular contribution to the kernels which we set equal to $\pm i\gamma_\mu$ in the following. We consider $0\leq\alpha<1$ and $0<r<\infty$ fixed, while $0\leq\lambda\leq\infty$ denotes the interaction parameter driving the transition.

Eqs. (1) and (2) represent a closed set of nonlinear integral equations for the determination of $\phi_\mu(t)$. Fig. 2 shows numerical results for $\alpha=0.5$, $r=2.5$, $K_\mu^{rest}(z)=i\Omega_+$ and $\Omega_-=0.4\ \Omega_+$.

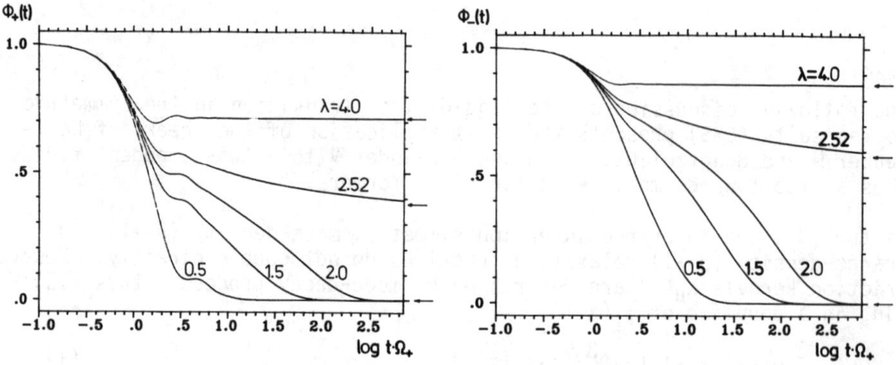

Fig. 2: Mass- and charge-density relaxation functions of SMS for various interaction parameters.

For $\lambda>\bar{\lambda}=2.52$ the relaxation functions do not vanish for $t\to\infty$ but relax to $f_\mu=\phi_\mu(t=\infty)\neq 0$, in other words the model has formed a glass at the transition point $\bar{\lambda}$. Inserting Eq. (2) into Eq. (1) leads to

$$\frac{f_+}{1-f_+} = \lambda(\alpha f_+^2 + (1-\alpha)f_-^2) \ ; \ \frac{f_-}{1-f_-} = \lambda r f_+ f_- \tag{3}$$

for f_μ. The solutions of Eq. (3) follow from the roots of a polynomial of degree five. There is only one physically acceptable solution $f_\mu(\lambda)$, which is zero for $\lambda<\bar{\lambda}$ and $0<f_\mu\leq 1$ for $\lambda\geq\bar{\lambda}$. (Some values of $\bar{\lambda}$ and $\bar{f}_\mu:=f_\mu(\bar{\lambda})$ for different α are shown in table 1.) Two complex conjugate roots (for $\lambda<\bar{\lambda}$) develop at $\lambda=\bar{\lambda}$ into a pair of real roots. As a consequence for $\lambda=\bar{\lambda}(1+\varepsilon)$, $\varepsilon>0$, we have

$$f_\mu = \bar{f}_\mu + c_\mu\ \varepsilon^{1/2} + 0(\varepsilon) \text{ with } c_- = c_+/(\lambda r \bar{f}_+^2), \tag{4}$$

showing the Debeye-Waller factor f_+ and the polarizability $(1-f_-)/f_-$ to approach their asymptotic limits at the glass transition with critical exponent 1/2. The same is true for Lamb-Mößbauer factor, localization length of a tagged particle and elastic moduli[6)].

The dynamical critical exponent β (table 1) differs slightly from that of the corresponding one-component-fluid model (β=1.266) and changes with the choice of the parameter α. The critical dynamics is described by one scaling function $\hat{g}(\zeta) = i\int_0^\infty dt\ e^{i t \zeta} g(t)$ for both mass- and charge-density relaxation obeying:

$$-a\frac{\sigma}{\zeta} + 2b\zeta\hat{g}^2(\zeta) + ci\int_0^\infty d\tau\, e^{i\tau\zeta}\, g^2(\tau) = 0 , \qquad (5)$$

where $\phi_\mu(t) = \bar{f}_\mu + \varepsilon^{1/2} g_\mu^{(\varepsilon)}(\varepsilon^\beta t)$,

$\hat{g}_-^{(0)}(\zeta) = \hat{g}(\zeta)$, $\hat{g}_+^{(0)}(\zeta) = \bar{f}_+/(1-\bar{f}_-)\,\hat{g}(\zeta)$, $\zeta = z/\varepsilon^\beta$,

$\sigma = +(-)$ glass (fluid) and a,b,c are constants.

The exponent β is determined from

$$2b/c = \Gamma(1-2x)/\Gamma^2(1-x) \text{ and } \beta = 1/2x,$$

solutions of which are given in table 1 for various parameter values.

α	λ	$\bar{f}_+$	$\bar{f}_-$	2b/c	x	β
0.00	2.216	0.398	0.546	1.9985	0.395	1.266
0.25	2.335	0.389	0.559	1.9883	0.394	1.269
0.50	2.520	0.379	0.580	1.9509	0.392	1.276
0.75	2.914	0.372	0.631	1.8296	0.381	1.312
0.95	3.698	0.451	0.760	1.766	0.375	1.332

TABLE 1

Note that our model (eqs. (2)) presents a special case of the general problem discussed by Götze[7]. Our solutions of the present model are consistent with the general results derived in Ref. 7.

References

1) E. Leutheußer, Phys.Rev. A29, 2765 (1984); Z.Phys. B55, 235 (1984)
2) U. Bengtzelius, W. Götze, A. Sjölander, J.Phys. C17, 5915 (1984)
3) J. Bosse, T. Munakata, Phys.Rev. A25, 2763 (1982)
4) J.P. Hansen, J.R.McDonald, Phys.Rev. A11, 2111 (1975)
5) J. Bosse, T. Munakata, Phys.Rev. A24, 2261 (1981)
6) W. Götze, this conference proceedings.
7) W. Götze, preprint (1985), submitted to Z. Phys.

LOCAL ORDER AT THE MANGANESE SITES IN IONIC SOLUTIONS BY XANES (X-ray absorption near edge structure)

M. BENFATTO, C.R. NATOLI
INFN Laboratori Nazionali di Frascati, 00044 Frascati, Italy
J. GARCIA[x], A. MARCELLI, A. BIANCONI
Dipartimento di Fisica, Università "La Sapienza", 00185 Roma, Italy,
I. DAVOLI
Dipartimento di Matematica e Fisica, Università di Camerino, 62032 Camerino, Italy

1. INTRODUCTION

The idea that important geometrical informations about local structure are contained in XANES spectra has stimulated theoretical and experimental studies in this field(1,4). The low energy part of the x-ray absorption spectra, in fact, is sensitive to the geometrical arrangement of the environment around the absorbing atom because of the strong scattering power of these low energy photoelectrons, favoring multiple scattering (MS) processes. Going to the high energy region of the spectra a gradual turn-over occurs from the MS regime to the single scattering (SS) regime where the modulation in the absorbtion coefficient (EXAFS) is substantially due to the interference effect of the outgoing photoelectron wave from the absorbing atom and the backscattered wave from each surrounding atom (5,6). Hence, while this part of the spectrum provides information about the pair correlation function, the first one contains information about higher order distribution functions by MS pathways which begin and end at the absorbing atom.

By making the expansion of the total cross section in the energy region where it is possible, it has been shown(7) that the contributions of consecutive scattering orders can be identified. To go deeper into the question, we report on the expressions for the first three contributions of scattering orders to the totel cross section valid for the K-shell.

In the framework of the multiple scattering theory we can write for the absorption coefficient

$$\mu_c = A \cdot \hbar \cdot \omega \cdot M_{01}{}^2 \cdot \alpha^{(1)} \qquad (1)$$

where M_{01} is the radial atomic dipole element which varies smoothly with energy whereas the quantity $\alpha^{(1)}$ contains all the geometrical informations around the absorbing atom. It can be expanded into the contributions due to successive orders of scattering pathways as $\alpha^{(1)}=1+\Sigma_{n-2}X_n{}^{(1)}$. This latter quantities are classified according to the number of scattering events n, therefore the corresponding number of neighbour atoms participating in the scattering processes are n-1. The following expression can be obtained for $X_2{}^{(1)}$:

$$X_2{}^{(1)}=\Sigma_j \mathrm{Im}\{f_j(k,\pi)\exp(2i(\delta_1+kr_j))/kr_j{}^2\}, \qquad (2)$$

where δ_1 is the partial phase shift of the absorbing atom, r_j is the distance between the central atom and the neighboring atom j, k is the photoelectron wave number and $f_j(k,\pi)$ is the backscattering amplitude.

This is the usual EXAFS term. It contains only information about the pair distribution function. The first MS contribution is the $X_3^{(1)}$ term which can be written as:

$$X_3^{(1)}=\Sigma_{i=j}\mathrm{Im}\{P_l(\cos\Phi)f_i(\omega)f_j(\Theta)\exp(2i(\delta_l+k(r_i+r_{ij}+r_j)))/kr_ir_{ij}r_j\}, \quad (3)$$

where r_{ij} is the distance between atoms i and j, $f_i(\omega)$ and $f_j(\Theta)$ are the scattering amplitudes which now depend on the angles through the Legrende Polynomials $P_l(x)$. They are expressed in the triangle which joins the absorbing atom to the ones i and j. In this expression $\cos\Phi = -\hat{r}_i\cdot\hat{r}_j$, $\cos\omega = -\hat{r}_i\cdot\hat{r}_{ij}$, $\cos\Theta = \hat{r}_j\cdot\hat{r}_{ij}$. All these quantities are obtained neglecting the angular dependence of the Hankel function in the free propagator. Here, we report on a study of the site geometry of manganese ions by XANES.

2. EXPERIMENT AND CALCULATION

The experiment has been performed at the Frascati Synchrotron Radiation Facility. The aqueous solutions at concentrations of 50mM have been prepared and a 1 mm thick cell has been used for the transmission experiments. In the data analysis the pre-edge absorption back-ground is subtracted. The MS calculation for the octahedral and tetrahedral clusters formed by a central Mn and neighbouring oxygens are made with the usual X-α potential and Mattheiss prescription. Hydrogen atoms were neglected.

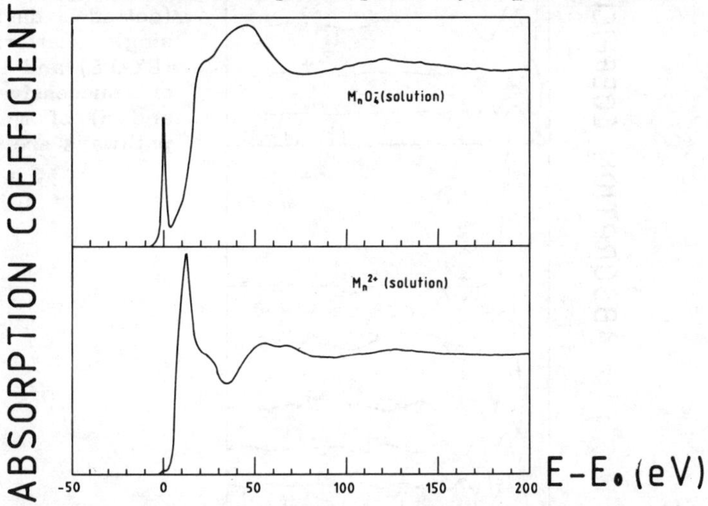

FIGURE 1. X-ray absorption spectra of MnO_4^- (upper panel) and Mn^{2+} (lower panel in acqueous solution.The zero of the energy scale is fixed at the 1s-3d excitation at threshold.

3. RESULTS AND DISCUSSION

In Fig.1 the X-ray absorption spectrum of Mn ions in solution are reported. The comparison between both configurations shows the different shape in the energy region between 0 and 160 eV due to the different geometrical structure around manganese atoms. Using the multiple scattering theory, briefly described above, we have obtained the theoretical spectra for the two clusters. Reported results are reproduced in Fig.2. The agreement with the experimental data is good with respect to the shape of the spectrum but it is less good concerning the location of the high energy maxima due to the energy independence of the X-α potential used. We also reporte on the breakdown of the total cross section in terms of the partial contributions X_n classified according to the number of scattering events n. In the octahedral geometry we observe that the MS contributions extend up to 150eV but, because of the destructive interference effect between the n=3 and n=4, the EXAFS signal becomes the main contribution above 40 eV. On the contrary, the expansion of the total cross section in tetrahedral geometry shows that the sum $1+X_2+X_3$ is enough to get good agreement between 50-140 eV and in this range X_n (n≥4) is negligible. For this reason, we have extracted from the MnO_4^- solution data the experimental X_3 contribution by substracting from the total experimental signal, normalized EXAFS-like, the n=2 theoretical

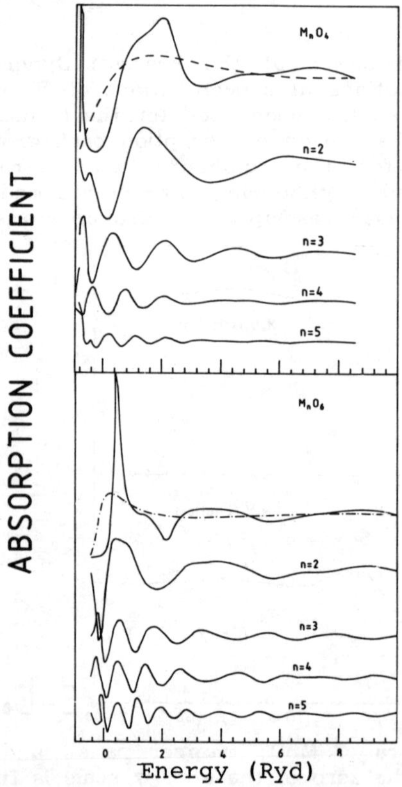

FIGURE 2. Theoretical absorption coefficient for the Mn atom in tetrahedral (upper panel) coordination. In each panel going from the top to bottom the total cross section, the atomic contribution (dotted- dashed line), the single scattering contribution (EXAFS) and the contributions of successive order (n=3,4 and 5) of multiple scattering pathways are shown.

contribution (see Fig. 3). Good agreement is obtained between the calculated X_3 and the experimental one. Fourier transforms of this two signals shows the same peak centered at 2.4 Å. This value coincides with the optical path length for the third order MS pathway for this cluster.

In conclusion we have shown that XANES data analysis has reached the level of quantitative agreement with experimental spectra. We also discussed the way to obtain higher order correlation functions.

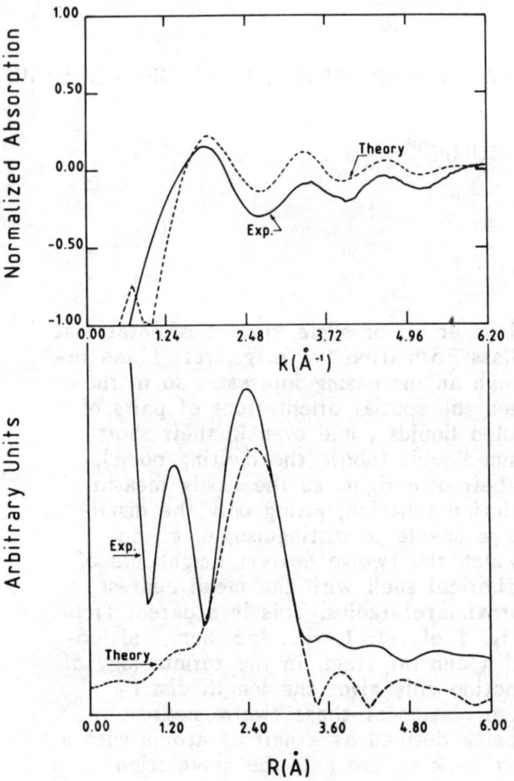

FIGURE 3. Upper part: comparison between the theoretical X_3 signal (dashed line) and the X_3 contribution (full line) obtained experimentally for the MnO_4^- cluster. The threshold energy E_0 is fixed at the first absorption peak. Lower part: Fourier transforms of the theoretical (dashed line) and experimental (full line) X_3 signals.

REFERENCES

1. A. Bianconi, Appl.Surface Science 6, 392 (1980)
2. F.W. Kutzler, C.R. Natoli, D.K. Misemer, S. Doniach, H.O. Hodgson, J.Chem.Phys. 73, 3274 (1980)
3. H.O. Hodgson, B. Hedman, J.E. Penner-Haln eds. "EXAFS and Near Edge Structure III", Springer Proc.Phys. Vol.2 (1984)
4. A. Bianconi, L. Incoccia, S. Stipchich eds. "EXAFS and Near Edge Structure", Springer Series in Chem.Phys. Vol. 27 (1983)
5. P.A. Lee, J.B. Pendry, Phys.Rev. B11, 2795 (1975)
6. D.E. Sayers, E.A. Stern, F.W. Lytle, Phys.Rev.Lett. 27, 1204 (1971)
7. M. Benfatto, C.R. Natoli, A. Bianconi, J.Garcia, A.Marcelli, I.Davoli Phys.Rev.B, to be published and Frascati L.N.F. Report (1985)
8. W.L. Schaich, Phys.Rev. B29, 6513 (1984)
9. G. Bunker, E.A. Stern, Phys.Rev.Lett. 52, 1990 (1984)

"BOND"-ORIENTATIONAL SHORT RANGE ORDER IN LIQUID ALUMINIUM

H. HAHN,* M. MATZKE,** U. ZUMPE*

1. INTRODUCTION

With the suggestion of D. Nelson et al. of some kind of orientational long range order underlying Glass Formation (v., e. g., ref. 1 and references therein), there has been an increasing interest also in the finite range correlation between the spatial orientations of pairs of nearest neighbours in undercooled liquids , and even in their short-range correlations in equilibrium liquids (above the melting point). The latter are interesting in their own right, as the easily measurable and calculable pair correlation function, giving only the distribution of atom pair distances, is unable to distinguish, e. g., between the different ways in which the twelve nearest neighbours of an atom are distributed over a spherical shell with the mean nearest neighbour distance as its (approximate) radius. This is apparent from the three examples given in Fig. 1 of ref. 1, viz., fcc, hcp, and icosahedral close packings around a central atom. In the terminology of ref.1, the pair distribution function only gives the length distribution, not the orientational correlation of these twelve nearest neighbour "bonds", a "bond" being defined as a pair of atoms with a distance falling within the first peak of the pair the correlation function.

Of the distance distribution for pairs of atoms, the closest analoga for pairs of bonds are two particular correlation functions which could in principle be calculated from the three- and four atom correlation functions: the distribution of the distances of the bond centres, and the distribution of the "angular distances " (i. e., the angles between two "bond" directions) for all pairs of bonds with a given distance between their centres. Since the direction of a dumbbell formed by two atoms is only defined up to a sign, it suffices to consider the distribution of the cosines of those relative angles, $\cos\vartheta$.

2. WHAT WAS CALCULATED FROM MOLECULAR DYNAMICS

Our figure 1 gives, as a function of the distance between bond centres, the values of the l=4 and l=6 coefficients of an expansion in terms of Legendre Polynomials, P_l $(\cos\vartheta)$, of the abovementioned distribution of cosines, for a Molecular Dynamics simulation of a sample of 4000 aluminium atoms in a cubic periodicity volume of (42,32 Å)^3, corresponding to the number density of molten aluminium at 984 K under atmospheric pressure. (The melting point is at 933 K). To model the interaction between atoms, the same Duesbery-Taylor local pseudopotential was taken as had been used in earlier simulations with a smaller

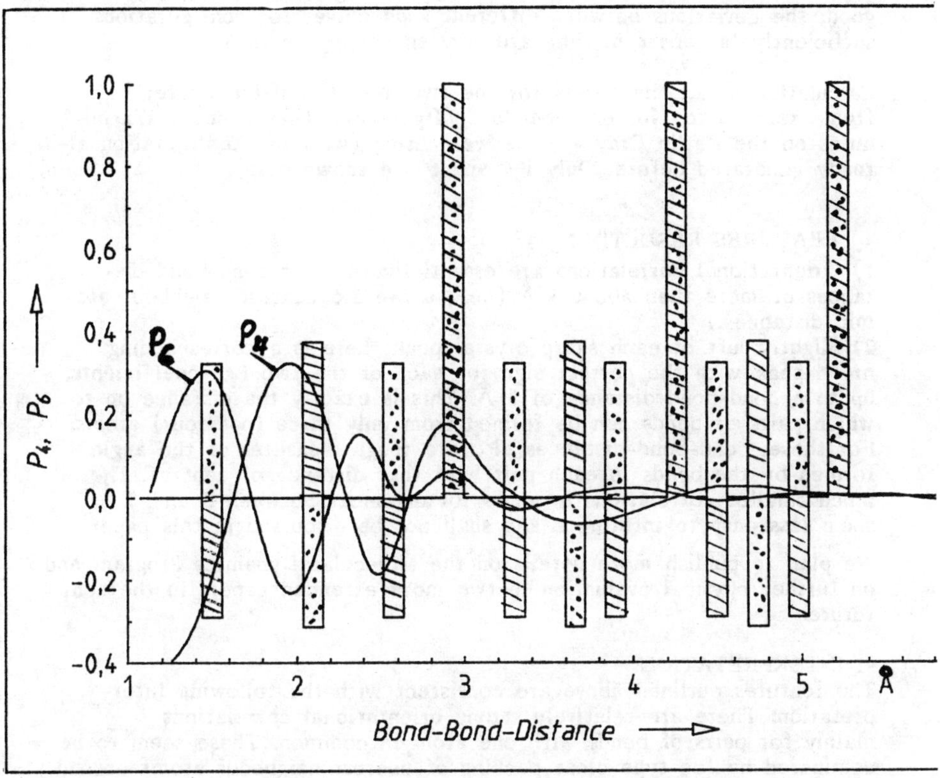

Fig. 1. Legendre Polynomial expansion coefficients for l=4 and l=6, as a function of bond-bond distance, for the bond-bond angle distribution of all bond pairs with their centres having distances in one of successive 0.12 Ångstrom - wide intervals. Ends of dotted bars mark values of P_6 for those intervals containing a bond -bond distance occuring in fcc lattice of same density ; hatched bars: same for P_4. (For the crystal, these values are identical with the coefficients, v. text).

number of particles, cf. references 2 and 3. The nearest neighbour distance in a simple cubic latice of the same density would be 3 Å, the peak of the pair correlation function lies at 2.85 Å. All pairs of atoms with a distance less than 3.2 Å were considered as forming a "bond".

For any given number of bonds, the distribution of cosϑ is just a sum of δ-functions at the occurring cosϑ values, divided by the number of bond pairs. Thus, the requested expansion coefficients are just the averages of the values the corresponding Legendre polynomial takes on the set of bond pairs considered (v. figure caption). Already with a single configuration of 4000 particles, the number of bond pairs is so enormous that statistics on the curves in fig. 1 are extremely good; the deviations between different such curves for configurations sufficiently far apart in time are correspondingly small.

Calculation of all the curves for the five even l coefficients for l from zero to ten for one sample configuration takes about 2 1/2 minutes on the Berlin Cray - 1 we were using (with the configuration already generated before. Only l=4 and 6 are shown here).

3. FEATURES RESULTING

1) Orientational correlations are essentially zero for bond-bond distances of more than about 5 Å (i.e.., above 1.6 nearest neighbour atomic distances.)
2) Slightly left of each sharp crystal peak there is a corresponding broad peak with the correct sign for each of the two P_ℓ coefficients, up to a bond-bond distance of 3 Å. This is exactly the distance up to which pairs of bonds can be formed from only three (not four) atoms. For these, bond-bond-distances also are rough measures of the angle formed by the bonds in each pair with that distance of centres. The much smaller correlations in angle for distances between 3 and 5 Å seem less easy to interprete and shall not be discussed in this paper.

We plan to publish more details on the Molecular Dynamics Program and on further physical evaluation in two more extended papers in the near future.

4. INTERPRETATION

The features outlined above are consistent with the following Interpretation: There are relatively strong orientational correlations mainly for pairs of bonds with one atom in common. These seem to be dominated by fcc-type close-packing of nearest neighbour atoms around a central atom. The clear and pronounced second peak at 2 Angstroms, corresponding to a 90° angle between three-atom pairs, should be absent in an icosahedral close-packing of "one-plus twelve" close-packed clusters. This precludes at least any dominant contribution of such clusters. An approximate calculation of the distribution function for the angle from a superposition of the Legendre polynomial constributions up to l=10 for this particular bondbond distance does give just one peak with a maximum at 90 degrees of bond-bond angle, thus corroborating the interpretation given from the distance dependence.

5. REFERENCES

1) STEINHARDT, P. J., NELSON, D. R., RONCHETTI, M., Phys. Rev. B 28, 784 (1983)
2) EBBSJÖ, I., KINELL, T., WALLER, I., J. Phys. C 13, 1865 (1980)
3) HAHN, H., MATZKE, M.,PTB - Bericht FMRB - 105, (ISSN 0341-6666) Physikalisch-Technische Bundesanstalt Braunschweig (January 1984)

6. ACKNOWLEDGEMENTS

We thank the Konrad Zuse Zentrum, Berlin, for the opportunity to perform the calculations on the CRAY-1 at Berlin.

*) Technische Universität Braunschweig, **) PTB, BRAUNSCHWEIG, FRG

3. STRUCTURE OF AMORPHOUS MATERIALS

ATOMIC SHORT RANGE ORDER IN AMORPHOUS MATERIALS

S. METHFESSEL

Ruhr-Universität, D-4630 Bochum, FRG

ABSTRACT:

1. INTRODUCTION

In the last NATO-ASI on "Liquid and Amorphous Metals" 1979 in Zwiesel my paper on "Ferromagnetism in Amorphous Metals came to the conclusion: "The similarity of intrinsic magnetic properties in the amorphous and crystalline state indicates, that the situation in the first coordination shell is more important than long range order. The directional d-d bonds and the screening by attached metalloid atoms stabilize the atomic structure in the first coordination shell so strongly that it survives melting and the usual preparation methods for amorphous materials. The most probable fundamental cluster is a closely packed tetrahedron of transition atoms." (1) The Bernal-Model and Radial Distribution Functions were then very fashionable for describing the structure of amorphous metals.

Five years later, in the 5th International Conference on "Rapidly Quenched Metals" 1984 in Würzburg, the concept of atomic short range order was much more accepted and discussed in many papers. One finds often statements such as that of Hafner: "The existence of chemical short range order (CSRO) in many amorphous alloys can now be considered as firmly established by diffraction experiments. The evidence for the existence of a certain degree of topological short range order

TSRO) is admittedly less direct, and in many cases the particular type of TSRO present in a given amorphous alloy can only be guessed by reference with the characteristic building blocks of the corresponding crystalline structures." (2), or by Sommer: "The CRSO, therefore, represents a necessary prerequisite for obtaining glassy alloys by rapid quenching from the molten state with attainable cooling rates" (3), or by Ellner et al.: "The similarities in the behaviour of crystalline and non-crystalline solids agree with the assumption that the structural unit has the same importance for the non-crystalline solid as the unit cell has for the crystalline solid. From a chemical point of view the stability of a structural unit results from conditions similar to those for the crystalline phases (valence electron concentration, atomic size, chemical affinity)." (4)

Now the problem remains to be solved how the structure of such "structural units" or atomic coordinations, which stabilize the glassy state against crystallization, can be described theoretically and determined experimentally. The big handicap lies here in the lack of transitional symmetry by definition. A glass is defined as a solid (i.e. the viscosity is above 10^{13} poise) without any translational symmetry which can be detected by interference methods: it is "a-morphous". Diffraction of waves in a glass gives no sharp spots at definite angles as one obtains from crystalline materials, but broad "Radial Distribution Functions" only and all information about the local Short Range Order is averaged out. However, many physical properties of metallic glasses, such as EXAFS (Extended X-ray Absorption Fine Structure), Hyperfine Fields in Mössbauer or NMR-measurements as well as magnetic properties, which depend in alloys mainly on the next neighbor coordination of the magnetic atoms, show very often a remarkable similarity in glasses and crystals of the same chemical composition. Then the importance of the atomic coordinations seems to become obvious, but we have still the problem to define in a very large array of many densely packed atoms the right size and symmetry of the "structural units", which are mainly responsible for the properties and stabilities of the glassy or crystalline state.

In general, one speaks about a "Topological Short Range Order" (TSRO) when only the size of the atoms is considered when packing them together as densely as possible. However, the "Chemical Short Range Order" (CSRO) is aware of the different chemical character of different atoms and brings into consideration chemical bonds and can, therefore, be related to the stability of the glass up to a certain temperature and to the "Glass Forming Ability" (GFA), that means how fast the melt must be quenched in order to remain in the metastable glassy state.

There are many modern books and survey articles available now about glasses in general (e.g. 5), glassy metals (e.g. 6 and 7), magnetic glasses (e.g. 8) and a digest of literature for specific problems (9). Therefore, this short lecture shall concentrate on some general aspects of atomic arrangements in glasses and their significance to the stability of the glassy state. We will find, that the conditions for the formation of metallic and oxide glasses are qualitatively less different than one would expect from the different character of their chemical bonding.

2. TOPOLOGY OF DENSELY PACKED ATOMS

In this approach all atoms shall have equal size and chemical properties such as in a simple metal and we consider only size effects. At very high temperatures there is enough "free volume" in the melt to allow a nearly "Random Distribution" of pointlike atoms. Approaching the point of solidification the melt becomes so dense that atoms cannot move freely anymore, but have to fill space as effectively as possible. The demand of space filling dictates the Topological Short Range Order which one can describe as a "Dense Random Packing of Hard Spheres" (DRPHS).

Statistical geometry describes TSRO by Voronoi polyhedra which contain only one atom in the center and are formed by bisecting planes perpendicular to the vectors running to the next neighbor atoms. The analogy to the Wigner cells in crystals is obvious, but the Voronoi cells in glasses have, of course, irregularly shaped planes, often with 5 corners. They form a space filling network, their shape reflects the TSRO and their average size the density of the glass (10,11).

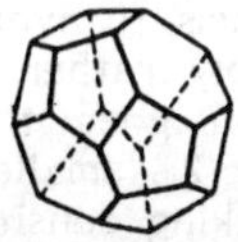

Another way to describe TSRO is concerned with the "Coordination Polyhedra", which are complementary to the Voronoi cells as the unit cells of crystals are to Wigner cells and appeal more to experimentalists for visualizing atomic arrangements.

2.1. The Tetrahedra Problem

Which coordination polyhedra provide the densest packing of atoms of equal size? In a plane of 2 dimensions the densest unit is the triangle. Since we can fill the plane without any gaps by using triangles only and hexagons as Voronoi polyhedra the situation is here much simpler than in 3 dimensions. In space the tetrahedron is the densest unit. It has a packing density, i.e. relative part of volume filled by matter, of about 78% compared to 72% for octahedral units. Serious problems arise now for the DRPHS from the fact that tetrahedra can be packed continously only in 4 but not in 3 dimensions. When one starts, for instance, to pack 5 tetrahedra densely together around a common edge, a gap is left with an angle of 7,5°. In general, there are no regular polyhedra besides the cube which can fill 3 dimensional space alone; one always needs at least two kinds of polyhedra packed together in the right proportions and sequence to reach optimal overall density. Only the space group of the crystal knows this recipe for space filling, but it is neglected by the melt.

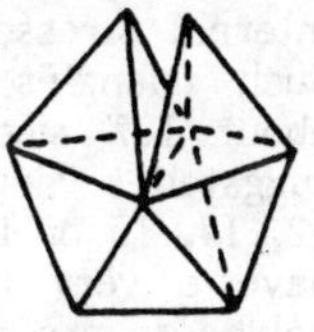

Close packed crystals of simple structure, such as f.c.c. ot h.c.p., use for space filling 1 octahedron per 2 tetrahedra and reach in this way an average packing density of 74%. Therefore, the local density in the tetrahedra is higher than the average density of the close-packed crystal, but cannot be maintained for steric reasons when a larger volume has to be filled. However, the melt has at higher temperatures not such rigid steric restrictions and can form any number of tetrahedral configurations. As soon as the melt solidifies, suddenly a large number of new complementary polyhedra such as octahedra have to be formed and packed together in the right fashion. This needs time and leads to steric frustrations and undercooling.

2.2. The Bernal Model for DRPHS

It is difficult to understand by theories how the melt manages to fill space nearly as dense as the crystal, but without using long range order. Therefor, Bernal (13) made a rather simple, but famous experiment. He packed 3000 steel balls into a balloon, kneaded them to an irreglular shape of maximum density. His ingenious technician, J. Mason, poured then paint into the balloon which coagulated in all gaps between balls, which were not wider than 5% of the diameter of the balls. Counting the paint dots gave the following results:

* The distribution of coordinations of touching or almost touching balls has a maximum at about 8 neighbor atoms.
* The canonical or "Bernal Polyhedra" formed by the disordered balls are to 73% tetrahedra, 20% octahedra and a small rest of trigonal prism (3%), tetragonal octahedra (3%), Archimedian antiprism (1%).
* The Voronoi polyhedra have an average of 13.6 faces with 5 corners each.

Cohen and Turnbull (14) suggested to apply the Bernal model also to solid glasses and it became here a leading concept because the Radial Distribution Functions, which were computer simulated in this model, were in good agreement with experimental observations (15). However, the Bernal model can have limited validity only for most metal glasses, which must contain more than one kind of atoms for stability reasons. Whittaker (17) reexamined the Bernal model and found even 86% tetrahedra and the rest larger holes surrounded by 8 to 10 neighbors. Polk (18) suggested that the stabilisation of the glass by foreign atoms results from "stuffing" them into the nearly 20% of large Bernal holes. The reality is probably more complicated because chemical interactions between the glassforming and -stabilizing atoms must be important. The observed tendency of the stabilizing atoms to avoid each other and surround themselves mainly with metal atoms is probably rather a chemical than a topological size effect only. In the same direction goes the observation, that the density measured of metal glasses is usually only 2% smaller than that of comparable crystals, while Bernal's DRPHS predicts a packing density of 63% , that means reduction of about 15%.

2.3. Close-Packed Tetrahedral Crystals

In f.c.c. crystals the tetrahedra are packed around the octahedra in such a way that adjacent tertahedra have common edges and vertices, but never common faces. In h.c.p. crystals the tetrahedra have common vertices and share also faces, but never with more than one other tetrahedron. In order to reach a higher local packing density, more tetrahedra have to share faces with one another, e.g. by piling them together along a spiral or when 12 atoms are coordinated around a central atom in form of an icosahedron. Unfortunately, such units cannot be packed continously in 3 dimensions and produce finally a lower overall density and lage internal stresses. The surface of such icosahedron consists of 20 regular triangles. Such triangles can form regular tetrahedra with the central atom, when this is about 10% smaller than the atoms in the shell (19). Frank and Kasper (20, 21) suggsted 4 "normal coordination polyhedra" or "Kasper Polyhedra" which coordinate 12, 14, 15 or 16 atoms of different size. They consist exclusively of tetrahedra and have a very high packing density. In 3 dimensional crystal lattices the Kasper polyhedra are arranged in "skeletons" and, in general, only layered structures must result. The Frank-Kasper phases can be described by a sequence of layers, which are tesselated by hexagons, pentagons and triangles and then stacked together in such a way that only tetrahedral interstices form between the planes.

The most abundant form of Frank Kasper phases in transition metal compounds are the Laves phases with the composition AB_2 and the structure types $MgCu_2$:C15, $MgZn_2$:C14 and MgN_2:C36. The ideal structure forms, when the A-atoms are 1.225 times larger than the B-atoms, which are imbedded in form of tetrahedra in a tetrahedral skeleton of A-atoms. We find then only B-B and A-A contacts, but no A-B contacts; the crystal seems to be separated into two interpenetrating lattices of A and B atoms, respectively. The B-B distances are often smaller than the interatomic distances in the elements, but the A-A distances are usually much larger. This leads to the conclusion (22, 23), that the small B-tetrahedra dominate and stabilize the structure. Laves and Witte (24) replaced Cu by Zn in $MgCu_2$ and found, that the form of the skeleton, in which the B-tetrahedra are packed, depends on the electron/atom ratio N. The structure changes from C15 to C36 at N=1.8 and to C14 when N approaches 2.0. This suggests that the skeleton of longe range order is stabilized by the conduction electrons in the band states formed by A-A overlapping, while the short range B-tetrahedra are hold together by directional d or p bonds.

The large flexibility of tetrahedra packing allows to form many more Frank Kasper by stacking the layers in different ways. Examples are A15, Delta, Mu, Nu, Sigma, Chi, M, P, R phases and many modifications of them as described in detail by (21). Many of such crystals have in their electronic structure sharp and high peaks of the density of states, what leads often to unusual properties, such as superconductivity at relatively high temperatures and magnetic fields, ideal itinerant magnetism etc. It was pointed out (25), that very often glasses form, when such close-packed tetrahedral crystals are destabilized by adding extra A-atoms.

Instead of stacking Frank Kasper polyhedra together in alternating layers, Schubert (26) uses the "Tetraederstern" of the A15:Cr_3Si structure as the "Building Block". All close-packed tetrahedra phases can be derived by symmetry operations such as translations, rotations, reflections and intergrowth (27).

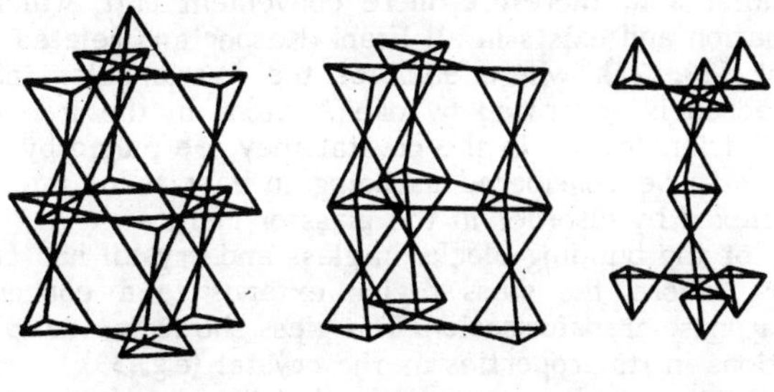

Arrangement of B-Tetrahedra in Laves Phases.

3. CHEMICAL SHORT RANGE ORDER

Solidification is driven by attracting interatomic forces. The resulting atomic configurations must minimize the energy of the valence electrons as well as satisfy steric conditions, because everything must fit into the Point Symmetry of one of the 14 Bravais lattices. The stability of the configurations depends then on the electron orbitals, which are available for hybridization into new orbitals having the right point symmetry. The van der Waals forces have purely radial character and can stabilize any close packed configuration. The metallic bond by nearly free conduction electrons is also rather unspecific about symmetry, but the "tightly bond" d-electrons in transition metals are very much restricted by the character of the local atomic orbitals they come from. The d-electrons with t_{2g} symmetry prefer tetrahedral configurations as in f.c.c. and h.c.p. structures, but the one with eg stabilize the less densely packed cubes of the b.c.c. structure. Another very famous tetrahedral bond is the sp^3 electron configuration which stabilizes diamond lattices. The ionic forces have again radial character, but ions of equal charge must be kept as far apart as possible.

Solid state physicists like to describe crystals as infinitely large entities where the bonding electrons interact as waves in reciprocal space with all atoms simultaneously. The crystallographers, when they have to design very complicated structures, believe that nature starts a crystal by forming small and rather simple building blocks, which represent the relevant chemical bonds. Then the crystal is built by joining together the building blocks by crystallographic operations, such as translation, rotation, reflection and intergrowth. The large family of Frank-Kasper-Phases is a typical example for this approach (27). Such building blocks are not identical with, but parts of the crystalline unit cell, which is much larger and guarantees agreement with one of the 230 space groups, so that later infinitely large crystals can be built by translations only.

Since the form of the atomic units depends so critically on the concentration, size and electron configuration of the involved atoms, it seems reasonable to assume, that the building blocks in a given alloy are basically the same in the crystalline, glassy and molten state (28). Perhaps, they may be somewhat distorted by the Madelung forces in the crystal or by the entropy of the melt. The amount of such distortions depends, of course, on the relative strength of the chemical bonds, which is small for Van der Waals forces and nearly free conduction electrons but much larger for localized d-electrons, covalent p-electrons and ionic bonds.

In the Laves phases AB_2, as an example, the 12fold coordinated Kasper icosahedron around the smaller B-atom can be used as a building block. This unit contains two regular B-tetrahedra and 2 triangles of larger A-atoms at its surface and includes all important information about B-B as well as A-A and A-B bonds. It is similar to the icosahedral "Frank units" and icosahedral chains, which are supposed to occur in eutectic melts (44).

An even smaller and, therefore, more convenient unit, which contains the same chemical information and exists in all Frank-Kasper and related phases, is Schubert's "Tetraederstern" (26, 27), where each of the 4 triangular faces of the strongly bound B-tetrahedron is decorated by one A atom. In the free cluster the A atoms may touch the B triangle, but in the crystal they are pulled by the A-A interactions and the crystal may be considered as being in a state of internal strain (38). This strain can be relaxed by disorder in the glass or melt.

The identity of the building blocks in glass and crystal has been demonstrated in ternary systems, where the glass region extends to a concentration, where the crystal makes a phase transformation. The glass shows for comparable stoichiometry the same variations in its properties as the crystal (e.g. 39).

When solidification shall get stuck in the disordered glassy state, impediments must exist against arranging the building blocks in the proper crystalline order. The energy meeded to overcome them must be comparable in size with Madelung energy to be gained by long range ordering. Since building blocks and bonding energies are the same in glass and crystal, such impediments can have only steric reasons and must be explained by the nature of the CSRO (37).

The free energy of molten alloys is dominated by the entropy of mixing, particularly at high temperatures. The survival of building blocks in the melt depends on their local enthalpy in comparison to the entropy one has to pay for their short range order. When temperature decreases, the restraints by space filling become stronger with increasing density and the Bernal structure develops its many tetrahedra for purely steric reasons, which are, of course, not solid particles but fluctuate in time and space. The life time of CSRO depends then on its symmetry. When it is not tetrahedral, entropy must be paid for deviations from the Bernal structure. CSRO with tetrahedral symmetry, however, can gain the local enthalpy by chemical bonds without paying any configurational entropy.

The topological frustration that happens during solidification, when the many Bernal tetrahedra have to be reorganized and complemented by other polyhedra, gets now an energy barrier by CSRO, which is the higher the stronger the chemical bonds are. In most simple metals this barrier is not high enough for glass formation under reasonable conditions. Glass stabilizers have to be added, which are the more effective the larger the cohesive energy or heat of sublimation of the alloy is (29). A good glass stabilizer should have a large heat of reaction, when added to the melt, and the resulting chemical bonds "associate" the atoms with one another in the melt. At cooling the associations grow into the CSRO, which is finally quenched into the glass. (3,30).

4. KINETICS OF GLASS FORMATION

Whether a crystal or a glass forms depends not only on the existence of atomic associations in the melt but also on kinetic processes during cooling. Crystallization is not a homogeneous transformation of the melt, but proceeds with a finite growth rate starting from a finite number of nuclei, which must contain the characteristic elements of the crystal structure to be formed, that means at least one unit cell. The rate of nucleation and the speed of linear growth increase first with decreasing temperature and then decrease again because of diffusion damping when the viscosity increases exponentially with temperature.

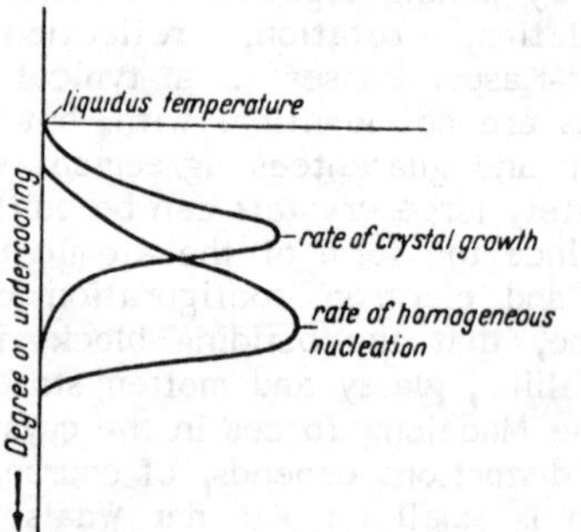

It is a characteristic property of glass forming melts, that their viscosity follows a Fulcher law (51) with $\exp(A/k_B(T - T_o))$. The Fulcher temperature T_o defines an "ideal" glass temperature where all atomic motions are perfectly frozen-in. The real "Glass Temperature" T_g has a somewhat higher value and is defined by the temperature where the viscosity reaches 10^{13} poise and the glass behaves practically like a solid.

A glass will form, when one succeeds to suppress crystallization while crossing the range between solidus and glass temperature. Therefore, in a glass forming system the Fulcher or glass temperature must lie so closely to the solidus temperature T_s that the atomic motions become here already so slow, that crystalline nuclei have no chance to form during the short quenching time. A quantitative relation between critical quenching rate and glass temperature has been worked out (31, 32) using the Johnson-Mehl theory of crystallization kinetics (33). The resulting Time-Temperature-Transformation (TTT-) Diagrams show that the reduced glass temperature $T_r = T_g/T_s$ controlls the critical cooling speed for glass formation. It varies from 10^{10} K/s for pure metals with $T_r = 0.25$ to 10^6 K/s for most metal glasses with $T_r = 0.5$ and 10^2 K/s for easy glass formers with $T_r = 0.7$ and higher. Since the glass temperature is connected with viscosity and viscosity with the size and stability of atomic "associations" in the melt, one can expect also a relation between bonding strength or enthalpy of mixing and the critical speed of cooling (34, 35). The glass temperature is related to the stability of the CSRO and the solidus temperature to the additional gain of energy by the long range order.

The kinetics of glass formation suggests the following general rules for glass making:

1. Go to sufficiently high quenching rates. Every melt, that undercooled, should be able to form a glass after sufficiently fast quenching.

2. Increase the glass temperature, that means the viscosity and CSRO in the melt by adding suitable components, which must not form solid solutions but stable compounds with high heat of reaction and melting temperatures at another stoichiometry than the glass has (29, 30). Rules for selecting suitable glass stabilizers for improving the Glass Forming Ability (GFA) are discussed in the next chapter 5.

3. Lower the melting temperature, that means the stability of the crystalline state, according to the Raoult-Van't Hoff law by adding any impurities, which are soluble in the liquid, but not in the crystalline state (36). This rule does not work for ideal, but only for anomalously deep entectics due to a high enthalpy of mixing, which means again stabilisation of the CSRO in the melt.

Since the crystal is always the most stable state of lowest energy, the glass can only be metastable. It is kept in its relative minimum of energy by steric frustrations and impediments. It forms crystalline nuclei immediately, when at the "cristallisation temperature" T_k the viscosity is reduced sufficiently for the atoms to overcome local frustrations and impediments by diffusion. The crystallisation temperature lies usually very closely above the glass temperature. Since the amount of built-in frustrations depends on the quenching rate also, the temperatures T_g and T_k are not controlled by chemistry alone but depend also on the speed of cooling or heating, respectively.

5. STRUCTURE MODELS FOR GLASSES

The concepts suggested sofar cover continously all grades of atomic order an amorphous material may have without showing it in diffraction experiments.

5.1. Microcrystalline Models suggest as building blocks little single crystals containing only so few unit cells that they cannot be detected by wave diffraction when packed without any regularity (40). This model seems to be applicable only to some special cases. An important and difficult question is, how to match the small crystallites together without using too much filling material in a lower state of order.

5.2. Random Networks have also very stable, well defined building blocks which, however, contain only a small section of the crystalline unit cell, which reflect the chemical situation. Such units are then netted together in an irregular manner but with about the same density as the crystal. The model was proposed by Zachariasen

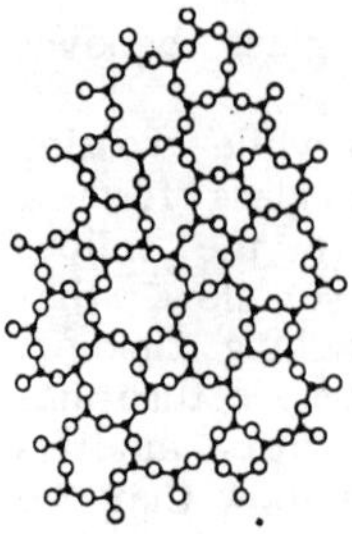

more than 50 years ago(28) for transparent silicate glasses and became quite useful for their technical design and production. Building blocks are the SiO tetrahedra, which have a large gap between valence and conduction electron states and therefore high chemical stability. Polk (41) applied this concept also to amorphous and semiconducting Si by replacing the SiO tetrahedra by Si-atoms. The resulting Si-tetrahedra are rigid, but netted together at their corners with enough freedom for rotation and little bending to form a glass with a Radial Distribution Function undistinguishable from that expected for DRPHS. The rigidly bond Si tetrahedra reproduce the semiconducting properties of the crystal.

5.3. Agglomerate Glass (42, 43) contains in melt and glass building blocks, which are stabilized by strong directional bonds but packed together by non directional forces, such as conduction electron bonds, van der Waal's forces etc. A typical example are the Laves phase compounds of d-metals with Rare Earth elements, where the 3d-atoms form rather stable tetrahedra which are then packed in a three dimensional lattice by the 5d conduction electrons from the Rare Earth.

5.4.Correlated Glass has no directional bonds, but strong central forces between metal and metalloid ions or between the ions of different polarity in a salt. This modulates the Bernal structure in such a way that equal ions are kept away from one another. In alkalioxide glasses, as an example, the Cs or Rb ions surround the oxygen ions (46) and form a stable unit not suitable for building crystals. Even for metal glasses, which have more than one kind of atoms, one must expect stereochemical modifications in the Bernal structure, which may are difficult to be treated in theory, but can produce atomic coordinations similar to that in the crystal (47).

5.5. Dense-packed Random Glasses form a random mixture of atoms which must be so similar in their size and chemistry that they can form perfect solid solutions. The Bernal structure can then be stabilized only by the gain in Madelung energy by space filling. Such glasses can exist only at very low temperatures or not at all, but the model has fundamental significance because of its relative simplicity.

More stable glasses are formed when the electronic bond is strengthened by adjusting the free electron concentration just so, that the double Fermi wave vector fits the wave vector of the first peak in the structure factor derived from the Radial Density Distribution by Fourier transformation. This Nagel-Tauc rule (48) corresponds to the Hume - Rothery Rule for stability of certain crystal phases in alloys. It comes now in the version that conduction electrons should stabilize the next neighbor coordination of the Bernal structure instead of the unit cell in a crystal. It applies to noble metal alloys such as Au-Si, Cu-Sn etc (49) with an electron concentration per atom around 2.

6. RULES FOR IMPROVING GFA

What are the properties, which characterize an element as an good glass stabilizer? This question aims directly to the problem of the structure and chemical interaction in the melt. It has, at present, many answers which may some times converge in their meaning while the understanding of the liquid and glassy state improves. It is even not yet sure, wether there will finally some general rules valid for all disordered materials or if we have to distinguish always several different groups.

6.1 Covalent Oxide-Glasses. Most stable and technically important are the glassy oxides of Bor and elements of the IV. and V. group. Goldschmidt (50) believed that the formation of O-tetrahedra around the metal atoms is the necessary condition for glass stability and derived simple rules for the atomic size ratio.

The Zachariasen Rules for the formation of three-dimensional networks (28) are more or less universally accepted now for oxide glasses and have significantly contributed to the understanding and development of technical glasses. He concluded from similar physical properties of glass and crystal that both must have the same atomic coordination, such as the SiO-tetrahedra in the Cristobalite structure, which must form 3 dimensional networks of comparable energy but without or with long range order, respectively. What are the topological conditions for such atomic configurations? From his studies on SiO_2-glasses he arrived at the following rules:

1. an oxygen atom is linked to not more than two A atoms (A = glass forming metal)
2. the number of oxygen atoms surrounding A atoms must be small
3. the oxygen polyhedra share corvers with each other, not edges or faces
4. at léast three corners in each oxygen polyhedron must be shared

The rules 1 - 3 make sure that the bonding energy is not much reduced, when long range order is given up, but also sufficient flexibility is obtained for building an irregular network without distorting the first coordination shell. A "small" number of oxygen atoms means 3 or 4, i.e. triangle or tetrahedral coordination. Rule 3 makes sure that the oxygen tetrahedra are not too compactly packed; rule 4 ensures the spatial coherence of the network, which must make all topological rearrangements difficult.

Cooper (52) has discussed several aspects of the rules in detail and reaches the conclusion "that these rules may be considered to be the necessary and sufficient conditions for glass formation in binary compounds". Since they are purely topological rules without any concern about chemistry, one cannot see any reason to confine their validity to oxide glasses alone. Cooper points out also, that tetrahedra are the only simple polyhedra, which provide in contrast to cubes and octahedra sufficient topological freedom for packing them densily together without the danger of structural repetitions after a certain number of steps.

Zachariasen suggested to generalized his rules as follows:

"Glass may be formed

1. if the sample contains a high percentage of cations which are surrounded by anion tetrahedra or triangles
2. if these anion tetrahedra or triangles share only corners with each other
3. if some anions are linked to only two such cations and do not form further bonds with any other cations."

The words "cation" and "anion" are used here just as names for the opposite partners in any chemical bond.

Smekal (53) pointed out that the mixture of chemical bonding mechanism in the same sample could improve the GFA. Sun (54) demonstrated that bond strength and dissociation energy are important for the GFA of oxide glass. The stronger the bonds, the more difficult become atomic rearrangements and crystallization.

6.2. Metallic Glasses

Pure simple metals have an extremely low GFA. Most metal glasses produced and investigated so far are alloys. Therefore, the rules for improving GFA are mainly concerned with the selection of suitable partners, which can be two metals or a metal and a metalloid. From the many recommendations made over the years, the following seem to have some relevance:

* The atomic size ratio and/or the difference in electronegativity should be so large, that solid solubility is excluded
* The resulting eutectics must be irregularly deep compared to the situation in melts with non-interacting atoms.
* The crystalline phases neighboring the eutectic should have high stability and melting temperatures.
* The alloy should show a strong enthalpy of mixing and a relatively small volume of the melt compared to neighboring concentrations.

Details, discussions and references of those rules are given in many review articles (e.g. 3 and 6) and shall not be repeated here. Obviously, the Bernal model of DRPHS has to be corrected by introducing CSRO as "a necessary prerequisite for obtaining glassy alloys by rapid quenching with attainable cooling rates" (3). Therefore, the situation becomes here quite similar to that for oxide glasses and it is an interesting question whether Zachariasen's topological rules can be applied to metallic glasses also.

7. EXAMPLES OF METAL - GLASSES

Several hundred metallic alloy systems are already known to form glasses in some regions of their phase diagram when they undergo melt spinning, clutch cooling, jet spraying, laser glazing, electron or ion bombardment, surface welding, sputtering or evaporation (55). Since each process has a different speed of quenching, the width of the glass region depends on the choice of the preparation method. We like to discuss here only two glass forming systems, which are very popular due to their potential for technical applications.

7.1. 3d-metal + Rare Earth Glass

Films of such glasses produced by high rate sputtering can have a magnetic anisotropy useful for "Bubble Memories". A wide glass region extends on the 3d-rich side of the phase diagram, which contains several Laves phases and related tetrahedral compounds. The "Confusion Principle" for glass formation claims that for so many structures of comparable stability the usual quenching time is too short for making a decision about the best crystal structure.

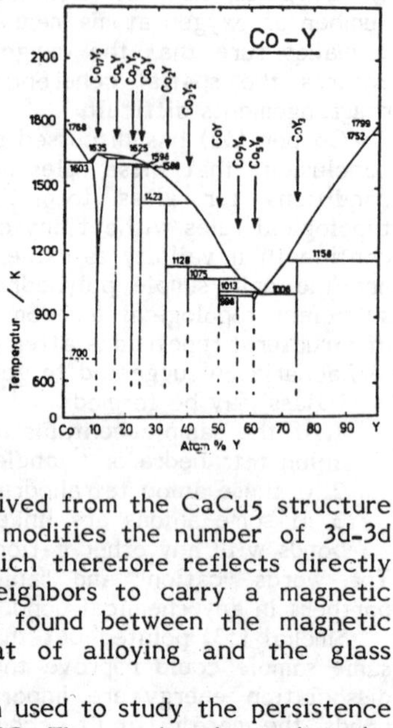

The topological similarity between the structure of glass forming Cristobalite, pure Si and the Laves phases, e.g. the cubic C15 version, becomes obvious when one takes instead of the simple tetrahedron the "Tetraederstern" (4 Rare Earth atoms around the small 3d-tetrahedron) as a structural unit, which should obey the Zachariasen rules. The other Co-rich phases have configurations, which can be derived from the $CaCu_5$ structure by using different stacking sequences (26, 59). This modifies the number of 3d-3d neighbors and the magnetic moment per 3d-atom, which therefore reflects directly CSRO. As an example, Co needs at least 8 Co neighbors to carry a magnetic moment (57). For Ni-glasses a correlation has been found between the magnetic moment, as indicator for the CSRO, and the heat of alloying and the glass temperature (59).

The magnetic moment of Co in Co_2Y has also been used to study the persistence of CSRO above the melting point (60) of about 1270°C. The results show, that the Co atoms coordinate in the melt up to about 1360°C in the building blocks of the Co_7Y_2 structure and assume above 1380°C the paramagnetic moment and Curie temperature of pure Co.

The variation of CSRO with temperature in the melt explains why in some cases the properties of the glass depend not only on the quenching rate but also on the temperature of the melt before quenching (e.g. 61).

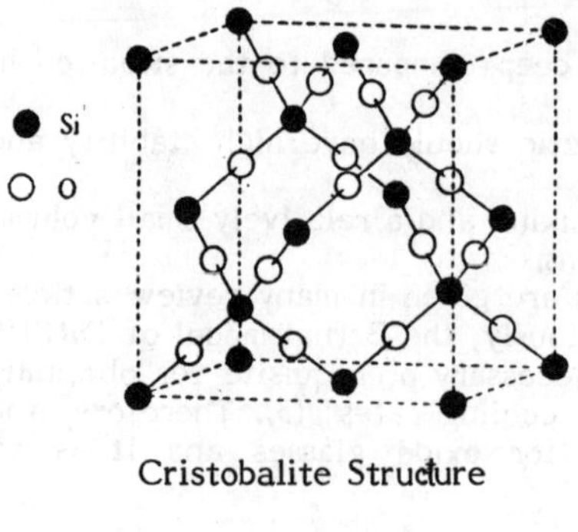

Cristobalite Structure

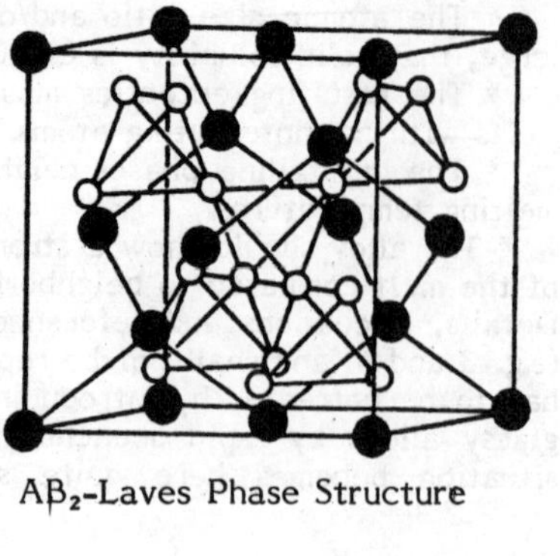

AB_2-Laves Phase Structure

7.2. Fe + B Glass

Glasses of this type find applications as soft magnetic materials and are therefore extensively investigated. Stable glasses can be prepared easily by melt spinning in the eutectic region around the 80:20 composition. The bordering phases, where to search for suitable building blocks for the glass, are pure Fe and Fe_2B with melting points at 1535°and 1389°C, respectively. The metastable Fe_3B-phase is of special interest, because it does not come from the melt but segregates from the glassy state in a narrow temperature range.

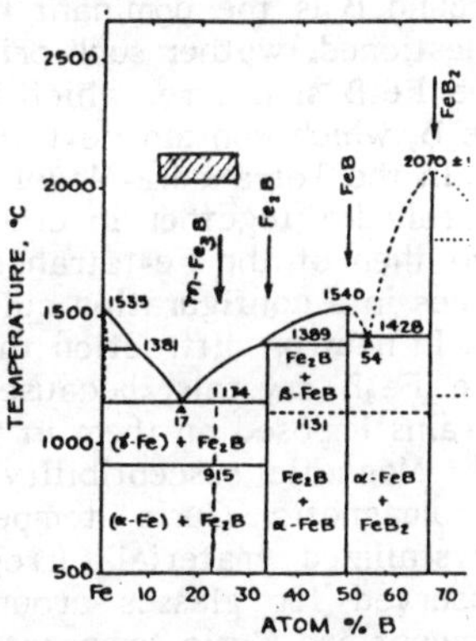

The compound Fe_2B:C16 belongs to the family of tetrahedral structures we have discussed before (26, 27). The central Fe-Fe tetrahedron within the "Tetraederstern" seems to be here the structure stabilizing unit, because the Fe-Fe distances are about 10% shorter than in b.c.c. Fe. However, the whole "Tetraederstern" is built here of 8 Fe-atoms only and that is unfavourable for glass formation because there are no "bridging" atoms of the other component available as requested by the first generalized Zachariasen Rule.

The B-atoms form chains parallel to the c-axis in the channels left by the Fe-stars. The B-B distances are about 25% longer than in the icosahedron of elemental B. This explains the much lower melting point compared to the other Fe-B compounds richer in B.

The compound Fe_3B contains still chains of Fe-Fe stars, but replaces also the B-chains by chains of mixed Fe-B "Tetraedersterne", which are suitable for glass formation. In addition they destabilize the crystalline state by pushing apart the chains of Fe-Fe stars. The B-"anions" at the corners of the Fe-B stars link to only 2 and not more Fe-tetrahedra, as requested by the 3. Rule. Obviously, the 2. Rule is satisfied also. The "B-B avoidance" in glasses has been confirmed by many experimental results, which suggest that each B is coordinated by 6+3 Fe-atoms in configurations similar of a triangular prism.

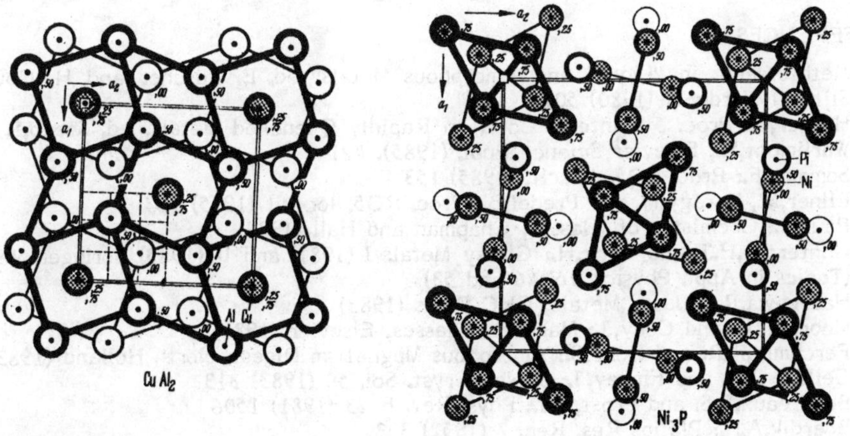

K.Schubert (26):
Crystal Structures of Fe_2B (left) and Fe_3B (right) with "Tetraedersternen".

Gaskell (45) concluded from the B-Fe configuration, that the triangular Fe prism around B is the dominant building block for stabilizing the glassy state. It may be questioned, wether such prism have sufficient stability, because the melting point of the Fe_3B structure, which is built by such units, is much lower than that of Fe and Fe_2B, which contain Fe-tetrahedra.

In the Tetrahedra-Model the B atoms take the role to net always two neighboring tetrahedra together in order to form a stable 3 dimensional network. The B-atoms see then of the Fe-tetrahedra, which they have to connect, only the two triangular faces in a configuration quite similar to a triangular prism.

In neutron diffraction the glass shows better defined inter-atomic distances than the Fe_3B crystals, because the "Tetraedersterne" can relax in the glass from the strains imposed on them in the crystal by longe range ordering (63).

Magnetic susceptibility measurements in the Fe-B melt give about the same paramagnetic Curie temperatures as in the glass, but smaller than in the crystallized material. Irregularities in density and crystallization temperature observed for glasses around 20%B can be related to irregularities in magnetic moment and Curie temperature at the same concentrations (62).

8. CONCLUSIONS

The CSRO controls the glass forming and many physical properties of a glass in the similar way as the unit cell the properties of the crystal.

The TSRO in glasses is controlled by DRPHS, which produces many more tetrahedra than can be used in any crystal structure.

The CSRO provides the topological rearrangements necessary for crystallizytion with an energy barrier, which is the higher, the stronger the tetrahedral bonds: the melt undercooles.

The similarity of densities and properties of glasses and crystals suggests, that the same CSRO is used as building block in both cases.

Therefore, the tetrahedra, which stabilize the glass by chemical bonds must, in principle, also be found in the crystals, which form in or around the glass forming region.

The Zachariasen Rules, which have been developed for oxide glasses, consider only topological and no chemical aspects and may have a more general validity, including metallic glasses.

3d+Rare Earth and Fe+B glasses are discussed from Zachariasen's point of view, taking "Tetraedersterne" as stabilizing building blocks instead of simple tetrahedra.

REFERENCES

1. Methfessel,S. in "Liquid and Amorphous Metals" ed. E. Lüscher and H. Coufal, Sijthoff-Nordhoff (1980) 501
2. Hafner,J.: Proc. 5th Intern. Conf. on Rapidly Quenched Metals, ed. Steeb,S. and Warlimont,H., Elsevier Science Publ. (1985), 421
3. Sommer,F.: Proc. RQ5, loc.cit. (1985) 153
4. Ellner,M., Boragy,M. and Predel,B.: Proc. RQ5, loc cit (1985) 183
5. Paul,A.: Chemistry of Glasses, Chapman and Hall (1982)
6. Günterodt,H.J. and Beck,H.: Glassy Metals I (1981) and II (1983) Springer Verlag (Topics in Appl. Physics Vol 46 and 53)
7. Hasegawa,R.: Glassy Metals, CRC-Press (1983)
8. Moorjani,K. and Coey,J.: Magnetic Glasses, Elsevier (1984)
9. Ferchmin,A.R. and Kobe,S.: Amorphous Magnetism Digest. North Holland (1983)
10. Gellatly,B.J. and Finney,J.: J. Non-Cryst. Sol. 50 (1983) 313
11. Boudreaux,D.S. and Frost,H.J.: Phys. Rev. B 23 (1981) 1506
12. Boerdjk,A.H.: Philips Res. Rep. 7 (1952) 303
13. Bernal,J.D.: Nature 185 (1960) 68; Proc. Roy. Soc. London A 280 (1964) 299
14. Turnbull,D. and Cohen,M.H.: J. Chem. Phys. 52 (1970) 3038
15. Finney,J.L.: Proc. Roy. Soc. London A 319 (1970) 479; Nature 266 (1977) 309

16. Cargill,G.S.: J. Appl. Phys. 41 (1970) 2248; Solid State Phys. 30 (1975) 227
17. Whittaker,F.J.W.: Journ. Non-Cryst. Solids 28 (1978) 293
18. Polk,D.E.: Acta Metall. 40 (1972) 485
19. Boerdjik,A.H.: Philips Res. Rep. 7 (1952) 303
20. Frank,F.C. and Kasper,J.S.: Acta Cryst. 11 (1958) 184
21. Sinha,A.K.: Topol.Close-Packed Structures, Pergamon Press 1972
22. Schulze,G.E.R.: ZS.Elektrochem. 45 (1939) 849
23. Berry,R.L. and Raynor,G.V.: Acta Cryst. 6 (1953) 178
24. Laves,F. and Witte,H.: Metallwirtschaft 15 (1936) 840
25. Hafner,J.: in Ref. 6, Vol. I, pg 93 with more references.
26. Schubert,K.: Kristallstrukt. zweikomp. Phasen, Springer Verlag (1964) 150
27. Andersson,S.J.: Solid State Chem. 23 (1977) 191
28. Zachariasen,W.H.: J. Am. Chem. Sol. 54 (1932) 3841
29. Donald,I.W. and Davies,H.A.: J. Non-Cryst. Solids 30 (1978) 77
30. Sommer,F.: Z. Metallkunde 72 (1981) 219; 73 (1982) 72, 77
31. Uhlmann,D.R.: J. Non-Cryst. Solids 7 (1972) 337
32. Davies,D.: J. Non-Cryst. Solids 17 (1975) 266
33. Johnson,W.A. and Mehl,R.F.: Trans AIME 135 (1939) 416
34. Rawson,H.: Inorganic Glass Forming Systems, Academic Press 1967
35. Turnbull,D. and Cohen,M.H.: J. Chem. Phys. 34 (1961) 120
36. Marcus,M. and Turnbull,D.: Mat. Sci. Eng. 23 (1976) 211
37. Phillips,J.C.: Phys. Today (Feb. 1982) 27
38. Berry,R.L. and Raynor,G.V.: Acta Cryst. 6 (1953) 178
39. Durand,J.: IEEE MAG 12 (1976) 945
40. Hornbogen,E.: J. Mat. Sci. 13 (1978) 666
41. Polk,D.E.: J. Non-Cryst. Solids 5 (1971) 365
42. Hoare,M.R. and Pal,P.: Adv. Phys. 20 (1971) 161
43. Predel,B.: Ber. Bunsen Ges. 80 (1976) 695
44. Hume-Rothery,W. and Anderson,E.: Phys. Mag. 5 (1960) 383
45. Gaskell,P.H.: Proc. RQ5. loc cit (1985) 413
46. Bauhofer,W. and Simon,A.: Phys. Rev. Lett. 40 (1978) 1730
47. Gaskell,P.H. in Ref. 6 (1983) 5
48. Nagel,S.R. and Tauc,J.: Pys. Rev. Lett. 35 (1935) 380
49. Häussler,P.: Proc. RQ5, loc. cit. (1984) 797, 1007
50. Goldschmidt,V.: Akad. Skr. Oslo 8 (1926) 7
51. Fulcher,G.S.: J. Am. Ceram. Soc. 8 (1925) 339, 789
52. Cooper,A.R.: Phys. Chem. of Glasses 19 (1978) 60
53. Smekal,A.: J. Soc. Glass Technol. 35 (1951) 411
54. Sun,K.H.: J. Am. Ceram. Soc. 30 (1947) 277
55. Wang,R.: Bull. Alloy Phase Diagr. 2 (1981) 269
56. Cromer,D.T. and Larson,A.C.: Acta Cryst. 12 (1959) 855
57. Buschow,K.H.J.: J. Appl. Phys. 53 (1982) 7713
58. Buschow,K.H.J.: Proc. RQ5, loc cit (1985) 163
59. Buschow,K.H.J.: Rep. Progr. Phys. 40 (1977) 1181
60. Kamp,A.: priv. comm., Uni Bochum, 1985
61. Alounian,Z. and Strom-Olsen,J.O.: Proc. RQ5, loc cit (1985) 447
62. Takahashi,M., Sato,F. and Ishio,S.: J. Mag. Mag. Mat. 49 (1985) 145
63. Lamparter,P., Sperl,W., Steeb,S. and Bletry,J.: ZS. Natf. 37 (1982) 1223

ELECTRONICALLY DRIVEN STRUCTURAL EFFECTS IN LIQUID AND AMORPHOUS METALS AND ALLOYS

J.HAFNER, G.KAHL, and A.PASTUREL[x]

Institut für Theoretische Physik, Technische Universität Wien
Karlsplatz 13, A 1040 Wien, Austria

1. INTRODUCTION

It is now generally agreed that the crystal structures formed by metals and alloys are determined by the electron-mediated atomic interactions. The increasing electron density results in a change from close-packed metallic structures in groups I to III to more open covalent lattices in groups IV and V with decreasing coordination numbers. The enhanced non-locality of the electron-ion interaction with increasing atomic number A produces a progressive distortion of the hexagonally close-packed structure in the group II elements, culminating in a unique structure in Hg. In group III one finds a similar phenomenon with a cubic close-packed structure in Al and a unique low-symmetry structure in Ga, but then a return to a less distorted one for In and a close-packed metallic structure in Tl. This tendency to form a more metallic structure with increasing atomic number prevails in groups IV and V and is most pronounced in the series Si - Ge - Sn - Pb. For binary alloys, it is again the electron-mediated part of the interatomic potentials which determines whether a system forms a substitutional alloy or an intermetallic compound and whether the structure of this compound is of a close-packed metallic or of a more open, at least partially covalent or ionic character.

The crystal structures and the cohesive properties of the metallic elements are now thoroughly well explained in terms of perturbative and non-perturbative total energy calculations based on the local-density approximation (see /1,2/ for a review of the state-of-the-art and for a list of additional references). However, we have to remember that when we say that "a calculation has determined the correct crystal structure of element X", we mean only that the total energy of the element has been calculated for a certain discrete set of structures (which has been arbitrarily chosen and is generally quite restricted) and that the lowest energy has been found for the experimentally determined structure. This is also the very reason which makes any attempt to develop an electronic theory of the structures of binary compounds so difficult - the available choice of possible competing structures and compositions is immense. Nonetheless, both perturbative /3-5/ (some of them including higher-order corrections /6/) and non-perturbative /7-9/ total energy calculations have treated successfully a number of alloys ranging from terminal solid solutions to Laves-phases (close-packed and metallic) and to Zintl-phases (with an at least partially covalent character of the chemical bond) - so progress in this difficult field is sure to come but will be necessarily slow.

---

[x] *On leave of absence from: Laboratoire de Thermodynamique et Physico-Chimie Métallurgiques, Ecole Normale Supérieure d'Electrochimie et d'Electrométallurgie de Grenoble, F 38402 Saint Martin d'Hères, France.*

The generally accepted picture of the structure of simple liquids and of simple liquid mixtures on the other hand is much more simplistic: it is agreed that the liquid has its volume determined by the volume forces and by the attractive part of the pair interactions, but once the equilibrium volume has been established the liquid may be considered as a hard-sphere fluid confined within that volume. It is true that it has been suggested very early /10,11/ that most of the trends in the crystalline structures persist in the liquid state. In the monoatomic systems the distortions of a close-packed crystalline structure are reflected in the form of distortions of the first few peaks in the static structure factor $S(q)$ and/or the pair correlation function $g(R)$ from their hard-sphere-like form (see Fig.1). In the binary systems, the formation of strictly ordered intermetallic compounds is paralleled by similar ordering phenomena in the liquid state which are best described in terms of the Bhatia-Thornton number-concentration structure factors $S_{NN}(q)$, $S_{NC}(q)$, and $S_{CC}(q)$ (and their Fourier-transforms, the number-concentration correlation functions $g_{NN}(R)$, $g_{NC}(R)$, and $g_{CC}(R)$). In an ordered liquid mixture (which is also often called "compound-forming") the concentration-fluctuation structure factor $S_{CC}(q)$ has a peak at small momentum transfers which is very similar to the superlattice peak observed in the diffraction pattern of the corresponding solid alloy /12,13/, see Fig.2. This close analogy suggests that the trends in the liquid structures are driven by electronic effects - in a way quite similar to the trends in the crystalline structures.

From one point of view the study of the liquid systems is even more instructive than the investigation of the crystalline systems: there all we could do was to compare the ground-state energies of a small set of discrete structures - at most we could study their quasi-continuous variation as a function of some structural parameter (such as an axial ratio for example), thereby exploring a small region in configuration space. For a liquid system on the other hand we can calculate the structure (at the level of a pair correlation function) from the interatomic force law, based on statistical-mechanical theory of the liquid state.

Thus the problem represented by the liquid structures is clearly divided into two distinct parts: (1) the derivation of appropriate expressions for the interatomic forces, based on an electronic theory of the chemical bond; and (2) the calculation therefrom of the pair correlation functions and of the static structure factors.

This review is organized as to follow this strategy rather tightly. First we turn to the monoatomic metallic melts: in Sec.2 we recapitulate very briefly the trends in the interatomic potentials and the physical mechanisms from which they originate. Sec.3 contains a systematic discussion of thermodynamic perturbation theory - the main advantage of thermodynamic perturbation theory being that it allows to treat each part of the complex interatomic potentials in an appropriate way. At the level of a thermodynamic variational calculation we find that we do not even have to know the interatomic forces explicitly - an expression for the ground-state energy in terms of the pair correlation functions is sufficient. This is important as it will allow us to extend our study to transition metal systems where expressions for the interatomic forces are not readily available. Sec.4 contains a discussion of the structures of the elements in the liquid state. The following sections bring a generalization of both parts of the theory to binary systems (Secs. 5 and 6) - this prepares the discussion of chemical ordering in liquid (Sec.7) and amorphous (Sec. 8) simple-metal alloys. Chemical ordering in transition-metal alloys is treated in Sec.9, our conclusions are presented in Sec. 10.

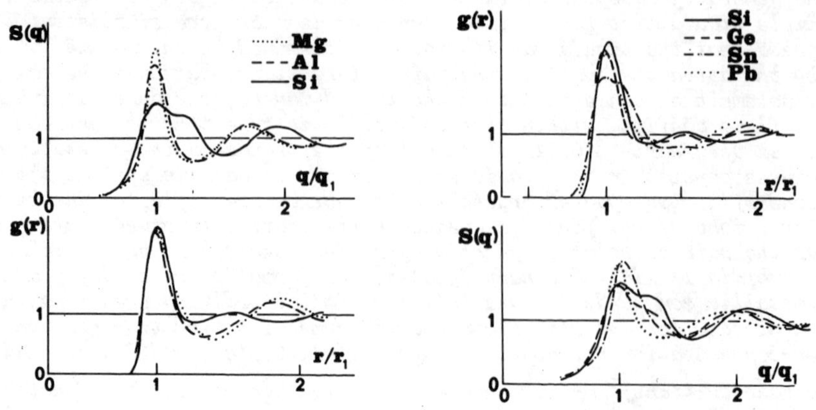

FIGURE 1. The trend from metallic, hard-sphere-like to more covalent, "open" structures in the series Na - Mg - Al - Si (a), and from open structures with low coordination numbers to more hard-sphere-like (high) coordination numbers in the series Si - Ge - Sn - Pb (b), cf. text. After Waseda /14/, and Gabathuler and Steeb /15/.

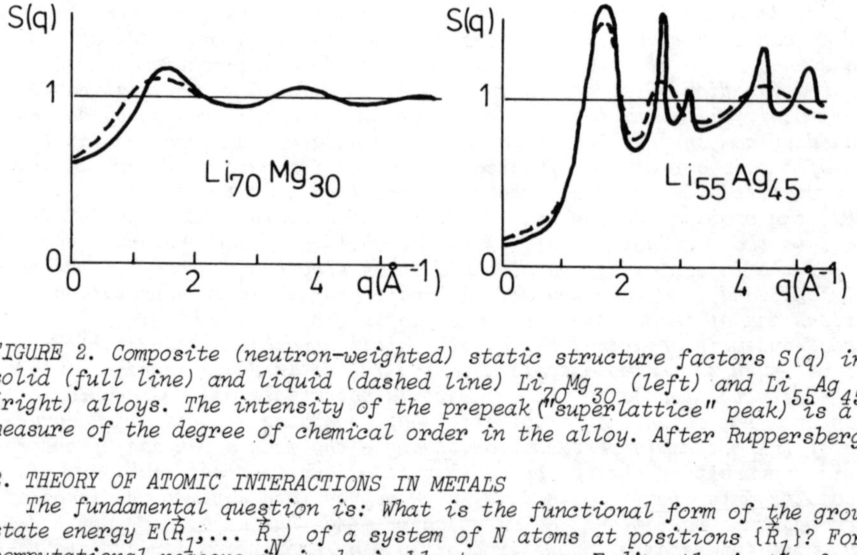

FIGURE 2. Composite (neutron-weighted) static structure factors S(q) in solid (full line) and liquid (dashed line) $Li_{70}Mg_{30}$ (left) and $Li_{55}Ag_{45}$ (right) alloys. The intensity of the prepeak ("superlattice" peak) is a measure of the degree of chemical order in the alloy. After Ruppersberg/16/.

2. THEORY OF ATOMIC INTERACTIONS IN METALS

The fundamental question is: What is the functional form of the ground state energy $E(\vec{R}_1,\ldots \vec{R}_N)$ of a system of N atoms at positions $\{\vec{R}_i\}$? For computational reasons it is desirable to express E directly in the form of an expansion in terms of volume-, pair-, and multi-body forces. The most convenient scheme for obtaining these interactions in metallic systems is based on the linear response of the conduction electrons to the electron-ion potential, which is valid only in systems where the ions scatter the electrons weakly and are describable by pseudopotentials /2, 17/. Only the so-called simple (or (s,p)-bonded) metals confirm with these requirements. For the transition metals approximate expressions for the interatomic potentials can be derived from a tight-binding approach /18/.

There are many equivalent ways to derive a pseudopotential from first principles (for a recent discussion see e.g. /19/), for the use in perturbative total energy calculations optimized pseudopotentials (the optimization serves to minimize the importance of higher order perturbation contributions) turned out to be particularly fruitful. The pseudopotential resulting from such a first-principles construction is a rather complicated non-local and energy-dependent operator. Even if the non-local character of the pseudopotential is important for obtaining quantitatively reliable results, experience with local ("on-Fermi-sphere") approximations suggests that even the simplest local pseudopotential (the empty-core model of Ashcroft /20/) is sufficient to explain the main trends and this was indeed confirmed by the work of Hafner and Heine /1/. Our discussion follows their work rather closely.

To second order in the pseudopotential, the ground state energy E of a metal is composed of a volume energy E_V and a term expressible as a sum over central, volume-dependent pair interactions $\Phi(R;V)$, i.e.

$$E = E_V + \frac{1}{2N} \sum_{i \neq j} \Phi(|R_i - R_j|; V) \quad . \tag{1}$$

For a local pseudopotential $v_{ps}(q)$ the pairpotential $\Phi(R)$ and the volume energy E_V are given by (see Ref./2/ p.276 ff and /21/, we use atomic units)

$$\Phi(R) = \frac{Z^2}{R} \{1 - \frac{2}{\pi} \int v_N(q)^2 \, (\varepsilon_p(q)^{-1} - 1) \, \frac{\sin qR}{q} \, dq\} \tag{2}$$

and

$$E_V = Z \, (E_{eg} - \frac{1}{2} V_a B_{eg}) + \frac{1}{2} \Phi_{ind}(R=0). \tag{3}$$

Here Z is the valence of the atom, $v_N(q)$ is a normalized pseudopotential matrix element with the Coulomb part factored out: we have

$$v_{ps}(q) = \frac{4\pi Z}{V_a q^2} v_N(q) \tag{4}$$

with V_a the atomic volume. The function $\varepsilon_p(q)$ describes the screening of the interaction between two atomic charges, including corrections for exchange and correlation effects among the screening electrons[x]. Equ.(2) shows that the effective pair interaction $\Phi(R)$ consists of the direct Coulomb repulsion between the ionic charges and the attraction that the ionic charge exerts on the screening charge induced by another ion and vice-versa. In equ. (3) E_{eg} and B_{eg} are the ground-state energy and the bulk modulus of the homogeneous electron gas, and $\Phi_{ind}(R)$ is the non-Coulombic part of $\Phi(R)$. Thus $\Phi_{ind}(R=0)$ represents the self-energy of a pseudo-atom, i.e. the interaction of an ionic core with its own screening electron cloud.

---

[x] Much effort has been spent to find the best way for including exchange and correlation effects in the dielectric function (for a recent review see /22/) and to investigate the apparently very drastic effects on $\Phi(R)$ /23,24/. What has often been overlooked is that the dielectric function has to obey numerous consistency conditions relating $\varepsilon_p(q)$ to the correlation energy, the compressibilty of the electron gas, the electron -electron pair correlation function etc.. After eliminating all approximations which violate one or more of these requirements, one is left with a rather small choice /25-27/ and one finds that the difference in $\Phi(R)$ produced by the different acceptable $\varepsilon_p(q)$'s is quite small and irrelevant in the present context. Again the local density approximation /28/ represents an attractive compromise between accuracy a. simplicity.

If we approximate the pseudopotential matrix element by its empty-core form (i.e. $v_N(q) = \cos qR_c$ where R_c is the empty-core radius /20,2/), we find that the interatomic potential $\Phi(R)$ depends on three essential parameters: the atomic valence Z, the electron density R_s ($R_s = (3V_a/4\pi Z)^{1/3}$), and the core radius R_c. As pointed out by Hafner and Heine /1/, the valence enters only the amplitude factor Z^2/R of the atomic interaction, its form is determined by the term in the curly brackets in equ. (2) (which is appropriately termed the reduced interaction $\Phi_{red}(R)$). $\Phi_{red}(R)$ depends only on R_c (via $v_N(q)$) and on R_s (via $\varepsilon_p(q)$). The general form of $\Phi_{red}(R)$ is well known: it consists of a strongly repulsive core plus an oscillatory potential at long and intermediate distances. In a Thomas-Fermi approximation (which is realistic only at high electron densities) the repulsive part takes the form

$$\Phi_{red}(R) \propto \cosh^2(R_c/\lambda_{TF}) \exp(-R/\lambda_{TF}) \tag{5}$$

(λ_{TF} is the Thomas-Fermi screening length, $\lambda_{TF} = (\pi/12)^{1/3} R_s^{1/2}$ in atomic units), the large R-behaviour is dominated by the Friedel oscillations

$$\Phi_{red}(R) \propto \cos^2(2k_F R_c) \cos(2k_F R)/(2k_F R)^3 \tag{6}$$

(with the Fermi momentum $k_F = (9\pi/4)^{1/3}/R_s$). The oscillations extend under the repulsive core and may be exposed by a variation of its radius. Hafner and Heine[x] have shown that around the nearest neighbour distance the trends in the form of $\Phi(R)$ shown in Fig.3 are dominated by two main effects: (1) The "core-effect". At constant R_c/R_s the oscillations are shifted to smaller distances (their characteristic wavelength is the Friedel length $\lambda_F = 2\pi/2k_F \propto R_s$), but the diameter of the repulsive core changes more slowly ($\lambda_{TF} \propto R_s^{1/2}$), altogether this results in a repulsive core moving over the first attractive wiggle with increasing electron density.

(2) The "amplitude-effect". The variation of $\Phi_{red}(R)$ with R_c/R_s at constant R_s is characterized by a drastic change in the amplitude of the oscillations. At $2k_F R_c = \pi/2$, i.e. at $R_c/R_s = 0.4092$ the leading term in the Friedel oscillations is completely suppressed by interference with the ionic pseudopotential (see equ. (6)). Consequently the damping of the wiggles is accompanied by an increasing phase shift - it is very large for $R_c/R_s = 0.41$ where one really sees the non-Friedel component of the potentials. Altogether one finds that the position of the main minimum in $\Phi(R)$ is determined by the core radius R_c and not by the Friedel oscillations, the strength of the minimum is related to the expansion of the repulsive core (and hence to the electron density R_s) and to the strength of the pseudopotential at $q = 2k_F$ (depending on the ratio R_c/R_s) setting the amplitude of the oscillations.

The next step must consist of locating the individual elements in the two-dimensional parameter space spanned by R_s and R_c/R_s. Here is advantageous to follow the strategy developed by Hafner and Heine /1/: the equilibrium atomic volume (which is in reality determined by the balance between the pressures arising from the volume- and the pair-forces) is taken as given - this is in fact equivalent to introducing a second parameter modelling the non-locality of the pseudopotential. But even taking V_a as given, we must remember the important trends in the size of the atoms through the periodic table (see e.g. Austin and Heine /29/): (1) The increase in atomic volume with the principal quantum number (dominating in

---

x *Ref./1/ contains a numerical investigation of the variation of $\Phi(R)$ as a function of R_c and R_s, this has been supplemented in a later paper /24/ by an analytic investigation.*

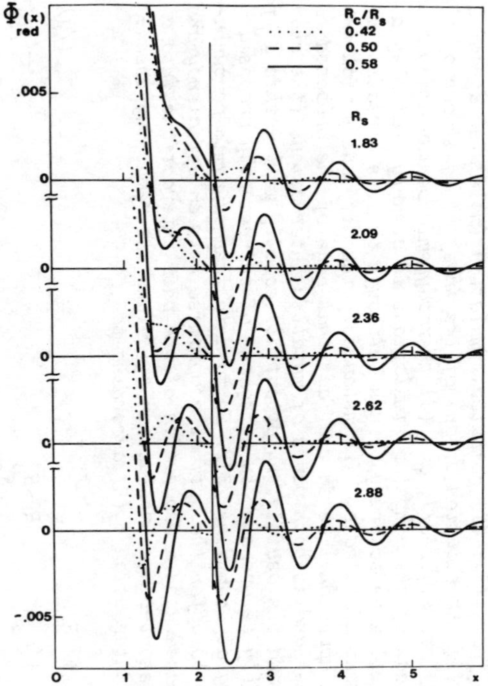

FIGURE 3. Variation of the reduced interatomic interaction $\Phi_{red}(R)$ as a function of R_c and R_s. Distance has been expressed in terms of the scaled variable $x=R/\lambda_F=2k_FR/2\pi$ which places the maxima of the ideal Friedel oscillations at integer values of x so that phase shifts due to the pseudopotential stand clearly out.

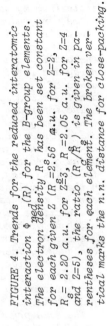

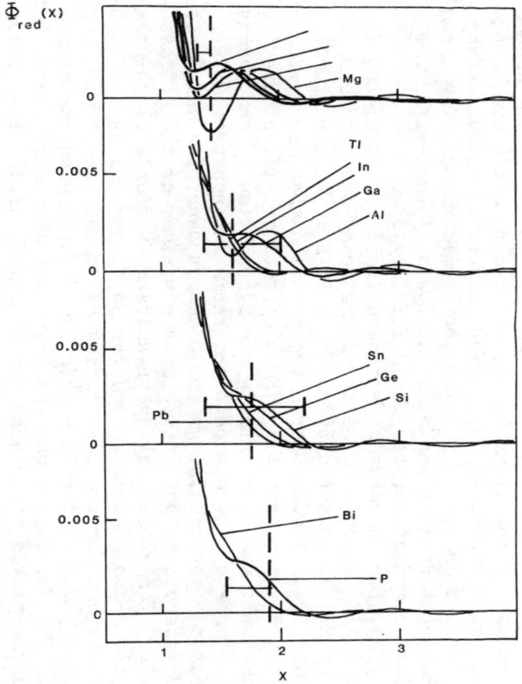

FIGURE 4. Trends for the reduced interatomic interaction $\Phi_{red}(R)$ for the B-group elements. The electron density R_s has been set constant for each given Z (R_s=2.56 a.u. for Z=2, R_s= 2.20 a.u. for Z=3, R_s=2.05 a.u. for Z=4 and Z=5), the ratio (R_c/R_s) is given in parentheses for each element. The broken vertical marks the n.n. distance for close-packing.

groups Ia and IIa); (2) the existence of a d-electron core makes the pseudopotential strongly attractive (hence the atomic volume varies only very little from Mg to Hg, from Al to Tl, and from Si to Pb); and (3) the atoms become more hydrogenic as the valence Z increases (consequently the atomic volume shrinks in the series Na - Mg - Al - Si). R_c is to a first approximation the actual radius of the atomic core, shifted slightly due to the non-locality and to relativistic effects which confere an extra binding energy and result in a decrease of R_c with the atomic number. The actual values of R_c can be taken from an empirical analysis of band-structure data (for a complete compilation, see Table I in /1/).

The important effect for the change from metallicity to covalency in going from group I to IV is found to be the decrease of the atomic volume (and a forteriori of R_s) at essentially constant R_c/R_s. The structural changes within groups II to V with increasing atomic number are dominated by the decrease of R_c/R_s at constant R_s. Both effects are illustrated in Fig.4. This diagram summarizes very nicely our description of the trends in the crystal structures of the elements and their interpretation in terms of the interatomic potentials. At low electron densities and large core radii we find interatomic potentials with a minimum at the nearest neighbour distance D_{cp} for close packing, resulting in stable close-packed structures for the A-group elements and for Be,Mg and Al. As the electron density increases and /or the core radius decreases, the minimum is shifted relative to D_{cp} and flattened - this yields distorted structures for Zn, Cd, Hg, Ga, In, Si, Ge, Sn and the V-group elements at first, but then a return to close-packed structures for Tl and Pb as the last trace of an oscillation in $\Phi(R)$ near D_{cp} has vanished.

The same physical effects also dominate the trends in the liquid structures. To show this we must first discuss the statistical mechanical theories which allow us to pass from the pair potential to the pair correlation function g(R).

3. FROM THE INTERATOMIC FORCE LAW TO THE STRUCTURE OF THE LIQUID

We dispose essentially of three different techniques which enable us to calculate g(R) from $\Phi(R)$: (1) computer simulation (molecular dynamics or Monte Carlo), (2) the solution of one of the integral equations of the theory of liquids (see e.g./30/), and (3) thermodynamic perturbation theory. The disadvantage of the computer simulations is clearly that they are too expensive to allow us to study trends (although this is certainly a subjective statement dictated by circumstances), the difficulty with the integral equation approach lies in the fact that none of the proposed integral equations handles both parts of metallic interactions (the short-range repulsions and the oscillatory potential) equally well - this difficulty could possibly be overcome by using one of the recently proposed parametrized integral equations /31/. So to date thermodynamic perturbation theory seems to offer the optimal compromise between efficiency and accuracy. It is based upon a separation of the pair interactions into a purely repulsive short-range-interaction $\Phi_0(R)$ and the remaining essentiall long-range interaction $\Phi_1(R)$ according to

$$\Phi(R) = \Phi_0(R) + \Phi_1(R) \tag{7a}$$

$$\Phi_0(R) = \begin{cases} \Phi(R) - \Phi(R_0) & R < R_0 \\ 0 & R > R_0 \end{cases} \tag{7b}$$

$$\Phi_1(R) = \begin{cases} \Phi(R_0) & R < R_0 \\ \Phi(R) & R > R_0 \end{cases} \tag{7c}$$

where R_o is the distance at which the pair interaction has its first minimum (or an inflection point, if the first minimum has been covered by the repulsive core).

Any integral equation for the correlation function may be cast into the form of an Ornstein-Zernike equation (n is the number density)

$$h(R_{12}) = c(R_{12}) + n\int c(R_{13})\ h(R_{32})\ d^3R_3 \qquad (8)$$

for the total correlation function h(R) (g(R)=h(R)-1) and the direct correlation function c(R), supplemented by a closure condition relating h(R), c(R) and the pair potential Φ(R). In analogy with equ.(7) we separate the correlation functions into two parts

$$h(R) = h_o(R) + h_1(R) \qquad (9a)$$

$$c(R) = c_o(R) + c_1(R). \qquad (9b)$$

Assuming for the moment that we know the solution of the Ornstein-Zernike equation of the Φ_o-system,

$$h_o(R) = c_o(R) + n\int c_o(R_{13})\ h_o(R_{32})\ d^3R_3, \qquad (10)$$

we can subtract the purely repulsive Φ_o-couplings from both sides of equ. (8) to derive a residual Ornstein-Zernike equation describing the effect of the long-range interactions

$$\begin{aligned} h_1(R_{12}) = c_1(R_{12}) &+ n\int c_o(R_{13})\ h_1(R_{32})\ d^3R_3 \\ &+ n\int c_1(R_{13})\ h_o(R_{32})\ d^3R_3 \\ &+ n\int c_1(R_{13})\ h_1(R_{32})\ d^3R_3\ . \end{aligned} \qquad (11)$$

An entire hierarchy of thermodynamic perturbation approaches may be built upon equs. (7) to (11):

(1) The simplest approach consists in replacing the correlation functions $c_o(R)$ and $h_o(R)$ by those of a reference system with a simple interaction $\Phi_{ref}(R)$ for which an analytic solution of (1o) is available. The effect of the long range interactions is included in a perturbative way, for the free energy we have

$$F = F_{ref} + \frac{n}{2}\int g_{ref}(R)\ \Phi_1(R)\ d^3R. \qquad (12)$$

The parameters of the reference system are determined by a minimization of the right hand side of (12) in the spirit of the Gibbs-Bogoljubov inequality /32/. Appropriate reference systems for pure liquid metals are a hard-sphere liquid /33-35/ or a one-component-plasma /36-38/, but recently reference systems with an intermediate character (charged hard spheres with Coulomb or Yukawa tails) have also been proposed /39,40/.

(2) In the next step one could correct for the difference between the exact Φ_o and the reference potential Φ_{ref}. If $\Phi_{ref}(R)$ is a hard-core interaction, this is most conveniently done by applying the so-called "blip-function" expansion (the blip-function measures the softness of the potential), which amounts to /41/

$$g_o(R) = g_{ref}(R)\ \exp(\beta\Phi_{ref}(R)-\beta\Phi_o(R))\ . \qquad (13)$$

The free energy is again given by (12).

(3) The long-range-interactions are most conveniently treated in the "optimized random phase approximation" (ORPA) defined by the following closure conditions to the residual Ornstein-Zernike equation (11)

$$h_1(R) = 0 \qquad R < \sigma \tag{14a}$$

$$c_1(R) = -\beta\Phi_1(R) \qquad R > \sigma\,, \tag{14b}$$

where σ is an effective hard-core diameter of a reference interaction replacing Φ_o /42/. Equ.(14a) states the impenetrability of the hard-core, (14b) assumes the random-phase approximation for the long-range interactions. Note that the ORPA is equivalent to the mean spherical approximation (MSA) for the residual Ornstein-Zernike equation (11) with $\Phi_o(R)$ replaced by a hard-sphere potential /43/, soft-core effects are included via (13). The consistency of the separate treatments of $\Phi_o(R)$ and $\Phi_1(R)$ is achieved by an iterative algorithm /44,45/. Attempts have been made the develop an ORPA based on the OCP reference system /46,47/ - however, there are a lot of problems associated with this particular approach (the form (14) of the closure conditions applies only to hard repulsive interactions, for soft potentials this should be replaced by the more general form of the MSA /43/, the approach is not self-consistent - for detailed discussions see /48/).

In its form based on a hard-sphere reference system, the ORPA offers a very accurate description of the structure and of the thermodynamics of pure liquid metals, and it is applicable to all liquid metals - from the very soft alkali metals to the polyvalent metals with rather harshly repulsive short-range interactions. Therefore most of the following discussions will be based upon the ORPA.

4. TOPOLOGICAL SHORT-RANGE ORDER IN LIQUID METALS

We begin by illustrating the relative role of the short-range and the long-range forces in determining the liquid structure at two extreme examples, the alkali metal Rb and the polyvalent metal Ge (Ge is metallic when molten!). Fig.5 shows the pair potential (based on an empty-core model - for these interactions molecular dynamics simulations by Rahman /49/ are available), the pair correlation function and the static structure factor of liquid Rb. The results clearly demonstrate the effects of the different components in the atomic interaction potential: the softness of the repulsive part rounds (via the "blip-function" procedure) the first peak in $g(R)$, the effect of the long-range oscillations $\Phi_1(R)$ is essentially to enhance $g(R)$ at distances where $\Phi_1(R)$ is attractive, and to reduce it where $\Phi_1(R)$ is repulsive. Thus we find that the primary effect of the Friedel oscillations is to order the liquid (by localizing the positions of the nearest neighbours) to a somewhat greater extent than it would be the case in their absence. In $S(q)$ this is mainly reflected by a progressive damping of the higher-order oscillations. However, near the melting point these effects are not strong enough to induce a qualitative change in an overall still hard-sphere like structure. Only at higher temperatures the attractive interactions are found to be responsible for the long-wavelength density fluctuations resulting in a pronounced increase in $S(0)$ (note however that - as the liquid-gas transition is coupled to a metal-insulator transition(see Prof. Hensels lecture in these proceedings) - the breakdown of the linear screening approximation prevents an extension of the theory up to the critical point/44/)

The situation is quite different for the tetravalent element Ge (see Fig.6). Here we find that the hard-sphere picture represents a rather poor approximation, and although the "blip-function" expansion does a rather good job in softening the core (as indicated by the left-hand side of the first peak in $g(R)$), agreement between theory and experiment is reached only after including the long-range oscillatory forces. Again we

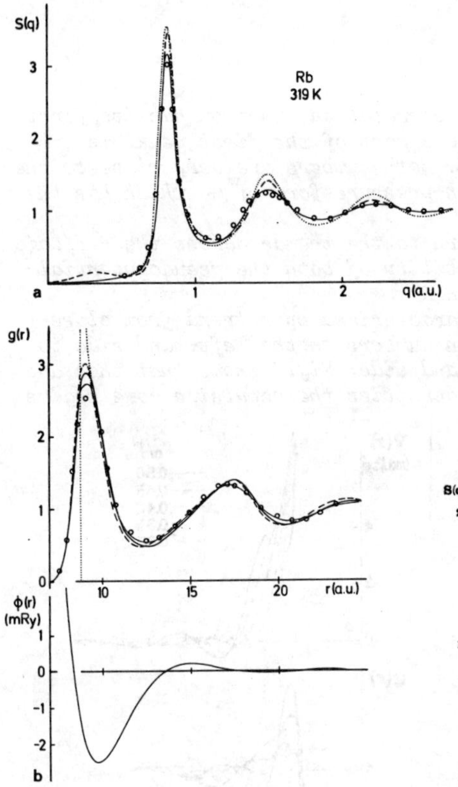

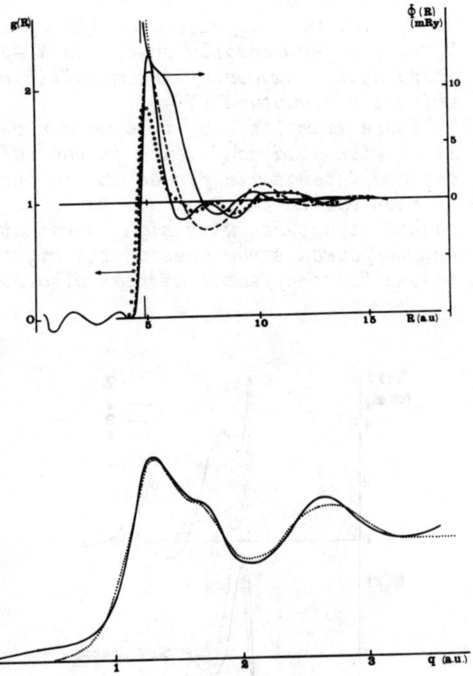

FIGURE 5. Comparison of the calculated pair correlation function g(R) and of the static structure factor S(q) for liquid rubidium with the results of a computer simulation using the same pair potential Φ(R). Open circles - molecular dynamics /49/, dotted - hard sphere approximation, dashed - blip-function approach, full line - ORPA without (bold) and with (thin) Jacobs-Andersen corrections. After /48/.

FIGURE 6. The structure of liquid germanium close to its melting point (T=1250 K): (a) effective pair potential Φ(R) and pair correlation function g(R), (b) static structure factor S(q). The symbols have the same meaning as in FIG. 5, the open circles representing the experimental results of Gabathuler and Steeb /15/. After /45/.

find that the effect of the Friedel oscillations is to force the atoms into the attractive wells of the potential. Thus in real space the structure of molten Ge is determined by the interplay of two charactersitic distances: the effective hard-core diameter σ expressing the geometrical requirements of sphere packing and the Friedel wavelength λ_F *setting the periodicity preferred by the intermediate and long-range interactions - in the static structure factor they show up as the peak near* $q \sim 2\pi/\sigma$ *and the high-angle shoulder at* $q \sim 2k_F$*. The fact that these shoulders appear preferrentially close to* $q=2k_F$ *has already been noticed by Beck and Oberle /50/, similarly encouraging results for other polyvalent metals have been obtained by dif-*

ferent groups /51-54/. It is interesting to note that the coordination number N_c *defined by*

$$N_c = 2n \int_0^{R_{max}} g(R)\, 4\pi R^2\, dR \qquad (15)$$

(R_{max} *is the distance at which the pair correlation function has its first peak, i.e. we integrate over the symmetric part of the first peak) is* N_c*=6.6, experimentally one finds* N_c*=6.5. Both numbers are very close to the coordination number of the metallic high-pressure form of Ge which has the white-tin structure (*N_c*=6).*

These examples show that we can return to the trends across the Periodic Table with good confidence in the reliability of both the pseudopotential-derived interatomic forces and in the ORPA.

Each row in the Periodic Table is characterized by a trend from close-packed structures with high coordination numbers on the left-hand side to loosely packed structures on the right-hand side. Fig.7 shows that this due to the "core-effect": at high electron densities the repulsive core begins

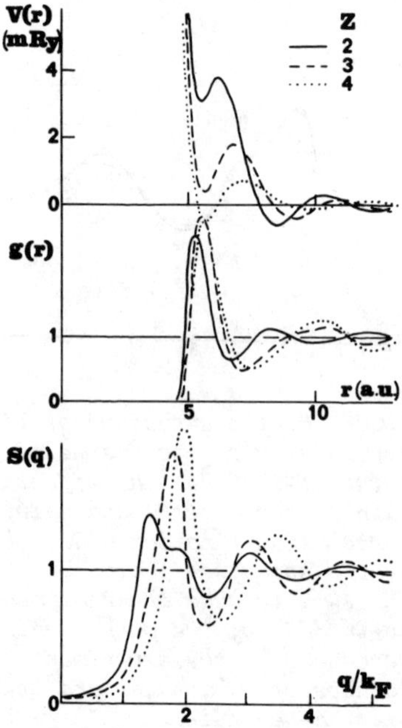

FIGURE 7. Variation of the interatomic potential Φ(R) and of the liquid structure with the valency Z at constant R_c/R_s*=0.45, reflecting the influence of the "core-effect" upon the transition from close-packed to losely packed structures, see text. After /54,55/.*

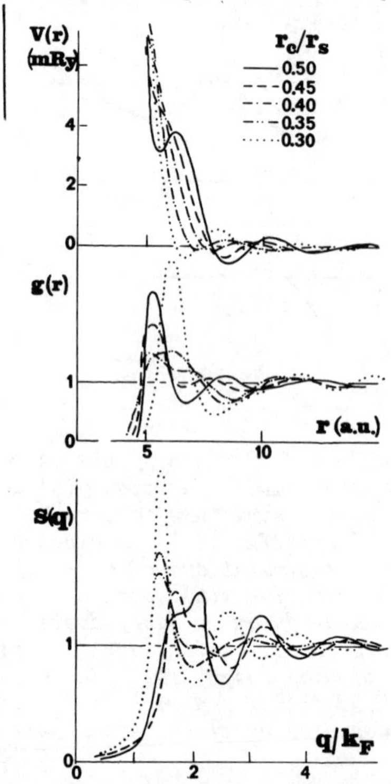

FIGURE 8. Variation of the interatomic potential Φ(R) and of the liquid structure with the ratio R_c/R_s *and constant Z=4 and constant atomic volume - this shows the return to hard-sphere-like structures for high atomic numbers (Si-Ge-Sn-Pb) is triggered by the "amplitude-effect". After /54,55/.*

to cover the attractive minimum. For Na,Mg and Al the first peak of a hard-sphere-like $g(R)$ falls into the main attractive minimum of the pair potential, and the higher-order maxima fit quite well into the minima of $\Phi(R)$ - clearly such a structure will be energetically favourable and hence stable. For Si, the hard-sphere peak has moved onto a repulsive hump of $\Phi(R)$ and the atoms rearrange as described above for Ge. Using equ.(15) we calculate N_c = 8.8, 9.5, 9.6, and 6.6 for Na, Mg, Al, and Si respectively.

The vertical trend within groups II to IV is again related to the "amplitude effect". This is illustrated in Fig.8 at the example of the tetravalent elements. With increasing atomic number the oscillations in $\Phi(R)$ are even stronger damped. For Pb finally the potential is purely repulsive at the nearest neighbour distance, the higher order oscillations are overdamped and therefore not effective. Note that using an inversion of the experimental structure data (based essentially on a blip-function +RPA procedure), Meyer et al /56/ found the same type of purely repulsive,non-oscillatory potential for liquid Pb.

To summarize: the main effect determining the liquid structure is the compatibility of the geometric and the electronic factors. At low electron densities and large core radii both are compatible with HS-like structures. As the electron density increases and/or the core radius decreases, this "constructive interference" between the pair interaction and the HS-pair correlation function disappears. The diameter of the repulsive core and the Friedel wavelength now set two characteristic distances - the complex structures of Zn, Cd, Hg, Ga, In, Si, Ge, and Sn result from the interplay of the two different distance scales. But finally the last trace of oscillations around the nearest-neighbour distance has disappeared and only the geomentrical factors prevail - this explains the return to a HS-like packing in Tl and in Pb.

5. ALLOYING EFFECTS ON INTERATOMIC INTERACTIONS

Formally the theory for deriving the atomic interactions is easily generalized to binary systems: they are now described by a set of three pair potentials which may be chosen to represent the interactions between AA, AB, and BB pairs (the defining equations are straightforwardly generalized from (2), $\{v_N(q)\}^2$ being replaced by $v_N^i(q)v_N^j(q)$, $i,j=A,B$) - alternatively they may be expressed in terms of an average potential coupling to the fluctuations in the mean number density,

$$\Phi_{NN}(R) = c_A^2\Phi_{AA}(R) + c_B^2\Phi_{BB}(R) + 2c_Ac_B\Phi_{AB}(R) \quad , \tag{16a}$$

an ordering potential coupling to the concentration fluctuations

$$\Phi_{CC}(R) = c_Ac_B(\Phi_{AA}(R) + \Phi_{BB}(R) - 2\Phi_{AB}(R)) \quad , \tag{16b}$$

and a cross-term

$$\Phi_{NC}(R) = 2c_Ac_B(c_A\Phi_{AA}(R) - c_B\Phi_{BB}(R) + (c_B-c_A)\Phi_{AB}(R)) \quad . \tag{16c}$$

Unfortunately it turns out that for binary alloys it is no longer possible to base a discussion even of the rough trends upon simple local model potentials . This would yield entirely misleading results. The reason for the failure of local model potentials to describe binary alloys is rather easily understood. According to Pauling's /57/ early interpretation the chemical bond in binary metallic systems is dominated by two competing charge transfer effects: First, electrons are transferred into the direction expected from the electronegativity difference. This creates relatively strong potential gradients which enhance the electron-electron potential energy. The system will try to reduce the potential energy by redistributing the electrons accordingly. This is what Pauling calls the electro-

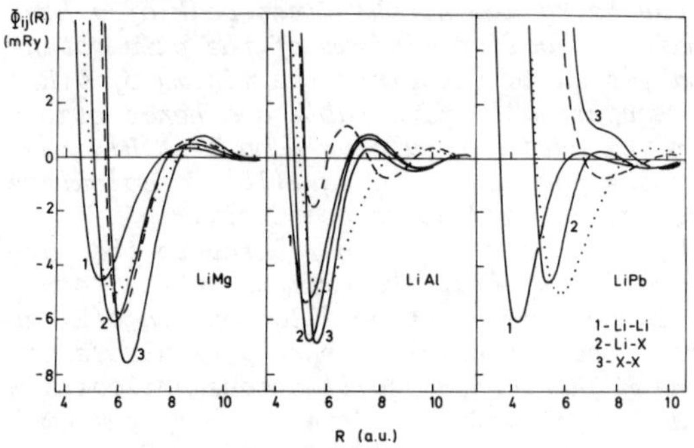

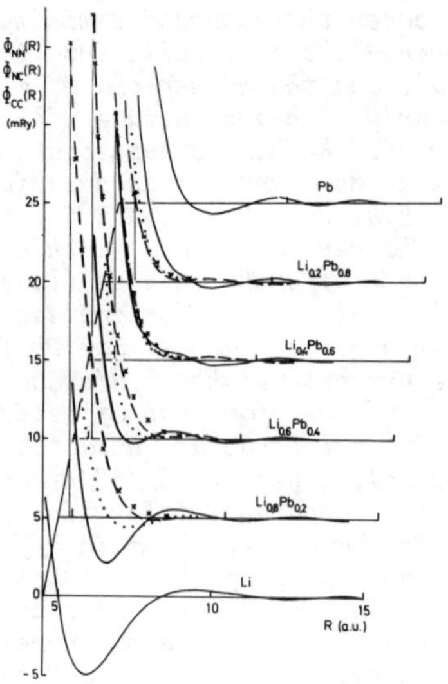

FIGURE 9. Interatomic pair potentials in binary alloys.
(a) Equiatomic alloys of Li with Mg, Al, and Pb. The dotted and dashed lines show the pair interactions in pure Li and Mg, Al, and Pb metals.
(b) Average pair potential $\Phi_{NN}(R)$ ———, ordering potential $\Phi_{CC}(R)$ - - - -, and cross term $\Phi_{NC}(R)$ ····· for Li_xPb_{1-x} alloys. The crosses show the reference-ordering interaction determined from a variational ansatz for a hard-sphere-Yukawa reference system. After /61/.

neutrality principle and it causes a second electron transfer mechanism opposite to the first one.

Hafner /3-5/ has shown that Pauling's ideas are directly reflected in a first-principles pseudopotential formalism. If we introduce the A and B ions forming an AB alloy into the homogeneous electron gas of the average density forming the unperturbed ground state of the system, the one which sees an increased electron density (relative to the pure metal, say A) will have to repell more electrons from the core region to satisfy the Pauli principle. This means that the A-ions have to form a larger orthogonalization hole and that the repulsive component of the A-pseudopotential will be stronger in the alloy than in the pure metal, the cancellation effect being altogether stronger. On the B-sites, the changes of the pseudopotential and of the orthogonalization hole are just the opposite. Thus the orthogonalization effect creates a charge transfer from A to B - which is just the direction expected from the electronegativity difference. In the next step the effective charges (=valence charge plus orthogonalization hole charge) are screened by the electron gas. Now the A ions will be screened more tightly, the B ions more loosely, than in the pure metals in order to level out too strong gradients in the local potential energy. In an actual calculation we find that the additional screening charge necessary to compensate for the increased orthogonalization hole charge on the A-ions is accumulated near the core region - as we would expect it according to the electroneutrality principle. For a more detailed discussion of these charge transfer effects in crystalline intermetallic compounds, see /58,59/. Here we are more interested in their manifestations in the interatomic potentials. We have to remember that both effects affect charge densities of a rather different character: the orthogonalization hole charges are localized, the screening charges are extended and *have a large overlap. Fig. 9 shows that the alloying effect in the interatomic potentials explains two important phenomena: (1) the "chemical compression" of an electropositive ion upon alloying with a more electronegative component /59,60/, and (2) the increasing tendency to chemical ordering with increasing difference in valence and/or electronegativity /12,13,16/. The chemical compression arises from the concentration of the screening charge around the electropositive ion - the Coulomb repulsion is screened over much shorter distances. It has been demonstrated that this effect is very important in understanding the stability of many topologically close-packed intermetallic compounds with a strongly non-ideal nominal radius-ratio of the components /5/. Because the ions are not completely screened over the average nearest-neighbour distance, the charge redistribution also causes a progressive non-additivity of the pair interactions - the interaction between AB nearest neighbour pairs is more attractive than the average interaction in AA and BB pairs. Clearly this results in an ordering potential $\Phi_{CC}(R)$ which is repulsive for these distances (with the sign-convention adopted in equ. (16b) a positive $\Phi_{CC}(R)$ favours heterocoordination). Note that the ordering potential has nearly an exponentially screened form - this is a consequence of the fact that all the three pair potentials $\Phi_{AA}(R)$, $\Phi_{AB}(R)$, and $\Phi_{BB}(R)$ oscillate assymptotically with the same wavelength and phase - and not an expression of an ionic character in the bond. That the compensation is complete already at the distance of the second nearest neighbours is however quire surprising.*

Qualitatively, this suggests that the pseudopotential approach explains the observed trends towards chemical short-range order, but this should of course be assessed by an actual calculation of the structure of the liquid alloy.

6. THERMODYNAMIC PERTURBATION THEORY FOR BINARY MIXTURES

Formally, the thermodynamic perturbation theories sketched in Sec.3 are straightforwardly generalized to multicomponent systems, but an actual calculation for a non-ideal mixture turns out to be quite difficult.

At the level of a thermodynamic variational calculation the difficulty consists in finding a reference system capable of simulating chemical short-range order. Clearly a mixture of hard spheres of different sizes is not an appropriate reference system because it is - apart from certain weak size-effects - always chemically random. Copestake et al /62-63/ have shown that the structural manifestations of ordering in molten salts, liquid semiconductors, and even in liquid metallic alloys may be modelled by a mixture of hard spheres having all the same diameter, but opposite charges - however respecting the overall charge neutrality condition

$$c_A Q_A + c_B Q_B = 0 \qquad (17)$$

and interacting through Coulomb or screened Coulomb (Yukawa) forces. For such a reference system, an analytical solution of the mean spherical approximation is available /64/. Hafner et al /65-66/ have demonstrated that, combined with a very simple treatment of the electronic contributions to the free energy of the alloy, this model may be used to fit the thermodynamic excess functions and the partial static structure factors of a large series of Li- and Na-based alloys with good success. Very recently, Pasturel et al /61/ have shown that a thermodynamic variational approach to chemical short-range order may be based upon this reference system.

In terms of the average and ordering potentials, the reference interactions are given by /61/

$$\Phi_{NN}(R) = \begin{cases} \infty & R < \sigma \\ 0 & R > \sigma \end{cases} \qquad (18a)$$

$$\Phi_{CC}(R) = \begin{cases} \infty & R < \sigma \\ c_A c_B \Delta Q^2 \; exp(-\kappa(R-\sigma))/R & \\ = -\varepsilon\sigma \; exp(-\kappa(R-\sigma))/R & R > \sigma \end{cases} \qquad (18b)$$

$$\Phi_{NC}(R) = 0, \qquad (18c)$$

where σ is again the hard-sphere diameter, $\Delta Q = Q_A - Q_B$ the difference in the component-charges, and κ the screening constant of the Yukawa potential. ε defined by the second equality in (18b) measures the strength of the ordering potential at hard contact. Because of (18c) the three coupled integral equations of the mean spherical approximation (MSA) decouple into two independent equations. One, with the closure conditions (to the Ornstein-Zernike equations)

$$h_{NN}(R) = -1 \qquad R < \sigma \qquad (19a)$$

$$c_{NN}(R) = 0 \qquad R > \sigma \qquad (19b)$$

describes the fluctuations in the mean number density and is identical to the Percus-Yevick equation for hard spheres, hence we know its analytical solution /67/. The second, with the closure conditions

$$h_{CC}(R) = 0 \qquad R < \sigma \qquad (20a)$$

$$c_{CC}(R) = \beta\varepsilon\sigma \; exp(-\kappa(R-\sigma))/R \qquad R > \sigma \qquad (20b)$$

describes the local fluctuations in the concentrations. Its analytical solution has been given by Waisman /64/. Within this reference system, the variational conditions assume the form

$$\left.\frac{\partial \overline{F}(\sigma,\varepsilon,\kappa)}{\partial \sigma}\right|_{T,n,\varepsilon,\kappa} = 0 \qquad (21a)$$

$$\left.\frac{\partial \overline{F}(\sigma,\varepsilon,\kappa)}{\partial \varepsilon}\right|_{T,n,\sigma,\kappa} = 0 \qquad (21b)$$

$$\left.\frac{\partial \overline{F}(\sigma,\varepsilon,\kappa)}{\partial \kappa}\right|_{T,n,\sigma,\varepsilon} = 0 \qquad (21c)$$

with the variational upper bound to the exact free energy /61/

$$\overline{F}(\sigma,\varepsilon,\kappa) = \tfrac{3}{2}k_BT + E_o + E_p(\sigma,\varepsilon,\kappa) - T\,(S_{HS}(\sigma) + S_{ord}(\sigma,\varepsilon,\kappa)), \qquad (22)$$

where E_o is a volume-energy, and E_p the pair energy

$$E_p(\sigma,\varepsilon,\kappa) = 2\pi\, n \int \{g_{NN}^{HSY}(R;\sigma)\Phi_{NN}(R) + g_{CC}^{HSY}(R;\sigma,\varepsilon,\kappa)\Phi_{CC}(R)\}R^2 dR \qquad (23)$$

expressed in terms of the reference-system correlation functions. Explicit expressions for these as well as for the HS- and the ordering contributions to the entropy (S_{HS} and S_{ord}) are given in /61/. Note that the complete decoupling of number-density and concentration fluctuations is an artefact of this reference system and of the MSA in particular.

The properties of the reference system, and especially the variation of the thermodynamic functions and of the partial structure factors have been investigated by Pasturel et al /61/ and by Holzhey et al /68/. The essential point is that at a given strength of the ordering potential at hard contact (i.e. for a fixed ε), the ordering effects do not vary monotonically with the strength of the screening: the strongest ordering effects appear if the ordering potential is restricted just to the nearest-neigh-

bour shell. For a longer-ranged ordering potential the competition between nearest- and next-nearest neighbour interactions tends to reduce the ordering effects. An application of this particular variational method is described below.

The blip-function expansion has also been applied to binary alloys /69-71/. Again the main problem is that realistic pair interactions tend to be non-additive, whereas the available reference system (a mixture of hard spheres of different diameters) is additive. A possible solution of the resulting problem is indicated in /71/. Only first attempts have been made to apply the ORPA to a binary system /72/. So at the present stage the HSY-variational method remains - apart from a few attempts to use molecular dynamics /73,74/ - the only viable way to calculate the structure of chemically ordering alloys.

7. CHEMICAL SHORT RANGE ORDER IN LIQUID SIMPLE-METAL ALLOYS

We begin by studying systems with a rather weak tendency to CSRO, for which the approximations inherent in the choice of a HSY-reference system should be reasonably well satisfied. The first alloy selected was Mg_7Zn_3 - for this composition X-ray scattering experiments indicate a rather pronounced CSRO /75,76/. The pair potentials for this alloy (Fig.10) show that this system is indeed ideally suited for the HSY-variational method: $\Phi_{NC}(R)$ is close to zero for all distances. The variational problem (2o-22) has a well defined solution: the hard-sphere diameter σ satisfies the condition

$$\Phi_{NN}(\sigma) = \Phi_{min} = \frac{3}{2} k_B T \qquad (24)$$

(Φ_{min} is the value of $\Phi_{NN}(R)$ at the first attractive minimum). This simple rule expresses the fact that the variationally determined hard-sphere diameter σ is just an average collision distance /35,77/. The parameter ε obeys the rule

$$\Phi_{CC}(\sigma) = - \varepsilon \quad , \qquad (25)$$

altogether we find that the variationally determined reference ordering potential models the exact $\Phi_{CC}(R)$ very closely (Fig.10).

Fig. 11(a) shows the partial structure factors $S_{NN}(q)$ and $S_{CC}(q)$ at two different temperatures - as one would intuitively expect, the ordering effect is rather strongly temperature dependent. Fig. 11(b) shows the composite (X-ray weighted) static structure factor S(q)

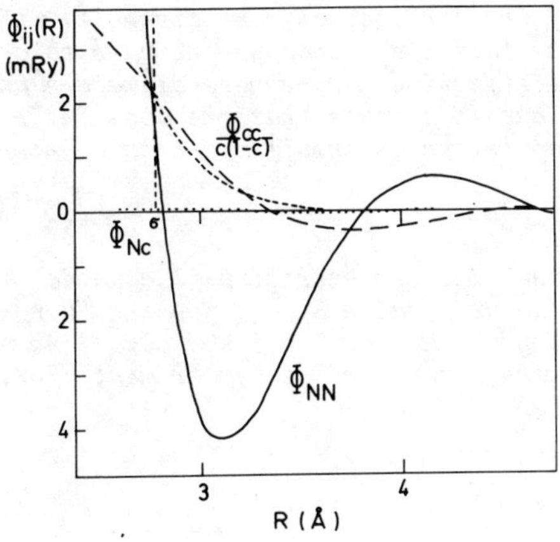

FIGURE 10. Pair potentials $\Phi_{NN}(R)$ (full line), $\Phi_{NC}(R)$ (dotted line) and $\Phi_{CC}(R)$ (long dashed line) for a liquid Mg_7Zn_3 alloy at T= 673 K. The vertical line marks the effective hard-core diameter, the short dashes show the variationally determined reference ordering potential $\Phi_{CC}^{HSY}(R)$. After ref. /61/.

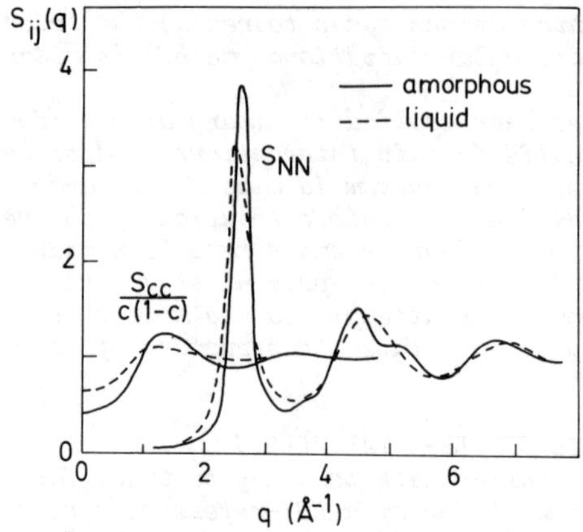

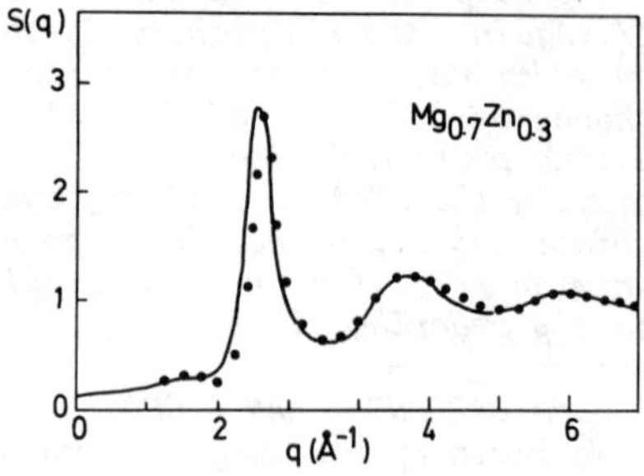

FIGURE 11. (a) Partial structure factors $S_{NN}(q)$ and $S_{CC}(q)$ for liquid (T=673 K) and for supercooled liquid (T= 373 K) Mg_7Zn_3 alloys (left). (b) Composite (X-ray weighted) static structure factor S(q) for liquid Mg_7Zn_3. Solid lines - theory, dots - experiment (Refs. 75,76). After /61/.

$$S(q) = \{\langle f(q)\rangle^2 S_{NN}(q) + \Delta f(q)^2 S_{cc}(q)\}/\langle f(q)\rangle^2 \quad , \tag{26}$$

where $\langle f\rangle = c_A f_A + c_B f_B$, $\Delta f = f_A - f_B$, the $f_i(q)$ are the X-ray scattering form factors of the components. The observed agreement between theory and experiment is really quite good.

The next system studied was Li-Mg: the difference in the atomic volumes is quite small, so that the HSY-model should still be applicable. However, in this case $\Phi_{NC}(R)$ is not small compared to $\Phi_{CC}(R)$. Again the variational problem has a well defined solution which complies with rules (24) and (25). The composition Li_7Mg_3 is a "zero-alloy" (i.e. $\langle f\rangle = 0$) for neutron scattering, thus a diffraction experiment measures directly $S_{CC}(q)$. We note a really good agreement between theory and experiment (see Fig.12), and the same is found for the thermodynamic excess functions /61/. Thus we conclude that pseudopotential theory, combined with the HSY variational method yields a good microscopic description of weakly ordering liquid alloys.

We now turn to the strongly ordering systems such as Li-Pb. In this case a straightforward application of the variational conditions (21) yields no physically acceptable solution: the total free energy continues to decrease in the direction of very small effective hard-sphere diameters and very large ordering interactions ε. Thus it appears that the loss in free energy from the mean interparticle repulsions is insufficient to counter-

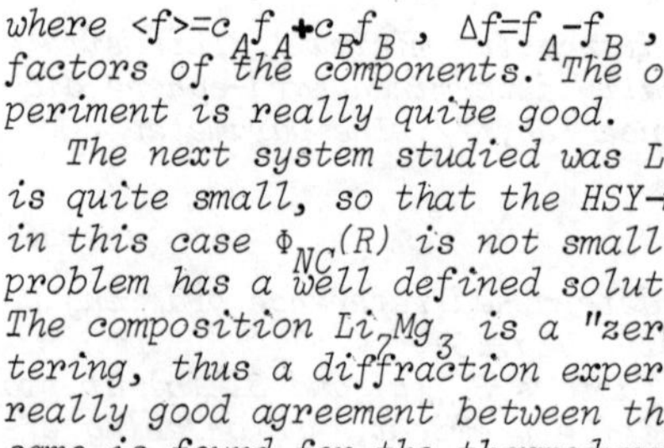

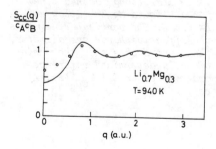

FIGURE 12. Concentration-fluctuation structure factor $S_{CC}(q)$ for the "zero-alloy" Li_7Mg_3 at T=940 K. Solid line - theory, open circles -experiment /78/. After /61/.

balance the energy gain from an ordering potential which increases strongly with decreasing distance, and that the difficulties arise from the complete decoupling of the number-density- and the concentration-fluctuations which is an artefact of the MSA-HSY model. Numerical solutions of the hypernetted chain equations for both hard- and soft-core Yukawa particles by Copestake et al /63/ and molecular dynamics calculations by Ronchetti et al /73/ for the same system have shown that even in the equal-diameter case the NC-correlation is non-negligible once the ordering potential is strong enough (unfortunately the solutions are only numerical and hence not useful as a basis for variational calculations). However, we find that we arrive at a physically realistic solution if we impose equ. (25) as an additional constraint and vary only σ and κ, i.e. the variational conditions are now given by

$$\left.\frac{\partial\bar{F}(\sigma,\varepsilon,\kappa)}{\partial\sigma}\right|_{T,n,\varepsilon,\kappa} = 0 \tag{27a}$$

$$\left.\frac{\partial\bar{F}(\sigma,\varepsilon,\kappa)}{\partial\kappa}\right|_{T,n,\sigma,\varepsilon} = 0 \tag{27b}$$

$$\Phi_{CC}(\sigma) = -\varepsilon \quad . \tag{27c}$$

The resulting reference-system interactions are compared with the exact potentials in Fig. 9(b), again equ. (24) is well obeyed and the variationally determined screening constant reproduces the decay of $\Phi_{CC}(R)$ very well. The calculated static structure factors for a series of Li_xPb_{1-x} alloys are shown in Fig.13 - the agreement with experiment is surprisingly good. The only slight discrepancy is that for the Li-rich alloy the higher order oscillations are a bit too strong. However, this seemingly small difference reflects a larger discrepancy in the thermodynamic excess functions /61/ which could be due to an overestimate of either ε or κ. Copestake et al /63/ and Ruppersberg and Schirmacher /80/ have attempted to derive an "experimental" ordering potential by inverting the MSA-equation (20).

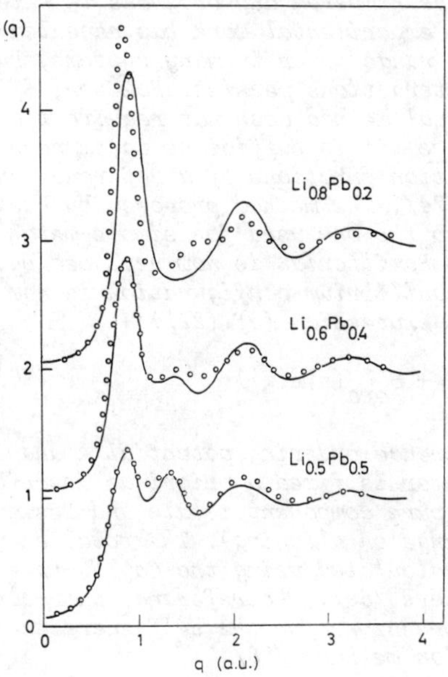

FIGURE 13. Composite (neutron-weighted) static structure factor S(q) for several Li_xPb_{1-x} alloys at T=1000 K. The alloy $Li_{80}Pb_{20}$ is very close to the composition of the "zero-alloy" so that S(q) is essentially identical to $S_{CC}(q)/c_Ac_B$, for larger Pb-concentrations there is an increasing contribution from $S_{NN}(q)$. Solid lines - theory, open circles - experiment /79/. After /61/.

The results correspond again quite closely to a Yukawa-type ordering potential with $\varepsilon=-20$ *mRy and* $\kappa=0.58$ *a.u. for* Li_8Pb_2*, from our variational calculation we had found* $\varepsilon=-30.2$ *mRy and* $\kappa=1.8$ *a.u.. This would suggest that our ordering potentials have about the right strength at hard contact, but are too strongly screened. Ruppersberg and Schirmacher argue that the small value of the screening constant is related to a reduced electronic density of states at the Fermi level compared to the free-electron value. Another manifestation of the reduced density of states would be the high value of the electrical resistivity at precisely this "stoichiometric" composition (see e.g. /81/).*

Thus it seems that our treatment of Li-based liquid alloys is a first, not unsuccessful step towards a microscopic theory of CSRO in metallic alloys, but two improvements are urgently called for: (1) Effects of a finite electronic mean free path must be incorporated in the screening function, perhaps along the lines proposed some time ago by de Gennes /82/. (2) We need a more realistic reference system, allowing for different effective hard-sphere diameters. A generalized MSA has been solved for such a system /83/, but the solution appears to be still too complicated to serve as the basis for a four-parameter ($\sigma_A, \sigma_B, \varepsilon, \kappa$) *variational calculation.*

Neverteheless, this is the direction which we will have to follow. Recent results on liquid alloys of the heavy alkali metals (K,Rb,Cs) with polyvalent metals indicate that in these systems thermodynamic, structural and electronic anomalies appear at two characteristic compositions (close to the "stoichiometric" composition, e.g. K_4Ge*, and close to the equiatomic composition KGe) /84,85/, and this is most likely due to an interference between size- and electronic effects.*

8. CHEMICAL SHORT RANGE ORDER IN LIQUID TRANSITION METAL ALLOYS

Up to now we have restricted our attention to simple metals and their alloys. The reason is twofold: first, for the pure liquid transition metals deviations from the hard-sphere like structures are very small /14/, secondly the difficulties associated with the construction of reliable interatomic potentials for transition metals have hampered any progress in this field. In transition-metal alloys recent experimental work has revealed the existence of a pronounced chemical short range order in many systems, but the difficulties with the interatomic interactions persist. However, if one realizes that the thermodynamic variational method does not require the explicit knowledge of the interatomic forces (it is sufficient to express the free energy as a function of the correlation functions of a reference system the way to progress is wide open again /74/. The method proposed by Pasturel and Hafner /74,88/ is very similar to the one used for simple-metal alloys, excepted that the pair potential Hamiltonian is now replaced by a d-band thight-binding Hamiltonian. The equilibrium configuration is now calculated by minimizing the variational expression (cf.(22))

$$\overline{F}(\sigma,\varepsilon,\kappa) = \frac{3}{2} k_B T + E_T(\varepsilon,\kappa) - T\,(S_{HS}(\sigma) + S_{ord}(\sigma,\varepsilon,\kappa)) \tag{28}$$

with respect to the strength of the reference-ordering potential ε *and the screenign constant* κ *(the hard-core diameter is fixed a priori as described in /74/) by fitting the entropies of the pure component metals and assuming that the hard-sphere volume does not change on alloying). The total energy* $E_T(\sigma,\varepsilon,\kappa)$ *for a given configuration is calculated using the Cayley-tree method introduced by Falicov and co-workers /89/. The reference pair-correlations* $g_{ij}(R)$ *enter for example the expressions for the self-energies* Δ_A, Δ_B. *In the Bethe-Cluster approximation we have /88/*

$$\Delta_A(z) = \frac{n\, c_A \int t_{AA}^2(R)\; g_{AA}(R)\; d^3R}{z - E_A - \Delta_A(z)} + \frac{n\, c_B \int t_{AB}^2(R)\; g_{AB}(R)\; d^3R}{z - E_B - \Delta_B(z)} \qquad (29)$$

and an ***analogous expression*** *for $\Delta_B(z)$. The $t_{ij}(R)$ are the transfer integrals and the E_i are the atomic energy eigenvalues. For each configuration represented by a pair of values ε and κ, the shift of the energy bands and the electron transfer are calculated self-consistently, the equilibrium configuration is determined by minimizing (28).*

This method is similar to other methods to calculate the CSRO in disordered systems /90/ which are also based on the Cayley-tree method and the HSY-reference system, but where the coupling was achieved by equating the spectrally defined electron transfer and the ionic charges of the HSY reference system. First of all this is necessarily somewhat ambigious (see the discussion in /74/) and furthermore this excludes the conceivable solution of a charge transfer without chemical ordering. The variational method avoids these problems and provides a sound thermodynamical basis.

First results for Ni-Ti are shown in Fig.14. As most available experimental results for partial structure factors refer to amorphous alloys, the calculation has been performed at a temperature of T=600 K, corresponding roughly to the glass-temperature of the alloy. A comparison of the theoretical result on a supercooled liquid at this temperature with an experiment on the glass presupposes of course that the degree of CSRO is about the same above and below the glass transition.

The electronic density of states shows the familiar shift of the d-band of the late transition metal (Ni) to higher binding energies /92/, the structure in the DOS is somewhat more pronounced in the ordered than in the disordered state. This is related to the fact that the CSRO tends to keep like atoms apart - the reduced overlap results in a narrowing of the

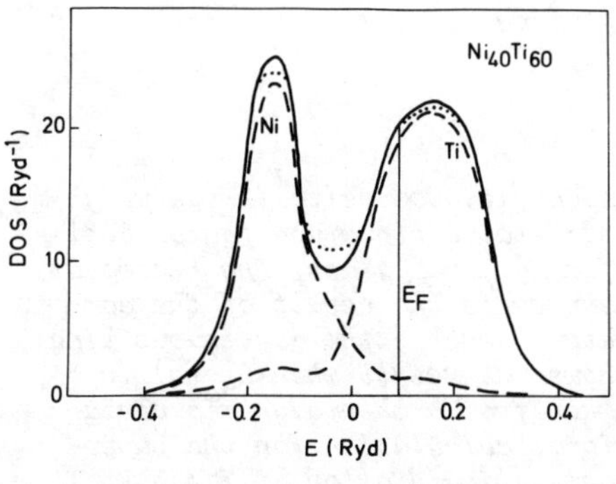

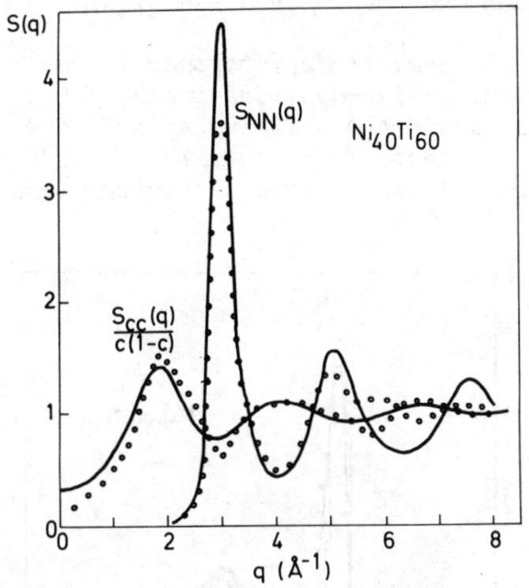

FIGURE 14. (a) Electronic density of states in supercooled (amorphous) $Ni_{40}Ti_{60}$: full line - including the variationally determined CSRO, dotted line - without CSRO, dashed lines - partial DOS's at Ni and Ti sites. (b) Partial structure factors $S_{NN}(q)$ and $S_{CC}(q)$ for amorphous $Ni_{40}Ti_{60}$. Full lines - variational calculation, open circles - isotope substitution experiment /91/. After Pasturel and Hafner /74/.

partial A- and B-type DOS. The partial structure factors agree very well with experiment - similarly encouraging results have been obtained for other alloys.

For the simple metal alloys we had found that the ordering potential depends on $|v_{ps}(q,E)|^2$ (non-local pseudopotentials are energy-dependent). For the transition metal alloys preliminary results on various Ni-based alloys /88/ suggest that, in accordance with the results of Gautier et al /93-95/ on crystalline alloys, the strength of the ordering interaction depends on the difference ΔE_d of the d-electron energies and on the band filling. The strongest effect appears for an approximately half-filled band and at large ΔE_d.

9. CHEMICAL SHORT RANGE ORDER IN AMORPHOUS ALLOYS

In practice the structure of an amorphous alloys depends on the interatomic forces and on the thermal history of the sample. A theoretical model will depend on the interatomic forces and on the modelling algorithm - the algorithm creates the "thermal history" of the model cluster. For a proper description of CSRO, we have virtually only two alternatives: (1) we can try to perform a computer-quench experiment, or (2) we can use the information on the CSRO in the supercooled liquid phase just above the glass transition to define a proper starting structure for a static relaxation calculation (supposing that the CSRO does not change during the glass-transition).

Hafner and Pasturel applied a molecular-dynamics quench-procedure to $Mg_{70}Zn_{30}$ (their quench procedure is rather similar to the one proposed by Weber and Stillinger /97/). The result is shown in Fig.15. The result of the computer experiment is very close - at least as far as CSRO is concerned - to the result of a thermodynamic variational calculation at T=373 K (which is just 20 K above the glass transition) and in good agreement with the experimental results. Thus it appears that the assumption that the glass and the supercooled liquid have the same degree of CSRO is legitimate.

This result is important in two respects. It assesses the usefulness of the molecular-dynamics quench for treating CSRO in glasses and it shows that again the ordering effects are first of all related to the form of the interatomic potentials. These investigation are now extended to other systems showing more pronounced ordering effects.

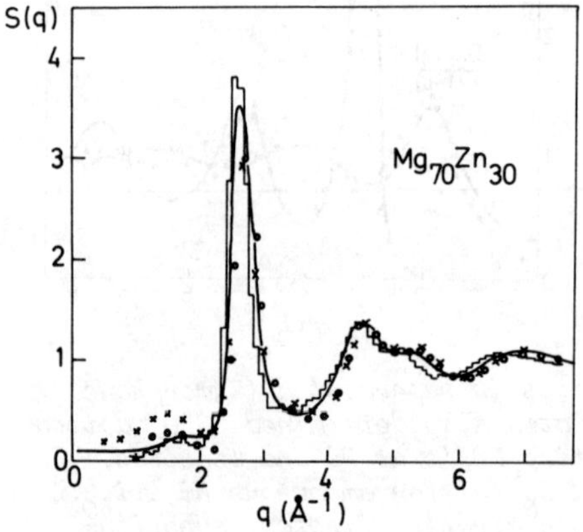

FIGURE 15. Composite (X-ray weighted) static structure factor S(q) of a $Mg_{70}Zn_{30}$ glass. The histogram represents the result of the computer quench , the continuous line shows the result when $S_{cc}(q)$ is taken from a thermodynamic variational calculation for the supercooled liquid alloy at T=373 K. The crosses and circles give the results of two different X-ray diffraction experiments /75,76/. After /74/.

10. CONCLUSIONS

In this paper we have presented a comprehensive microscopic theory for the topological short-range order in pure liquid metals and for the chemical short-range order in liquid and in amorphous alloys. The trends in the short-range topology through the Periodic Table are found to be driven by the interplay of geometrical (space-filling) effects and the electron-induced oscillatory form of the atomic interactions.

The microscopic interpretation of the origin of CSRO in simple-metal and in transition-metal alloys is based in two seemingly rather different concepts: one emphasizes the properties of the mobile s and p electrons such as the interstitial electron density and the chemical potential, the other focuses on the covalent bonding between free-atom-like d-states. In one case the electronegativity would be associated with the chemical potential (which is not necessarily to be identified with the Fermi energy), in the other with the position of the atomic energy eigenvalues.

The key to a reconciliation of both concepts lies in a consideration of the changes of the partial densities of states on alloying, even in the simple-metal case. Generally, one will expect the following: the lowest state in the alloy lies at the self-consistent mean potential (or pseudopotential) $\overline{W}$ in the alloy. The, if B is the more electronegative element (the one with the more attractive pseudopotential), states near the bottom of the band will have large amplitudes in the B-cells and small amplitudes in the A-cells, i.e. $n_B(E)>n_A(E)$ near $\overline{W}$, but both will converge to their free-electron form at large kinetic energies /98/. This means that the pseudopotential, averaged over the atomic cell, plays a role similar to the atomic energy eigenvalue; the concentration-dependent changes in the pseudopotentials parallel the role of the hybridization splitting of the energy levels. In a real material of course the pseudopotential is non-local, i.e. electrons with different angular momenta experience different potentials and the analogy becomes more difficult to follow. Here again we have made only a first step towards the exploration of a vast field, but it seems worthwhile to proceed.

ACKNOWLEDGEMENTS

This work has been supported by the Fonds zur Förderung der wissenschaftlichen Forschung in Österreich and by the Österreichische Forschungsgemeinschaft. A.P. thanks the Centre National de la Recherche Scientifique for a leave of absence.

REFERENCES

1. J.Hafner and V.Heine, J.Phys. F13, 2479(1983).
2. V.Heine and D.Weaire, Solid State Physics 24, 249(1970).
3. J.Hafner, J.Phys. F6, 1243(1976).
4. J.Hafner, Phys.Rev. B15, 617(1977).
5. J.Hafner, Phys.Rev. B21, 406(1980).
6. A.B.Maknovetskii and G.L.Krasko, phys.stat.solidi (b) 80, 341(1977).
7. P.C.Schmidt, Phys.Rev. B31 5015(1985).
8. N.E.Christensen, Phys.Rev. B32, 207(1985).
9. J.Hafner and W.Weber, Phys.Rev. B (submitted).
10. V.Heine and D.Weaire, Phys.Rev. 152, 603(1966).
11. D.Weaire, J.Phys. C1, 210(1968).
12. P.Chieux and H.Ruppersberg, J.Physique (Paris) 41, C8-321(1980).
13. W.van der Lugt and W.Geertsma, J.Non-cryst.Solids 61/62, 187(1984).
14. Y.Waseda, The Structure of Non-crystalline Materials - Liquids and Amorphous Solids (McGraw-Hill, New York 1980).

15. J.P.Gabathuler and S.Steeb, Z.Naturforschung 34a,1314(1979).
16. H.Ruppersberg, Metallwissenschaft und Technik, 33,1281(1979).
17. W.A.Harrison, Pseudopotentials in the Theory of Metals (Benjamin, New York 1966)
18. W.A.Harrison and J.M.Wills, Phys.Rev. B28,4363(1983).
19. J.Hafner, in "Amorphous Solids and the Liquid State", ed. by N.H.March, R.A.Street, and M.P.Tosi (Plenum, New York 1985), p.91.
20. N.W.Ashcroft, Phys.Lett. 23,48(1966).
21. M.W.Finnis, J.Phys. F4,1645(1974).
22. S.Ichimaru, Rev.Mod.Phys. 54,1017(1982).
23. G.Jacucci and R.Taylor, J.Phys. F11,787(1981).
24. J.Hafner and V.Heine, J.Phys. F (in print).
25. S.Ichimaru and K.Utsumi, Phys.Rev. B24,7385(1981).
26. P.Vashishta and K.S.Singwi, Phys.Rev. B6,875(1972).
27. D.J.W.Geldart and R.Taylor, Can.J.Phys. 48,167(1970).
28. R.Taylor, J.Phys.F8,1699(1978).
29. B.J.Austin and V.Heine, J.Chem.Phys. 45,928(1966).
30. C.A.Croxton, Liquid State Physics - A Statistical Mechnaical Introduction (Cambridge University Press, Cambridge 1974).
31. F.J.Rogers and D.A.Young, Phys.Rev. A30,999(1984).
32. A.Isihara, J.Phys. A1,539(1968); T.Lukes and R.Jones, J.Phys. A1,29 (1968).
33. H.D.Jones, J.Chem.Phys. 55,2640(1971).
34. I.H.Umar, A.Meyer, M.Watabe, and W.H.Young, J.Phys. F4,1691(1974).
35. J.Hafner, Phys.Rev. A16,351(1977).
36. H.Minoo, C.Deutsch, J.P.Hansen, J.Physique-Lettres 38,L191(1977).
37. M.Ross, H.E.de Witt, W.B.Hubbard, Phys.Rev. A24,2145(1981).
38. K.K.Mon, R.Gann, D.Stroud, Phys. A24,2145(1981).
39. J.B.Hayter, R.Pynn, and J.B.Suck, J.Phys. F13,L1(1983).
40. H.B.Singh and A.Holz, Phys.Rev. A28,1108(1983).
41. J.D.Weeks, D.Chandler, and H.C.Andersen, J.Chem.Phys. 55,1497(1971); for a variant of this method which gives slightly better results for small momentum transfers see R.E.Jacobs and H.C.Andersen, Chem.Phys. 10,73(1975).
42. H.C.Andersen, D.Chandler, and J.D.Weeks, Adv.Chem.Phys. 34,105(1976).
43. W.G.Madden and S.A.Rice,J.Chem.Phys. 72,4208(1980); W.G.Madden, J.Chem.Phys. 75,1984(1981).
44. G.Kahl and J.Hafner, Phys.Rev. A29,3310(1984).
45. G.Kahl and J.Hafner, Solid State Comm. 49,1125(1984).
46. D.K.Chaturvedi, G.Senatore, M.P.Tosi, Lett.Nuovo Cim. 30,47(1981).
47. G.Pastore and M.P.Tosi, Physica 124B,383(1984).
48. G.Kahl and J.Hafner, Z.Physik B58,282(1985).
49. A.Rahman, Phys.Rev. A9,1667(1974).
50. R.Oberle and H.Beck, Solid State Comm. 32,959(1979); H.Beck and R.Oberle, J.Physique 41,C8-289(1980).
51. C.Regnaut, J.P.Badiali, and M.Dupont, Phys.Lett. 74A,245(1979).
52. C.Regnaut and J.L.Bretonnet, J.Phys. F14,L85(1984).
53. J.L.Bretonnet and C.Regnaut, Phys.Rev. B31,5071(1985).
54. J.Hafner and G.Kahl, J.Phys. F14,2259(1984).
55. J.Hafner, J.Non-cryst.Solids 61/62,175(1984).
56. A.Meyer, M.Silbert, and W.H.Young, Phys.Chem.Liqu. 10,279(1981).
57. L.Pauling, The Nature of the Chemical Bond, 2nd. ed., Sec. 12.5, (Cornell University Press, Ithaca 1952).
58. J.Hafner, J.Phys. F15,1689(1985).
59. See e.g. W.Biltz and F.Weibke, Z.Anorg.Allgem.Chemie 223,32(1935).
60. J.Hafner, J.Phys. F15,L43(1985).

61. A.Pasturel, J.Hafner, and P.Hicter, Phys.Rev. B (Oct.1985).
62. A.P.Copestake, R.Evans, and M.M.Telo da Gama, J.Physique 41,C8-145 (1980).
63. A.P.Copestake, R.Evans, H.Ruppersberg, and W.Schirmacher, J.Phys. F13,1993(1983).
64. E.Waisman, J.Chem.Phys. 59,495(1973).
65. J.Hafner, A.Pasturel, and P.Hicter, J.Phys. F14,1137(1984);ibid. 2279.
66. J.Hafner, A.Pasturel, and P.Hicter, Z.Metallkunde 76,432(1985).
67. M.S.Wertheim, Phys.Rev.Lett. 10,321(1963).
68. Ch.Holzhey, J.R.Franz, and F.Brouers, J.Phys.F14,2475(1984).
69. S.I.Sandler, Chem.Phys.Lett. 33,351(1975).
70. S.Sung and D.Chandler, J.Che,Phys. 56,4989(1972)
71. G.Kahl and J.Hafner, J.Phys. F15,1627(1985).
72. G.Kahl, poster contribution at this school.
73. M.Ronchetti, G.Jacucci, and W.Schirmacher, in Proc. 3rd. Intern. Conference on the Structure of Non-crystalline Materials, ed. by Ch.Janot (Les Editions de Physique, in print).
74. J.Hafner and A.Pasturel, in Proc. 3rd. Intern. Conference on the Structure of Npn-crystalline Materials, ed. by Ch.Janot (Les Editions de Physique, in print).
75. H.Rudin, S.Jost, and H.J.Günterodt, J.Non-cryst. Solids 61/62,303(1984).
76. E.Nassif, P.Lamparter, W.Sperl, and S.Steeb, Z.Naturforschung 38a, 142(1983).
77. F.Igloi, phys.stat.solidi (b) 92,K85(1979).
78. P.Chieux and H.Ruppersberg, J.Physique 41,C8-321(1980).
79. H.Ruppersberg and H.Reiter, J.Phys. F12,1311(1982).
80. H.Ruppersberg and W.Schirmacher, J.Phys. F14,2787(1984).
81. V.T.Nguyen and J.E.Enderby, Phil.Mag. 35,1013(1977).
82. P.G. de Gennes, J.Physique 23,630(1962).
83. M.C.Abramo, C.Caccamo, and G.Pizzimenti, J.Chem.Phys. 78,357(1983).
84. W.van der Lugt, these proceedings.
85. J.A.Meijer, W.Geertsma, and W. van der Lugt, J.Phys. F15,911(1985).
86. S.Steeb and P.Lamparter, J.Non-cryst.Solids 61/62,237(1984).
87. C.N.J.Wagner, in Proc. Intern. Workshop "Amorphous Metals and Semiconductors", San Diego 1985, ed by R.I.Jaffee and P.Haasen (Acta Met. Press, in print).
88. A.Pasturel and J.Hafner, to be published.
89. L.M.Falicov and F.Yndurain, Phys.Rev. B12,5664(1975); R.C.Kittler and L.M.Falicov, J.Phys. C9,4259(1977).
90. Ch.Holzhey, F.Brouers, and W.Schirmacher, J.Non-cryst. Solids 61/62, 65(1984).
91. T.Fukunaga, N.Watabe, and K.Suzuki, J.Non-cryst.Solids 61/62,343(1984).
92. See e.g. P.Oelhafen et al., J.Non.cryst.Solids 61/62,1067(1984) and the contribution by P.Oelhafen in these proceedings.
93. F.Ducastelle and F.Gautier, J.Phys. F6,2039(1976).
94. A.Bieber and F.Gautier, Z.Physik B57,335(1984).
95. A.Bieber and F.Gautier, Solid State Comm. 38,1219(1981).
96. J.Hafner, in "Amorphous Solids and Non-Eqilibrium Processing", ed. by M.von Allmen, p.219 (Les Editions de Physique, Paris 1984).
97. T.A.Weber and F.H.Stillinger, Phys.Rev. B31,1954(1985).
98. M.Schlüter and C.M.Varma, Phys.Rev. B

THE ORIGIN OF MICROSTRUCTURES OF RAPIDLY SOLIDIFIED ALLOYS

Erhard Hornbogen

Institut für Werkstoffe, Ruhr-Universität Bochum, D-4630 Bochum, West-Germany

1 INTRODUCTION

At high temperatures the homogeneous, disordered, entropy stabilized structures are favored, such as a liquid alloy with random distribution of atoms. At low temperatures pure crystalline components or highly ordered intermetallic compounds will exist in stable, most frequently, heterogeneous equilibrium. This highly ordered and decomposed state may form during cooling of a liquid from $T > T_{f\alpha}$ (Fig. 1).

Glass formation is identical with freezing-in the disordered high temperature structure, i. e. cooling fast enough to a temperature at which diffusion is negligeable, to avoid nucleation and growth of crystals. In most alloys there exist a wide range of structures between glass and the structures which represent full equilibrium: ordered, relaxed, decomposed glass → icosahedral structures → metastable, homogeneous crystal structures → metastable heterogeneous structures → stable heterogeneous structures (i. e. two- or multiphase-structures). [1, 2]

Which structure forms depends on the particular cooling system. Atomizing of powders [3], melt spinning [4], and laser heating [5] are the most popular methods. A certain cooling rate $dT/dt = \dot{T}$ limits the number of atomic jumps. The cooling rate depends on the temperature difference which decreases with time

$$\dot{T} = dT/dt = \kappa\ (T-T_o) \quad (1)$$

and on the position x inside the material

$$\dot{q} = dq/dt = \lambda\ dT/dx \quad (2)$$

a particular temperature ($T^* < T_{kf}$) moves with a velocity dx/dt through the cross-section of the material (Fig. 1).

The cooling rate can be approximated by an average value $\bar{\dot{T}} = \Delta T/\Delta t$. The range between splat cooling and cooling in a parent body of a meteorite comprehends many orders of magnitude.

$$10^8\ K/s > \bar{\dot{T}} > 10^{-8}\ K/a\ . \quad (3)$$

For the analysis of relaxation, decomposition and crystallization reactions, it is more convenient to use isothermal heat treatment, which implies rapid down- or up-quenching and holding for a period Δt at T = const. The average number of atomic hops which take place during this period is

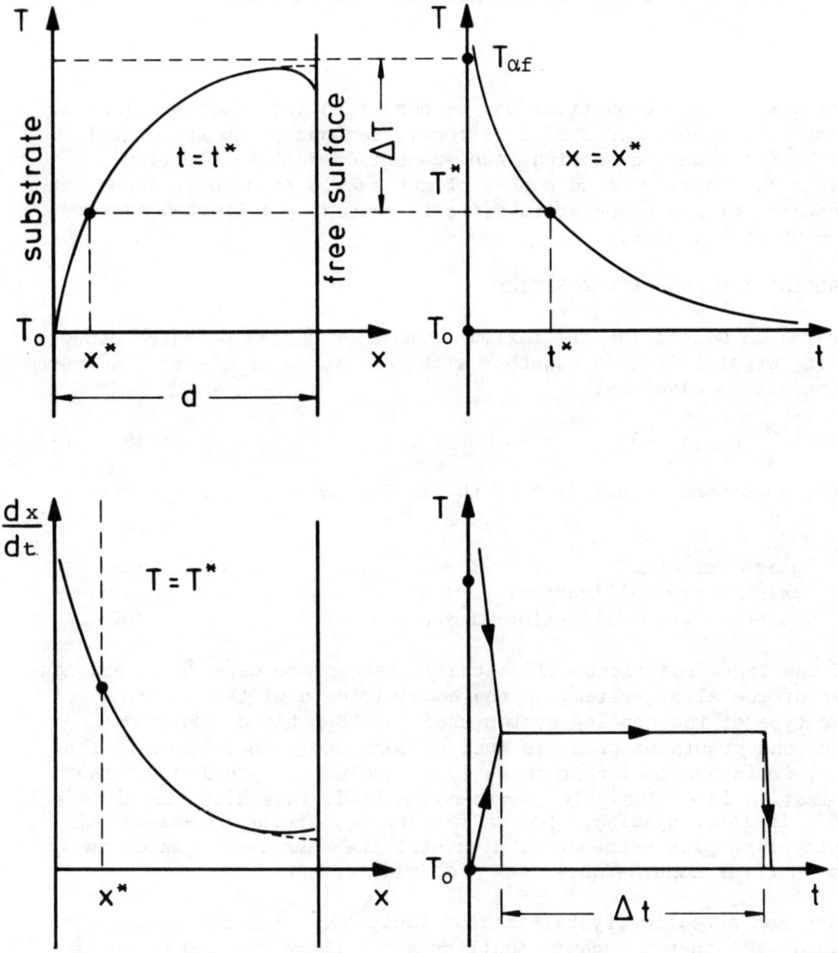

Fig. 1 Rapid cooling by heat transfer through one substrate, schematic.

A Temperature gradient at constant time t^*,

B Cooling rate at a site x^* inside a ribbon of thickness d ,

C Velocity of the temperature T^* as it moves across the ribbon ,

D Rapid cooling or heating followed by isothermal heat treatment.

controlled by the diffusion coefficient $D = D_O \exp - Q/RT$ and the atomic spacing b. $n < 1$ defines the requirement for freezing-in.

$$n = \frac{\sqrt{D\ t}}{b} \quad . \tag{4}$$

Continuous cooling conditions can be converted into isothermal ones, if the temperature dependence of D is considered and if no structural changes take place during cooling. Achievement of full equilibrium requires $n \to \infty$. The very large number of metastable structures found after rapid quenching or low temperature aging is caused by a limited number $1 < n << \infty$ of atomic hops.

2 CRYSTALLIZATION AND DECOMPOSITION

According to Gibb's law the maximum number of phases p, which can coexist during crystallization together with the liquid or glass in a system of k components is given by:

$$p \leq k + 1 = 3 \quad . \tag{5}$$

For two component systems($k = 2$) the following reactions may take place:

glass formation:	$l \to a$	(6A)
primary crystallization:	$l \to a + \alpha$	(6B)
eutectic crystallization:	$l \to a + \alpha + \beta$.	(6C)

Which of the three reactions will actually take place depends on the constitution of the alloy system, on the composition c of the alloy as well as on the type of the cooling system used [6]. For glass formation nucleation and growth of crystals must be completely inhibited. In the following, it is presumed that there is no nucleation problem, so that crystallization is exclusively growth-controlled. This situation is idealy fulfilled in laser-glazing. For melt spinning, it can be assumed that high undercooling plus presence of a crystalline substrate reduces the nucleation barrier considerably at $x = 0$ (Fig. 1 A, C).

Primary and eutectic crystallization imply two types of constitution of the alloy. The thermo-dynamic equilibria are characterized by phase diagrams(T-c) and free energy-diagrams (F-c, T = const., Fig. 2).

A necessary (existence of a driving energy) and a sufficient condition (mobility of the reaction front) are required for the occurence of crystallization. If competing reactions are existing, the fastest reaction will take place (v is the velocity of the reaction front $dx/dt = v$),

$$v \to v_{max} \, , \tag{7A}$$

i.e. the reaction doesn't take place necessarily which leads to a maximum decrease in free energy. The velocity of a reaction front v is determined by the product of a driving energy for crystallization f_c and the mobility of the reaction front m. If the mechanisms of the reactions are known v can be calculated:

$$v = f_c \, m = \frac{1}{A} \frac{dF}{dx} m \tag{7B}$$

Fig. 2 Constitution and reaction kinetics of a binary alloy with complete miscibility in the liquid state and limited mutual solubility of the crystalline phases α and β, schematic.

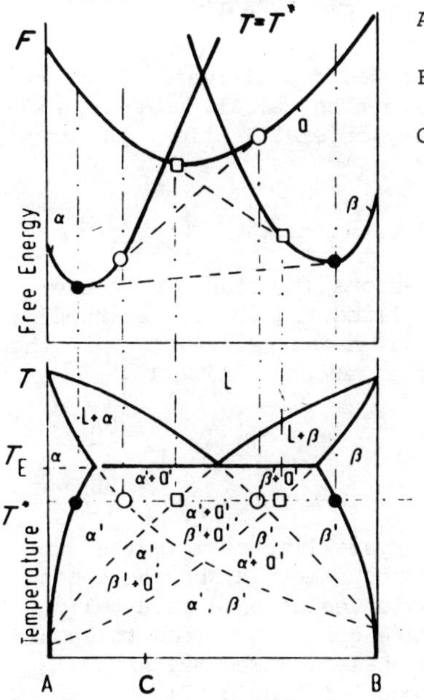

A Free energy-chemical composition diagram (F-c, T = T*).
B Equilibrium-phase diagram (T-c), and metastable equilibria (dotted).
C Time-temperature-crystallization diagram (t-T-c) of alloy C (Figure 2 B), l = liquid, a = amorphous, a', α', β = phases in metastable equilibria.

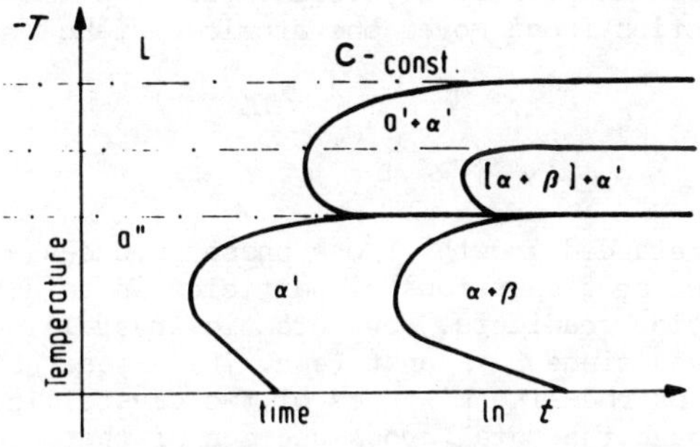

Fig. 3 Ranges of chemical composition in which massive (M), primary (P), and eutectic (E) crystallization can occur.

A F-c-diagram for $T < T_E$ (eutectic temperature).

B Type of phases which form by the different crystallization reactions in seven different concentration ranges. Two or three reactions may compete (equ. 7A).

3

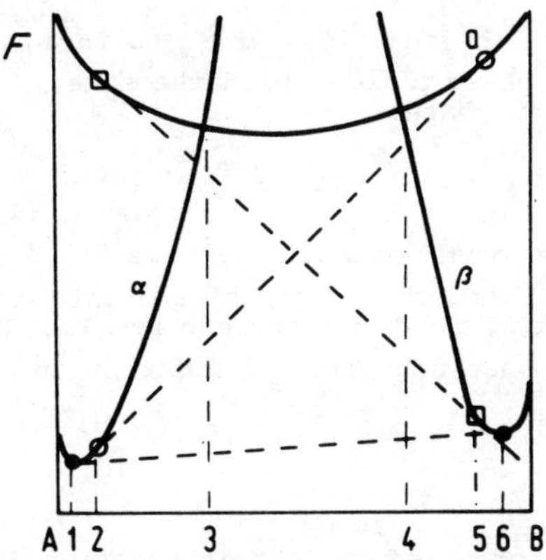

Reaction	Composition						
	A–1	1–2	2–3	3–4	4–5	5–6	6–B
M	α	α'	α'		β'	β'	β
P	–	–	α' + a'	a' + α', β'	β' + a'		
E	–	α + β	α + β	α + β	α + β	α + β	

$$\text{one crystalline structure} \quad v = f_c \frac{V_m}{RT} \frac{D}{\Delta x} \tag{7C}$$

$$\text{two crystalline structures} \quad v = (f_c - f_i) \frac{V_m}{RT} \frac{D}{S} = \text{const.} \tag{7D}$$

The velocity v is constant if the driving force f_c, diffusion distance Δx, or the lamellar spacing S do not vary with time t. If, however, Δx varies with time, as it is implied for primary crystallization, the growth velocity becomes retarded (compare equ. 7C):

$$\Delta x = \sqrt{Dt}\,. \tag{7E}$$

A constant velocity is caused by a homogeneous structure and consequently by a constant mobility of the crystallization front. Retarded growth in primary crystallization implies that n > 1 (equ. 4), while the reaction front moves one atomic distance at a period of time τ^x:

$$\tau^x \gg \tau \tag{8A}$$

$$b/v \gg b^2/D\,. \tag{8B}$$

Retarded growth of one phase produces a crystalline seam in the surface, or dispersions of particles in an amorphous matrix showing concentration gradients. For eutectic crystallization equation 8 is usually not obeyed since S = const (equ. 7D). Steady state conditions lead to lamellar or rod-like mixtures of two crystalline phases, which may grow through the total cross-section of the undercooled liquid [7].

3 THE MASSIVE CRYSTALLIZATION REACTION

Maximum mobility m is provided for $\Delta x \rightarrow b$ (equ. 7C), or $S \rightarrow b$ in equation 7D. This implies that a crystalline phase α forms with the same composition as that of the amorphous parent-phase:

$$l \rightarrow a \rightarrow \alpha\,. \tag{9}$$

The term "massive" is used in analogy to the established term "massive transformation". In this case one crystal structure β transforms into α by individual atomic hops from β into α across the α-β-interface [8]. The velocity of the reaction front increases when substituting b for Δx in equ. 7C:

$$v = f_c \frac{V_m}{RT} \frac{D}{b} = \text{const.} \tag{10}$$

Following the principle of maximation of v (equ. 7), this reaction will be found as a consequence of the high mobility m, if the necessary conditions (i. e. existence of a driving force) for its occurrence are fulfilled (Fig. 5).

Massive crystallization leads to homogeneous columnar or equiaxed microstructures, depending on whether it starts by heterogeneous or homogeneous nucleation. The crystalline phases can be stable or metastable solid solutions or compounds in a wide range of compositions, if the undercooling is high (Fig. 6). It is worth noting now, that homogeneous solid solutions may form either by ultra-slow or ultra-fast cooling. Very slow

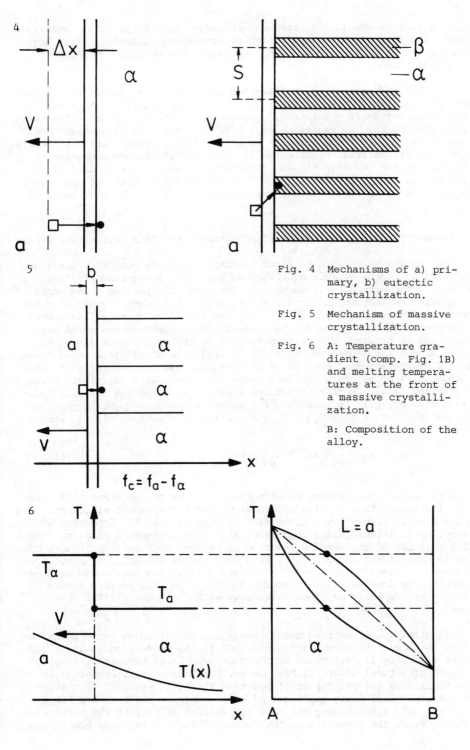

4

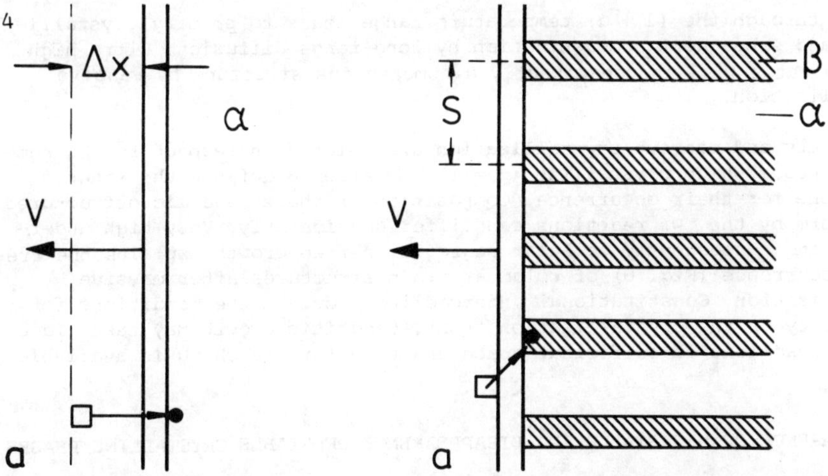

5

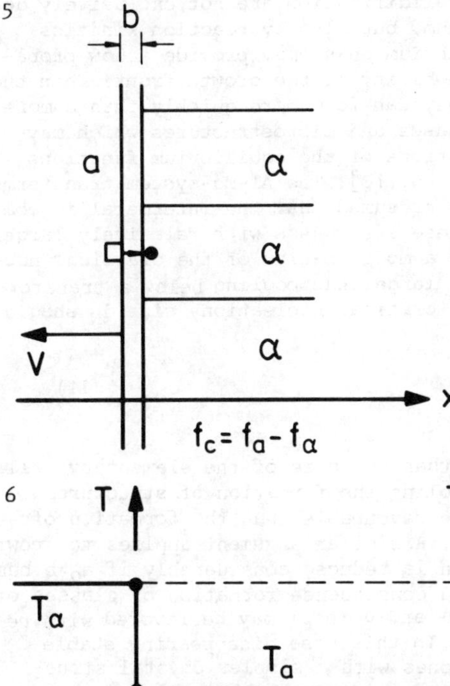

Fig. 4 Mechanisms of a) primary, b) eutectic crystallization.

Fig. 5 Mechanism of massive crystallization.

Fig. 6 A: Temperature gradient (comp. Fig. 1B) and melting temperatures at the front of a massive crystallization.

B: Composition of the alloy.

6

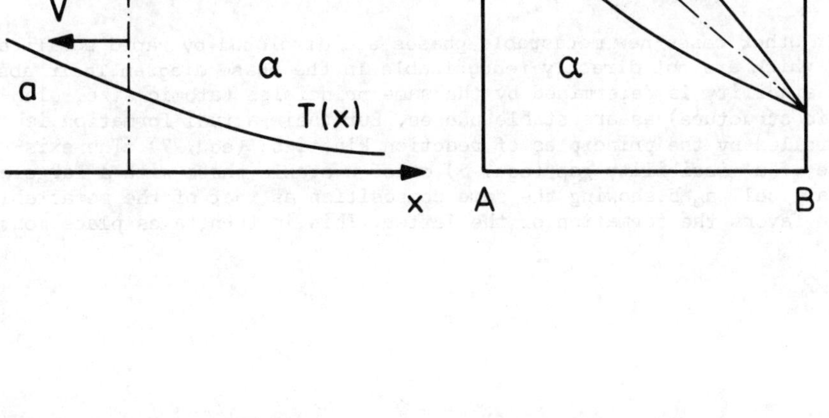

cooling through the $(l + \alpha)$-temperature range leads to primary crystallization and subsequent homogenization by long-range diffusion. Ultra-high cooling rates will induce directly a homogeneous structure by massive crystallization.

Primary and massive crystallization are related in respect to the number of phases which crystallize $(p = 1)$. Equation 8 defines the kinetic conditions for their occurrence. Composition of phases and microstructures which form by the two reactions can differ considerably. Very high undercooling and consequently cellular segregation-free growth explains the frequent occurrence (Fig. 6) of columnar grain structures after massive crystallization. Constitutional undercooling provides the conditions for an instability of the crystallization front. Dendritic growth may take place in both reactions if sufficient space and time for branching is available [6].

4 APPEARENCE OF METASTABLE AND DISAPPEARENCE OF STABLE CRYSTALLINE PHASES

The phases which form by rapid solidification are not exclusively determined by thermo-dynamic equilibrium, but also by reaction kinetics (equ. 7). The structure of an equilibrium phase may provide a low probability for nucleation and/or a low mobility of the growth front. Then the metastable phases will appear, if they can form more quickly than a more stable one. An indication for the phases and microstructures which may form, can be obtained from extrapolations of the equilibrium functions given in the phase diagramm (Fig. 7) [9, 10]. The Al-Mg-system transforms into a simple eutectic one, if it is presumed that the intermetallic compounds β, γ do not form. In fact these are phases with relatively large elementary cells. In this connection a modification of the classical nucleation theory [11] is required. For large undercooling below a transformation temperature $\Delta T = T_{f\alpha} - T$, the critical nucleation size d_N should not approach zero.

$$d_N - a_o = \frac{2\gamma_{a\alpha}}{\Delta T \; S_{a\alpha}} > a_o . \qquad (11)$$

It cannot, however, become smaller than the size of the elementary cell a_o. It follows that at high undercooling the formation of structures with a small elementary cell is more favourable than the formation of complicated intermetallic compounds. A similar argument applies to growth. The velocity of the massive reaction is reduced considerably if $a_o > b$ has to be substituted into equ. 10. As a consequence formation of glasses of the simple crystal structures α (fcc) and δ (hcp) may be favored with respectto the intermetallic compounds. In this case disappearing stable phases are replaced by other known ones with a simpler crystal structure.

In other cases new metastable phases are disclosed by rapid solification, which are not directly recognizable in the phase diagram.Their absolute stability is determined by the same principles (atomic size, electronic structure) as are stable phases. But their actual formation is controlled by the principles of reaction kinetics; (equ. 7): The existence of a miscibility gap (equ. 5) or of a stable phase with a large elementary cell $a_o > b$,showing the same composition as that of the metastable phase favors the formation of the latter. This,in turn,takes place most

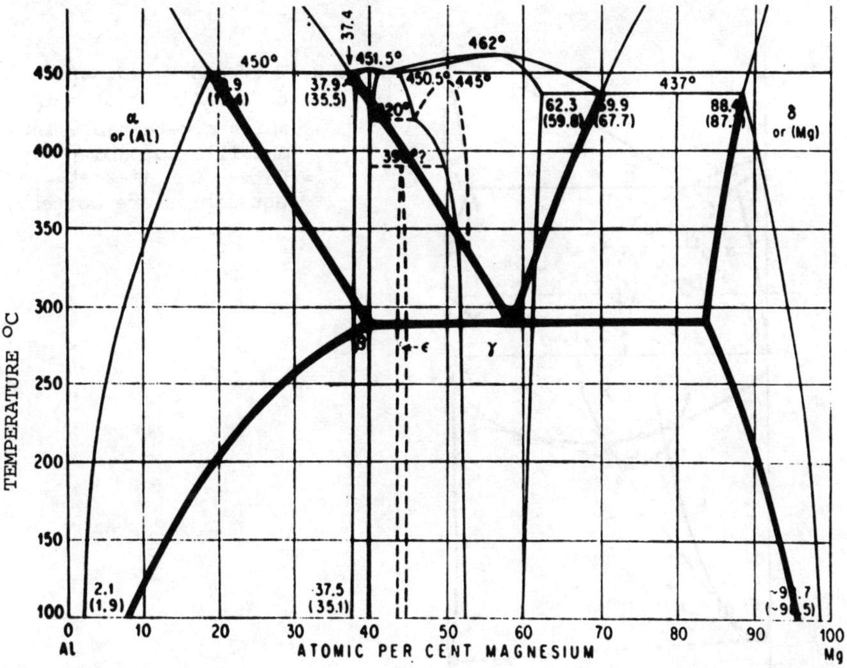

Fig. 7 Phase diagrams to which metastable equilibria have been added (strong lines):

A Al-Mg: non-formation of the compounds (β, ε, γ) leads to a metastable eutectic at about 300 °C, or primary, or massive formation of α or δ.

B Al-Mn: the intermetallic compounds which are situated above 12 % Mn are replaced by icosahedral structures at intermediate or by glasses at high cooling rates.

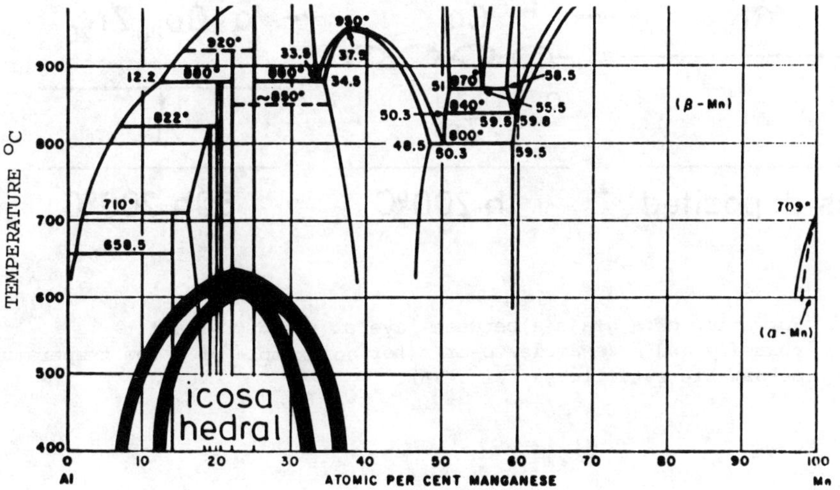

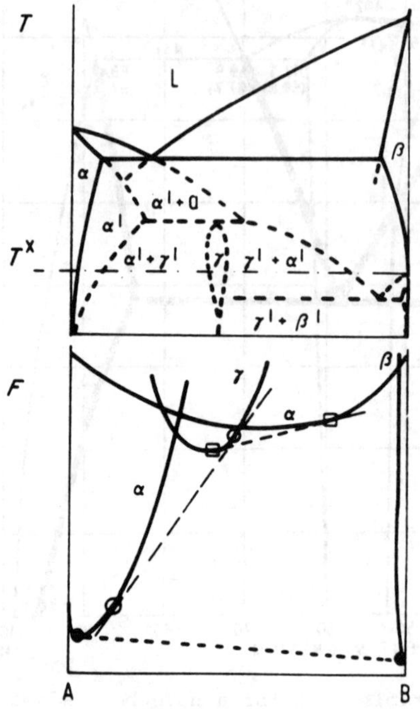

Fig. 8 T-c- and F-c-diagrams of an alloy system in which a metastable **intermetallic** compund γ may form (metastable equilibria are dotted).

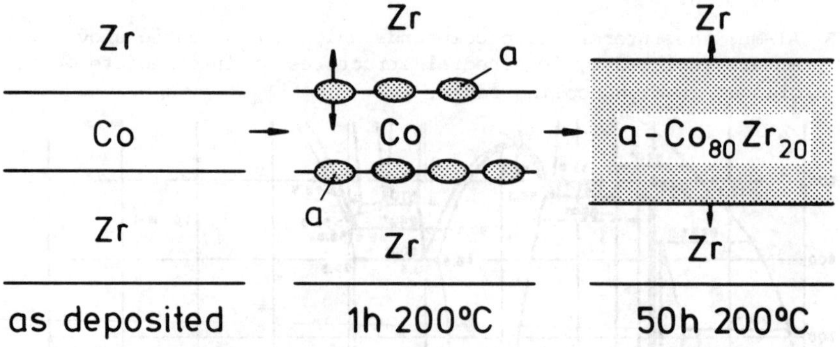

Fig. 9 Formation of a glass(a)between layers of two crystalline phases α and β (example Co-Zr) (for an example of a low temperature metastable eutectic see Fig. 7A).

rapidly by the massive reaction. Figure 8 indicates the role of a metastable phase in a schematic F-c-, and T-c-diagram. The Al-Mn-diagram shows a large number of intermetallic compounds in the range between 10 and 30 at. % Mn. They are replaced by icosahedral crystal structures, if glasses will not form at very high cooling rates [2]. The Mg-Li-diagram discloses that a bcc-ordered metastable phase could possibly exist at about 25 at. % Li. Its formation by rapid cooling is however not likely because the disordered simple crystal structures bcc (Li-rich) and hcp (Mg-rich) can grow rapidly by the massive reaction.

Metastable phases will not only form a liquid which is highly undercooled (glass), but also from crystalline phases. At the first sight it may appear paradox that an amorphous phase a may form from two crystalline phases [12, 13]:

$$\alpha + \beta \rightarrow a \, . \tag{12}$$

This type of reaction can however be explained using known principles: The amorphous atomic mixture should possess a lower energy as the two crystalline phases (existence of a driving force), and the amorphous phase should form more rapidly than the more stable crystalline compounds. For example in Zr-Co, Hf-Ni alloys the α-β interface provides the nucleation sites for the formation of the amorphous phase. The reaction is similar to melting of a metastable eutectic (Fig. 7). The growth of the glass into the lamellar crystals follows the $\sqrt{t}$-law as for primary crystallization (equ. 7C, D, Fig. 9).

5 SUMMARY AND CONCLUSIONS

Rapid cooling of liquid alloys can provide the following new phases: amorphous, known crystal structures in an extended range of composition, new crystalline solid solutions and compounds.

These phases in turn can be arranged in form of a large number of microstructures. Types of phase mixtures are: glass-glass (a' + a"), glass-crystals (a + α), metastable crystal (α' + β'). One-phase microstructures can be ultra-fine equiaxed and random grain structures (homogeneous nucleation), or columnar and textured (heterogeneous nucleation). Homogeneous and heterogeneous microstructures which contain metastable phases are susceptible to secondary reactions:

$$\alpha' \rightarrow \alpha + \beta , \tag{13A}$$

precipitation of β in supersaturated solid solutions α' which had formed by the massive reaction:

$$\alpha' \rightarrow \beta + \gamma , \tag{13B}$$

eutectoid decomposition of α',

$$a + \alpha \rightarrow (a \rightarrow \beta) + \alpha \rightarrow \beta + \alpha , \tag{13C}$$

massive crystallization of the amorphous matrix a in a microstructure which originally has formed by primary crystallization and

$$\beta \rightarrow \alpha_M, \tag{13D}$$

martensitic transformation of β, that had first formed by massive crystallization.

It can be concluded that in addition to the free energy of the amorphous and crystalline phases, crystallization kinetics and the cooling system are essential for an understanding of the microstructure of alloys which was produced by rapid cooling as well as by aging at large undercoolings below the melting temperature.

6 REFERENCES

[1] Hornbogen, E., in: Rapidly Quenched Metals, S. Steeb, H. Warlimont eds., North Holland Publ. (1985) p. 785.

[2] Shechtman, D., et al., Phys. Rev. Lett. 53 (1984) p. 1951.

[3] Grant, N. J., ref. 1, p. 3.

[4] Heineman, W. A., ref. 1, p. 27.

[5] Mukherjee, K and Mazumder eds., Laser in Metallurgy, AIME Conf. Proc. 1981.

[6] Chalmers, H., Principles of Solidification, J. Wiley, New York 1964.

[7] Hornbogen, E and I. Schmidt, in: Liquid and Amorphous Metals, NATO Adv. Study Series E 36, Sijthoff & Nordhoff (1980) p. 353.

[8] Gilbert, A. and W. S. Owen, Acta Met. 10 (1962) p. 45.

[9] Hornbogen, E., J. Mat. Sci. 18 (1983) p. 127.

[10] Müller, B.A., J.J. Richmond and J.H. Perepezko, ref. 1, p. 43 and 47.

[11] Zettlemoyer, A. C., eds., Nucleation, M. Dekker, New York 1969.

[12] Schwarz, R. B. and W. L. Johnson, Phys. Rev. Lett. 51 (1983) p. 415.

[13] van Rossum, M., M. A. Nicolet and W. L. Johnson, Phys. Rev. 1329 (1984) p. 5498.

[14] Blank, E., Dr. rer. nat.-thesis, University Göttingen (1971).

7 LIST OF SYMBOLS

A	m^2	amy area
a_o	m	size of elementary cell
b	m	atomic spacing
c	1	chemical composition
C_p	$J\ mol^{-1}K^{-1}$	specific heat
D	$m^2\ s^{-1}$	diffusion coefficient
d_N	m	size of nucleus
f_c	Jm^{-3}	free energy of crystallization ($f_a - f_\alpha$)
f_i	Jm^{-3}	energy density due to α-β-interfaces
F	J	free energy $dF/dx \cdot A = f$

$\gamma_{a\alpha}$	Jm^{-2}	interface energy between glass and crystal
k	1	number of components of alloy systems
κ	s^{-1}	$\kappa = \lambda/c_p\rho$
λ	$J\,K^{-1}s^{-1}m^{-1}$	thermal conductivity
m	$m^4\,J^{-1}s^{-1}$	mobility of crystallization front
n	1	number of atomic hops
p	1	number of phases
$\dot{q}$	$Jm^{-2}s^{-1}$	current density of heat flow
ρ	$g\,m^{-3}$	density
S	m	lamellar spacing
$S_{a\alpha}$	$Jmol^{-1}k^{-1}$	entropy of crystallization
ΔT	K	undercooling $\Delta T = T_{\alpha f} - T$
$\dot{T}$	Ks^{-1}	cooling rate dT/dt
$\bar{\dot{T}}$	Ks^{-1}	average cooling rate $\Delta T/\Delta t$
Δt	s	period of time
T_o	K	ambient temperature
$T_{\alpha f}$	K	melting temperature
τ	s	time for one atomic hop
τ^*	s	time which the $\beta\rightarrow\alpha$-reaction front needs to progress by one atomic distance b
V_m	m^3mol^{-1}	molar volume
v	ms^{-1}	velocity of crystallization front
v_{max}	ms^{-1}	maximum velocity of crystallization front
Δx	m	diffusion path in x-direction after n jumps

ACKNOWLEDGEMENT

The support of this work by the German Science Foundation (DFG - Schwerpunkt Mikrokristalline Legierungen Ho 325-19) is gratefully acknowledged.

DISORDER IN METALS - SELECTED PROBLEMS

K.H. Bennemann
Institute for Theoretical Physics, Freie Universität Berlin
D-1000 Berlin 33

Various properties of disordered metals are discussed. In particular, the effects of atomic disorder on superconductivity, magnetism and surface reconstruction are discussed. Furthermore, results on clusters and atomic disorder occuring at interfaces are presented.

1. INTRODUCTION

Disorder in metals has been studied with increasing intensity for the last 40 years (1). Disordered small particles, amorphous and liquid like glassy metals exhibit many properties which are very useful for technical applications. In particular, the effect of atomic disorder on superconductivity, magnetism and corrosion is of great interest. Therefore, we present a few typical examples illustrating how disorder effects the superconducting transition temperature, the magnetic moments and spin-ordering and surface segregation. Atomic disorder consisting of reduced atomic coordination is very important for properties of small clusters and surfaces. The effect of reduced atomic coordination on structural stability, recrystallization and electronic structure of small clusters, interfaces and surfaces is discussed. In the following various important cases of disorder are studied in detail.

2. SELECTED PROBLEMS

2.1 Granular- and Amorphous Superconductors

That atomic disorder may have dramatic effects on properties of superconductors like the superconducting transition temperature T_c and the critical magnetic field H_c has been shown clearly (2). While the effects on H_c are still not clearly understood on an electronic basis, the effects on T_c result from changes of the phonon spectrum, atomic coordination and electronic structure. Using the McMillan formula for T_c,values for the

Table 1: Experimental and theoretical values of the superconducting transition temperature T_c(K) for the disordered state and other characteristic quantities. (For Pd we use $\lambda_o = 0.6$).

	metal	$T_c^{exp.}$(K)	$T_c^{calc.}$(K)	$(N(o)/N_o(o))_{calc.}$	$(\langle\omega_{ph}^2\rangle/\langle\omega_{ph}^2\rangle_o)^{1/2}$ calc.	$(\theta_D/\theta_D^o)_{exp.}$	$\lambda_{cal.}$	$\lambda_{exp.}$
granular	Al	~3	~3.4	~1	0.87		0.5	
	Zn	1.3	1.6	~1	0.92		0.6	
	$Sn_{0.9}Cu_{0.1}$	~4.6	~4.6	~1	~0.9		0.9	
metallic glasses	Al	~5.8	~6.4	0.94	~0.8		0.6	0.66
	Ga	8.4	~11	1.6	~0.6	0.3	2.04	~2
	$Sn_{0.9}Cu_{0.1}$	7	~7(11)	~1.18	~0.85(1)		~1.3(1.4)	1.8
	La(β)	-	4.3	~1	~0.7		~1	
	$La_{0.7}Cu_{0.3}$	3.5	~3	~1	$\lesssim$0.8		≈1	
	$La_{0.78}Ni_{0.22}$	3	~3	$\gtrsim$1	$\gtrsim$0.8		≈1	
	Zr	~3	~4	~1.2	~0.6		~0.9	
	$Zr_{0.67}Cu_{0.33}$	2.27	~2.7	~1.4	$\gtrsim$0.6	0.6	~0.8	~0.9
	$Zr_{0.65}Pd_{0.35}$	3.5	~4	$\gtrsim$1.6	~0.7	0.64	~0.9	0.6
	$Zr_{0.7}Be_{0.3}$	2.8	~3.5	~1.4	>0.7		~0.8(?)	
amorphous	Pd(?)	~3.2	~2.5	<<1	~0.7		~0.6	
	Nb_3Ge	3.9	$\lesssim$4.5	~0.4	~0.8		~0.7	0.6
	Nb_3Si	3.8	$\lesssim$4.5		~0.8		~0.7	
	Nb_3Sn	~3	~4	~0.5	$\lesssim$0.8		~0.7	

superconducting transition temperature have been calculated successfully for various granular- and amorphous superconductors (3). Results are shown in table 1 for various granular superconductors, strongly disordered metals and metallic glasses (1) and in Fig. 1 for vapour quenched transition metals (4). These results can be understood by determining the effects of atomic disorder on $\langle\omega_{ph}^2\rangle$, the average squared phonon frequency, or on the Debye-temperature Θ_D, on the electron-phonon coupling constant $\lambda=2\int d\omega\, \alpha^2 F(\omega)/\omega$ and on μ_{sp}, the coupling constant for the interaction between electrons and spin excitations. α is the r.m.s. electron-phonon coupling constant and $F(\omega)$ describes the phonon density of states. Note that T_c is approximately given by

$$T_c = \frac{\langle\omega_{ph}^2\rangle^{1/2}}{1,2} \exp\{- \frac{1+\lambda+\mu_{sp}}{0.96\lambda-(\mu^*+\mu_{sp})(1+0.6\lambda)}\}$$

and λ by $\lambda\approx\langle N(o)\langle J^2\rangle/M\ \langle\omega_{ph}^2\rangle$, where $N(o)$ is the electronic density of states at the Fermi-energy ε_F, M is the ionic mass and $\langle J^2)$ is the average over the Fermi-surface of the squared matrix element of the electron-ion deformation potential $J(q)$. For transition- metals one finds (5) $\langle J^2\rangle\ \alpha\ W\ \alpha\ N^{-1}(o)$, W is the band-width, and approximately $N(o)\langle J^2\rangle\approx q_o E_{coh}$ Here, E_{coh} denotes the cohesion-energy and $q_o R\approx 3$ with R denoting the interatomic spacing. Primarily atomic disorder will change $N(o)$, the electron-phonon matrix element, e.g. $\langle J^2\rangle$, the phonon density of states and decrease μ_{sp} for the late transition metals like Ni, etc. Typical results are illustrated in table 2. For simple metals we use $\mu^*\approx 0.1$ and for $N(o)$ the free-electron value, $\delta\ln\langle\omega_{ph}\rangle\simeq-0.05\ \gamma_G$ for granular superconductors (γ_G = Grüneisen constant) and $\delta\ln\langle\omega_{ph}\rangle\simeq-\gamma_G\delta\ln v+(\delta\ln\langle\omega_{ph}\rangle/\delta\ln N(o))\delta\ln N(o)$ for amorphous superconductors. For transition metals we use $\mu^*\simeq0.13$, $\lambda=\eta(v)/M\langle\omega_{ph}^2\rangle$, $\mu_{sp}=0$, and for $N(o)$ the value obtained from a parabolic tight-binding like density of states. The results shown in table 1 and Fig. 1 are then explained as follows:

(1) for granular superconductors T_c increases mainly as a result of lattice softening at the surface. One expects that T_c first increases rapidly as the size of the granules decreases, that T_c nearly saturates as the surface area becomes comparable to the size of the granules and then that T_c ($T_c \lesssim T_c^a$) decreases again as the diameter of the granules becomes smaller than the coherence length ξ.

(2) For amorphous simple metals like Al, Ga and Sn the superconducting

Table 2: Summary of main factors determining changes in λ and T_c due to atomic disorder.

metal	granular superc.	amorph. superc.
simple metals	latt. softening $\lambda > \lambda_o$ $T_c \gtrsim T_{co}$	latt. softening $N(o) \hat{=}$ free el. $\langle J^2 \rangle \gtrsim \langle J^2 \rangle_o$ $\lambda \gtrsim \lambda_o$ $T_c \gtrsim T_{co}$
transition metals	latt. softening $\lambda > \lambda_o$ $T_c > T_{co}$	latt. softening $N(o) \gtrless N_o(o)$ $N(o)\langle J^2 \rangle \sim E_{coh}$ $\mu_{sp} \rightarrow o$ $\lambda \gtrless \lambda_o$ $T_c \gtrless T_{co}$
A-15 compounds etc.	latt. softening struct. changes	latt. softening $N(o) < N_o(o)$ struct. changes $\lambda < \lambda_o$ $T_c < T_{co}$

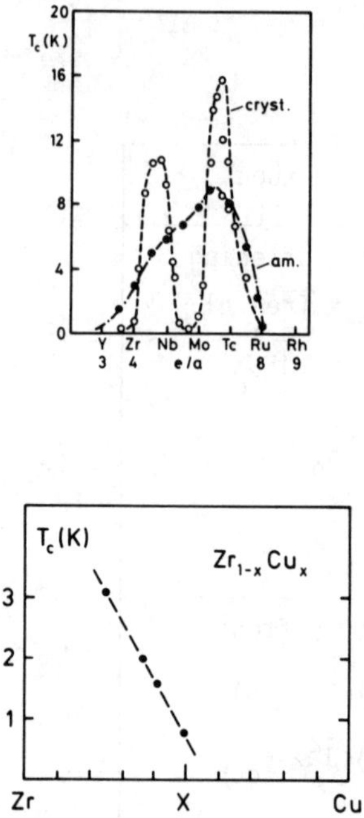

Fig. 1

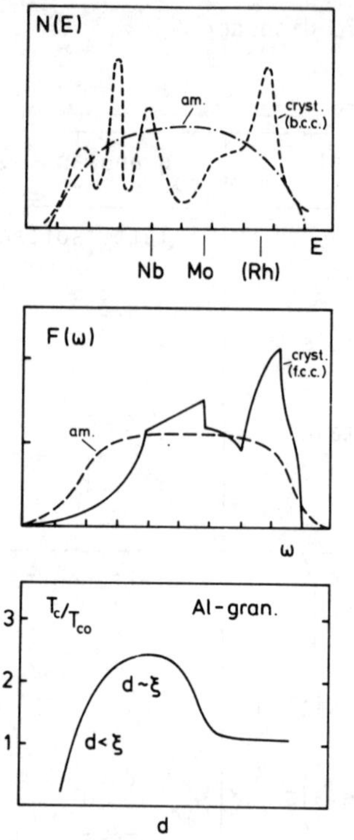

Fig. 2

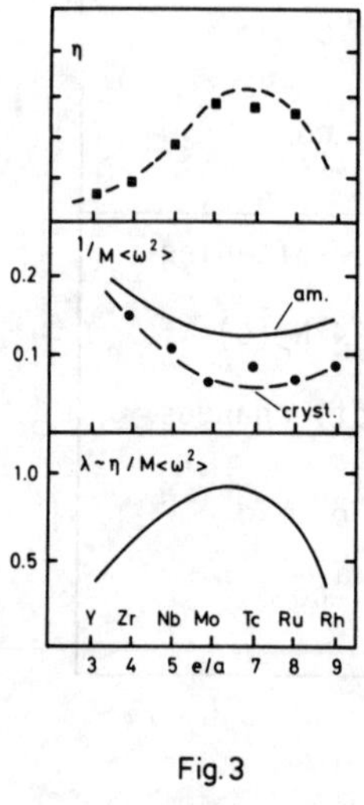

Fig. 3

Fig. 1: Results for T_c of transition-metals and their alloys.

Fig. 2: Illustration of the behaviour of N(E), F(ω) and T_c in disordered metals.

Fig. 3: Valency dependence of η, $\langle\omega_{ph}^2\rangle$ and λ.

transition temperature increases due to lattice softening. Also an increase of N(o) and $\langle J^2\rangle$, in particular, can be observed if the atomic structure coordination changes (see Sn, Ga,...). For N(o) one may use in amorphous simple metals the free-electron value, compare Fig. 2.
(3) For transition-metal glasses like La_xM_{1-x}, Zr_xM_{1-x}, etc. the changes in T_c are mainly due to lattice softening, alloying with a quasi non-superconducting element (s. $La_{0.7}Cu_{0.3}$, etc.) and $\mu_{sp} \to 0$ ($Zr_{0.65}Pd_{0.35}$). Note, that for the late transition-metals like Co, Rh, Ir and Ni, Pd, Pt N(o) decreases strongly upon atomic disorder and consequently $\mu_{sp} \to 0$.(6). This explains superconductivity in highly disordered Pd (7).
$N(o)\langle J^2\rangle \approx q_0 E_{coh}$ will not depend strongly on disorder as long as the atomic coordination is not changed and strongly anisotropic crystals are not considered. For example, in the case of $Zr_{0.67}Cu_{0.33}$ we estimate the changes $N(o)/N_0(o) \approx 1.4$ and $\langle J^2\rangle/\langle J^2\rangle_0 \approx 0.6$ from tight-binding approximations using the typical distance dependence of the d-electron hopping integrals and a parabolic density of states (7,8) as well as (9) $\theta_D \approx 200$ K and $\theta_D^0 \simeq 315$ K. Thus, one obtains approximately $\lambda/\lambda_0 \approx 0.7(\theta_D^0/\theta_D)^2 = 1.7$, e.g. $\lambda \approx 0.7$ and $T_c \approx 3 \div 4$ K for amorphous Zr and its alloys. For $\mu_{sp} = 0$, Ni, Rh, Pd, etc. are expected to show nearly the same λ as Zr. Note, $T_c(Zr_{0.65}Pd_{0.35}) \approx T_c(Zr)$ suggests already $T_c \approx 3.5$K for amorphous Pd.(4).

For amorphous A-15 compounds and other compounds with anisotropic atomic structure favourable for superconductivity, the changes in T_c arise essentially from the change in the local atomic structure (decrease of N(o), change of atomic coordination, etc.) which is accompanied frequently by lattice softening. Thus, one expects that disordered Nb_3Ge etc. resembles amorphous Nb. This explains, why T_c of disordered Nb_3Ge, etc. is smaller (alloy effect), but nearly equal to $T_c \approx 6$K of amorphous Nb. (5).

The valency dependence of T_c for amorphous transition metals shown in Fig. 1 results essentially from the valency dependence of E_{coh} and $\langle\omega_{ph}^2\rangle$, respectively θ_D (9,10), Figs.2,3. For amorphous transition metal alloys A_xB_{1-x} one obtains in accordance with alloy theory (11) that $T_c(x)$ is described by the valency dependence. This is shown in Fig. 1 in the case of alloys for which rigid band theory is valid. However, $T_c(x) \simeq T_c(B) + x(T_c(A)-T_c(B))$ in the case of alloys for which split band behaviour is valid. If the A-element is non-superconducting one expects a critical concentration at which superconductivity vanishes. Results for $Zr_{1-x}Cu_x$,... are shown in Fig. 1. A non-linear dependence on x is also possible.

In summary, the effect of atomic disorder on T_c is relatively well understood. Of course, the calculated values for T_c could be inaccurate due to the use of the McMillan formula and of the approximated expressions for λ and $<\omega_{ph}^2>$, etc. But the observed trends for the effect of disorder on T_c are believed to be calculated reliably. The main goal of the theoretical studies was to clarify the important physical changes produced by atomic disorder.

2.2 Magnetism in Amorphous Metals

Atomic disorder in Fe, Co, Ni and alloys thereof exhibit interesting changes of their magnetic behaviour (12). Due to the effect on atomic coordination and interatomic distance, disorder affects the magnitude of the local magnetic moments μ_i (see local atomic environment effects in CuNi- and AuV alloys, Fe_3O_4, etc.), the exchange interaction J(r), and thus the long-range- and of course, also short-range-spin-order and magnetic anisotropy. Regarding μ_i one observes in the amorphous state only a small change for Co, for Ni a 20% to 30% decrease and for Fe even larger changes. Due to the fluctuations in the local atomic environment which are present in metallic glasses one expects a distribution of atomic magnetic moments $\{\mu_i\}$ and a distribution of J(r) for next nearest atoms due to variations in their distance. This has been observed using NMR and Mößbauer experiments. Note that crystalline FeSi and CoSn are non-magnetic, but magnetic when amorphous. If in amorphous Fe-alloys the distance d between Fe atoms varies from 2.4 Å to 2.5 Å, one expects $J(r) < 0$ (antiferromagnetic) to change to $J(r)>0$, since in antiferromagnetic γ-Fe $d \cong 2.4$Å and in ferromagnetic α-Fe $d \cong 2.5$ Å. The resulting spectral distribution p(J) is illustrated in Fig. 4. This fact causes a spin system with competing interactions. For transition metal-metalloid glasses ($(TM)_{0.75}B_{0.25}$, B = P, B, C, Si, Al, etc.) one observes similar changes. The moment reduction seems to be proportional to the number of s,p-electrons of the metalloid (largest for $P(S^2p^3)$, then $C(S^2p^2)$ which causes hybridization effects, etc.). Furthermore, one observes a flattening of M(T), $M(T)<M_0(T)$, as a result of the distribution of molecular fields (see variation of J(r)).

In the following part of this paper we discuss in particular the effects of atomic disorder on the magnetic moments μ_i, on the magnetization and the Curie-temperature, and on the magnetic phase-diagram.

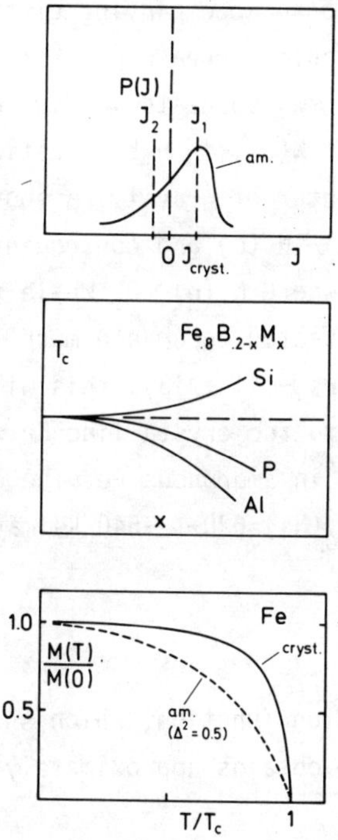

Fig.4

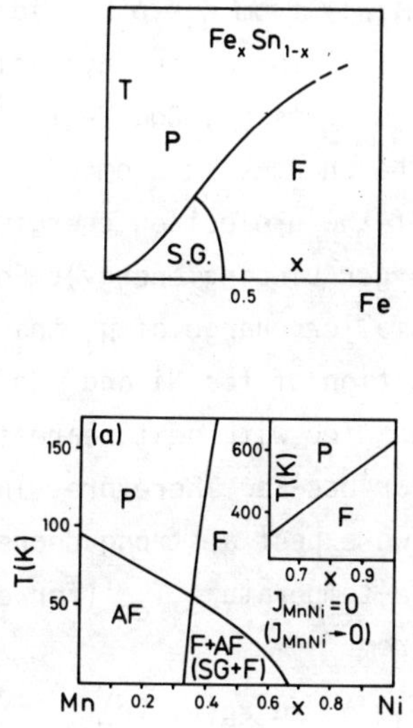

Fig.5

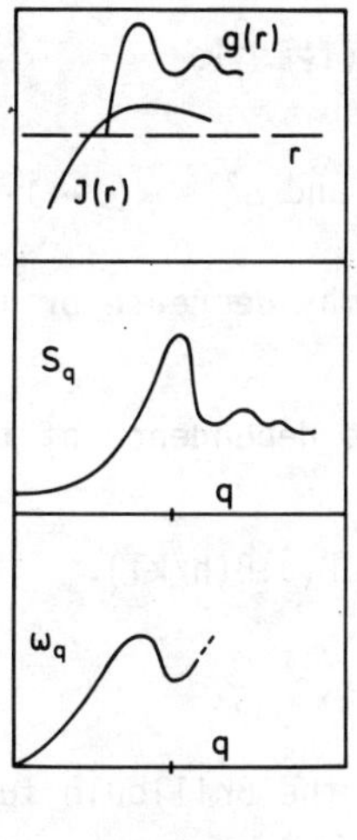

Fig.6

Fig. 4: P(J), Curie-temperature T_c and magnetization M(T) in amorphous metals.

Fig. 5: Phase-diagrams for spin-systems with competing spin-interactions.

Fig. 6: g(r), S_q and spin-wave frequency ω_q for amorphous metals.

First, for transition metal-metalloid glasses, changes in μ_i are expected to result mainly from d-s,p hybridization and an accompanying electron charge transfer. Increasing hybridization should decrease μ_i. For the itinerant ferromagnets Ni, Co, Fe the changes in μ_i upon atomic disorder are caused by the changes in atomic structure, in N(o), in hybridization, and in shifts of the d-electron energies (the center of gravity ε_d should shift to larger binding energy). For Co $N_\uparrow(o) \approx N_\downarrow(o)$ and consequently one expects a smaller change of μ_i than for Ni, where $N_\uparrow(o) \approx 0$. While the atomic coordination of fcc-Ni and -Co is not expected to change much (amorphous state ≈ fcc with next nearest neighbours ≈ 11 ÷ 13), this will be different for bcc-Fe. Therefore, in analogy to fcc crystalline Fe with no magnetism, we expect a strong decrease in μ_i in amorphous Fe. The change of the Curie-temperature T_{co} (for example $T_{co}(Ni)=620 \rightarrow T_c \simeq 540$ K) may be estimated from

$$T_c \sim \sum_j J_{ij} \sim \int d^3r g(r) J(r),$$

where g(r) is the usual atomic pair correlation function, which should be similar to the one observed for liquids. One obtains approximately

$$T_c{}^a = T_c' \cdot (1 - a\Delta^2),$$

with $T_c' = \alpha \langle J \rangle$ and $\Delta^2 = \langle (J - \langle J \rangle)^2 \rangle / \langle J \rangle^2$.

Therefore, T_c may decrease or increase depending on ⟨J⟩, compare Fig. 4 for examples.

The temperature dependence of the magnetization M(T) can be obtained from

$$M(T) = \int dJ P(J) B(h/kT),$$

with $h = \sum_j J_{ij} \langle S_z \rangle$,

where B is the Brillouin function. Results are given in Fig.4 . At small temperatures one finds $\Delta M = aT^{3/2} + bT^{5/2}$ which is reminiscent of spin waves.

Phase diagrams of metallic glasses are shown in Fig. 5. The observed spin-glass phases result from the variation in J(r) for next nearest and next next nearest atoms, e.g. from competing spin interactions. Approximating P(J) by a spectrum with narrow peaks at $J_1 > 0$ and $J_2 < (0)$ (Fig.4) one may model the phase-diagrams by the one of a fictitious alloy A_xB_{1-x} with $J_{AA} > 0$ and $J_{BB} < 0$. For such a fictitious alloy one may use Ni_xMn_{1-x}, for example (13). Results obtained by the Migdal renormalization procedure are shown in Fig.5. Similar results were obtained by more realistic and improved calculations which were performed by Moorjani et al. (14.)

Finally, regarding spin-waves in metallic glasses, one observes (see experiments for $Co_{80}P_{20}$) a softening of the spin wave frequencies ω_q at wavelengths where the structure factor S(q) peaks, see Fig.6. Following early studies by Beeby and Hubbard one has approximately

$$\omega_q \simeq \int d\overset{3}{r}(1-e^{i\underline{q}\cdot\underline{r}})J(r)g(r),$$

which for $q \to 0$ yields $\omega_q = D_q^{\;2}$

with $D = \int d\overset{3}{r}\; r^2 J(r)g(r)$.

Thus,calculated values for ω_q agree quantitatively with experiment (15).

In summary, it has been shown that atomic disorder causes variations in magnetic moments and molecular fields. These facts are primarily responsible for the magnetic properties of metallic glasses. Additional studies are still needed to understand the electronic origins of the magnetic properties of amorphous metals more clearly.

2.3 Structural Disorder at Interfaces and Surfaces

Interfaces of two metals and surfaces of metals, in particular transition metals, are of importance for many problems in physics and chemistry. Due to broken bonds, e.g. reduced atomic coordination, and bond-disorder (at interfaces of two different metals) and surface defects (steps, etc.) one expects structural disorder or changes in the lattice structure as compared to bulk. This has been observed in the form of commensurate and incommensurate interfaces (16), relaxation and reconstruction at surfaces (17). Note, chemisorption induced reconstruction may be viewed parallel to the impurities needed for stabilizing the amorphous state (see Cu-impurities for obtaining amorphous Sn, for example). Changes of atomic coordination will not only affect the atomic structure, but in general also significantly the electronic structure. This can be clearly seen from the behaviour of small metallic clusters M_n(M=Pb, Ni,...) with the number of atoms n ranging from a few to a few hundred atoms. In the

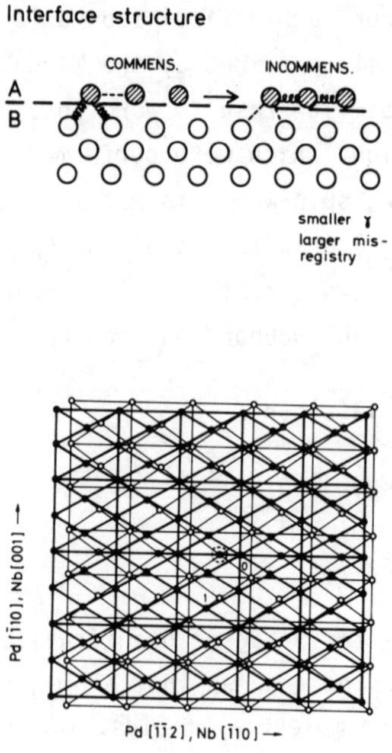

Fig. 7

Fig. 7: Commensurate - incommensurate structure at interfaces.

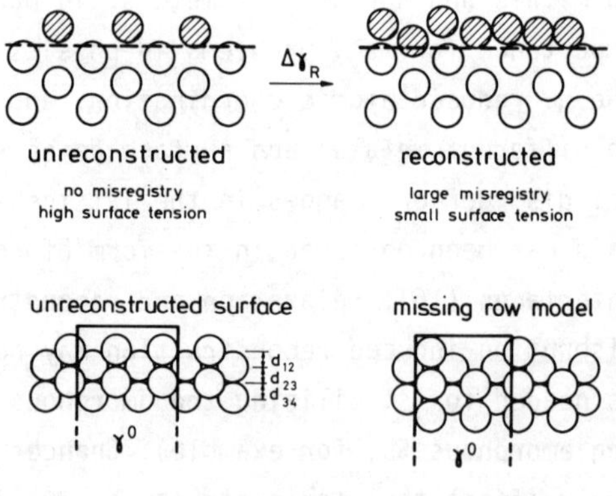

Fig. 8

Fig. 8: Illustration of reconstruction at transition-metal surfaces.

following we study in more detail the effects on the atomic- and electronic structure due to reduction of atomic coordination.

First, at interfaces of two different metals of type A and B, in particular transition metals, one expects competing interatomic interactions due to different energy bonds between A-atoms and B-atoms and A- and B-atoms. As long as the energy gain resulting from A-B bonds is larger than the one due to B-B bonds, one expects a commensurate interface. The B-atom will prefer positions in registry with the substrate lattice of A-atoms. Such a behaviour has been observed for Pd layers on Nb(110), see Fig. 7 for illustration. Performing a tight-binding calculation we find, in agreement with LEED experiments, a transition Pd(110) → Pd (111) for increasing coverage θ of the Nb-(110) surface with Pd. Pd forms a (111) overlayer structure when the energy gain from Pd-Pd bonds dominates ($\theta > 1-2$). The energetic reason of the lattice transformation is illustrated in Fig. 7. While the uptake of H is small for Pd(110) it is large for incommensurate Pd(111).

This lattice transformation is expected to be typical for metal/metal interfaces. At free surfaces atomic disorder caused by the reduced coordination occurs in the form of surface relaxation and reconstruction. This is illustrated in Fig. 8. The atomic redistribution seems strongest for transition-metals. Typically, the surface layer of atoms relaxes outwardly. Roughly speaking surface reconstruction occurs when the gain in surface energy is larger than the increase in strain energy due to misregistry of the surface atoms with the underlying atomic layers of the bulk lattice. Since chemisorption changes the surface energy, chemisorption induced reconstruction becomes understandable. At T = 0 one may use the general criterion

$$\Delta\gamma = \Delta\gamma_0 - \Delta\gamma_{ads} < 0$$

for reconstruction, see Fig.8. Here, $\Delta\gamma$ denotes the difference in surface energy between the reconstructed and unreconstructed surface. $\Delta\gamma_{ads}$ denotes the contribution to $\Delta\gamma$ due to chemisorption. Surface relaxations are determined by $\partial\gamma_0/\partial r_{ij} = 0$, where r_{ij} refers to the distance between atoms i and j. The surface energy γ_0, including strain energy, may be calculated with sufficient accuracy using a Born-Mayer potential. For the electronic energy the formula $E_{el} = \int d\varepsilon(\varepsilon-\varepsilon_0)N(\varepsilon)$ applies. Here ε_0 refers to the center of gravity of the electronic band with density of states $N(\varepsilon)$. Thus, we have shown for transition metals that missing-row

Table 3: Values for the ratio of the shear modulus and of the cohesive energy for fcc transition-metals.

Co 11.5	Ni 18.5	Cu 14.7	↑ difficult
Rh ?	Pd 11.4	Ag 10.8	reconstruction
Ir	Pt 8.2	Au 7.4	easy ↓

Table 4: Values for the structure energy $E_{bs}=E_{coh}(bcc)-E_{coh}(fcc)$, the bulk cohesive energy $E_{coh,b}$, and their ratio for different bcc metals. Also included is the critical cluster size N_{crit}, at which the structur transformation is expected.

cluster	E_{bs}(eV)	$E_{coh,b}$(eV)	$E_{bs}/E_{coh,b}$	N_{crit}
V	0.286	5.33	$5.37x10^{-2}$	2950
Cr	0.381	4.12	$9.25x10^{-2}$	580
Nb	0.286	7.48	$3.82x10^{-2}$	8190
Mo	0.381	6.83	$5.58x10^{-2}$	2630
Ta	0.286	8.11	$3.53x10^{-2}$	10380
W	0.381	8.81	$4.32x10^{-2}$	5660

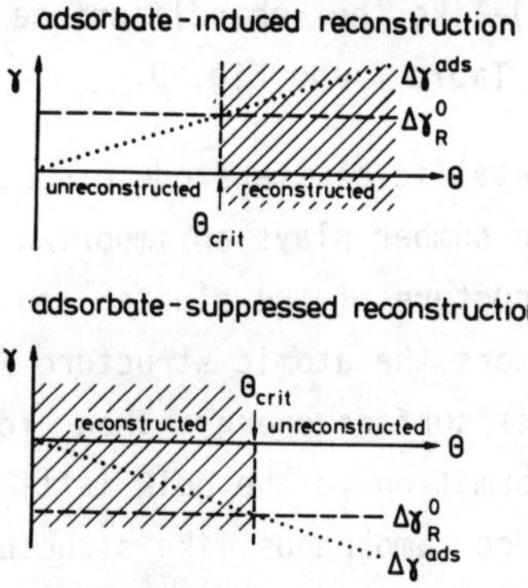

Fig. 9

Fig. 9: Illustration of chemisorption effects on reconstruction.

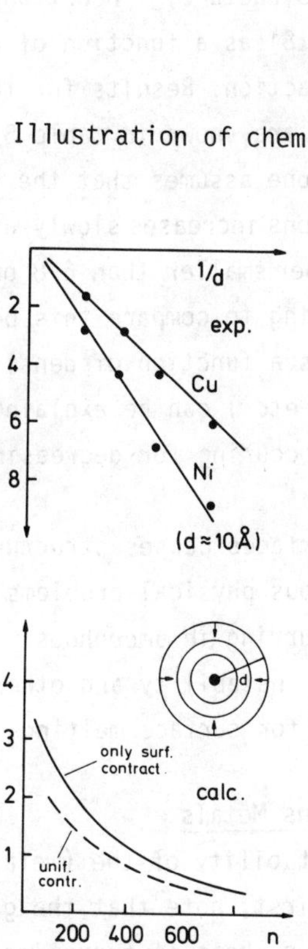

Fig.10

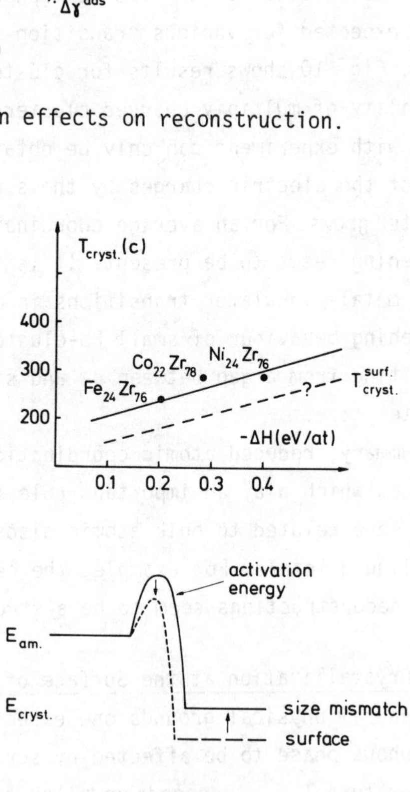

Fig. 11

Fig. 10: Results for contraction of small transition-metal clusters.

Fig. 11: Illustration of the effect of surfaces on recrystallization of glassy metals.

reconstruction, see Fig. 8, occurs most likely and also why the (1oo)-surface transforms to a (111)-like (hex-phase) surface (17). Chemisorption effects are summarized in Table 3 and Fig. 9.

Finally, for small metallic clusters one expects again that the reduced average coordination number plays an important role for various properties like atomic structure of the cluster and also its electronic structure. For small clusters the atomic structure will correspond to the one which yields the lowest surface energy. Then, for increasing cluster size one expects a transformation to the bulk lattice structure. Table 4 contains results for the fcc (amorphous like structure) → bcc transformation expected for various transition-metals (18) as a function of cluster size. Fig. 10 shows results for cluster contraction. Results for the stability of multiply charged clusters $Pb_n^{(m+)}$ are shown in table 5. Agreement with experiment can only be obtained if one assumes that the screening of the electric charges by the s,p electrons increases slowly as the cluster grows. For an average coordination number smaller than 6÷8 only poor screening seems to be present. It is interesting to compare this behaviour with metal- insulator transitions in metals as a function of density. The screening behaviour of small Pb-clusters (Hg, etc.) can be explained as resulting from a gap between p- and s-states occuring for decreasing cluster size.
In summary, reduced atomic coordination at surfaces causes structural changes which play an important role for various physical problems. They are related to bulk atomic disorder occurring in amorphous and liquid metals. For example, the hex-phase instability and other surface reconstructions seem to be a strong hint for surface melting.

2.4 Crystallization at the Surface of Amorphous Metals

On general physical grounds one expects the stability of the (metastable) amorphous phase to be affected by surfaces. First, note that the glass temperature $T_{cryst.}$ depends on alloy composition, heat of formation, etc. (1,8). (For illustration see Fig.11.) Now, at surface one expects in general atomic segregation to occur which should affect T_{cry}. Furthermore, if the amorphous state is stabilized relatively to a large extent by strain energy $T_{cryst.} \propto E_{sef-diff}$) one expects recrystallization of amorphous metals to occur first at the surface due to surface lattice softening. The decrease of $T_{cryst.}$ is expected to be strongest for amorphous metals AB with

Table 5: Results are given for the minimal number of atoms, n_{cr}, for which m-fold positively charged spherical clusters $M_n^{(m+)}$ are stable. n_{cr} is calculated from $E_{coh,s}(n)=\delta E_c(n)$. The spherical cluster is built within a fcc lattice. Point charges are put at equal maximal distance at the surface. Results are obtained using (a) bare Coulomb-forces ($\lambda=0$), (b) using for the inverse screening length $\lambda=0$ for $n<30$ and $\lambda\propto(n-30)$ for $n>30$ with $\lambda=\lambda_b$ for $n\leq300$. The interatomic distance is taken to be d=3 Å and $\lambda_b=1.8$ Å^{-1} for Pb_n.

M_n^{m+}	n_{cr}(calc.)		n_{cr}(exp)
	a($\lambda=0$)	b($\lambda\to\lambda_b$) 30<n→300	
Pb_n^{2+}	30	31	30
Pb_n^{3+}	130	49	45
Pb_n^{4+}	325	61	~72
Pb_n^{5+}	500	~70	-
Ni_n^{2+}	~8	~8	-
Ni_n^{3+}	30	30	-
Ni_n^{4+}	≳100	~40 (to 50)	-

large Δv. Here, Δv refers to the difference in atomic volume. Surface segregation results from the interplay of surface-energy and strain energy-(size-mismatch energy) lowering. If the strain-energy change dominates, then one expects smaller surface segregation for the glassy state. One may assume that the size mismatch energy is smaller in the glassy than in the crystalline state. We expect in general a smaller surface energy for the surface of an amorphous metal as compared to the surface energy of the corresponding crystalline metal. (see γ_{100}, γ_{110}, γ_{111}). This tends to stabilize the amorphous state. Note, however, for the segregation kinetics also the activation energy barrier for atomic motion needs to be considered. We expect that this energy barrier is reduced due to reduction in the average coordination number or aquivalently due to $E_{coh,s} < E_{coh,b}$. Presently, we perform model calculations to study this problem.

Regarding corrosion behaviour, note that this is strongly affected by surface defects and segregation. For example, for $Fe_{1-x}Cr_x$ we estimated that corrosion resistance will occur at a smaller Cr-concentration than in the crystalline case. Vacancies favouring oxidation should be less present at amorphous surfaces.

In summary, we have studied various problems related to strong atomic disorder. This shows that atomic disorder is a widespread phenomenon and plays a significant role for understanding material properties.

Many helpful discussions with Dr. Ghatak, Dr. Tománek and S. Mukherjee are gratefully acknowledged.

References

1. Glassy Metals I, Topics in Applied Physics, 46, ed. H.J. Güntherodt and H. Beck, Springer Verlag Berlin, 1981.
2. Buckel, W. Physica 126B, (1984) 1.
3. Garland, J.W. et al., Phys. Rev. Lett. 21 (1968) 1315.
4. Collver, M.M. and Hammond, R.H.,Phys. Rev. Lett. 30 (1973) 92.
5. Barisic, S., Labbé, J. and Friedel, J.,Phys. Rev. Lett. 25 (1970) 919.
6. Berk, N.F. and Schrieffer, J.R., Phys. Rev. Lett. 17 (1966) 433.
7. Stritzker, B., J. de Phys. Lett. 21 (1978) L 398.
8. Garoche, P. and Veyssie, J.J., J. de Phys. Lett. 42 (1981) 365; Kübler, J., Bennemann, K.H., Lapka, R., Rösel, F., Oelhafen, P. and Güntherodt, H.J., Phys. Rev. B 23 (1981) 5176.
9. Bennemann, K.H. and Garland, J.W.. Superconductivity in d- and f-Band Metals, AIP Conf. Proc. No. 4, ed. D. Douglass, Am. Inst. of Physics, New York, 1972.
10. Kerker, G. and Bennemann, K.H., Z. Physik 264 (1973) 15.
11. Kerker, G. and Bennemann, K.H., Sol. St. Comm. 14 (1974) 399 and 15 (1974) 29. For experimental results see Willer, J., Fritsch, G., Lüscher, E., J.Less.Com.Met. 77 (1981) 191, Appl.Phys.Lett. 36 (1980) 859.
12. Durand, J. lassy Metals: Magnetic Chemical and Structural Properties, ed. Hasegawa, R., CRC Press (1983) 109; Moorjani, K. and Coey, J.M.D., Magnetic Glasses, Elsevier, Amsterdam 1984 .
13. Schlottmann, P. and Bennemann, K.H., Phys. Rev. B 25 (1982) 6771.
14. Moorjani, K., Ghatak, S.K., Rao, K.V., Kramer, B. and Chen, H., J. Phys. 41 (1980) C8-718; Mitra, A. and Ghatak, S.K., J. Magn. Magn. Mat. 51 (1985).
15. Ghatak, S.K. (to be published).
16. Kumar, V. and Bennemann, K.H., Phys. Rev. B 28 (1983) 3138.
17. Tománek, D. and Bennemann, K.H., Surf. Sci. (1985), to appear.
18. Tománek, D., Mukherjee, S. and Bennemann, K.H., Phys. Rev. B 28 (1983) 665.

DEFECTS IN AMORPHOUS ALLOYS

W. TRIFTSHÄUSER and G. KÖGEL

Institut für Nukleare Festkörperphysik, Universität der Bundeswehr München, 8014 Neubiberg, Federal Republic of Germany

1. INTRODUCTION

The existence of localized defects is one of the key concepts for the understanding of many properties of crystalline materials. The verification and identification of similar defects in amorphous alloys will possibly facilitate the interpretation of the specific properties like diffusion, embrittlement, corrosion, magnetic hardening in these alloys. Amorphous metallic alloys are metastable solids with atomic arrangements which are not spatially periodic. Metallic glasses can be produced in a variety of ways: evaporation, sputtering, or fast quenching from the liquid state.

Most investigations have been performed on amorphous alloys which are stable at room temperature. Several structural models have been proposed. The dense random packing of hard spheres'model was derived from the Bernal model, conceived to explain the structure of simple liquids [1]. Such a structure contains different kinds of empty spaces, the so-called Bernal holes. The empty-space type holes with the largest volume are the dodecahedron with 0.19, the trigonal prism with 0.31 and the Archimedean antiprism with 0.38 of the volume of a single vacancy in the corresponding crystalline matrix [2]. Such a dense random packing of hard spheres is in good agreement with the experimental results for most metallic glasses.

In the microcrystalline models, the structure is made up of an assembly of uncorrelated crystalline regions of sizes around 10 Å to 30 Å. For the explanation of the experimental results, packing faults, local strain and dislocations have to be introduced [3,4]. Experimental evidence for chemical and configurational short-range order [5-7] and for local density fluctuations [8] have been found in many amorphous alloys.

Positron annihilation results provided the means for great progress in understanding the nature of point defects in crystalline metals. Positrons detect vacancy-type defects even at a concentration of a few ppm. The lifetime values and the parameters derived from the Doppler broadening studies of the annihilation radiation of trapped positrons give information on the size of the submicroscopic vacancy agglomerates and the microvoids. Positrons are also trapped with high probability at the core of dislocations and dislocation loops. The metallic glasses may be classified conventionally into two groups according to their constituents: metal-metalloid (M1 type) and metal-metal (M2 type) systems.

For positrons, amorphous alloys look like irregular arrays of potential wells with different strength. There are several possible states for a thermalized positron:

i) The positron is in a free-particle state, if the scattering of the positron by the irregular array of potentials is weak. Then the annihilation characteristics reflect the bulk properties.

ii) The scattering is strong and the mean free path of the positron is of the order of the average atomic spacing. The annihilation characteristics are then statistically averaged quantities over the annihilation sites.

iii) The positron is localized with its wave function confined to a small region of space. Then the annihilation characteristics reflect the local environment of the annihilation sites.

2. EXPERIMENTAL

The Doppler broadened annihilation spectra have been measured with a high-purity germanium diode. The γ-line of 497 keV from ^{103}Ru is used as a reference and is recorded simultaneously for stabilization of the system. The intrinsic resolution of the detector system is 1.14 keV FWHM at 497 keV. From the measured energy distribution the lineshape parameters I_v and I_c are evaluated [9].

The system for the positron lifetime measurement has a resolution of about 200 ps FWHM using ^{60}Co with energy windows set for ^{22}Na [10]. 10^6 to 10^7 events are collected in each spectrum resulting in a statistical uncertainty of better than 1 ps for most of the spectra.

A sandwich-type source-sample arrangement has been used. A 25 µCi ^{22}Na source between two 5 µm thick titanium foils is faced from both sides by three to four layers of metallic glass ribbons (each layer about 25 µm to 30 µm thick).

Lifetime and Doppler broadening measurements have been carried out simultaneously on the same specimen using this sample-source geometry.

Some metallic glasses have been also investigated with monoenergetic positrons of variable energy [11]. Doppler broadening measurements have been performed as a function of the incident positron energy between 450 eV and 28 keV [12].

The electron irradiation of $Ti_{50}Be_{40}Zr_{10}$ has been performed with a 1.5 MeV Van de Graaff at the Hahn-Meitner Institut in Berlin. During the irradiation the specimen has been cooled by liquid nitrogen resulting in an effective specimen temperature of about 80 K. The irradiation of $Ti_{60}Cu_{40}$ has been carried out with 3 MeV electrons at the Kernforschungsanlage Jülich. The specimens have been immersed in liquid helium during the irradiation.

After the irradiation the specimens are transferred into the measuring device (cryostat of variable temperature) at liquid nitrogen temperature.

One amorphous alloy ($Ni_{76}Si_{12}B_{12}$) has been exposed to helium ions of 8 keV energy. These helium implanted specimens have been investigated with the positron beam system at various positron energies.

3. RESULTS AND DISCUSSION

3.1 Intrinsic Defects in Metal-Metalloid Alloys

For iron based amorphous metal-metalloid (M1) alloys of different composition the Doppler profiles are compared in Figure 1 to pure crystalline iron and to the spectrum obtained for annihilations at single vacancies in crystalline iron. Independent of the constituents, the Doppler curves are very similar, but clearly different to the curve due to vacancies. In Figure 2 difference curves of the measured Doppler spectra with respect to pure crystalline iron are compared in more detail with those for crystalline Fe_2B and with deformed (50%) crystalline iron. The Doppler profiles are very similar to those of the deformed

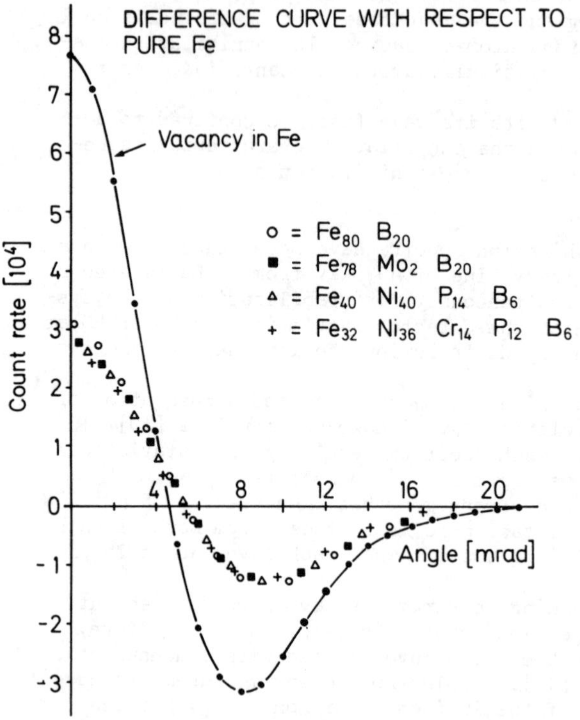

FIGURE 1. The Doppler-profile difference curves for various iron-based amorphous (M1) alloys are compared with the corresponding curve for vacancies in crystalline iron.

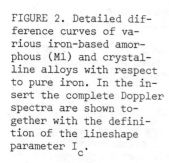

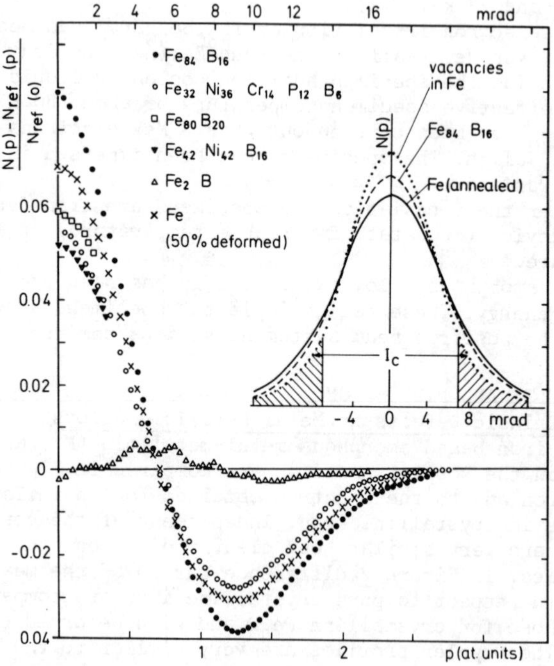

FIGURE 2. Detailed difference curves of various iron-based amorphous (M1) and crystalline alloys with respect to pure iron. In the insert the complete Doppler spectra are shown together with the definition of the lineshape parameter I_c.

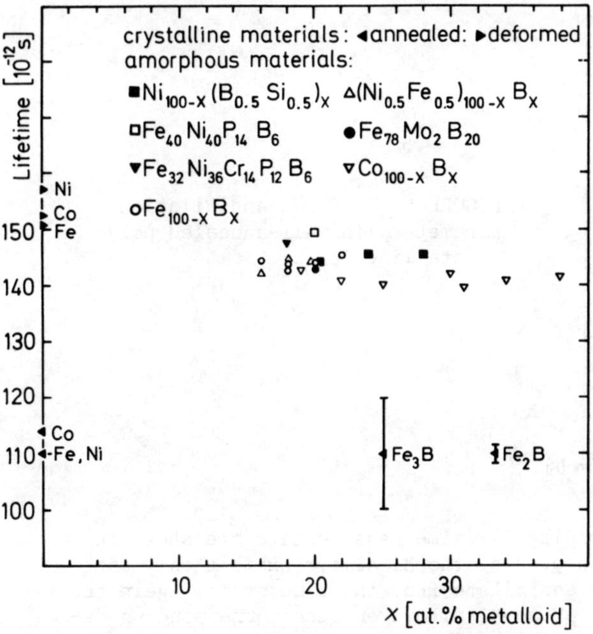

FIGURE 3. Positron lifetimes of various crystalline and amorphous (M1) materials.

crystal (dislocations) but reveal distinct differences with respect to annihilations in the bulk of iron and at vacancies in iron [12,13].

In Figure 3 the most sensitive lineshape parameter I_c (7-14 mrad) is plotted for a series of M1 alloys together with the corresponding crystalline metals. The strong correlation between the Doppler lineshape of the amorphous M1 alloys and the corresponding deformed crystals is valid over the whole range of the metalloid concentration investigated and independent of the chemical composition.

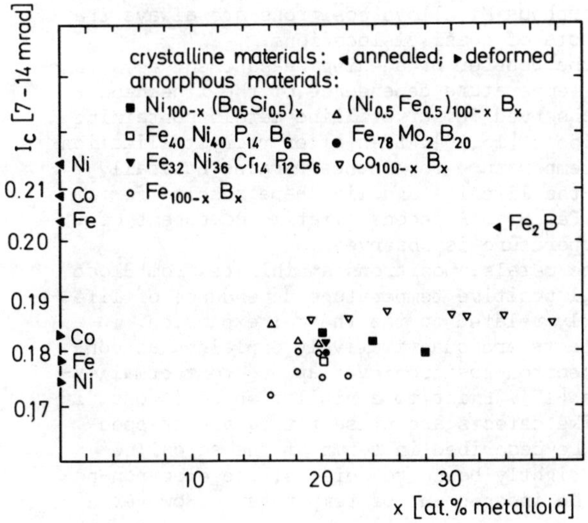

FIGURE 4. Doppler lineshape parameter I_c (see Figure 2 for definition) of various crystalline and amorphous (M1) materials.

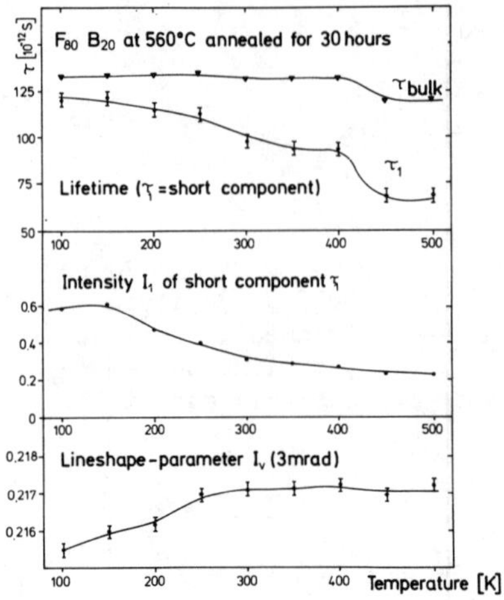

FIGURE 5. Positron annihilation parameters in well-annealed polycrystalline $Fe_{80}B_{20}$.

The results of the corresponding lifetime measurements are shown in Figure 4. Since all lifetime spectra of the M1 alloys exhibit only one single lifetime, all positrons annihilate from the same state. Again the strong correlation between amorphous M1 alloys and the corresponding deformed crystal is obvious and striking. Thus not only the electron momentum distribution but also the electron density sampled by the positron is almost identical in amorphous M1 alloys and in deformed crystals. The lifetime of the positron is inversely proportional to the electron density at the site of the positron. The electron density experienced by positrons in the amorphous M1 alloy is only 75% of that one in the corresponding crystals. However, since the average electron densities are almost identical in amorphous and crystalline alloys, this is a strong indication, that positrons in amorphous M1 alloys annihilate from regions with lower electron density. The only consistent picture for this behaviour is, that in amorphous M1 alloys positrons are always trapped at small vacancy-like defects of quasi dislocations.

This is also supported by the Doppler broadening measurement as a function of temperature [14]. The temperature dependence of the lineshape parameter is very similar as observed for crystalline metals containing dislocations and dislocation loops [15,16]. Even after recrystallization of the amorphous alloys, the temperature dependence remains basically the same [14]. Figure 5 shows the lifetime and lineshape results for well-annealed polycrystalline $Fe_{80}B_{20}$. A second lifetime component of 157 ps, but independent of temperature is observed.

In well annealed crystalline metals, positrons annihilate from Bloch states. There is always a small positive temperature dependence of lifetime τ and lineshape I_v linearly related to the thermal expansion. In simple metals the observed effects are quantitatively explained as consequences of a reduction of electron-positron overlap due to thermal expansion and lattice vibrations [17]. Prior to annihilation positrons diffuse in a metal. If vacancy-like defects are present they are trapped and the trapping kinetics can be described in terms of the so-called trapping model. For positrons tightly bound to defects, the electron-positron overlap is expected to be independent of temperature. However a

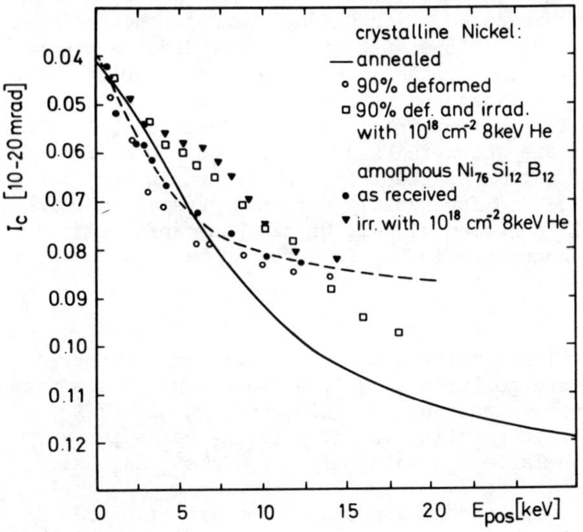

FIGURE 6. Doppler lineshape parameter I_c as a function of the positron implantation energy. The solid and the dashed line, respectively, are fits according to ref. [21].

variation of the annihilation characteristics with temperature is possibly by thermally activated detrapping. At a sufficiently high temperature T, positrons which are bound to defects with a binding energy E_b will escape at a rate proportional to (kT/h) exp $(-E_b/kT)$ [18]. As shown in Figures 3 and 4 there is complete positron trapping at vacancy-like defects in M1 alloys. Since these defects can have a distribution in size and binding energy, detrapping with temperature applies very well to this situation.

At very low temperature positrons are trapped by all defects with equal probability. If at higher temperatures positrons are thermally detrapped from the weakly binding defects they are still being trapped at defects with larger binding energies. Therefore the observed positron lifetime increases with rising temperature and the results can be explained by defects with binding energies $E_b \geq 0.3$ eV and with a total defect concentration $\geq 10^{-3}$. This model also explains the small changes introduced by crystallization [14,19].

As long as the grain sizes remain small compared to the diffusion length of the positron, positrons can be trapped at dislocations and grain boundaries. If by further annealing the grain sizes have grown sufficiently, a large fraction of positrons will be trapped at low temperature at grain boundaries or interfaces which are weakly binding defects. With increasing temperature positrons can escape from these defects and annihilate from Bloch states. Thus the observed positron lifetime decreases with increasing temperature (Figure 5). The second lifetime τ_2 is the indication of defects possibly either already present intrinsically and becoming visible after annealing or are produced during the annealing and recrystallization procedure.

All experimental results are consistent with the fact that in amorphous M1 alloys positrons are always trapped at vacancy-like defects with an open volume like the core of a dislocation. The concentration of these defects can be investigated with positrons of variable energy [12,13]. Positrons of energies between 1 to 20 keV are thermalized near the surface of the specimen. For a given energy of the implanted positrons, the fraction of the positrons is measured which diffuse back to the surface. Trapping of positrons at defects in the specimen and the defect concentration is directly related to the probability to reach the surface. From the measurement of the Doppler lineshape parameter as a function of the incident positron energy this positron fraction can be deduced [21]. The results of the measurements are shown in Figure 6. The positron diffusion length and hence the concentration of the trapping defects is very similar for amorphous $Ni_{76}Si_{12}B_{12}$ and for heavily deformed crystalline nickel, but quite different for well annealed nickel. On the basis of a simple diffusion model the defect concentration c_d is obtained by

$$c_d = (\ell_o^2 - \ell_d^2)/(\ell_d^2 \varkappa_d \tau_b). \qquad (1)$$

ℓ_d and ℓ_o are the diffusion lengths for the crystal with and without defects, respectively. τ_b is the positron bulk lifetime and $\varkappa_d$ is the trapping rate for unit defect concentration. For well annealed nickel ℓ_o is 870 nm [21]. The most uncertain quantity in Eq. 1 is the value for $\varkappa_d$. For the results presented in Table 1 a value of $\varkappa_d = 10^{14}s^{-1}$ is used.

TABLE 1: Diffusion length ℓ_d and the corresponding concentration $c_{v\ell}$ of vacancy-like defects in amorphous alloys.

material	ℓ_d(nm)	$c_{v\ell}$
$Ni_{76}Si_{12}B_{12}$	22 ± 2	1.4 x 10^{-3}
$Ni_{40}Fe_{40}B_{20}$	20 ± 2	1.6 x 10^{-3}
$Co_{75}B_{25}$	7.3 ± 0.6	1.3 x 10^{-2}

From the results we conclude that in all amorphous M1 alloys investigated there is definitely a large density of intrinsic defects. The detailed structure of those defects, however, remains unclear.

According to simple estimates, it is possible to explain the observed lifetimes by the annihilation of positrons trapped in an association of large Bernal holes [13,19]. But it seems unlikely that such configurations of empty holes exist with a density of 10^{-3} for metalloid concentrations as high as 38 atomic %. Probably larger holes are present if in the amorphous M1 alloys there are random packings of rigid metal-metalloid aggregates as proposed by Gaskell [22] instead of only single atoms.

All positron annihilation results are even better consistent with a quasi-dislocation model. From the magnetization curves of ferromagnetic amorphous M1 alloys such a model has been proposed [23]. Since the structure of the larger Bernal holes is closely related to the structure of the dislocation cores both of the defect structure models seem equivalent from the point of view of positron annihilation.

The decoration of these intrinsic defects with hydrogen was unsuccessful. Charging with hydrogen in amorphous alloys of various boron composition resulted in no observable changes of the positron annihilation characteristics. Even in amorphous $Pd_{80}Si_{20}$ the local electron density of the intrinsic defects remains unchanged by the charging with hydrogen [20], although this alloy is known for its good hydrogen-storage capability [24]. This is a strong indication that hydrogen is not filling up the intrinsic vacancy-like defects, but suggests that hydrogen is either

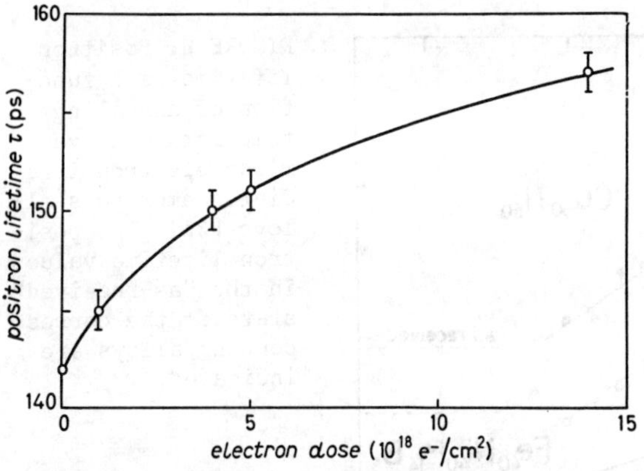

FIGURE 7. Positron lifetime as a function of the irradiation dose [20].

located at holes too small in order to be detected by positrons or located in the vicinity of the vacancy-like defects. A similar situation exists in copper [25] where hydrogen is not completely decorating a vacancy but is located close to the vacancy and in other fcc metals [26].

3.2 Irradiation Effects in Metall-Metalloid Alloys

Since amorphous M1 alloys contain the high density of intrinsic defects, as indicated in the preceding section, defects introduced into these alloys by irradiation have to be comparable in concentration in order to be seen by the positrons. This also explains why irradiation effects at room temperature and with relatively low doses have not been detected [27,28]. The effect of the electron-irradiation dose on the positron lifetime in amorphous $Fe_{80}B_{20}$ is shown in Figure 7 [29]. The lifetime spectra are well described by a single lifetime component increasing continuously with the electron dose [28]. At high doses a saturation value close to the positron lifetime for vacancies in crystalline iron is observed.

The change of the positron lifetime for various electron irradiated amorphous alloys as a function of the annealing temperature is shown in Figure 8 [20]. The lifetime decreases continuously with annealing temperature and at around room temperature the values before the irradiation are obtained again. Irradiation of amorphous $Fe_{80}B_{20}$ with high energy neutrons show the same saturation behaviour at a dose of 0.5 x 10^{17} neutrons/cm^2 as obtained for electron irradiation (Figure 7) [20]. The recovery behaviour of the positron lifetime agrees exactly with that in Figure 8 after electron irradiation. In $Ni_{76}Si_{12}B_{12}$ a heavily damaged surface layer of about 40 nm has been produced by irradiation with 8 keV helium ions to a dose of $10^{18}cm^{-2}$. The same irradiation has been applied to a heavily deformed nickel specimen. The radiation-damage effect in both materials is clearly visible (Figure 6) if positrons of the appropriate energy are used, so that the positrons are stopped within the damaged region of the specimens. Again the amorphous M1 alloy and the deformed nickel crystal behave very similar.

3.3 Intrinsic Defects in Metal-Metal Alloys

In the amorphous metal-metal (M2) alloys we always observe two lifetime components [12,13]. In contrast to the M1 alloys, the Doppler profiles and the observed positron lifetimes are now correlated to those of

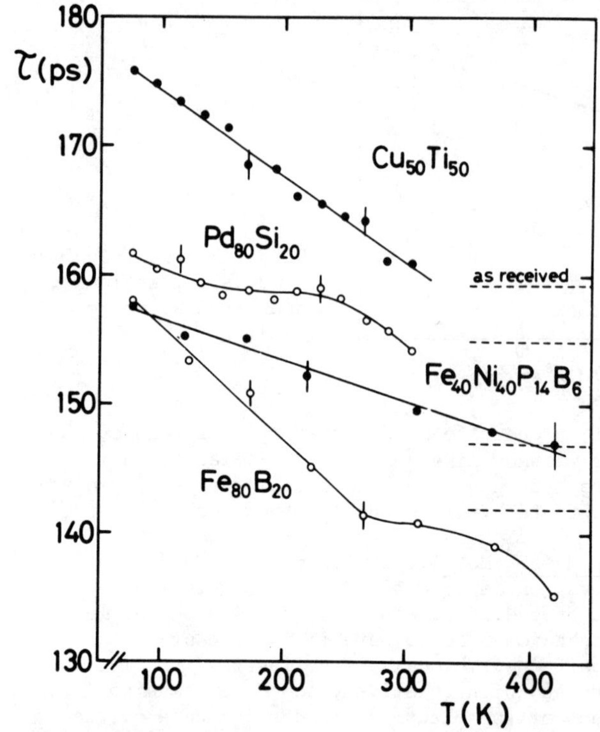

FIGURE 8. Positron lifetime as a function of annealing temperature in various electron irradiated amorphous alloys [31]. The positron lifetime values in the "as-received" state of the corresponding alloys are indicated.

the corresponding annealed crystalline constituents. The lifetime result for various amorphous M2 alloys are summarized in Table 2 compared with the corresponding pure crystalline constituents.

TABLE 2. Positron lifetimes in picoseconds (ps) for Cu, Ti, Zr, Be and amorphous copper-zirconium, $Ti_{50}Be_{40}Zr_{10}$ and $Ti_{60}Cu_{40}$ alloys.

crystalline	annealed	deformed	vacancy
Cu	125	170-180	190 - 200
Ti	143	168	222(ref. [30])
Zr	165	185	240 - 250
Be	137	178(62%)460(38%) $\bar{\tau}$ = 285	-

amorphous	τ_1	I_1(%)	τ_2	I_2(%)
$Cu_{57}Zr_{43}$	104 ± 10	9.0 ± 0.5	169	91
$Cu_{60}Zr_{40}$	69 ± 10	2.5 ± 0.5	166	97.5
$Cu_{62}Zr_{38}$	71 ± 10	3.5 ± 0.5	165	96.5
$Ti_{50}Be_{40}Zr_{10}$	145	82	251 ± 5	18
$Ti_{60}Cu_{40}$	163	93.5	270	6.5

A good and acceptable fit of the measured lifetime spectra is only pos-

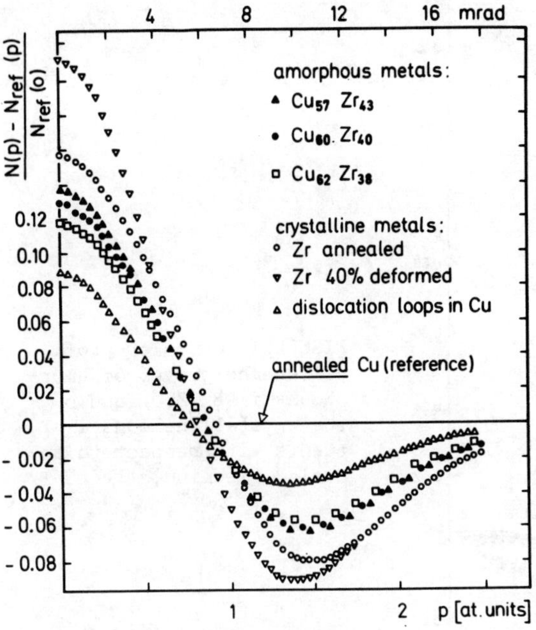

FIGURE 9. Doppler-profile difference curves of amorphous copper-zirconium (M2) alloys and of the corresponding crystalline components with respect to annealed copper.

sible with two lifetimes τ_1 and τ_2 indicating the existence of different positron-decay modes. In all specimens τ_1 is the result of annihilations from Bloch states in the amorphous matrix, whereas τ_2 is due to annihilations of positrons at defects. (In the copper-zirconium glasses τ_1 shows the well-known reduction due to defect trapping). These facts are also confirmed by the Doppler-profile results shown in Figures 9 and 10. The Doppler broadening difference curves of the amorphous copper-zirconium alloys are plotted and are compared with annealed and deformed zirconium and copper, respectively, (Figure 9).

Under the assumption that the Doppler profiles are also characteristic for defects in the hypothetically pure amorphous alloy, we estimate from the difference of the Doppler profiles of the annealed and the "as received" sample of $Ti_{50}Be_{40}Zr_{10}$ (Figure 10) that about 45% of the neighbours of the defects are Beryllium atoms in agreement with the expected value for a random alloy. The lifetime τ_1 of this amorphous alloy is similar to the bulk positron lifetime of Be and Ti, whereas τ_2 is similar to the lifetime at vacancies in Ti and to an average lifetime in deformed Be (Table 2). Annealing at 513 K for two hours reduces the intensity I_2 considerably (Table 3). We therefore attribute τ_2 to annihilations from quenched-in vacancy-like defects with a concentration of about 10^{-5}.

The positron mean lifetime and the Doppler lineshape parameter for $Ti_{50}Be_{40}Zr_{10}$ as a function of temperature are shown in Figure 11. The temperature dependence of lifetime and lineshape parameter in M2 alloys is characteristic for weakly binding defects and is just opposite to the corresponding behaviour in the M1 alloys.

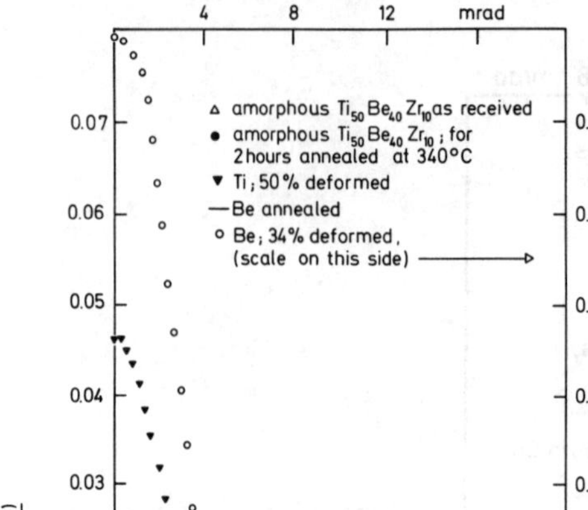

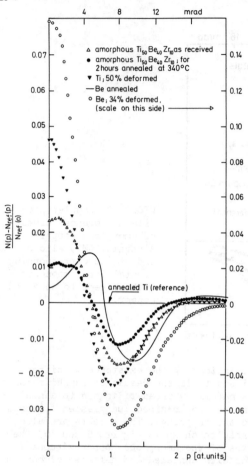

FIGURE 10. Doppler-profile difference curves of amorphous $Ti_{50}B_{40}Zr_{10}$ and of the crystalline constituents with respect to annealed titanium.

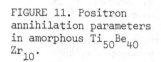

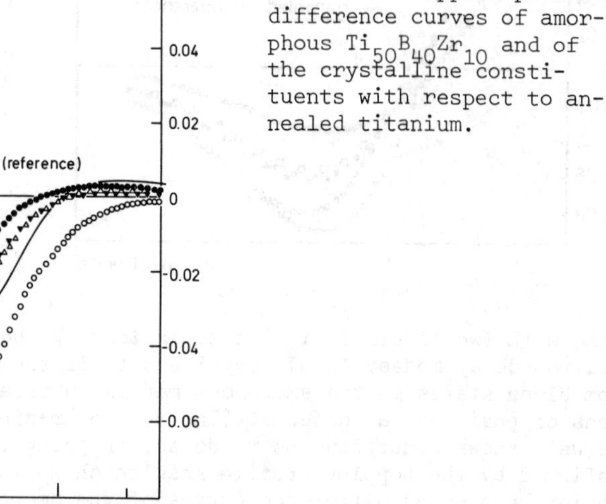

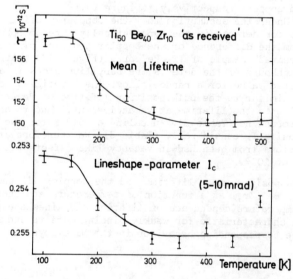

FIGURE 11. Positron annihilation parameters in amorphous $Ti_{50}Be_{40}Zr_{10}$.

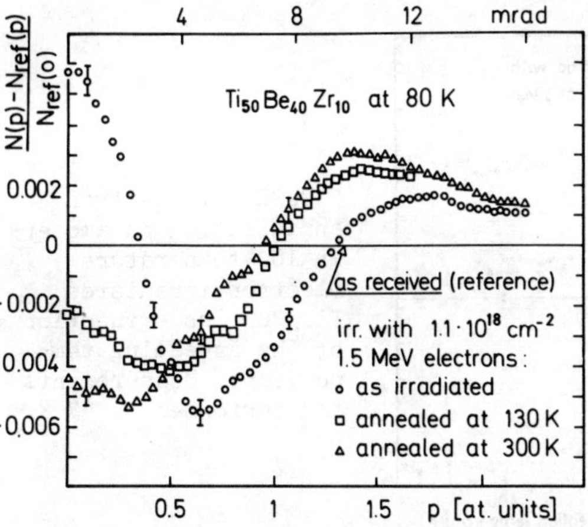

FIGURE 12. Doppler-profile difference curves for electron irradiated $Ti_{50}Be_{40}Zr_{10}$ at different annealing temperatures with respect to the "as received" alloy.

TABLE 3. Positron lifetimes for amorphous $Ti_{50}Be_{40}Zr_{10}$ at various temperatures.

temperature	τ_1	$I_1(\%)$	τ_2	$I_2(\%)$
300 K (as received)	145	82	251 ± 5	18
80 K (as received)	165	95.2	270 ± 5	4.8
300 K (after annealing at 513 K for 2 hours)	150	95.5	363 ± 5	4.5

3.4 Radiation Induced Defects in Amorphous Metal-Metal Alloys

The intrinsic defect concentration in the M2 alloys is sufficiently low to allow the study of isolated point defects produced by low-temperature electron irradiation. Amorphous $Ti_{50}Be_{40}Zr_{10}$ has been irradiated with 1.5 MeV electrons at about 80 K to a dose of $1.1 \times 10^{18} cm^{-2}$. After the irradiation the same two lifetimes are observed as for the unirradiated specimen but the intensity I_2 for the longer lifetime component is

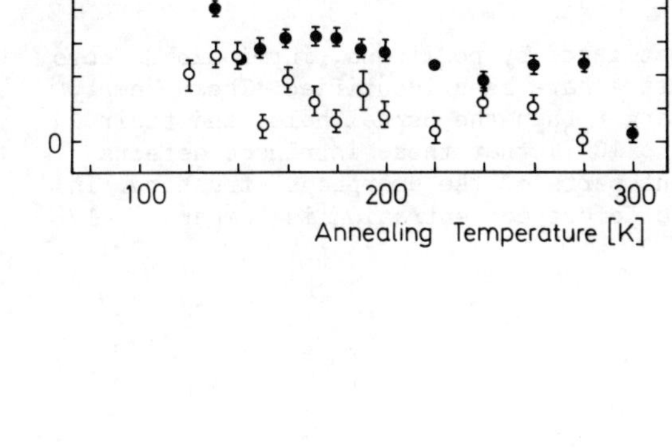

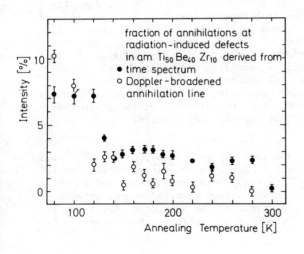

FIGURE 13. The fraction of annihilations at radiation-induced defects in amorphous $Ti_{50}Be_{40}Zr_{10}$. The contribution of the intrinsic defects is subtracted.

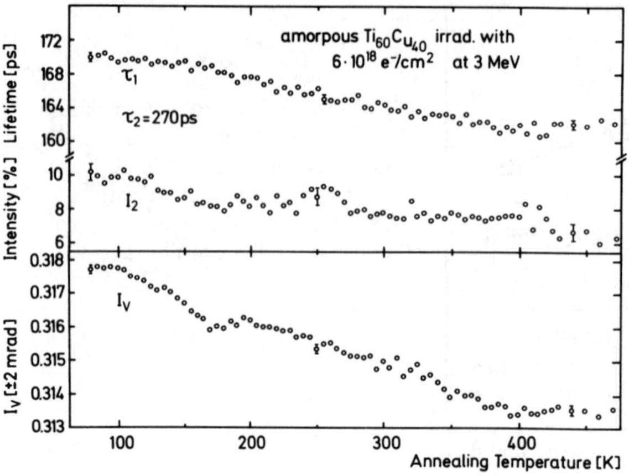

FIGURE 14. Positron annihilation parameters in low-temperature electron irradiated $Ti_{60}Cu_{40}$ as a function of the annealing temperature. Measurements are performed at 77 K.

enhanced indicating that more of these defect types have been produced during the irradiation. The irradiation effect is also clearly visible in the Doppler profile (Figure 12). Results of the isochronal (30 min) annealing study are shown in Figures 12 and 13. The defect responsible for the increase of the lifetime intensity I_2 and the Doppler profile anneals out at about 130 K (Figure 13). The changes of the Doppler spectra (Figure 12) indicate that at least two different radiation induced components are present. A broad component (from about 8 to 15 mrad) appears and remains even after annealing at 300 K. Probably this component is due to a radiation induced global relaxation of the amorphous structure. The narrow component (extending to about 4 mrad) which is typical for trapping at defects anneals out at 130 K. Thus both lifetime and Doppler broadening results reveal a radiation-induced defect of the same structure as the intrinsic quenched-in defects.

Amorphous $Ti_{60}Cu_{40}$ has been irradiated at 4.2 K with 3 MeV electrons to a dose of 6 x 10^{18}cm^{-2}. Prior to irradiation the specimen has been annealed at 473 K. Again the same two lifetimes are observed and long lifetime component I_2 is enhanced to about 10% after irradiation. The results of lifetime and Doppler lineshape parameter I_v as a function of the annealing temperature are shown in Figure 14. There is more or less continuous recovery of lifetime and lineshape with annealing temperature. There is, however, a weak indication of annealing stages at about 130 K and 170 K. The absence of distinct annealing stages is a clear demonstration that uncorrelated Frenkel pairs do not exist because neither long-range migration nor vacancy clustering is observed [20]. The hypothesis of close-correlation Frenkel pairs, widely distributed in configurations, explains the observed experimental results.

4. CONCLUSIONS

In all amorphous alloys investigated by positrons, intrinsic defects with a vacancy-like local structure have been identified. These "empty spaces" in M1 type alloys are larger than the Bernal holes and their concentration is so high (10^{-3} to 10^{-2}) that these intrinsic defects must be considered as constituent parts of the amorphous structure. In the M2 type alloys the intrinsic defect concentration is comparatively low.

In both types of amorphous alloys radiation induced defects have been observed. After low-temperature irradiation the created defects behave like close Frenkel pairs: the ejected atom stays not far from its initial position and is associated to an empty zone, the volume of which is approximately the volume of the missing atom. The strong elastic interaction between interstitial-like and vacancy-like defects makes the pair metastable, and recovery occurs by thermally activated collapse [20]. The continuous recovery demonstrates that there is a wide range of pair configurations.

Since in crystalline materials the technologically interesting properties like ductility and magnetic softness depend on defects, we expect that further studies, also with positrons, will contribute to the understanding of the aging and annealing behaviour of those properties in amorphous alloys.

5. ACKNOWLEDGEMENT

This work has been in part supported by the Deutsche Forschungsgemeinschaft. We would like to thank Dr. F. Dworschak (IFF der KFA Jülich) and Dr. R. Poerschke (Hahn-Meitner-Institut Berlin) for their valuable support for the low-temperature irradiation of the specimens. The help of Dr. S. Schuhbeck is appreciated.

REFERENCES

1. Bernal, JD: Proc. R. Inst. G.B. A37,335(1959); Nature(London),183 141(1959); 185,68(1960)
2. Cargill III GS: in Solid State Physics, Seitz F, Turnbull D, Ehrenreich, H: Vol. 30, New York, N.Y. 1975, p. 227
3. Cargill III GS: J. Appl. Phys. 41,12(1970)
4. Jensen EJ, Damgaard Kristensen W, Cotterill RMJ: J. Phys. Paris Coll. 36, C-2,49(1975)
5. Vincze I, Boudreaux DS, Tegze M: Phys. Rev. B19,4896(1979)
6. Waseda Y, Chen HS: Phys. Status Solidi (a) 49,387(1978)
7. Panissod P, Aliaga-Guerra D, Amamon A, Durand J, Johnson WL, Carter WL, Poon SJ: Phys. Rev. Lett. 44,1465(1980)
8. Nold E, Steeb S, Lamparter P, Rainer-Harbach G: J. de Physique 41, C-8,186(1980)
9. Mantl S, Triftshäuser W: Phys. Rev. B17,1645(1978)
10. Kögel G, in: Positron Annihilation, Proc. Fifth Intern. Conf., Hasiguti, R.R., Fujiwara K (eds.), The Japan Institute of Metals, Sendai, 1979, p. 383
11. Triftshäuser W, Kögel G: Phys. Rev. Lett. 48,1741(1982)
12. Kögel G, Triftshäuser W in: Positron Annihilation, Proc. Sixth Intern. Conf., Coleman PG, Sharma SC, Diana LM(eds.), North Holland Publishing Co., 1982, p.595
13. Triftshäuser W, Kögel G in: Metallic Glasses, Hargitai C, Bakonyi I, Kemeny T (eds.) Kultura Budapest 1980, p. 347
14. Kajcsos, Zs,Winter J, Mantl S, Triftshäuser W: Phys. Stat. Sol. (a) 58,77(1980)
15. Mantl S, Kesternich W, Triftshäuser W: J. Nucl. Mat. 69&70,593 (1978)
16. Nieminen RM, Laakonen J, Hautojärvi P, Vehanen A: Phys. Rev. B 19, 1397(1979)
17. Stott MJ, West RN: J. Phys. F8,635(1978)
18. Seeger A: Appl. Phys. 4,183(1974)

19. Kögel G, Winter J, Triftshäuser W, in: Metallic Glasses, Hargitai C, Bakonyi I, Kemeny T, (eds.) Kultura Budapest 1980, p. 311
20. Moser P, Corbel C, in: Positron Solid State Physics, Proc. Intern. School of Phys., Brandt W, Dupasquier A (eds.), North-Holland Publ. Co., Amsterdam 1983, p.697
21. Triftshäuser W, Kögel G, in: Positron Annihilation, Proc. Sixth Intern. Conf., Coleman PG, Sharma SC, Diana LM (eds.), North-Holland Publ. Co. 1982, p. 142
22. Gaskell PH: Journ. Non-Crystalline Solids 32,207(1979)
23. Kronmüller H: IEEE Trans. Magn. 15,1218(1979)
24. Chen HS: Rep. Progr. Phys. 43,353(1980)
25. Lengeler B, Mantl S, Triftshäuser W: J. Phys. F 8,1691(1978)
26. Bugeat JP, Ligeon E: Phys. Lett. A71,93(1979)
27. Chen HS: Phys. Stat. Sol. (a) 34, K127(1976)
28. Chen HS, Chuang SY: Appl. Phys. Lett. 31,255(1977)
29. Moser P, Hautojärvi P, Yli-Kauppila J, Corbel C, in: Positron Annilation, Proc. Sixth Intern. Conf., Coleman PG, Sharma SC, Diana LM (eds.), North Holland Publ. Co. 1982, p.592
30. Yli-Kauppila J, Moser P, Künzi H, Hautojärvi P: Appl. Phys. A 27, 31(1982)
31. Moser P, Hautojärvi P, Yli-Kauppila J. Van Zurk R, Chamberod A: Conf. Rapidly Qunenched Metals, Sendai (Japan), August 1981.

SUPERCONDUCTING AND STRUCTURAL PROPERTIES OF AMORPHOUS Zr(Ni,Mo,W)-ALLOYS

A. FLEISCHMANN, P. MÜLLER, W. GLÄSER
FAKULTÄT FÜR PHYSIK E21, TU MÜNCHEN
D-8046 GARCHING

1. INTRODUCTION

$Zr_{1-x}B_x$ compounds form glasses with a great variety of B-elements. Most of these alloys show glass forming ability even in a large concentration range. This favours amorphous (a-) Zr alloys for systematic investigations (1,2). In the present work we report on measurements of superconducting and structural properties of the isoelectronic and isomorphous alloys $Zr_{1-y}Mo_y$ and $Zr_{1-z}W_z$. We studied transition temperatures T_c and upper critical fields $H_{c2}(T)$ and discuss the results in terms of a non orthogonal tight binding approach (1,3). Furthermore tunneling experiments on a series of $Zr_{1-x}Ni_x$ alloys were performed in order to get a more direct picture of superconductivity in these compounds.

2. EXPERIMENTAL

The samples were prepared in a DC-Planar-Magnetron sputter facility working with Argon (6N) at 0.3 - 3 µbar and a starting vacuum of $2\cdot10^{-4}$ µbar. The deposition rates were varied between 1 and 5 Å/sec. The substrates were cooled by liquid Nitrogen. We used inhomogeneous sheet targets where the area ratio of the two elements determines the composition of the sample. To achieve a good homogeneity we chose a relatively large distance between target and substrate (100 mm) compared to the discharge area (r = 15 mm). The absolute composition of the samples was measured by Rutherford Backscattering or neutron activation respectively. The following metglasses were prepared: $Zr_{1-x}Ni_x$, $0.15 \leq x \leq 0.30$; $Zr_{1-y}Mo_y$, $0.27 \leq y \leq 0.42$; $Zr_{64}W_{36}$.

The superconducting junctions were of the type metglass - insulator - counterelectrode. The insulating barrier was prepared in situ either by glow discharge oxidation or by sputter deposition of a 20 Å Al layer and oxidation in air. Several thin strips of Al or Ag or Bi(Te) evaporated across the metglass served as counterelectrodes. T_c, ρ, $H_{c2}(T)$ were measured with the standard four terminal method. The X-ray scattering experiments were performed in a 6 kW rotating anode system using monochromatized MoKα radiation (4).

3. RESULTS AND DISCUSSION

The results of the T_c and $H_{c2}(T)$ measurements are summarized in fig. 1 and table 1. For $Zr_{1-x}Ni_x$ alloys we found a linear decrease of T_c with increasing x in agreement with refs. 2, 5. Compared to liquid quenched samples the glass forming region could be extended down to x = 0.15. This alloy showed T_c = 4.09K. The $Zr_{1-y}Mo_y$ compounds had transition temperatures of about 4.2K with no significant concentration dependence. All measured samples showed transition widths smaller than 20 mK.

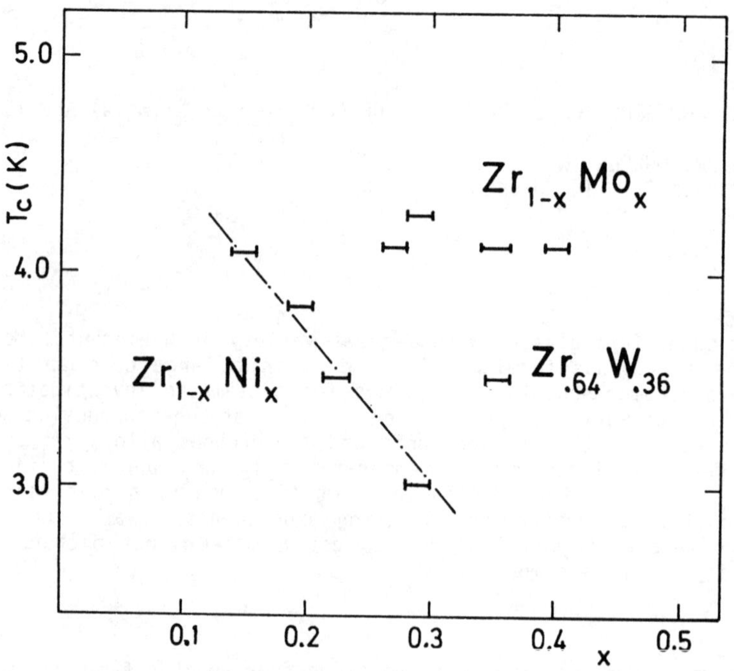

Fig. 1 Transition temperatures vs. composition.

For transition metal alloys Varma and Dynes (3) have argued using a non orthogonal tight binding analysis, that the parameter $\delta=\lambda/N(0) = W(1 \mp 2S)$ is approximately constant (the upper/lower sign belongs to less/more than half filled d-bands). W is the bandwidth and S the overlap integral.

Using McMillan's T_c formula (with $\mu^* = 0.13$) we obtain approximate values for the electron-phonon-coupling constant λ. The Debye temperatures were determined by scaling Θ_D of a-Zr_3Rh (191 K). The dressed density of states $N^*(0) = N(0)\cdot(1+\lambda)$ can be calculated with the Ginzburg - Landau - Abrikosov - Gorkov theory from the slope of the upper critical field at T_c. In the case of a-$Zr_{1-y}Mo_y$ and a-$Zr_{64}W_{36}$ this procedure leads to $\delta = 0.5 - 0.6$ in good agreement with the value of ref. 1 for alloys governed by early transition metals (ETM). Within this simplified bandstructure picture one also can give a crude explanation of the $T_c(y)$ dependence of $Zr_{1-y}Mo_y$ alloys (1): Firstly, δ should not change with y because Zr and Mo are both ETMs, and secondly N(0) is roughly constant, because the Zr and Mo subbands contribute a comparable amount to the total density of states at the Fermi energy.

Table 1

$Zr_{1-y}Mo_y$	y	T_c (K)	θ_D (K)	λ	$H'_{c2}(T_c)$ (kOe/K)	ρ (μΩcm)	N^* (states/eV atom)	δ
	0.27	4.15	193	0.67	25.3	155	1.9	0.59
	0.30	4.27	193	0.68	26.9	150	2.0	0.57
	0.35	4.12	192	0.67	25.1	145	2.0	0.56
$Zr_{1-y}W_y$	0.36	3.50	166	0.67	26.9	136	2.2	0.51

The results of our tunneling measurements are given in fig. 2. We find, that the whole series is well described by the BCS relation $2\Delta/k_BT_c = 3.52$ for weak coupling superconductors. This is in agreement with specific heat measurements (6). No deviations from the BCS conductivity were observed.

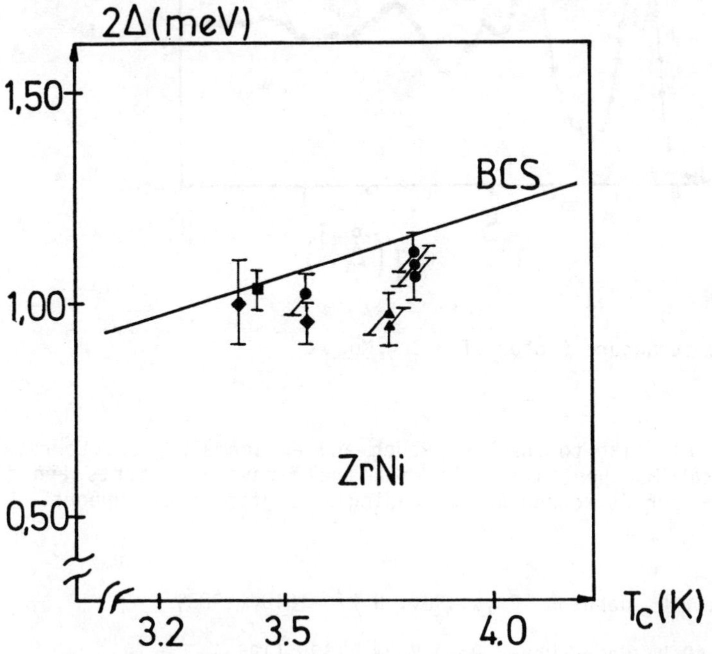

Fig. 2 Superconducting energy gaps vs. transition temperature.

The difference of the X-ray scattering lengths of Zr and Mo is very small. Therefore the weighing factor w_{NC} of the number-concentration correlation function $S_{NC}(q)$ is ten times smaller, and w_{CC} of the concentration-concentration correlation function $S_{CC}(q)$ is about 400 times smaller than the one of the number-number correlation function $S_{NN}(q)$ for $Zr_{1-y}Mo_y$. Hence, the scattered intensity is dominated by $S_{NN}(q)$.

Fig. 3 shows the static structure factor S(q) for $Zr_{70}Mo_{30}$. The maximum of the first peak occurs at $q = 2.56Å^{-1}$. An interesting feature of this curve is the slight splitting of the third peak which may be caused by a pronounced topological short range ordering.

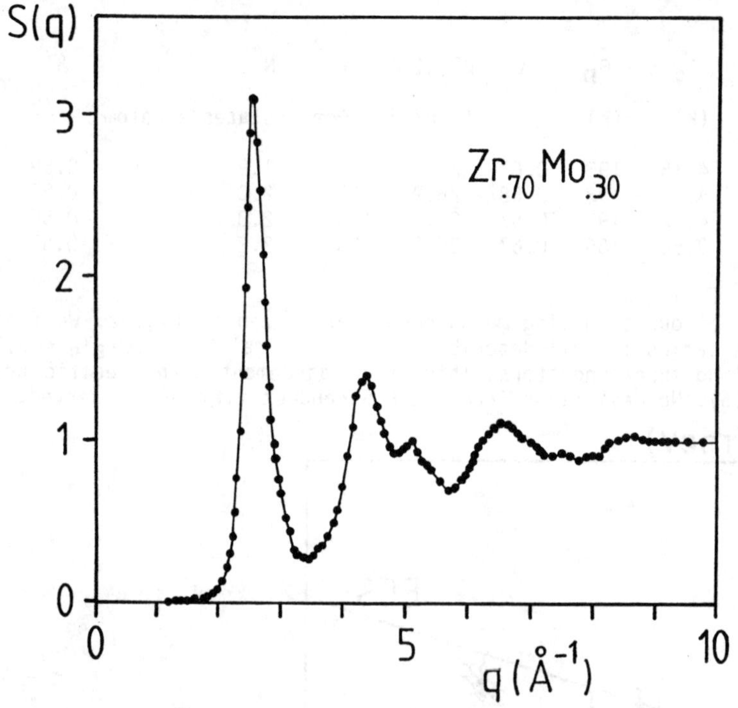

Fig. 3 Static structure factor of a-$Zr_{70}Mo_{30}$.

ACKNOWLEDGEMENTS

The authors would like to thank R. Rauch and A. Schmalzbauer, Lehrstuhl Peisl, Universität München, for help with the X-ray experiments and the Bundesministerium für Forschung und Technologie for financial support.

REFERENCES:

(1) M. Tenhover, W.L. Johnson, Phys. Rev. B 27, 1610 (1983)

(2) M.G. Karkut, R.R. Hake, Phys. Rev. B 28, 1396 (1983)

(3) C.M. Varma, R.C. Dynes, Superconductivity in d- and f-Band Metals edited by D.H. Douglass (Plenum, New York, 1976) p. 507

(4) D. Grasse, Dissertation, LMU München (1980)

(5) Z. Altounian, J.O. Strom-Olsen, Phys. Rev. B 27, 4196 (1983)

(6) D.E. Moody, T.K. Ng, Proceedings of the 17th International Conference on Low Temperature Physics ed. by N. Eckern, A. Schmidt, W. Weber, H. Wühl (North Holland, Amsterdam, 1984) p. 1323

4. DYNAMICS OF LIQUIDS AND AMORPHOUS MATERIALS

ATOMIC MOTIONS IN LIQUIDS

ALF SJÖLANDER

Institute of Theoretical Physics, S-412 96 Göteborg, Sweden

1. INTRODUCTION

At the Advanced NATO Study Institute at Zwiesel in 1979 I gave a few lectures on the atomic theory of liquid dynamics. The theory was based on the Zwanzig-Mori memory function approach. Since then the whole field has matured and we have seen some more applications of the theory. However, the conceptual ideas have not changed and I will here repeat several points, which I stressed at that time. This kind of theory can be viewed as a generalization both of the ordinary Boltzmann equation for dilute gases and of the Vlasov equation for classical plasmas. For both of them one introduces the density in the six-dimensional phase space, denoted by $f(\vec{r}\vec{p}t)$. Its evolution in time is in the appropriate situations governed by these equations. The linearized version of the Vlasov equation reads

$$\left(\frac{\partial}{\partial t} + \frac{1}{m}\,\vec{p}\cdot\vec{\nabla}_r\right)f(\vec{r}\vec{p}t) - \left\{\int d\vec{r}\,'d\vec{p}\,'\vec{\nabla}_r v(\vec{r}-\vec{r}\,')f(\vec{r}\,'\vec{p}\,'t)\right\}\cdot\vec{\nabla}_p f^{eq}(p) = 0, \quad (1)$$

where $v(r)$ is the interparticle (Coulomb) potential, m is the particle mass, and $f^{eq}(p) = n\,\phi_M(p)$ is the equilibrium distribution, with n being the uniform particle density and $\phi_M(p)$ the normalized Maxwellian momentum distribution. The linearized Boltzmann equation takes the form

$$\left(\frac{\partial}{\partial t} + \frac{1}{m}\,\vec{p}\cdot\vec{\nabla}_r\right)f(\vec{r}\vec{p}t) - \int d\vec{p}\,'K(\vec{p}\vec{p}\,')f(\vec{r}\vec{p}t) = 0 \quad (2)$$

with a kernel $K(\vec{p},\vec{p}\,')$, which contains information on the effect of a single binary collision. The interaction between the particles enters in $K(\vec{p},\vec{p}\,')$ through the binary collision cross section. Within the Born approximation it depends quadratically on the interaction potential. Already at this point we notice a significant difference between eq.(1) and eq.(2). The first one contains an interaction term linear in the potential and it depends explicitly on the interaction at a distance. This so called mean field term is crucial in plasmas but becomes insignificant in dilute classical systems with short range interactions.

In a dense system of atoms, both the aspects, which are contained in the two transport equations above, become important. One may consider that each atom experiences an average force from its surrounding as in plasmas and this can be incorporated in a mean field term. However, due to the strong short range repulsion between the atoms, aspects of binary collisions would also play a significant role and can be incorporated in a similar way as was done by Boltzmann. A well-defined model of this kind is the so called generalized Enskog model[1] and it has been analysed in quite some detail[2-4]. It treats the fluid as a collection of hard spheres of a certain radius and includes a proper mean field term as well as a collision term. Comparisons with computer simulations have shown that this model applies quite well to dense gases and also to liquids. One finds, however, that there are some very interesting discrepancies, particularly in the liquid region, which contain some new physics and which require a basic reformulation of the whole theory. These aspects turn out to be dominant when approaching the gas-liquid critical point and also when going into the supercooled region and approaching the liquid-glass transition.

In many of the experiments - light scattering, neutron scattering, NMR etc. - one measures directly certain correlation functions. Similarly, various hydrodynamic transport coefficients, as the viscosity coefficients and the heat conductivity, are also expressible as integrals over certain space- and time-dependent correlation functions[5]. The kind of theory I will describe in the following applies for the phase-space correlation function

$$F(\vec{p}\vec{p}';\vec{r}-\vec{r}',t-t') = \langle \delta f(\vec{r}\vec{p}t)\ \delta f(\vec{r}'\vec{p}'t')\rangle \ , \tag{3}$$

where here $\delta f(\vec{r}\vec{p}t)$ is the change in the microscopic distribution from its equilibrium value,

$$\delta f(\vec{r}\vec{p}t) = \sum_{\ell=1}^{N} \delta(\vec{r}-\vec{r}_\ell(t))\ \delta(\vec{p}-\vec{p}_\ell(t)) - n\phi_M(p) \ . \tag{4}$$

For simplicity we consider a one-component fluid and the summation runs over all the particles with positions and momenta denoted respectively by $\vec{r}_\ell(t)$ and $\vec{p}_\ell(t)$. The bracket in eq.(3) means an averaging over all initial configurations at t' according to the equilibrium distribution. If we integrate over $\vec{p}$ and $\vec{p}'$ we obtain the ordinary density correlation function. Fourier transforming this with respect to the space- and time-variables we get the dynamical structure factor $S(q,\omega)$, measured both in neutron and light scattering experiments. The initial value of $F(\vec{p}\vec{p}';\vec{r}t)$ is

$$F(\vec{p}\vec{p}';\vec{r}) = n\phi_M(p)\delta(\vec{p}-\vec{p}')\delta(\vec{r}) + n^2\phi_M(p)\phi_M(p')\{g(r)-1\} \ , \tag{5}$$

where $g(r)$ is the static pair-correlation function. It represents a situation where one particle is located at the origin and the other particles are in equilibrium around this particle. The microscopic time evolution describes then how the central particle diffuses out from the origin and how the surrounding particles move in. It is evident that this motion becomes strongly collective in dense media. We may, if desired, separate the motion of the central particle, the so called self-motion. We do this by simply considering one term in eq.(4) and we denote the corresponding correlation function by $F^s(\vec{p}\vec{p}';\vec{r}t)$. This has the initial value

$$F^s(\vec{p}\vec{p}';\vec{r}) = n\phi_M(p)\delta(\vec{p}-\vec{p}')\delta(\vec{r}). \tag{6}$$

The single particle aspects are obviously inherent in F^s, whereas the hydrodynamic aspects show up only in F. Neutron scattering revolutionalized our understanding of liquid dynamics and, since events over distances of the order 1-20Å and times of the order 10^{-14}-10^{-11} sec. are explored, both single particle and collective aspects appear. For large momentum transfers in the scattering experiments the neutrons literally see the individual atoms and the scattering cross section depends essentially only on F^s. The opposite is true in Brillouin scattering where the spatial resolution is of the order 5000Å or so.

I shall assume that the interaction potential is a pair-potential. This is quite appropriate for discussing the essential aspects of the atomic motions and is satisfactory also quantitatively for liquids like argon and simple metals like sodium and rubidium. In the latter cases one would have to introduce an effective ion-ion potential, which depends on the density. It is, of course, true that through these assumptions we loose all those contributions to the transport coefficients which depend explicitly on the free motion of the conduction electrons. In order to incorporate this one would have to treat the ions and the electrons on the same footing. To my knowledge nobody has fully developed the present general approach to this situation and I have to refrain from doing that here.

The equilibrium pair-correlation function enters in the theory as an input. In this respect the theory is complementary to all those theories which deal with the static properties of liquids. Even though it would be highly desirable to have a treatment which determines both the static and the dynamic properties on the same footing, such a theory is still lacking.

2. GENERAL TRANSPORT EQUATIONS

I begin by stating the general result and pointing out its connexion to the classic transport equations. I will emphasize the new aspects, which enter in the following exact equation for $F(\vec{p}\vec{p}';\vec{r}t)$[6]:

$$\left(\frac{\partial}{\partial t} + \frac{1}{m}\,\vec{p}\cdot\vec{\nabla}_r\right)F(\vec{p}\vec{p}';\vec{r}t) - \left\{\int d\vec{r}''d\vec{p}''\,\vec{\nabla}_r v_{eff}(\vec{r}-\vec{r}'')F(\vec{p}''\vec{p}';\vec{r}''t)\right\}\cdot\vec{\nabla}_p f^{eq}(p)$$
$$+ \int_0^t dt''\int d\vec{r}''d\vec{p}''\;\Gamma(\vec{p}\vec{p}'';\vec{r}-\vec{r}'',t-t'')\;F(\vec{p}''\vec{p}';\vec{r}''t'') = 0\;. \qquad (7)$$

The memory function $\Gamma(\vec{p}\vec{p};\vec{r}t)$ can also be written as[7]

$$\Gamma(\vec{p}\vec{p}';\vec{r}-\vec{r}',t-t') = -\,\vec{\nabla}_p\cdot\,\mathbf{L}(rpt;r'p't')\cdot\{\vec{p}'+(1/m\beta)\vec{\nabla}_{p'}\}\;, \qquad (8)$$

where $\beta=1/k_BT$ and where we have introduced a new memory function of tensor character. The mean field potential is given by

$$v_{eff}(r) = -\,\frac{1}{\beta}\,c(r)\;, \qquad (9)$$

where $c(r)$ is the direct correlation function and is uniquely determined by the static pair correlation function $g(r)$.

Eqs.(7-9) contain up to now no approximations but are of not much practical use unless we can specify Γ or $\mathbf{L}$ more explicitly. However, already the form of eq.(8) guarantees that the continuity equation for the density is fulfilled. Approximating $\mathbf{L}$ by

$$\mathbf{L}(\vec{r}\vec{p}t;\vec{r}'\vec{p}'t') = \zeta\;\delta(\vec{r}-\vec{r}')\delta(\vec{p}-\vec{p}')\delta(t-t')\;, \qquad (10)$$

we recover the ordinary Fokker-Planck collision term. We recall that this is appropriate if we can consider the force on a single atom from the surrounding ones as a truly random force. However, the surrounding has its own dynamics with both rapid and slow motions and this invalidates the assumptions behind eq.(10). The full $\mathbf{L}$ is meant to incorporate the motion of any single particle as well as the simultaneous motion of its surrounding, which is not already taken care of through the mean field term. It should therefore be no surprise that the hydrodynamic transport coefficients - viscosity coefficients and heat diffusion constant - are directly related to certain integrals over Γ. We may visualize each individual atom performing some Brownian type motion, all the time disturbing its surrounding and at the same time being influenced by the same surrounding. For instance, the long wavelength disturbances in the medium around any single atom would propagate according to ordinary hydrodynamics and would decay rather slowly. This leads to a long lasting feedback effect entering as a long time tail in the memory function. The single binary collisions would last only for a very short time, of the order 10^{-13} sec. or less, and lead to a rapid initial decrease of Γ. It is by now well known and also well explained that the memory function goes asymptotically as $t^{-3/2}$ in three dimensions[8,9].

A phenomenological formulation along the above lines was given long time ago by Akcazu and Duderstadt[10] and by Lebowitz et al[1] with a simplified form for Γ. Forster and Martin[11] derived an equation of the same form for a weakly interacting system and they demonstrated how the various conservation laws enter in Γ. However, the full equation was first derived and analysed as far as its general structure is concerned by Mazenko[12]. He[13] showed how the ordinary Boltzmann equation follows from eq.(7), when considering a dilute system. Mazenko did not follow the famous projection operator procedure of Zwanzig[14] and Mori[15], but the quickest way of arriving at his results seems to be to just apply their scheme to the microscopic phase-space function $f(\vec{r}\vec{p}t)$. The general equation has been applied by various authors for practical calculations with some approximations for Γ.

Before going into any of these applications I would like to continue along an exact path and derive what is called generalized hydrodynamics. Following standard procedures we may expand $\delta f(\vec{r}\vec{p}t)$ in eq.(4) in terms of a complete set of 3-dimensional Hermite polynomials

$$\delta f(\vec{r}\vec{p}t) = \phi_M(p) \sum_{\nu=0}^{\infty} A_\nu(\vec{r}t) H_\nu(\vec{p}) , \tag{11}$$

where

$$\int d\vec{p}\; H_\mu(\vec{p})\, H_\nu(\vec{p})\, \phi_M(p) = \delta_{\mu\nu}. \tag{12}$$

The first few coefficients $A_\mu(\vec{r}t)$ represent the microscopic particle density, the three Cartesian components of the particle current density, the kinetic energy density etc. In this representation the correlation function $F(\vec{p}\vec{p};\vec{r}t)$ becomes a matrix

$$F_{\mu\nu}(\vec{r}t) = \int d\vec{p}\; H_\mu(\vec{p}) F(\vec{p}\vec{p}';\vec{r}t) H_\nu(\vec{p}')d\vec{p}' , \tag{13}$$

and its Fourier-Laplace transform is

$$F_{\mu\nu}(\vec{q}z) = \int_0^\infty dt\; e^{-zt} \int d\vec{r}\; e^{-i\vec{q}\cdot\vec{r}}\; F_{\mu\nu}(\vec{r}t) . \tag{14}$$

Eq.(7) takes then the matrix form

$$zF_{\mu\nu}(\vec{q}z) - \Omega_{\mu\lambda}(\vec{q})F_{\lambda\nu}(\vec{q}z) + \Gamma_{\mu\lambda}(\vec{q}z)F_{\lambda\nu}(\vec{q}z) = F_{\mu\nu}(\vec{q}), \tag{15}$$

where summation over repeated indices is implied and where

$$\Omega_{\mu\nu}(\vec{q}) = -i\int d\vec{p}\; H_\mu(\vec{p})(\vec{q}\cdot\vec{p}/m)H_\nu(\vec{p})\Phi_M(p) + iq(m\beta)^{-1/2}nc(q)\delta_{\mu 1}\delta_{\nu 0} \,, \qquad (16)$$

and c(q) is the Fourier transform of the direct correlation function and $\delta_{\mu\nu}$ is the Kronecker symbol. From the definition follows that $F_{\mu\nu}(q)$ is a strictly diagonal matrix. I here choose the μ=1 component of the current to be along the vector $\vec{q}$. From the form of eq.(8) follows that $\Gamma_{0\nu}(\vec{q}z)=0$ and the μ=0 component of eq.(15) is simply the continuity equation for the density. The μ=1 component represents the equation for the longitudinal current and the μ=2 and 3 components are the equations for the transverse current. $H_4(\vec{p})$ is defined in such a way that the μ=4 component of eq.(15) is the equation for the kinetic energy density and it goes in the hydrodynamic limit, i.e. $q,\omega\to 0$, over into the heat diffusion equation for the temperature.

Generalized hydrodynamics is obtained from eq.(15) by concentrating on the μ,ν=0-4 components and eliminating all the others. This is formally achieved by introducing the projection operator[12]

$$P = \sum_{\mu=0}^{4} |\mu\rangle\langle\mu| = I - Q \,. \qquad (17)$$

We can then deduce the equation

$$zF_{jk}(\vec{q}z) + D_{j\ell}(\vec{q}z)F_{\ell k}(\vec{q}z) = F_{jk}(\vec{q}) \;,\; j,k,\ell = 0\text{–}4 \,, \qquad (18)$$

where

$$D_{j\ell}(\vec{q}z) = \Sigma_{j\ell}(\vec{q}z) - \Sigma_{j\mu}(\vec{q}z)\{Q[z+Q\Sigma(\vec{q}z)Q]^{-1}Q\}_{\mu\nu}\Sigma_{\nu\ell}(\vec{q}z) \,, \qquad (19)$$

and

$$\Sigma_{\mu\nu}(\vec{q}z) = -\,\Omega_{\mu\nu}(q) + \Gamma_{\mu\nu}(\vec{q}z) \,. \qquad (20)$$

The Q-operator implies here that the summation over the intermediate components runs only over the non-hydrodynamic ones, $\mu>4$. One can now compare eq.(18) with the ordinary hydrodynamic equations and then identify generalized transport coefficients, which depend both on the wavenumber and the frequency. For the longitudinal current correlation function we obtain from eq.(18)

$$F_{11}(\vec{q}z) = z\{z^2+(q^2/m\beta S(q)) + zD_{11}(\vec{q}z) + \frac{zD_{14}^2(\vec{q}z)}{z+D_{44}(\vec{q}z)}\}^{-1}, \qquad (21)$$

where S(q) is the static structure factor. Similarly, for the transverse current correlation function we have

$$F_{22}(\vec{q}z) = F_{33}(\vec{q}z) = \{z+D_{22}(\vec{q}z)\}^{-1}. \qquad (22)$$

These two equations can now be compared with the known hydrodynamic ones;

$$F_{11}(\vec{q}z) = z\{z^2+(q^2/\rho\kappa_T)+zq^2(\tfrac{4}{3}\,\eta+\zeta)/\rho + \frac{T}{\rho}\,(\frac{\partial p}{\partial T})_v^2\,\frac{zq^2}{zC_v+q^2\lambda_T}\}^{-1}. \qquad (23)$$

Here, ρ=mn is the mass density, κ_T is the isothermal compressibility, η and ζ are the shear and bulk viscosities, C_v is the heat capacity at constant volume, λ_T is the heat conductivity, and finally $(\partial p/\partial T)_v$ is the temperature derivative of the pressure at constant volume. $m\beta S(q)$ is equal to $\rho\kappa_T$ for q=0 and we do obtain the correct isothermal sound velocity. That D_{11} becomes proportional to q^2 and D_{14} proportional to q follows from conservation of momentum. Similarly

$$F_{22}(\vec{q}z) = \{z+q^2\eta/\rho\}^{-1} . \tag{24}$$

It is quite revealing to see how the full heat capacity enters in our equation. $F_{44}(qz)$ is the correlation function for the kinetic energy density. However, $D_{44}(qz)$ incorporates the coupling between the kinetic and potential energy and it was shown by Mazenko[12] and Forster[16] that for $q,z\to 0$ it goes over to

$$D_{44}(\vec{q}z) \to (C_v^o)^{-1}\{z(C_v-C_v^o) + q^2\lambda_T\} , \tag{25}$$

where C_v^o is the free particle heat capacity. $F_{44}(qt)$ contains information on how the fluctuations in the kinetic energy are partly converted into potential energy.

The density correlation function is obtained from $F_{11}(qz)$ by using the continuity equation. We have

$$F_{oo}(\vec{q}z) = \frac{nS(q)}{z} - (q^2/m\beta z^2)F_{11}(\vec{q}z) , \tag{26}$$

and the dynamical structure factor $S(q\omega)$ is then found by inserting $z=i\omega$ and taking the real part of $F_{oo}(q,i\omega)$. Actually, following the conventional definition of $S(q\omega)$[17] we have also to devide by $n\pi$.

Let us pause for a moment and recollect what we have learned so far. From eq.(19) we see that there are two different contributions to our generalized transport coefficients. The first part, the direct part, consists of the hydrodynamic matrix elements of Ω and Γ, where Ω contributes only to the generalized pressure. The other part, the indirect part, involves the non-hydrodynamic matrix elements. This reflects what occurs when we solve the ordinary Boltzmann equation. In principle we then have to find the eigen states of the Boltzmann collision operator in order to determine its inverse. Only those eigen states which are orthogonal to the hydrodynamic ones enter in calculating the viscosity and the heat conductivity. For instance, the shear viscosity is within the Boltzmann approximation given by

$$D_{22}^B(\vec{q}z) = - \Omega_{2\mu}^o(\vec{q})\{Q\Gamma Q^{-1}\}_{\mu\nu}\Omega_{\nu 2}^o(\vec{q}), \tag{27}$$

where Ω^o is the kinetic part of Ω and Γ is here the Boltzmann operator. The direct part vanishes. One condition for the validity of ordinary hydrodynamics is that the time integral over the memory function Γ exists. This is not the case at the gas-liquid critical point due to the development of a long time tail in Γ and it requires a more sophisticated treatment.

Since the single particle aspects appear quite explicitly in the experiments on $S(q\omega)$ for larger q-values, the hydrodynamic form in eq.(21) is less suitable in this case. Furthermore, one has found that the coupling to the true heat modes becomes then quite unimportant [18-20]. It would therefore be more suitable to split off the self part $F^s(\vec{p}\vec{p}';\vec{q}z)$, which satisfies an equation similar to eq.(15) with another memory term Γ^s and without any mean field term. The corresponding current correlation function can be obtained and this describes the Brownian type motion of a single atom. Doing this one can for $F_{11}(qz)$ write[21]

$$F_{11}(\vec{q}z) = F^s_{11}(qz)/\{1-[(q^2/m\beta z)nc(q)-\Gamma^d_{11}(qz)]F^s_{11}(qz)\} \ , \tag{28}$$

where

$$\Gamma^d_{11}(\vec{q}z) = \Gamma_{11}(\vec{q}z) - \Gamma^s_{11}(\vec{q}z) \ . \tag{29}$$

It should be stressed that we have here made certain approximations by keeping only the Γ^d_{11}-component. Explicit calculations have shown this to be quite acceptable. However, this kind of simplification may not be permitted for calculating $F^s_{11}(qz)$. One has to keep all the $\mu\nu$-components of the free particle flow term. This is one reason why one may have to make different approximations for the self and the collective motion. The denominator in eq.(28) becomes essentially unity for large wavevectors and $F_{11}(qz)$ approaches $F^s_{11}(qz)$. One finds that the single particle aspect becomes rapidly less important for q-values below the first peak in the static structure and the equation goes automatically over to the hydrodynamic form, where the collective aspects are more clearly revealed. Retaining eq.(28), we miss the heat diffusion peak in $S(q,\omega)$. Since F^s_{11} contains that part of the indirect term in eq.(19) which refers to the self motion, we still get a finite viscosity in the low density limit.

It may be worth pointing out the connexion between eq.(28) and the ordinary mean field theory or random phase approximation. For weak interaction $F^s_{11}(qz)$ goes over into the free particle expression, $-c(q)/\beta$ approaches the Fourier transform of the bare potential, and $\Gamma^d_{11}(qz)$ becomes a higher order term to be neglected. This gives then directly the Vlasov, or RPA, result. In eq.(28) one is evidently replacing the free particle motion by the proper description of the single particle one, renormalizing the bare potential, and furthermore introducing non-mean field contributions through $\Gamma^d_{11}(qz)$. The latter is essential in order to obtain proper hydrodynamics.

The generalized hydrodynamics in eq.(18) can be obtained directly by applying the Zwanzig-Mori projection operator technique on the five hydrodynamical variables. Such an approach was used by Götze and collaborators[22,23] and applied for explicit calculations. The drawback is that it is then more difficult to illustrate the single particle aspects.

Having tried to explain the physical content of our general transport equation, it now remains to give a more explicit form for $\Gamma(\vec{p}\vec{p}\,';\vec{r}t)$ or the matrix $\Gamma_{\mu\nu}(\vec{r}t)$.

3. THE MEMORY FUNCTION

Following the procedure of Zwanzig and Mori we first write[6,14,15]

$$\delta f(\vec{r}\vec{p}t) = \exp(L_{op}t)\delta f(\vec{r}\vec{p}0) \ , \tag{30}$$

where L_{op} is the full Liouville operator for the system. It implies that as time proceeds $\delta f(\vec{r}\vec{p}t)$ couples to higher order distributions as in the familiar BBKY-hierarchy of equations. A part of $\delta f(\vec{r}\vec{p}t)$ is directly correlated to the initial fluctuation $\delta f(\vec{r}\vec{p}0)$ and the proportionality factor is essentially the correlation function introduced in eq.(3). In order to sort out this from all other contributions we can introduce projection operators P and Q - not the same as in eq.(17) - such that

$$P + Q = I \;, \tag{31}$$

and

$$\langle \delta f(\vec{r}\vec{p}t) Q \delta f(\vec{r}'\vec{p}'0) \rangle = 0 \,. \tag{32}$$

The other part, $\langle \delta f(\vec{r}\vec{p}t) P \delta f(\vec{r}'\vec{p}'0) \rangle$, is then by definition proportional to $F(\vec{p}\vec{p}';\vec{r}-\vec{r}',t)$ and it excludes those fluctuations in $\delta f(\vec{r}\vec{p}t)$ which are uncorrelated to the initial fluctuation. The basic idea is analogous to that contained in Langevin's equation for a Brownian particle, where one splits up the force into a friction term proportional to the particle velocity and a stochastic one statistically independent of the initial velocity.

The quickest way of arriving at eq.(7) is to formally split $\delta f(t)$ into two parts, $P\delta f(t)$ and $Q\delta f(t)$ and write eq.(30) in the form (the phase-space variables are suppressed)

$$\begin{aligned} \frac{d}{dt} P\delta f(t) &= PL_{op}P\delta f(t) + PL_{op}Q\delta f(t) \,, \\ \frac{d}{dt} Q\delta f(t) &= QL_{op}Q\delta f(t) + QL_{op}L\delta f(t) \,. \end{aligned} \tag{33}$$

Elimination of $Q\delta f(t)$ from the second equation leads to

$$\frac{d}{dt} P\delta f(t) = PL_{op}P\delta f(t) + \int_0^t dt' \; PL_{op}Q\exp\{QL_{op}Q(t-t')\}QL_{op}P\delta f(t') \;, \tag{34}$$

which is just another way of writing eq.(7). For more details I refer to two recent textbooks on the subject[5,6].

The first part in eq.(34) yields the mean field term and its explicit form is obtained by using eq.(32) and carrying out the statistical averaging. Similarly, one obtains an expression for the memory function from the second part of eq.(34). It is more suitable at present to have the matrix elements $\Gamma_{\mu\nu}(\vec{r}t)$ and they are[21,24]

$$\begin{aligned} \Gamma_{\mu\nu}(\vec{r}-\vec{r}',t-t') = -\frac{1}{n}\int d\vec{p}H_\mu(\vec{p})\{\int d\vec{r}_2 d\vec{p}_2 \; \vec{\nabla}_r v(\vec{r}-\vec{r}_2)\cdot\vec{\nabla}_p \langle \delta f(\vec{r}\vec{p},\vec{r}_2\vec{p}_2) \\ \times \; Q\exp\{QL_{op}Q(t-t')\}Q \;\delta f(\vec{r}'\vec{p}',\vec{r}_2'\,\vec{p}_2')\rangle \;\vec{\nabla}_{r'} v(\vec{r}'-\vec{r}_2')\cdot\vec{\nabla}_{p'}\} \;H_\nu(\vec{p}')d\vec{p}' \,. \end{aligned} \tag{35}$$

The corresponding memory function for the self motion has the same form with the change that the variables $(\vec{r}\vec{p})$ and $(\vec{r}'\vec{p}')$ refer to one and the same particle, the self particle. δf_2 is the fluctuation in the microscopic two-particle distribution function. It is obvious that $\Gamma_{\mu\nu}$ and also Γ^s incorporates simultaneous motion of a single particle and its surrounding. In the self part we single out one particular particle and follow its motion. For a dilute gas $\Gamma_{\mu\nu}$ would primarily contain the dynamics of the rapid binary collisions. However, it is found to contain also longer lasting collective feedback effects. Actually, it was recognized that the non-linear Boltzmann-Enskog equation incorporates such effects[25], but they are erased in the linearized version. One should notice that the non-linearities are in the present formulation hidden in the memory function.

In order to explore $\Gamma_{\mu\nu}$ further one has to analyse how the four point correlation function in eq.(35) evolves in time. The initial value of $\Gamma(\vec{p}\vec{p}';\vec{q}t)$ has the form[10],

$$\Gamma(\vec{p}\vec{p}';\vec{q},t=0) = \zeta\, \vec{\nabla}_p\cdot\{\vec{\nabla}_p+(\beta/m)\vec{p}\}\delta(\vec{p}-\vec{p}') - \phi_M(p)p^{\alpha}\Lambda^{\alpha\beta}(\vec{q})p'^{\beta}, \tag{36}$$

where

$$\zeta = (n/\beta)\int d\vec{r}\; g(r)\nabla^2 v(r)\ ,$$
$$\Lambda^{\alpha\beta}(q) = q^{\alpha}q^{\beta}nc(q)/m - \delta_{\alpha\beta}(n\beta/m^2)\int d\vec{r}\,\cos(\vec{q}\cdot\vec{r})g(r)(\partial^2 v(r)/\partial r_{\alpha}^2). \tag{37}$$

The first term is the Fokker-Planck operator and it arises from the self motion. The second term provides a coupling to the collective current of the medium and it gives the initial value of Γ_{11}^{d} in eq.(28). Akcasu and Duderstadt[10] and Lebowitz et al[1] took into account the time dependence by multiplying with an exponential decay factor. Sjögren[21] assumed instead that the binary collisions, giving rise to a rapid initial decrease in Γ_{11}, can be incorporated through a factor $\exp\{-\alpha(q)t^2\}$. The q-dependence was then determined from the known sixth moments of $S(q\omega)$ (or the fourth moment of the longitudinal current correlation function). For calculating the transverse current correlation one would use its fourth moment instead. This seems to be the best one can do at present, when considering a realistic interaction potential. For a system of hard spheres the situation is simpler. The binary collisions lead then to an instantaneous collision term and $\Gamma(\vec{p}\vec{p}';\vec{q}t)$ has a $\delta(t)$-singularity and this provides the Enskog operator[12]. In this case the fourth and higher order frequency moments of $S(q\omega)$ are infinite, reflecting the strong singularity of the potential at short distance.

The remaining part of Γ incorporates collective effects, where the motion of the medium surrounding any individual atom can be described in terms of the hydrodynamical variables. For instance, we may visualize the creation of a backflow current of the kind Feynman and Cohen[26] introduced for He and which Alder and Wainwright[8] observed in their computer simulations. The so called cage effect, where each atom is thought of being caught for some time in a potential well from surrounding atoms, would also be a collective effect. All these aspects are certainly contained in the correlation function in eq.(35) and they are usually taken care of in some mode-coupling approximation.

Following the physical arguments above we split up $\Gamma_{\mu\nu}$ into two parts, the first one containing essentially only binary collisions and decaying rapidly in time and the second collective part containing recollisions of the particles. Following Sjögren[21] I write

$$\Gamma_{\mu\nu}(\vec{q},z) = \Gamma_{\mu\nu}^{B}(\vec{q},z) + \frac{1}{2n}\int \frac{d\vec{q}'}{(2\pi)^3}\, T^{B+}_{\mu;\lambda\eta}(\vec{q}\vec{q}',z)\ \Delta_{\lambda\eta;\sigma\zeta}(\vec{q}\vec{q}',z)\ T_{\sigma\zeta;\nu}(\vec{q}\vec{q}',z) \tag{38}$$

where

$$\Delta_{\lambda\eta;\sigma\zeta}(\vec{q}\vec{q}',t) = F^{-1}_{\lambda\lambda}(\vec{q}-\vec{q}')F^{-1}_{\eta\eta}(\vec{q}')\{F_{\lambda\sigma}(\vec{q}-\vec{q}',t)F_{\eta\zeta}(\vec{q}',t) -$$

$$-F^{B}_{\lambda\sigma}(\vec{q}-\vec{q}',t)F^{B}_{\eta\zeta}(\vec{q}',t)\}F^{-1}_{\sigma\sigma}(\vec{q}-\vec{q}')F^{-1}_{\zeta\zeta}(\vec{q}'). \tag{39}$$

The middle term in the integral above contains a complete set of pair of modes, where we subtract off the short time contribution, which is supposed to be incorporated already in $\Gamma^{B}_{\mu\nu}$. It is expected that the non-hydrodynamic modes decay quite rapidly and for that reason contribute mainly to $\Gamma^{B}_{\mu\nu}$, whereas the hydrodynamic ones give rise to a much slower decay. This is consistent with our knowledge that the long wavelength transverse current modes lead to the asymptotic $t^{-3/2}$ behaviour. For these reasons we may restrict the summation in eq.(38) to the hydrodynamic modes. They now contain the corresponding correlation function and these should be determined self-consistently.

The coupling constants $T^{B}_{\mu;\lambda\eta}$ and $T_{\sigma\zeta;\nu}$ are wavevector and also in general frequency dependent. This reflects the fact that locally the disturbance around any single atom is a multi-mode excitation and is in the vocabulary of the many-body theorists called vertex corrections. One part is strictly instantaneous and therefore frequency independent and it always contains coupling to at least one density mode. This is simply a consequence of the fact that a potential couples directly to the density of particles. There is no way of calculating the coupling constants exactly. Using the superposition approximation one can express the instantaneous part in terms of the static structure factor. For instance (s stands for static),

$$T^{(s)}_{\sigma\zeta;1}(\vec{q}\vec{q}',z) = in(m\beta)^{-1/2}\{(q\cdot\vec{q}')S(\vec{q}-\vec{q}')[S(\vec{q}')-1] + $$

$$+ (q\cdot(\vec{q}-\vec{q}')S(\vec{q}')[S(\vec{q}-\vec{q}')-1]\}\delta_{\sigma o}\delta_{\zeta o}\, , \tag{40}$$

with q being the unit vector along $\vec{q}$. One finds that the longitudinal current couples instantaneously to a pair of density modes. The same is true for the transverse current. Up to this point T^{B} and T are identical and one has also

$$T^{B^{+}}_{1;\lambda\eta}(\vec{q}\vec{q}',z) = T^{B}_{\lambda\eta;1}(-\vec{q},-\vec{q}',-z)\, . \tag{41}$$

The vertices $T_{\alpha\beta;\nu}$, $T_{o\beta;\nu}$ and $T_{\alpha o;\nu}$ - α and β refer to the three current components - contain coupling to at least one current mode and they can be calculated exactly for t=0 and for q'=0 but arbitrary times. The interesting point is that

$$T_{o\alpha;1}(\vec{q},\vec{q}'=0,t) = nS(q)\Gamma_{11}(\vec{q},t)q^{\alpha}, \tag{42}$$

and it has the same time dependence as $\Gamma_{11}(qt)$ itself. This followed from the conservation of total momentum. Some similar results were obtained for other vertices, which couple to the current. In order to take into account properly the known initial value as well as the property in eq.(42), Sjögren made the simplifying ansatz

$$T_{\alpha\beta;\nu}(\vec{q}\vec{q}',z) = \{T_{\alpha\beta;\lambda}(\vec{q}\vec{q}',t=0)\Gamma^{-1}_{\lambda\mu}(\vec{q},t=0)\}\Gamma_{\mu\nu}(\vec{q}z) \tag{43}$$

and similarly for $T_{o\beta;\nu}$ and $T_{\alpha o;\nu}$. The binary part of $T_{\alpha\beta;\nu}$, denoted as $T^{B}_{\alpha\beta;\nu}$, is obtained by in eq.(43) replacing Γ by Γ^{B}. Inserted into eq.(38) it leads to an equation for $\Gamma_{\mu\nu}(qz)$. If on the right hand side we replace the full Γ by Γ^{B}, we obtain what is called the ring approximation. The complete solution corresponds to an infinite set of recollisions. The self-motion can be handled in an analogous way. It turns out that a proper treatment of recollisions is necessary in order to obtain the correct prefactor of the asymptotic $t^{-3/2}$-term[21,24]. The time dependence in the vertex function implies that the renormalization of the coupling to the transverse current, for instance, gradually changes as we couple to slower and slower current fluctuations.

The same considerations as above enter also for a system of hard spheres and some of the explicit calculations are then simplified. Treatments along the same line have been carried out for such a system by Furtado et al[2], by Cukier and Mehaffey[27] and by Leutheusser[20,28], and they have produced interesting numerical results. I would also like to mention other work of Cukier and collaborators[29] and of Sung and Dahler[30], where they have been concerned with the dynamics of an emulsion of microscopically large particles in a fluid. They show that a proper treatment of repeated recollisions is then crucial. There is no difficulty to extend the above ideas to multi-component liquids, but the explicit calculations become much more difficult and require quite a lot of simplifying approximations. The agreement between theory and experiments, including computer simulations, are therefore less striking in this case[31,32].

4. COMPARISON WITH EXPERIMENTS AND COMPUTER SIMULATIONS.

Even though the present kind of theory has certainly deepened our understanding of the dynamics of dense classical fluids compared to earlier more phenomenological models, it is essential also to demonstrate that it leads to significantly better numerical agreement between theory and experiments. Unfortunately, it requires quite heavy computations and this has certainly prevented it from being widely used.

I shall here present a selection of results, which I think prove some specific points. Extensive computer simulations have been done for a hard sphere system and comparisons were made both with calculations based on the generalized Enskog model and on the present kinetic theory. The Enskog model has no memory effects built in. Fig.1 shows the spectrum of the transverse current correlation function based on the two theories.

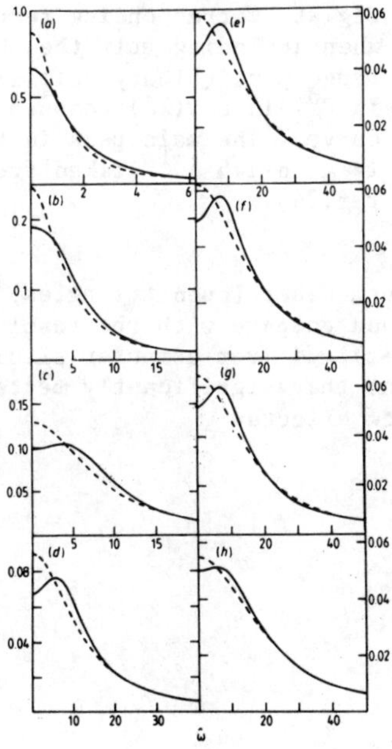

Fig.1. Normalized spectrum of transverse current correlation function from present theory (full curve) compared with solution of generalized Enskog theory (broken curve) at several wave numbers :(a) q=0.759, (b) q=1.52, (c) q=2.28, (d) q=3.04, (e) q=5.31, (f) q=6.83, (g) q=8.35, (h) q=15.2. The main peak in S(q) is at q=7.4 (taken from ref.20).

One finds that the latter one gives a weak resonance for an intermediate range of wavenumbers, implying a damped oscillatory behaviour in time. This is precisely what is observed in the computer simulations and one finds reasonable agreement with the theory. Concerning the longitudinal current correlation function and S(qω) the mean field term plays a quite dominant role and the difference between the two theories is less. However, with the feedback included one does obtain somewhat better agreement[4,20]. The same is true when the calculations of Sjögren are compared with experimental data on aluminium and rubidium[33,34].

Fig.2a shows S(qω) for rubidium both from theory and from computer simulations - the latter agree well with the neutron scattering data - and in Fig.2b we show curves with only binary collisions included (dashed curve).

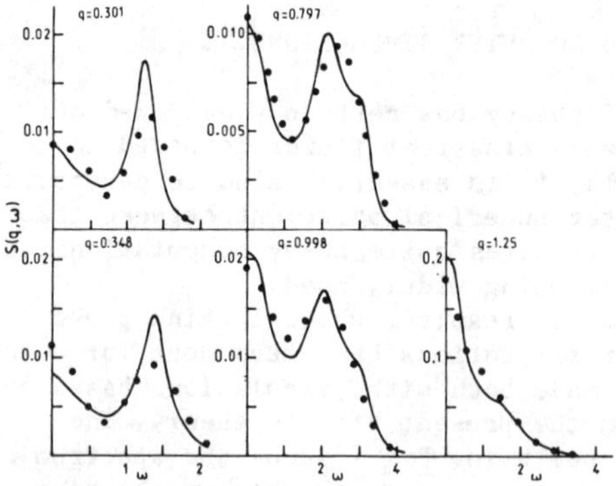

Fig.2a. S(qω) versus ω for liquid rubidium. Present theory (full curve) is compared with simulation data of Rahman and experiments of Copley and Rowe (dots).

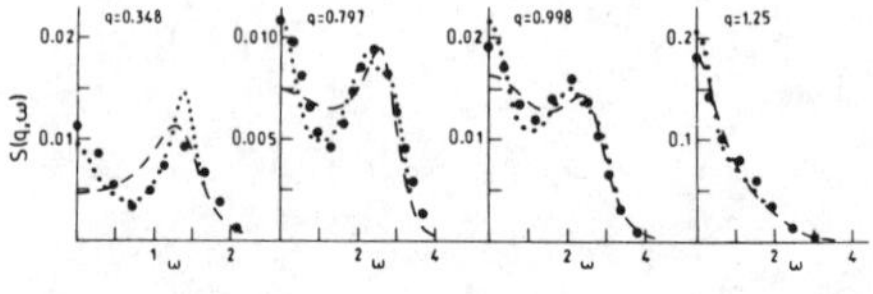

Fig.2b. Corresponding result when including only the short range part (binary collisions) in Γ^d_{11} in eq.(28) (dashed curve). The main peak in S(q) is at q=1.5 $Å^{-1}$ (taken from ref.34).

In Fig.3 I show again results from Leutheusser (open triangles) on the wavevector dependent shear viscosity and compare with the results from generalized Enskog model (open circles) and from computer simulations (closed circles). It is again evident that significantly better agreement is obtained by including feedback effects.

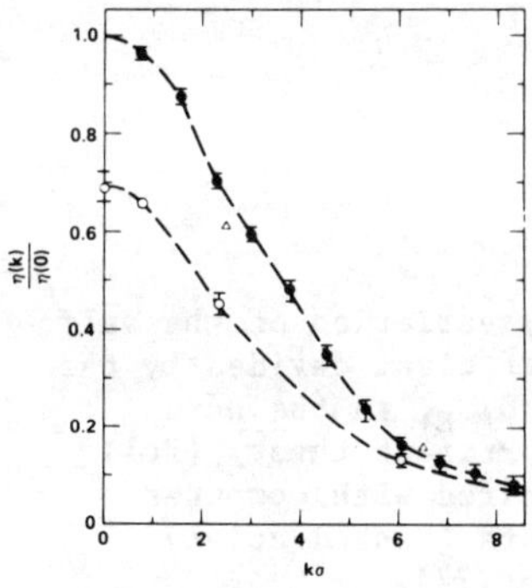

Fig.3. Wavevector dependent shear viscosity at zero frequency (divided by its value for k=0) versus kσ, σ being the hard sphere diameter, at a density corresponding to the freezing point. The computer simulation results (closed circles) and the results from the generalized Enskog theory (open circles) are compared with results obtained by Leutheusser with mode-coupling terms included (two triangles). The main peak in S(k) is at kσ=7.4 (taken from ref.3).

Larsson[35] has tried to extract the memory function $\Gamma_{11}^{d}(qt)$ in eq.(28) directly from his experimental data on lead and rubidium, and he finds very similar shape as that calculated by Sjögren. It begins with a rapid decrease and shows then an oscillatory time tail, depending strongly on the wavenumber. Fig.4 illustrates this.

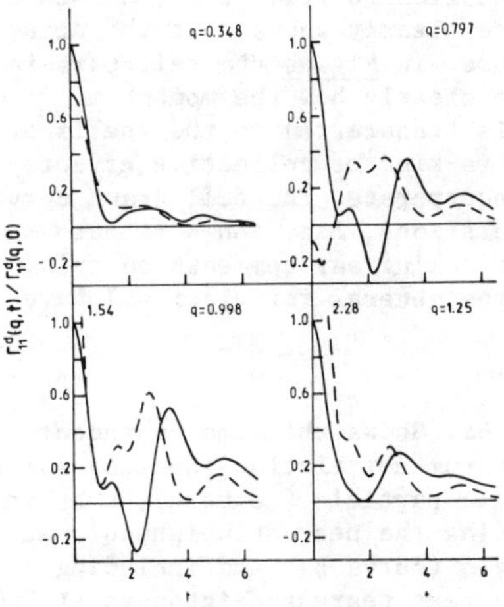

Fig.4. $\Gamma_{11}^{d}(qt/\Gamma_{11}^{d}(q,t=0)$ versus time (cf. eq.(28)). The full curves show the results from the present theory. The dashed ones, representing an earlier and incomplete version of the theory, should be ignored in this context (taken from ref.34).

Let us then turn to the self-motion. It has been found that the feedback effects have a significant influence on how the self diffusion constant varies with density. The theory shows that it is larger than the Enskog value for low density and smaller for high density. This is demonstrated in Fig.5, where comparison is made with computer simulations for a hard sphere system. The smaller value is due to the cage effect at high densities and would be more pronounced for supercooled liquids. The larger value at low densities can be understood as arising from the backflow current, which pushes the self particle forward.

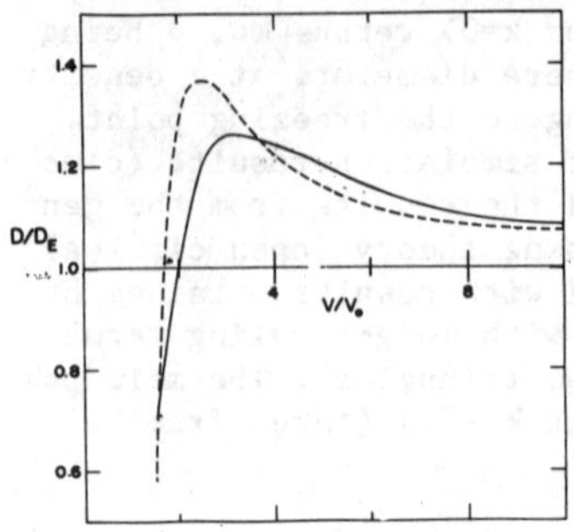

Fig.5. Density variation of the self-diffusion coefficient devided by the Enskog value, D/D_E, in the hard sphere fluid. Present theory (full curve) is compared with computer simulations data (dashed curve) (taken from ref.27).

Balucani et al[36] present some very interesting computer simulation results for rubidium. Besides the single particle momentum correlation function $\langle\vec{p}_1(t)\cdot\vec{p}_1(o)\rangle$ they also consider the correlations between the momentum of one particle at t=0 and the total momentum at a later time of the same particle plus a shell of nearest neighbour particles, as well as that including also next nearest neighbours. This is shown in Fig.6a. If we would include all neighbours, the correlation function stays constant due to conservation of momentum. The figure clearly shows that the decay becomes slower as we include more neighbours. In Fig.6b the self-part is subtracted and, therefore, it reveals more clearly how the momentum, initially confined to the self-particle, is transferred to the shells of neighbouring particles. It is precisely this kind of collective effects the mode-coupling terms are supposed to incorporate. The full drawn curve shows results from their theoretical calculations, based on a rather phenomenological treatment. I refrain from making any comments on this, since it would fall outside the scope of the general formulation I have given here.

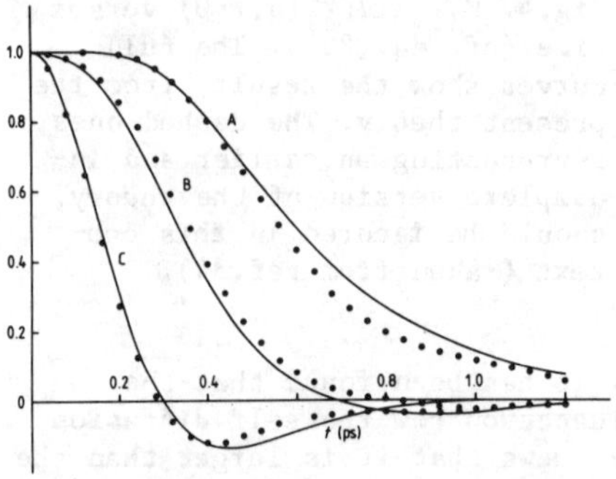

Fig.6a. Shows the time dependent momentum correlation function for a single particle (curve c), that including the nearest neighbours at time t (curve B), and including also next nearest neighbours at time t (curve A).

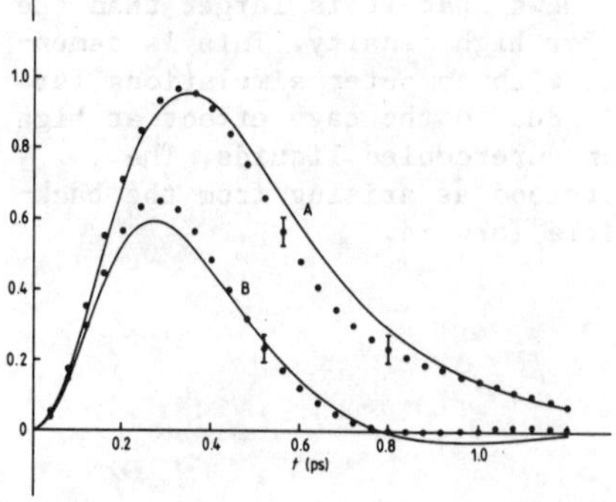

Fig.6b. Shows the same results with the self-part subtracted. The computer results are indicated by dots (taken from ref.36).

Fig.7 shows results for the velocity correlation function and the corresponding memory function for argon and rubidium. The dots are the computer simulation data and we notice slight oscillations in the tail of the memory function for rubidium. It is found to arise from the oscillatory behaviour of the density correlation function for intermediate wavenumbers, entering in the mode-coupling integral. Since no short wavelength sound waves exist for argon, no such oscillations are found in Γ(t) for argon.

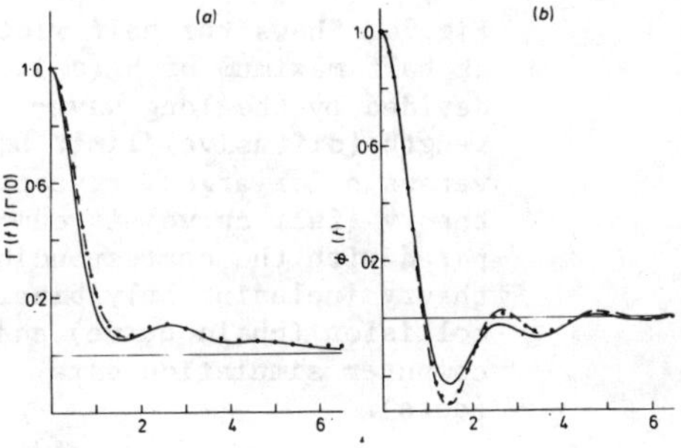

Fig.7a. Shows the memory function and the velocity correlation function for rubidium. Present theory (full curve) is compared with computer data (dots).

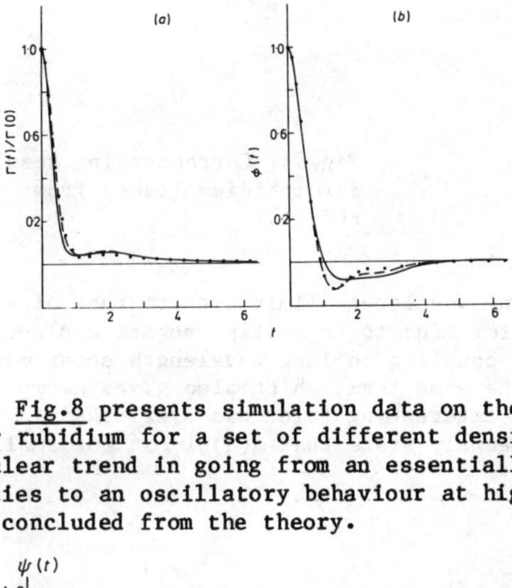

Fig.7b. Corresponding results for argon (taken from ref.37, see also a somewhat improved version in fig.2 in ref.34).

Fig.8 presents simulation data on the velocity correlation function for rubidium for a set of different densities and temperatures. There is a clear trend in going from an essentially exponential decay at low densities to an oscillatory behaviour at higher densities. This is also what is concluded from the theory.

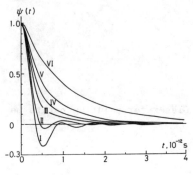

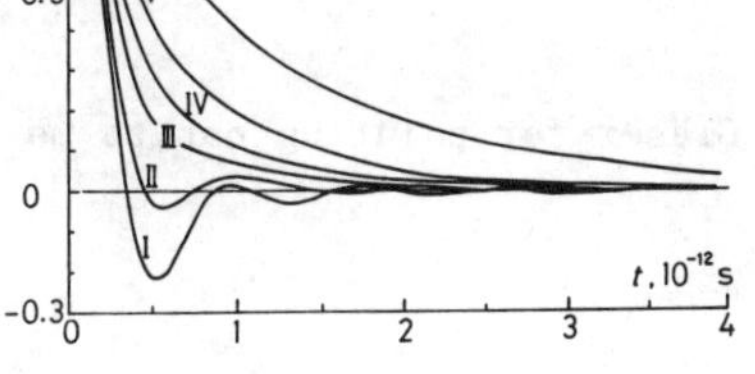

Fig.8. Shows simulation data for the velocity correlation function for rubidium at different densities: (1) ρ=1.502 gcm^{-3}, T=319 K, (2) ρ=1.336, T=600, (3) ρ=1.095, T=1143, (4) ρ=0.900, T=1523, (5) ρ=0.707, T=1876, (6) ρ=0.346, T=2093. They lie essentially along the liquid saturation curve up to the critical point (taken from ref.38).

As a last illustration I show results for the width of the quasi-elastic peak in the Van Hove self-correlation function $S^s(q\omega)$. It was found from computer experiments [39,40] that the width has a somewhat surprising q-dependence. Fig.9 shows comparison with the theory with and without the feedback effects, i.e. the tail in the memory function. The agreement is certainly quite striking and we notice that the feedback effect is important. When comparing the full and dashed curves one should lower the latter one until they overlap for q=0, since otherwise the widths are normalized differently. The expression for the width, as

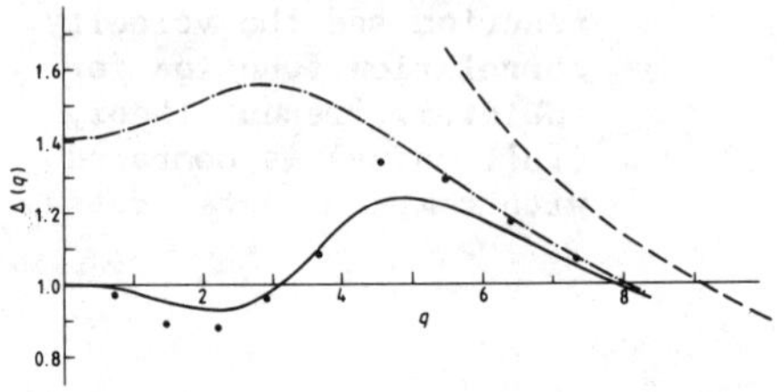

Fig.9a. Shows the half width at half maximum of $S^s(q\omega)$, devided by the long wavelength (diffusive) limit Dq^2, versus q for argon. Present theory (full curve) is compared with the corresponding theory including only binary collision (chain curve) and computer simulation data (dots).

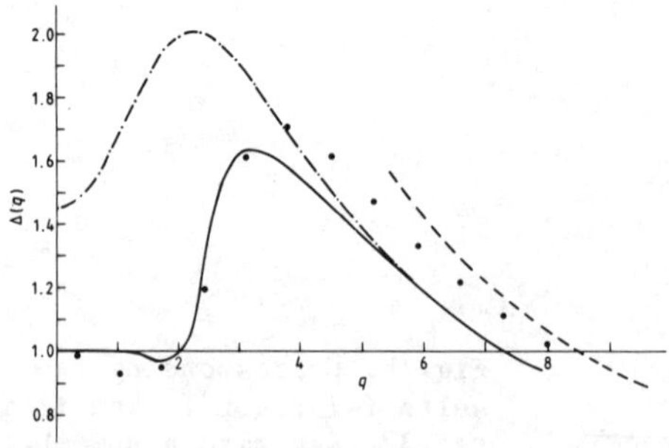

Fig.9b. Corresponding results for rubidium (taken from ref.41).

given in ref.[41], goes in the long wavelength limit over to that of Schepper et al [42] and it leads for $\Delta(q)$ to an initial negative slope linear in q. This originates from coupling to long wavelength shear modes in the medium and it comes from the same term, which also gives asymptotically the $t^{-3/2}$-behaviour. This interesting point was overlooked in ref.[41], where the authors erroneously state that $\Delta(q) \propto 1-cq^2$ for small q.*)

*) I am grateful to Prof. Bosse and Prof. Gläser for pointing out to me the wrong statement in ref.[41].

5. CONCLUDING REMARKS

In these lectures I attempted to clarify both the essential physics entering in the dynamics of monatomic liquids and how one nowadays tries to incorporate the various aspects in the theoretical calculations. In order to keep the essential steps free from all techniqualities I have purposely avoided any details of how the various authors carry out the explicit calculations. For instance, a fully self-consistent treatment is hardly ever done, but I do not think that this changes the results in any essential way. Comparisons have mostly been done against computer simulations. This just proves how crucial they have been for revealing the basic physics. A reason for this is the large flexibility one has in the computer simulations compared to the real experiments, which are mainly restricted to $S(q\omega)$. Secondly, in comparing with the real experiments lack of knowledge of the true interaction potential introduces an extra uncertainty. However, comparisons between simulations and real experiments are important in order to ascertain that the former contain the essential features of the real systems.

It was mentioned earlier as a passing remark that the feedback becomes a dominant effect at the gas-liquid critical point. The present approach with mode-coupling approximations goes then over to those of Kadanoff and Swift[43], Kawasaki[44] and others. Actually,it throws some doubts on the use of a static (zero frequency) coupling function in this connexion. It was also recognized that a dominant feedback effect should appear in supercooled liquids, when approaching the glass transition[45]. The reason of this is, however, entirely different from that at the critical point. Detailed calculations, based on the present kind of treatment have been carried out recently[46,47] and it confirms the earlier conjecture.

These lecture notes should by no means be considered as a review article. It has not been possible to give proper credit to all those who have contributed to the field of liquid dynamics. Nor have I included any comprehensive reference list. One may find a more complete list in the excellent book of Boon and Yip[6]. Instead, I have mainly referred to articles where in my mind important aspects are most clearly spelled out or quantitative results have clearly revealed certain points I wanted to explain.

ACKNOWLEDGEMENTS

It is my pleasure to acknowledge numerous and clarifying discussions with Dr. Lennart Sjögren on various aspects covered in these lectures.

REFERENCES

1. Lebowitz, J.L., Percus, J.K. and Sykes, J., Phys. Rev. 188, 487 (1969).
2. Furtado, P.M., Mazenko, G.F. and Yip, S., Phys. Rev. A12, 1653 (1975).
3. Alley, W.E. and Alder, B.J., Phys. Rev. A27, 3158 (1983).
4. Alley, W.E., Alder, B.J., and Yip, S., Phys. Rev. A27, 3174 (1983).
5. Forster, D., Hydrodynamics, Fluctuations, Broken Symmetry, and Correlation Functions, W.A. Benjamin Inc., London, 1975.
6. Boon, J.P. and Yip, S., Molecular Hydrodynamics, McGraw-Hill, New York, 1980.
7. Sjögren, L., Ann. Phys. (N.Y.) 113, 304 (1978).
8. Alder, B.J. and Wainwright, T.E., Phys. Rev. A1, 18 (1970).
9. Zwanzig, R. and Bixon, M., Phys. Rev. A2, 2005 (1970).
10. Akcasu, A.Z. and Duderstadt, J.J., Phys. Rev. 188, 479 (1969); A1, 905 (1969).
11. Forster, D. and Martin, P.C., Phys. Rev. A2, 1575 (1970).
12. Mazenko, G.F., Phys. Rev. 7, 209, 222 (1973), A9, 360 (1974).
13. Mazenko, G.F., Phys. Rev. A6, 2545 (1972).
14. Zwanzig, R., J. Chem. Phys. 33, 1338 (1960); Phys. Rev. 124, 983 (1961).
15. Mori, H., Progr. Theor. Phys. 34, 399 (1965).
16. Forster, D., Phys. Rev. A9, 943 (1974).
17. Egelstaff, P.A., An Introduction to the Liquid State, Academic Press New York, 1967.
18. Ailawadi, K.N., Rahman, A. and Zwanzig, R., Phys. Rev. A4, 1616 (1971).
19. Sjödin, S. and Sjölander, A., Phys. Rev. A18, 1723 (1978).
20. Leutheusser, E., J. Phys. C: Solid State Phys. 15, 2827 (1982).
21. Sjögren, L., Phys. Rev. A22, 2866 (1980)
22. Götze, W. and Lücke, M., Phys. Rev. A11, 2173 (1975)
23. Bosse, J., Götze, W. and Lücke, M., Phys. Rev. A17, 434. 447 (1978), A18, 1176 (1978).
24. Sjögren, L. and Sjölander, A., J. Phys. C: Solid State Phys. 12, 4369 (1979).
25. Ernst, M.H. and Dorfmann, J.R., Physica (Utrecht) 61, 157 (1972).
26. Feynman, R.P. and Cohen, M., Phys. Rev. 102, 1189 (1956)
27. Cukier, R.I. and Mehaffey, J.R., Phys. Rev. A18, 1202 (1978).
28. Leutheusser, E., J. Phys. C: Solid State Phys. 15, 2801 (1982).
29. Cukier, R.I., Mehaffey, J.R. and Kapral, R., J. Chem. Phys. 69, 4962 (1978); 74, 2494 (1980).
30. Sung, W. and Dahler, J.S., J. Chem. Phys. 78, 6264, 6280 (1983).
31. Sjögren, L. and Yoshida, F., J. Chem. Phys. 77, 3703 (1982)
32. Bosse, J. and Munakata, T., Phys. Rev. A25, 2763 (1982).
33. Larsson, K.E., Phys. Chem. Liq. 12, 273 (1983).
34. Sjögren, L., Phys. Rev. A22, 2883 (1980)
35. Larsson, K.E., preprint.
36. Balucani, U., Vallauri, R., Murthy, C.S., Gaskell, T. and Woolfson, M.S., J. Phys. C: Solid State Phys. 16, 5605 (1983).
37. Sjögren, L., J. Phys. C: Solid State Phys. 13, 705 (1980).
38. Tanaka, M., Progr. Theor. Phys. Suppl. 69, 439 (1980).
39. Nijboer, B.R.A. and Rahman, A., Physica 32, 415 (1966).
40. Levesque, D., and Verlet, L., Phys. Rev. A2, 2514 (1970).
41. Wahnström, G. and Sjögren, L., J. Phys. C: Solid State Phys. 15, 401 (1982).

42. De Shepper, I.M., Van Beyeren, H. and Ernst, M.H., Physica 75, 1 (1974); see also Verkerk, P., Builtjes, J.H. and de Schepper, I.M., Phys. Rev. A31, 1731 (1985).
43. Kadanoff, L.P. and Swift, J., Phys. Rev. 166, 89 (1968).
44. Kawasaki, K., Phys. Rev. 150, 291 (1966); Ann. Phys. (N.Y.) 61, 1 (1970).
45. Sjölander, A. and Turski, L.A., J. Phys. C: Solid State Phys. 11, 1973 (1978).
46. Leutheusser, E., Phys. Rev. A29, 2765 (1984).
47. Bengtzelius, U., Götze, W. and Sjölander A., J. Phys. C: Solid State Phys. 17, 5915 (1984).

EXPERIMENTAL STUDIES OF ATOMIC MOTIONS IN LIQUID METALS

W. Gläser

Physik-Department, Technische Universität München
8046 Garching, Germany

I. SINGLE PARTICLE MOTION

Much of our present knowledge on single particle motion in liquid metals was collected over the last two decades /1/. Recent progress mainly came through computer simulations /2,3/ and the development of kinetic theories /4,5/. On the experimental side besides tracer methods and NMR techniques the usefullness of neutron scattering techniques became aware. Whereas the former techniques yield information on the Brownian limit of diffusion, inelastic scattering of slow neutrons in principle can explore the entire time and space range of atomic motions.

The problem of understanding the properties of liquids starts with the problem of liquid structure which is illustrated in Fig. 1.

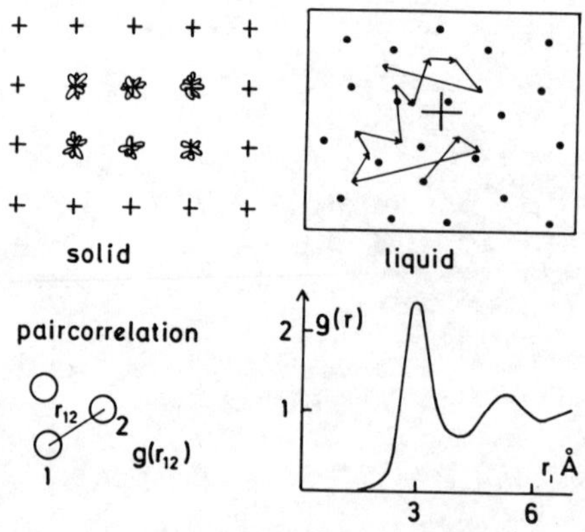

Fig. 1. Different concepts for crystalline and liquid structures.

The underlying center of mass lattice of a crystalline solid of moving atoms can be determined by a Laue diffraction experiment. However, in a liquid the atoms can eventually reach every available point if sufficient time is allowed. The time averaged centers of mass of all atoms are finally the same. So, the usual structure concept is good for nothing and has been replaced by the concept of pair correlation which describes the probability of finding an atom 2 at a certain distance from atom 1.
The Fouriertransform of g(r) the structure factor S(Q) can be measured with X-rays. If the atoms move, the correlation decays and the Fouriertransform of the time dependent correlation the van Hove scattering function can be measured with neutrons. What can be measured is a double diffential scattering cross section:

$$\frac{d^2\sigma}{d\Omega d\omega} = \frac{k}{k_o} \{ \frac{\sigma_{inc}}{4\pi} S_s(Q,\omega) + \frac{\sigma_{coh}}{4\pi} S(Q,\omega) \}, \tag{1}$$

$$S(Q) = \int_{-\infty}^{\infty} S(Q,\omega)\, d\omega, \qquad \omega = \frac{\hbar}{2m}(k_o^2 - k^2), \quad \bar{Q} = \bar{k}_o - \bar{k},$$

where $\hbar Q$ is the momentum transfer and $\hbar\omega$ the energy transfer of the neutron to the sample, σ_{inc} and σ_{coh} are incoherent and coherent scattering cross sections of the nuclei. The van Hove function $S(Q,\omega)$ reflects collective motions whereas its selfpart $S_s(Q,\omega)$ is a picture of single particle motions. The limits of $S_s(Q,\omega)$ for $Q \to 0$ and $Q \to \infty$ are well known:

$$Q \to 0: \quad S_s(Q,\omega) = \frac{\pi^{-1} DQ^2}{\omega^2 + (DQ^2)^2},$$
$$Q \to \infty: \quad S_s(Q,\omega) = \frac{1}{\sqrt{\pi} Q v_o} e^{-\frac{\omega^2}{Q^2 v_o^2}}, \quad \beta v_o = \sqrt{\frac{2k_B T}{M}}. \tag{2}$$

Most samples scatter both incoherently and coherently. If one wants to study the single particle motion one has to correct the data for coherent scattering which can in principle be done by realistic models. Recently a new technique became available which allows a model independent separation of incoherent and coherent scattering by neutron polarization analysis, if the incoherence is caused by spinflip scattering. It is easy to show that in this case, for incident polarized neutrons:

$$\sigma_s^F = \frac{2}{3}\sigma_{inc} \qquad \sigma_s^{nF} = \sigma_{coh} + \frac{1}{3}\sigma_{inc}, \tag{3}$$

where σ_s^F is the spinflip cross section and σ_s^{nF} is the nonflip cross section. The spin state after scattering can be determined by polarization analysis. The new spectrometer D7B /6/ at the high flux reactor at Grenoble which can be used for this type of experiments is illustrated in Fig. 2.

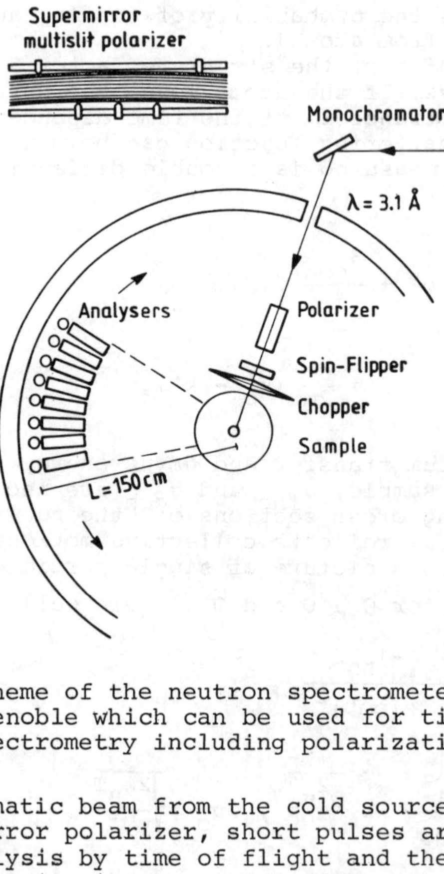

Fig. 2. Scheme of the neutron spectrometer D7B at the HFR Grenoble which can be used for time-of-flight spectrometry including polarization analysis.

A monochromatic beam from the cold source is polarized with a super mirror polarizer, short pulses are generated for energy analysis by time of flight and the polarization state after scattering is characterized by super mirror analysers in front of each TOF-detector. Fig. 3 shows some preliminary results for the selfpart $S_s(Q,\omega)$ of liquid sodium at 600K.

A drastic change in the ratio of both functions is observed which mainly is due to the static structure factor S(Q). There is also a Q-dependence of the ratio of widths. Up to now this data have been used only to check the parameters of a simple model for coherent scattering, namely Lovesey's /7/ "viscoelastic model". But certainly the application and improvement of this new technique will open the possibility to study single particle and collective motions in the same liquid simultaneously independent on any model assumption.

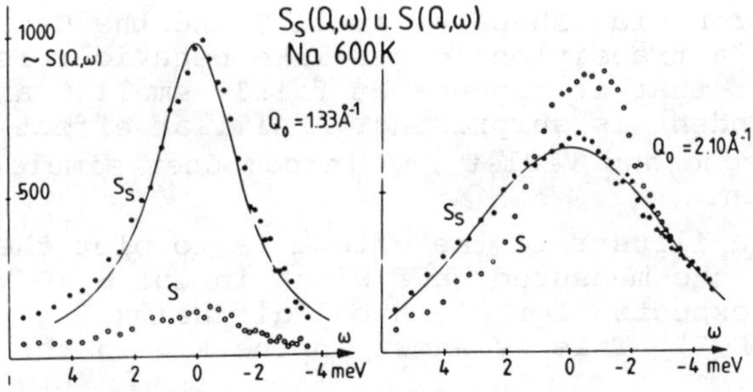

Fig. 3. Scattering functions $S(Q,\omega)$ and $S_s(Q,\omega)$ for liquid sodium as determined by polarization analysis.

In the following the results of a model analysis of scattering data on liquid sodium which are measured on the TOF-spectrometers IN 4 and IN6 at the HFR and are described more in detail in a paper by Chr. Morkel /8/ at this conference are summarized. If one tries to characterize the shape of the incoherent scattering function he derived, by a simple shape parameter one could use e.g. the product of line width and peak height: $\Delta Q = 2\omega_{1/2}(Q)S_s(Q,0)$.
The results for the 3 temperatures investigated are shown in Fig. 4.

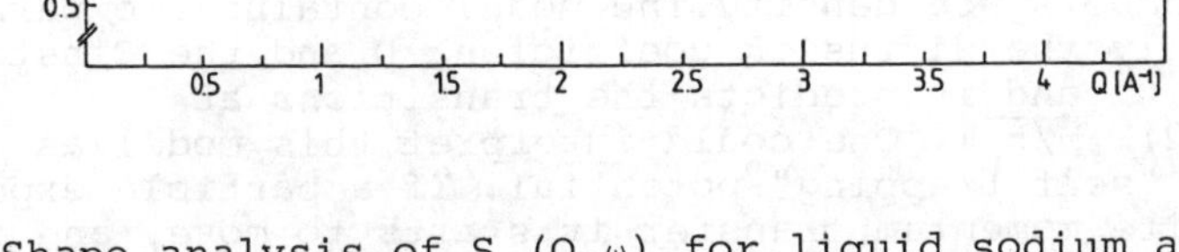

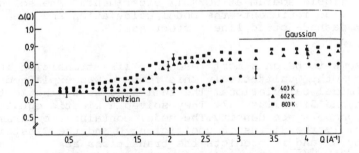

Fig. 4. Shape analysis of $S_s(Q,\omega)$ for liquid sodium at different temperatures using the shape parameter $\Delta Q = 2\ \omega_{1/2}(Q)S_s(Q,o)$.

The product of full width and height should be $2/\pi$ for a Lorentzian and $2\sqrt{\frac{\ln 2}{\pi}}$ for a Gaussian. The data seem to approach the Lorentzian shape at small Q and the Gaussian at large Q. Now, a transition to gas like behaviour is expected at large Q but that it happens at fairly small Q and is temperature-dependent is surprising. A similar effect has been found by Levesque and Verlet /9/ in computer simulations of liquid argon.

Another way to illustrate the effect is to plot the reduced width, namely the measured half width in ω of $S_s(Q,\omega)$ divided by the width expected for classical diffusion ($\gamma(Q) = \omega_{1/2}/DQ^2$). This is shown in the Fig. 5 for sodium at T = 602K.

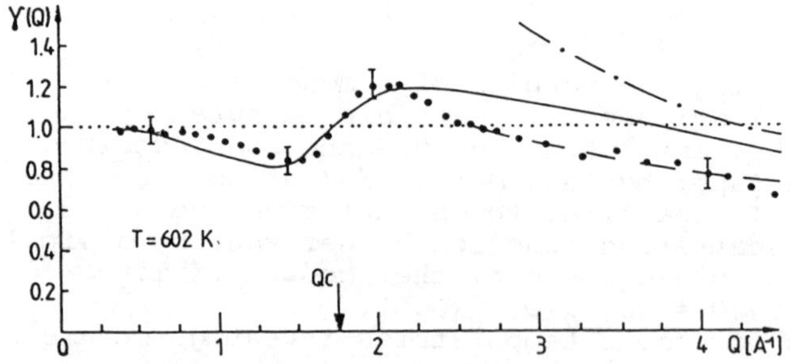

Fig. 5. Reduced width $\gamma(Q) = \omega_{1/2}(Q)/DQ^2$ of $S_s(Q,\omega)$ of liquid sodium at 602K (• experiment). The solid line represents the model calculation and the dashed-dotted line perfect gas.

There seems to be an almost resonance like transition in the reduced width. Qualitatively this feature was explained by kinetic model calculations. The full line is a kinetic theory by Götze and Zippelius /4/. They solved a kinetic equation for the phase space density. The model contains two parameters, namely the diffusion coefficient D and the Einstein frequency Ω_E and it predicts the transitions at $Q_c = (\Omega_E/2)\sqrt{M/k_BT}$. One could interpret this model as describing a "self trapping" potential. If a particle experiences a finite momentum transfer it starts to move, and polarizes its environment which leads to additional binding and therefore diffusion retardation. This effect vanishes at Q_c, above Q_c the mobility grows rapidly. So in a sense at low Q the liquid behaves more like a bound system and a high Q more gas like. At the highest Q measured the data are still away from perfect gas behaviour and are not in to good agreement with the model.

Large Q behaviour

However, if we consider the scattering function of a hard sphere fluid assuming a known hard sphere diameter $\sigma_{h.s.}$ for sodium and a mean free path

$$\ell_E = [\sqrt{2}\ \pi\ n\sigma^2_{h.s.}\ g(\sigma_{h.s.})\ ]^{-1}$$

(n = number density, $g(\sigma_{h.s.})$ = pair correlation at contact) according to Enskog's theory, then better agreement at high Q is achievable. The behaviour of $S_s(Q,\omega)$ for a hard sphere fluid at large Q can be expressed by the perfect gas behaviour corrected for hard core collisions by an inverse Q series /10/. The leading terms for $\omega_{1/2}(Q)$ and $S_s(Q,0)$ are:

$$\omega_{1/2}(Q) = \sqrt{\frac{2k_BT\ \ln 2}{M}}\quad Q\ \{\ 1 - \frac{\xi}{l_EQ} + \mathcal{O}(Q^{-2})\ \}, \qquad (4)$$

$$S_s(Q,0) = \sqrt{\frac{M}{2\pi k_BT}}\quad \frac{1}{Q}\ \{1 + \frac{s}{l_EQ} + \mathcal{O}(Q^{-2})\ \}.$$

ξ and s have been calculated /11/ to be: ξ = 0.449, s = 0.328. The results of this model for sodium at T = 602K with $\sigma_{h.s.}$ = 3.1 Å and l_E = 0.33 Å are compared to the experimental data in Fig. 6.

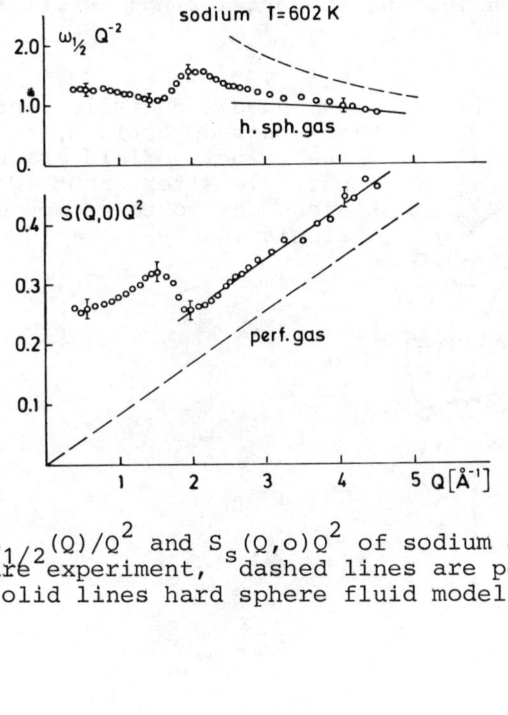

Fig. 6. $\omega_{1/2}(Q)/Q^2$ and $S_s(Q,o)Q^2$ of sodium at 602K. Circles are experiment, dashed lines are perfect gas and solid lines hard sphere fluid model.

The dashed lines give the perfect gas values. The agreement with this simple model at large Q is remarkable, especially if one notes that there is no free parameter left and that at high particle densities the classical mean free path l_E becomes comparable with the particles wave length (λ_{Na} = 0.37 Å at T = 602K).

Small Q behaviour

At low Q it may be of interest to see whether experimental techniques are accurate enough to see deviations from Brownian diffusion behaviour. A nonexponential $t^{-3/2}$ long time decay of the velocity autocorrelation function of liquid particles was found by Alder and Wainwright /12/ in computer simulations of hard sphere fluids. They concluded:

$$\langle v(0)v(t)\rangle \underset{t\to\infty}{\sim} t^{-3/2} \qquad f(\omega) = \frac{D}{\pi} \left| 1 - \sqrt{\frac{\omega}{\omega_0}} + \ldots \right| . \qquad (5)$$

Here $f(\omega)$ is the spectral density, $\omega_0 = 72(\pi nMD/k_BT)^2(D+\nu)^3$ and ν is the kinematic viscosity.

Whereas from the Langevin equation follows an exponential decay of $\langle v(0)v(t)\rangle$ and a horicontal tangent of $f(\omega)$ at the origin, this picture predicts a cusp at $\omega = 0$ and ω_0 determines its steepness. Zwanzig and Bixon /13/ explained this effect by long living shear excitations generated by the tagged particles leading to vortex rings as illustrated in Fig. 7.

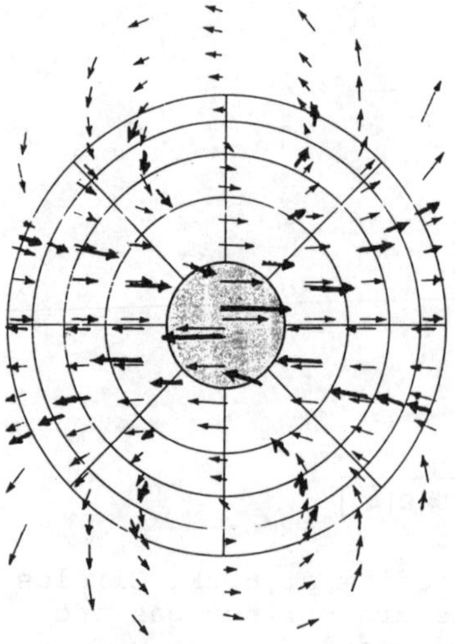

Fig. 7.
The shadowed circle represents the tagged particle, the arrows the velocity field around the particle after about 10 collisions due to the computer simulations.

The shadowed circle represents the tagged particle and the arrows the velocity field in its neighbourhood after about 10 collisions as derived from the computer simulations. In principle the spectral density $f(\omega)$ can be derived from $S_s(Q,\omega)$ in the limit $Q \to 0$. However, the extrapolation of existing neutron data /14/ was inconclusive because of the uncertainties involved.

Theoretically $S_s(Q,\omega)$ for small Q can be calculated in mode coupling theory /15/:

$$S_s(Q,\omega) = \frac{\pi^{-1} DQ^2}{\omega^2 + (DQ^2)^2} + \frac{\pi^{-1}}{DQ^* Q} \operatorname{Re} G\left(\frac{i\omega + DQ^2}{\delta DQ^2}\right) + \mathcal{O}(Q^{-1/2}), \quad (6)$$

where $Q^* = 16\pi n M D^2/k_B T$, $\delta = D/(D+\nu)$ and G is the complex function

$$G(z) = \tan^{-1}(z-1)^{-1/2} - (z-2)(z-1)^{1/2} z^{-2}.$$

The first term on the right hand side of equation (6) is due to Fick's diffusion law and the second term due to the coupling of the diffusion mode with the shear mode of the fluid. It is straightforward to calculate $S_s(Q,0)Q^2$ and $\omega_{1/2}(Q)Q^{-2}$ from equation (6). The results /16/ are:

$$S_s(Q,0)Q^2 = (\pi D)^{-1}\left[1 + G(\delta^{-1})Q/Q^* + \mathcal{O}(Q^{3/2})\right], \quad (7)$$

$$\omega_{1/2}(Q)Q^{-2} = D\left[1 - bQ + \mathcal{O}(Q^{3/2})\right],$$

where $b = 1.453 \frac{\delta^{3/2}}{Q^*}\left[1 - 0.728\,\delta - 0.152\,\delta^2 - \mathcal{O}(\delta^3)\right]$.

For liquid sodium at T = 602 and 803K the relevant numbers are given in table 1.

Table 1

T(K)	602	803
$n(Å^{-3})$	0.0229	0.0216
$D(Å^2/ps)$	1.28	2.10
$\nu(Å^2/ps)$	36.6	27.7
$G(\frac{1}{\delta})/Q^*$ (Å)	0.19	0.29
b(Å)	0.10	0.15

The calculated $S_s(Q,0)Q^2$ and $\omega_{1/2}(Q)Q^{-2}$ are compared with the experimental data in Fig.8.

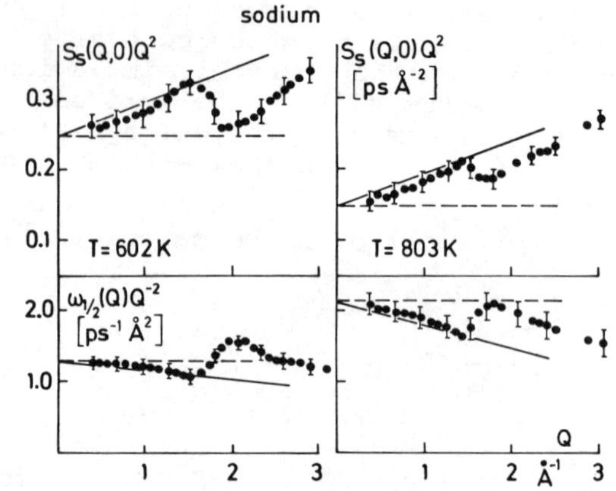

Fig. 8. Experimental values for $S_s(Q,o)Q^2$ and $\omega_{1/2}(Q)Q^2$ of liquid sodium at 602K and 803K(•) compared to the mode coupling results (solid lines).

For the two thermodynamic states of liquid sodium reported here the measured values of $S_s(Q,0)Q^2$ approach the hydrodynamic limit from above and the measured values of $\omega_{1/2}(Q)Q^{-2}$ approach this limit from below. The solid lines are calculated from mode coupling theory as described above. The reasonable agreement found can be interpreted as evidence for the long time tail behaviour. Surprisingly this behaviour seems to extend up to Q values of $Q \approx 1.4$ Å^{-1}, well above the validity range of hydrodynamics estimated before.

In conclusion one might say that single particle motion in a simple liquid metal can be understood in the low Q region within the mode coupling approach; in the high Q region a simple hard sphere fluid picture taking only binary collision into account may be sufficient. But in the transition region, where the change from one region to the other occurs probably more work is necessary to achieve a better understanding of the microscopic behaviour of a fluid.

II. COLLECTIVE MODES

In the preceding lecture we have seen, that some progress in understanding single particle motion in simple liquid metals has been made. The mode coupling theory probably is the right receipt for extending the hydrodynamic regime for increasing Q. On the other side for Q larger than the inverse mean free path simple binary collision models may suffice to describe single particle motions adequate. We begin also to understand the transition region but there certainly more detailled work is necessary.

The situation is very similar concerning collective excitations in liquids. Here the importance of the transition region has been stressed in many theoretical and experimental papers. Obvious experimental evidences for the existence of collective excitations in this region in liquid rubidium was found by Copley and Rowe /17/ a decade ago. Since then other cases have been observed. Fig. 9 shows some recent neutron scattering results of Bucher /18/ on liquid potassium.

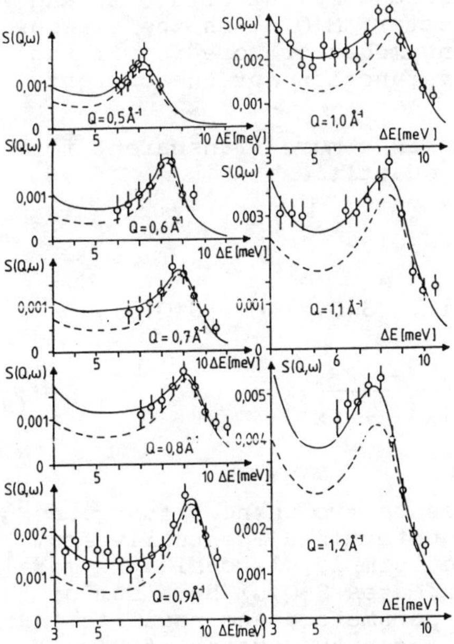

Fig. 9.
Neutron scattering data on liquid potassium at 338K in the small Q range. The data include incoherent scattering, the dashed lines are corrected for this effect /18/.

The measured peak energies are about 10% higher than those found in computer simulations by Jacucci and McDonald /19/.

A new methodical approach

In this lecture I want to present a new methodical approach /20/ for analysing the dynamical structure factor $S(Q,\omega)$ which may help to geht more insight in atomic relaxation processes. A similar but more theoretical oriented discussion was given by Bosse et al. /21/.
After introducing the basic ideas this approach will be illustrated for the liquid rubidium case. Although the method was designed for dense gases and simple liquids it may be extentable to liquid alloys and glasses. We start from a wellknown expression for the symmetrized dynamical structure factor, which is a generalization of the hydrodynamic theory of relaxation in liquids:

$$S(Q,\omega) = \frac{\pi^{-1} \frac{k_B T}{M} Q^2 \mathrm{Re}\, M(Q,\omega)}{|\omega_o^2 - i\omega M(Q,\omega) - \omega^2|^2}, \tag{8}$$

where $\omega_o^2 = k_B T Q^2/MS(Q)$ is the second moment of $S(Q,\omega)$ and $S(Q)$ is the static structure factor. $M(Q,\omega)$ is the viscous damping function for propagating modes or more formal the Laplace transform of the memory function for the current-current correlation function.

One can rewrite this expression in a more transparent form by using the following reduced quantities:

$$x = \omega/\omega_o\ , \quad c = \pi\omega_o S(Q,o)/S(Q),$$

$$\sigma = \sigma_1 + i\sigma_2 = \omega_o c/M^*(Q,\omega)\ , \quad S(Q,x) = \omega_o S(Q,\omega),$$

$$S(Q,x) = \frac{S(Q)}{\pi} \frac{\sigma_1/c}{|\frac{\sigma}{c}(1-x^2) + ix|^2}\,. \tag{9}$$

Here we see, that $S(Q,x)$ depends on two unknown functions σ_1/c and σ_2/c which have to be determined eventually from measured experimental data. The "simple visco-elastic model" suggested by Lovesey /7/ approximates $S(Q,\omega)$ by a sum of three Lorentzians with ω-independent parameters and requires that the zeroth, second and fourth frequency moments of $S(Q,\omega)$ are satisfied. In our notation this corresponds to $\sigma = 1 + i\omega\tau$ where τ is a unknown constant and the memory function is an exponential function of time.

In the general case it is easy to show that for a number of special x values which are reasonably well separated equation (9) can be solved for σ_1 and σ_2 without resorting to a Kramers-Kronig analysis which is of limited value because accurate experimental data are attainable for a restricted

x-range only.
If we use instead of the measured $S(Q,\omega)$ the reduced quantity $n = S(Q,0)/S(Q,\omega)$ then the equation for σ_1 and σ_2 can be written:

$$n\sigma_1 = (1 - x^2)^2\sigma_1^2 + \left[\sigma_2(1-x^2) + cx\right]^2. \qquad (10)$$

Because σ_2 is an odd function of x one sees immediately that:

$$\sigma_1(x = 0) = 1 \quad \text{and} \quad \sigma_1(x = 1) = c^2/n(x = 1).$$

Instead of the odd σ_2 we prefer to work with $\sigma_2^o = \sigma_2/x$.

We can formally solve equation (10) for σ_1 and σ_2^o:

$$\sigma_1 = \frac{n \pm \sqrt{n^2 - 4(1-x^2)^2x^2(\sigma_2^o + c - \sigma_2^o x^2)^2}}{2(1 - x^2)^2} \qquad (10a)$$

and

$$\sigma_2^o = \frac{-\,cx \pm \sqrt{n\sigma_1 - (1 - x^2)^2\,\sigma_1^2}}{x(1 - x^2)}. \qquad (10b)$$

The values of σ_2^o at $x = 0$ and $x = 1$ can be determined by extrapolation and interpolation respectively.
If both functions are real over the whole range of x then the radicants of the roots have to satisfy the following conditions:

$$y_1 = \frac{n}{(1 - x^2)^2} \geq \sigma_1, \qquad (11a)$$

$$y_2 = \frac{n \mp 2\,cx\,(1 - x^2)}{2x(1 - x^2)^2} \geq \pm\,\sigma_2^o. \qquad (11b)$$

We observe that y_1 and y_2 involve only experimental quantities. An obvious experimental procedure is to plot y_1 and y_2 as functions of x^2. The resulting curves will be convex downwards. The inequalities say that σ_1, σ_2^o cannot lie inside these curves. Unique solutions require the equality sign and if σ_1 and σ_2^o are slow varying functions of x they should be tangents to the curves close to their minima.

This procedure has been tested for two simple models, namely the perfect gas and a Brillouin line model. In the former case σ_1 and σ_2^o can be determined by numerical integration of the Laplace transforms. Fig. 10 shows a comparision of the exact functions for the gas model with the results of the "experimental inversion procedure".

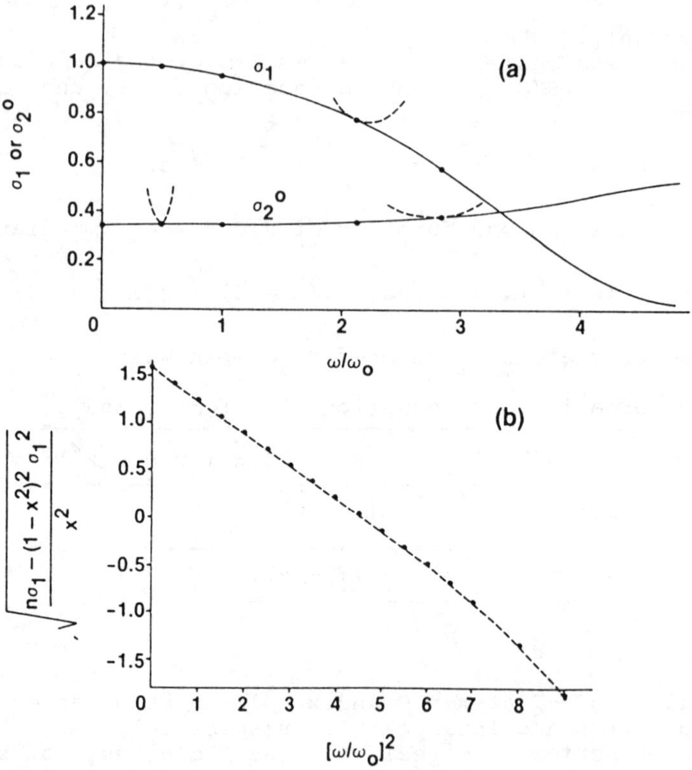

Fig. 10. (a) $\sigma_1(\omega)$ and $\sigma_2^o(\omega)$ for the perfect gas model (solid lines are exact). The dashed lines illustrate the minima of y_1 and y_2. (b) illustrates the extrapolation for determining $\sigma_2^o(0)$.

In the Brillouin line case, which may be a good approximation for $S(Q,\omega)$ of a liquid metal at low Q we started with a reasonable assumption for σ_1 and σ_2^o, calculated $S(Q,\omega)$ and reconstructed σ_1 and σ_2^o by the inversion procedure. The result is shown in Fig. 11.

Because these tests demonstrate the feasibility of the inversion procedure we applied it to real experimental data /22/. Of course experimental errors lead to additional uncertainties which my eventually request more accurate measurements. An example of a recent measurement on liquid rubidium by Gemperlein /23/ is illustrated in Fig. 12.

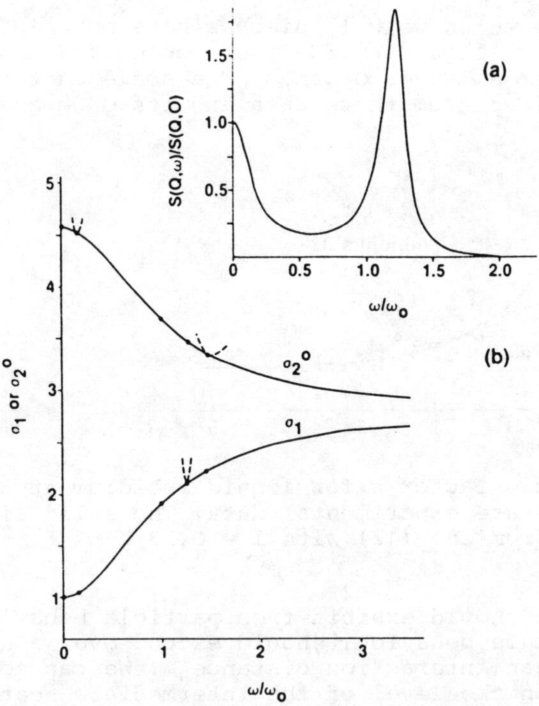

Fig. 11. (a) is a model Brillouin spectrum. (b) shows the corresponding $\sigma_1(\omega)$ and $\sigma_2^0(\omega)$ (solid lines) and the dots and dashed curves are derived using the described inversion procedure.

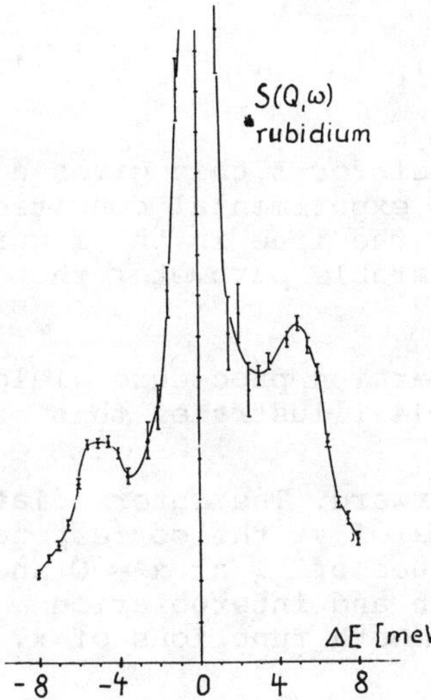

Fig. 12.
Example of neutron scattering data on liquid rubidium at $Q = 0.7$ Å^{-1} by Gemperlein /22/ (raw data).

However the data which we will discuss here more in detail are the results on liquid rubidium of Copley and Rowe /17/ which extend over a larger Q range. The scale factor c can easily be evaluated from these data and its Q dependence is shown in Fig. 13.

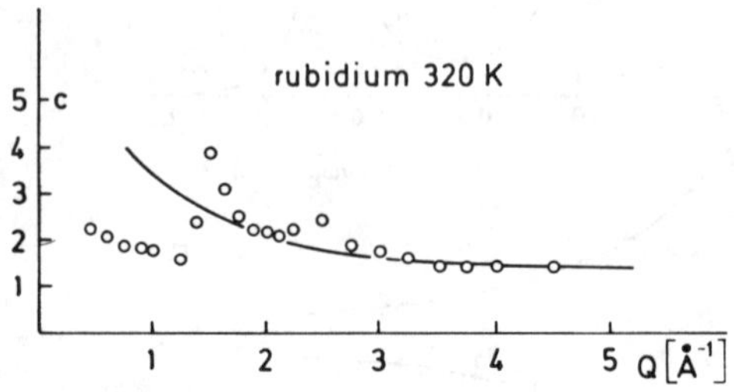

Fig. 13. The scale factor c for liquid rubidium at 320K. Circles are experimental data. The solid line results from eq. (12) with l = 0.33 Å.

At very high Q c should exhibit free particle behaviour. In a dense system this behaviour should extend over a distance range up to a mean interaction distance ℓ. We can modify the free gas model on the level of the intermediate scattering function to represent this behaviour at short times and a diffusion like behaviour at long times:

$$I(Q,t) = \exp - \{Q^2\ell|v_o^2t^2 + \ell^2|^{1/2} - Q^2\ell^2\}.$$

The result for c(Q) is:

$$c = Q\ell K_1(Q^2\ell^2) \exp (Q^2\ell^2). \tag{12}$$

If we use the Enskog hard sphere model for ℓ this gives a value ℓ = 0.33 Å for rubidium at the experimental conditions and a scale factor c as indicated by the line in the figure. Having in mind that there is no adjustable parameter the agreement at large Q is quite good.

Now we consider the experimental inversion procedure yielding σ_1 and σ_2 for liquid rubidium. Fig. 14 illustrates this procedure at Q = 3 Å^{-1}.

σ_1 at x = 0 and x = 1 are straightforward. The intermediate points for σ_2 follow from the extrema of y_2 the corresponding values of σ_1 follow from y_1. The values of σ_2 at x = 0 and x = 1 are determined by extrapolation and interpolation respectively. σ_1 and σ_2 seem to be smooth functions of x.

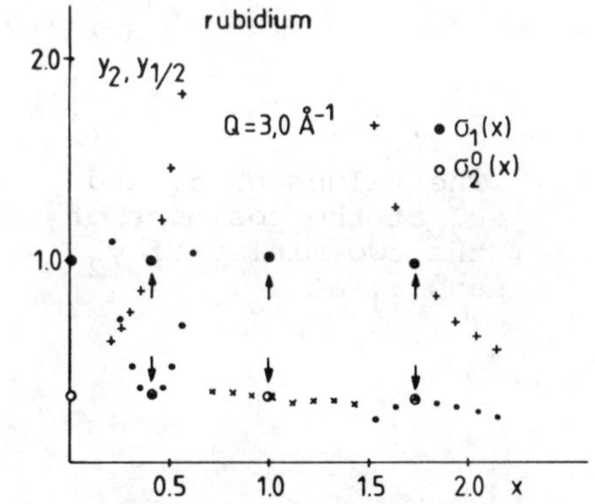

Fig. 14.

Application of the experimental inversion procedure on liquid rubidium data (320K) at $Q = 3\ Å^{-1}$. $\cdot$expression (11b), + (10a) and x shows the interpolation.

We can plot the ω-positions of the extrema of y_2 as functions of Q in comparison to ω_o and this is illustrated in Fig. 15.

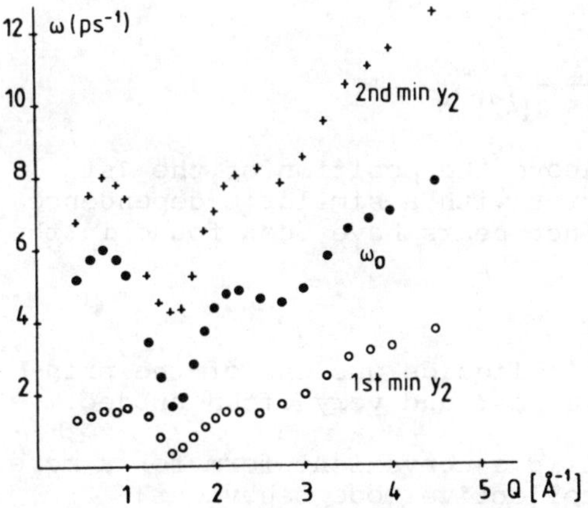

Fig. 15.

The positions of the two minima of y_2 and ω_o as functions of Q for liquid rubidium at 320K.

There is a surprising scaling of the Q-dependence of these energy values with the one of ω_o, which however, is not fully understood yet. Finally Fig. 16 shows the Q-depence of σ_1 and σ_2^o for these special ω values.

We recognize immediately that Loveseys "viscoelastic model" is only a crude approximation which seems to be reasonable for large $Q > 2.5\ Å^{-1}$. σ_1 and σ_2^o are ω-dependent at smaller Q. Below $Q \approx 1\ Å^{-1}$ the ω-dependence of σ_1 and σ_2^o agrees qualitative with that of the Brillouin model. This is the Q region where distinct excitation peaks in $S(Q,\omega)$ were found /17,23/.

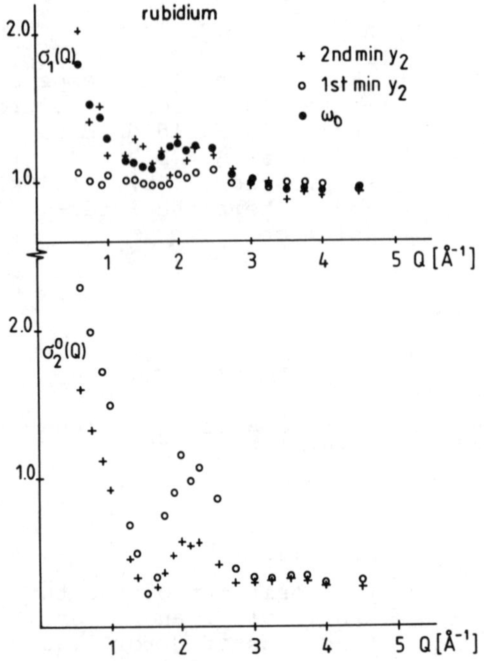

Fig. 16.

The values of σ_1 and σ_2^0 at the positions of the two minima of y_2 and σ_1 at ω_0.

But there is also a region above the position of the 1st maximum of the structure factor with a similar ω-dependence of σ_1 and σ_2^0 where no distinct peaks have been found in the experiments up to now.

Discussion of mode behaviour

Damping of collective modes in liquids was one of the principal points discussed in the past und very often subjective criteria were introduced.
Equation (10) can be written in a convenient form for a more quantitative discussion of collective mode behaviour:

$$n\sigma_1 = (1 - x^2)^2 \sigma_1^2 + (R - x^2)^2 \sigma_2^2 ,$$

$$R = \frac{\sigma_2 + \omega\tau_i}{\sigma_2} , \quad \tau_i = c/\omega_o . \qquad (13)$$

It is clear that a mode would correspond to a distinct minimum of the right hand side of equation (13). E. g., if $\sigma_2 >> \omega\tau_i$, then $R \rightarrow 1$ and $S(Q,\omega)$ would exhibit δ-function behaviour at $\omega = \omega_o$. For $R >> 1$ one can show that equation (13) predicts two distinct modes. To determine mode frequency ω_s and width Z_s we try to express the right hand side of equation (13) by

$$|\omega^2 - (\omega_s + iZ_s)^2|^2. \qquad (14)$$

We get different solutions for ω_s and Z_s in different x-regimes. For $x^2 \geq R \gg 1$ we can approximate equation (13) by the following expression:

$$\frac{n}{\omega^2\tau^2} \simeq \frac{\sigma_1}{\omega_o^4} \left| (\omega^2 - \omega_o^2 R)^2 + \omega^2/\tau^2 \right|, \quad \tau = \sigma_2/\sigma_1\omega . \tag{15}$$

Equating the latter bracket with expression (14) yields

$$\omega_s = \sqrt{R\omega_o^2 - 1/4\,\tau^2}\,, \quad Z_s = 1/2\tau . \tag{16}$$

The condition for this mode (mode 1) to propagate is $\omega_o \tau R > \sqrt{R/2}$ and the mode becomes the Brillouin peak as $Q^o \to 0$. Equation (13) can exhibit also a low frequency mode (mode 2) in the regime $R \gg x^2 \geq 1$, where we can write:

$$n \simeq \frac{1}{\omega_o^4} \left| (\omega^2 - \omega_o^2)^2 + \omega_o^4 \tau^2 R^2 \omega^2 \right| . \tag{17}$$

Equating with (14) yields:

$$\omega_s = \frac{\omega_o}{2} \sqrt{4 - \omega_o^2 R^2 \tau^2}, \quad Z_s = R\omega_o^2\,\tau/2 . \tag{18}$$

Propagation occurs for $\omega_o \tau R < 2$.
For large R the propagation conditions for both modes are normally exclusive. Finally for $\omega \ll \omega_o$ no propagating modes exist, instead a central peak of width $\sqrt{2}/R\tau$ is obtained.

Because of the ω-dependence of the solutions a clearer illustration of the mode behaviour can be achieved by resolving equation (13) in three modes:

$$\frac{n\omega_o^4}{\sigma_1} = (\omega^2 - \omega_o^2) + \tau^2\omega^2(\omega^2 - R\omega_o^2)^2$$

$$= \tau^2 |\omega + iZ_o|^2 |\omega - \omega_s + iZ_s|^2 |\omega + \omega_s + iZ_s|^2 . \tag{19}$$

Equating coefficients of equal powers of ω gives implicite equations for Z_o, Z_s and ω_s as functions of τ and R. If we define a parameter d by

$$\omega_s^2 + Z_s^2 = dZ_s^2 \quad \text{or} \quad \omega_s/Z_s = \sqrt{d-1}\,,$$

we can eliminate Z_o, Z_s and ω_s from these equations and derive a closed relation between σ_1/c and $\sigma_2^o/c = u^{-1}$ as a function of d:

$$\left(\frac{c}{\sigma_1}\right)^2 = \left(\frac{u}{u+1}\right)^2 \left\{\frac{4-d}{d} + \left(\frac{ud^2+8d-16}{8(u+1)d^2}\right)\left(ud+4 \pm \sqrt{u^2d^2-8ud-16(d-1)}\right\}. \qquad (20)$$

The ultimate boundary between propagating and nonpropagating solutions occurs for $d = 1$ ($\omega_s = 0$) and ω_s equals Z_s for $d = 2$. Fig. 17 illustrates these boundaries.

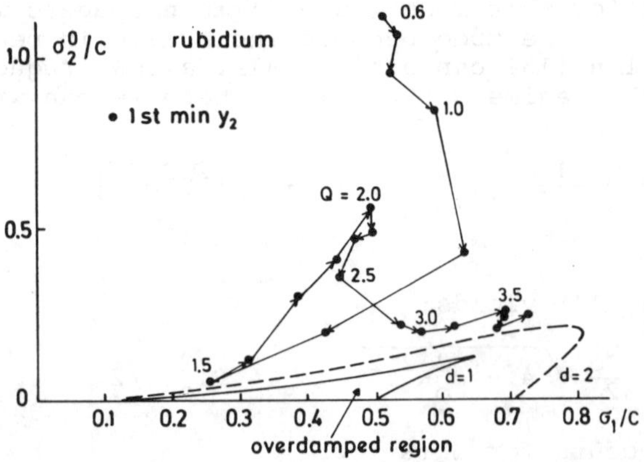

Fig. 17. Plot of σ_2^o/c vers. σ_1/c at the 1st minimum of y_2 for different Q. The solid line ($d = 1$) is the ultimate boundary between propagating and nonpropagating solutions of equation (19), whereas the dashed line holds for $\omega_s = Z_s$.

Certainly inside the $d = 1$ boundary modes are overdamped. In this figure also the 1st min of y_2-values for liquid rubidium in the experimental covered range are plotted. All the (σ_2^o/c , σ_1/c) pairs lie outside the $d = 2$ boundary, similarly all 2nd min of y_2 values which are not plotted lie outside this boundary. It is obvious from this plot, that in liquid rubidium at 320K propagating modes exist at all measured Q values.

Using the relations for Z_o, Z_s and ω_s following from equation (19) we vcan calculate the real part ω_s and the imaginary part Z_s of the mode frequency. The numbers calculated for σ_2^o and σ_1-values at the 2nd min of y_2 are plotted in Fig. 18.

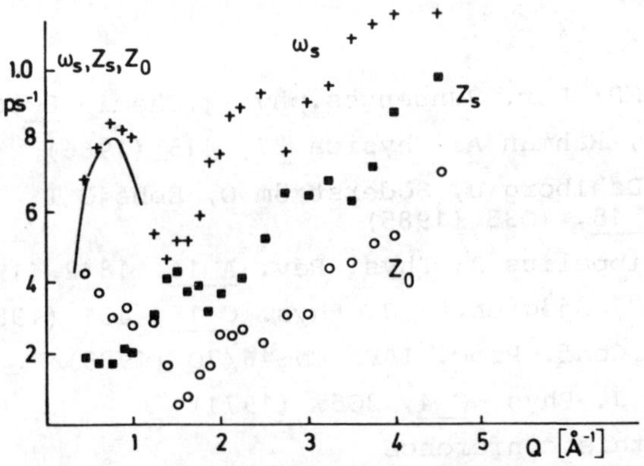

Fig. 18. Solutions of equation (19) for liquid rubidium at 320K (σ_1, σ_2^o at the 2nd min of y_2). The solid line at small Q represents the direct determined peak positions of the experimental S(Q,ω) data.

Using σ_2^o- and σ_1-values at the 1st min of y_2 gives very similar results for ω_s and it is remarkable, that this ω_s values are in quite good agreement with the loci of the 2nd min of y_2 on the ω-scale. The line drawn in this figure at small Q represent the direct determined peak positions of the experimental S(Q,ω) data. We realize that in this Q region the imaginary part is small compared to the real part of the mode frequency, making distinct peaks easy detectable.
The graph suggests that this might still be possible in the $Q \simeq 2 - 2.5\ \text{Å}^{-1}$ region and has to be checked experimentally. But the visibility of distinct peaks seems not to be a necessary criterium for the existency of propagating modes in liquids.

In summary in this lecture a new scheme of interpretation of the dynamical structure factor in terms of linear responce theory has been described. The scheme allows to determine a reduced memory function σ(Q, ω) at several values of ω in the experimentally accessible ω-region for each Q and consequently permits the selection of realistic models for the dynamic structure factor of a liquid.

ACKNOWLEDGEMENTS

It is a great pleasure to acknowledge many discussions with Prof. P. A. Egelstaff, Prof. W. Götze, Prof. J. Bosse and Dr. C. Morkel on the subject of this paper. This work was supported in part by the Bundesministerium für Forschung und Technologie.

REFERENCES

1. Nachtrieb NH: Ber. Bunsenges.physik.Chemie 80,678 (1976)
2. Nijboer, B, Rahman A: Physica 32, 415 (1966)
3. Kinell T, Dahlborg U, Söderström O, Ebbsjö I: J. Phys. F 15, 1033 (1985)
4. Götze W, Zippelius A: Phys. Rev. A 14, 1842 (1976)
5. Wahnström G, Sjögren L: J. Phys. C 15, 401 (1982)
6. Schärpf O: Conf. Proc. IAEA-CN-46/70 (1985)
7. Lovesey S: J. Phys. C 4, 3057 (1971)
8. Morkel C: this conference
9. Levesque D, Verlet L: Phys. Rev. A 2, 2514 (1970)
10. Sears V: Phys. Rev. A 5, 452 (1972)
11. Montfrooy W, Verkerk P, de Schepper I: submitted to Phys. Rev. A
12. Alder B, Wainwright T: Phys. Rev. A 1, 18 (1970)
13. Zwanzig R, Bixon M: Phys. Rev. A 2, 2005 (1970)
14. Carneiro K: Phys. Rev. A 14, 517 (1976)
15. de Schepper I, Ernst MH: Physica A 98, 189 (1979)
16. Montfrooy W, de Schepper I, Bosse J, Gläser W, Morkel C: submitted to PRL
17. Copley JRD, Rowe JM: Phys. Rev. A 9, 1656 (1974)
18. Bucher G: PhD thesis, Technische Universität München(1984)
19. Jacucci G, Mac Donald IR: J. Phys. F 10, L 15 (1980)
20. Egelstaff PA, Gläser W: Phys. Rev. A 31, 3802 (1985)
21. Bosse J, Götze W, Lücke M: Phys.Rev. A 17, 434 (1978)
22. Gläser W, Egelstaff PA: to be published
23. Gemperlein H: diplom thesis, Technische Universität München (1984)

ATOMIC DYNAMICS IN BINARY LIQUIDS WITH ATTRACTIVE A-B INTERACTION

D. QUITMANN

Institut für Atom- und Festkörperphysik, Freie Universität Berlin, D1000 Berlin 33, Fed. Rep. Germany

ABSTRACT

Experimental information on the relative dynamics of A-B particle pairs in metallic liquid alloys is discussed, using as examples quasielastic neutron scattering from $Li_{0.8}Pb_{0.2}$ and quadrupolar nuclear spin relaxation in In-Sb, In-Hg. Connection with thermodynamic data, and an anschaulich interpretation in terms of excess attractive A-B potential and of lifetimes of AB association, are stressed.

INTRODUCTION

The transition from monatomic metals to binaries with new properties follows from a differentiation between two species (A,B) with respect to mass, size or, most important, effective interaction. In the solid state, an exhiliarating wealth of new phenomena result from this generalization (formation of salts, molecular solids, semiconductors, various crystallographic phases, etc.). In the liquid state, one may assume that only the most important interactions survive at the elevated temperature and strong disorder. Thus one still has e.g. liquid salts, molecular liquids, liquid semiconductors, short range order. We shall be concerned here with atomic scale dynamics in some liquid metallic A-B systems formed from s-p-metals, covering two experimental methods where results have been obtained so far. For a review of a somewhat earlier status see e.g. /1/; see also /2/.

Strong AB attraction (compared to the average AA- and BB-interaction) in the liquid state is to be suspected whenever there occur $A_\mu B_\nu$ solid components with melting points markedly higher than the average of the constituents, like e.g. in Na-Cl, Li-Pb, Ga-Te, In-Sb; if in the liquid at intermediate concentrations a strongly (w.r.t.k_BT) negative heat of mixing is observed; if the structure of the liquid shows chemical short range order (heterocoordination, see e.g. /3/); if the viscosity increases; if the electron mobility is reduced. More than one of these effects are usually found in the same system, but they are not all necessarily interdependent. Chemical short range order by itself is not a proof of strong AB attraction, glass forming alloys like As-Se being counterexamples. However in many liquid alloys with interesting electronic properties, these may be a consequence of strong AB-attraction leading to (partial) localization of the electrons engaged in the AB-bonding; systems discussed by Hensel, van der Lugt, Nicholson, Schirmacher at this school are striking examples.

Atomic dynamics and transport properties are particularly sensitive to the attractive part of the interactions, as has been demonstrated e.g. by their dependence on ion polarizability in ionic melts or on the range of interaction in liquid Rb /4/. - Of extreme importance for our understanding of liquids are the molecular dynamics calculations. They are, however, difficult for complicated interactions (as may be present in AB ordered alloys), and sumptuous if reliable data for $\omega \to 0$ are required, as for diffusion. Experimental data are all the more indispensable. - An understanding of atomic dynamics is also central to a dynamical approach to glass formation /5/.

Description of Dynamics in Binary Liquids

The van Hove correlation function for a classical liquid

$$G(\bar{r},t) = \frac{1}{N} < \sum_{ij} \delta(\bar{r}+\bar{R}_i(0)-\bar{R}_j(t)) > = G_{self}(\bar{r},t) + G_{dis}(\bar{r},t) \quad (1)$$

which gives the probability density to find a particle at position $\bar{r}$ at time t, given that some particle was at position $\bar{r} = 0$ at t=0, is isotropic in the liquid with G(r,0) = g(r) the radial distribution function. It splits naturally into three parts for an AB-mixture

$$G(r,t) = G_{AA}(r,t) + G_{BB}(r,t) + G_{AB}(r,t) \quad (2)$$

in the present context we may assume $G_{AB}=G_{BA}$. As an example of extremely strong heterocoordination we may consider the $g_{ij}(r)$'s for alkali halides /6/. Similarly, the distance between a given atom and its neighbour i, $r_i(t)$, becomes species dependent, see fig. 1. The additional AB attraction stabilizes any spatially close AB-configuration over the effective range of the potential which causes the ordering.

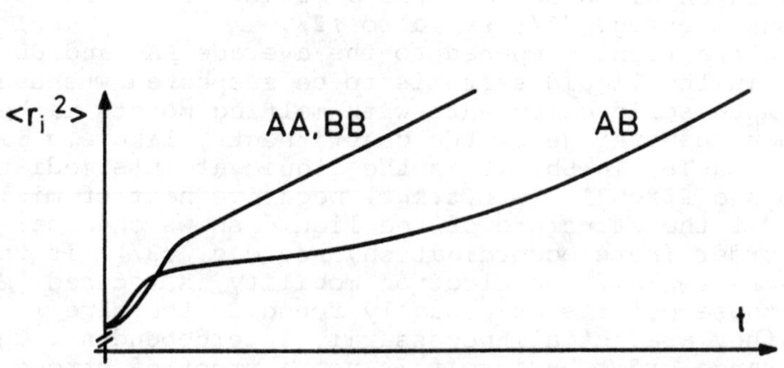

Figure 1: Qualitative behaviour of the time development of the magnitude of the distance between two individual particles in a system with strong AB-interaction; $D_{SA}=D_{SB}=D_S$ is assumed.

We start with a quasielastic/inelastic neutron scattering experiment. The double Fourier transform of G_{AB} etc. is

$$S_{\alpha\beta}(q,\omega) = \iint d^3r\, dt\, e^{i(\bar{q}\bar{r}-\omega t)}\, G_{\alpha\beta}(\bar{r},t) \qquad \alpha,\beta = \{A,B\} \quad (3a)$$

$$S_{\alpha\beta}(q) = \int d\omega S_{\alpha\beta}(q,\omega) = \int d^3r\, e^{iqr}\, G_{\alpha\beta}(r,0). \quad (3b)$$

Bhatia and Thornton /7/ introduced a different set of scattering laws S which allows to single out rather clearly the effect of the difference in interactions between AB and AA, BB pairs:

$$S_{NN}(q,\omega) = c_A S_{AA} + 2\sqrt{c_A c_B}\, S_{AB} + c_B S_{BB} \quad (4a)$$

$$S_{Nc}(q,\omega) = c_A c_B \left[S_{AA} + \frac{c_B - c_A}{\sqrt{c_A c_B}} S_{AB} - S_{BB}\right] \quad (4b)$$

$$S_{cc}(q,\omega) = c_A c_B \left[c_B S_{AA} - 2\sqrt{c_A c_B}\, S_{AB} + c_A S_{BB}\right]. \quad (4c)$$

These new S's still contain the full information. The effects of a specific AB interaction $w = w_{AB} - (w_{AA}+w_{BB})/2$ are contained in $S_{cc}(q,\omega)$ /1/, /8/. For a "normal" liquid with no difference between $G_{AB}(r,t)$ and $[G_{AA}(r,t)+G_{BB}(r,t)]/2$ eqs.(4) reduce essentially to one S viz. $S_{NN}(q,\omega)=S(q,\omega)$; $S_{Nc}=0$; and $S_{cc}^{ideal}(q,\omega)$, eqs. (6), (12), (13).

The q = 0 value of $S_{NN}(q)$ and $S_{cc}(q)$, i.e. the macroscopic averages of $G_{NN}(r,0)$ and $G_{cc}(r,0)$, are well known thermodynamic quantities: Just as $S_{NN}(0) = 1/N <\Delta N^2> = \rho\, k_B T \chi_T = N\, k_B T/(\partial^2 G/\partial N^2)$ is the stability against changes of density induced by thermal fluctuations or externally (χ_T = isothermal compressibility), so

$$S_{cc}(0) = N <(\Delta c)^2> = N\, k_B T \,/\, \left(\frac{\partial^2 \Delta G}{\partial c^2}\right)_{p,T,N} \quad (5)$$

is the stability of the system against changes of concentration; it is called the "stability". Systems with strong AB-attraction (which is working essentially on an atomic scale) resist macroscopic demixing, $S_{cc}(0) \ll S_{cc}^{ideal}$. In "ideal" alloys, $w/k_B T \to 0$,

$$S_{cc}^{ideal}(q) = c_A c_B \quad (6)$$

independent of q at all T. For the example Li-Pb fig. 2 shows the close connection between phase diagram with high melting compounds, thermodynamic function $\Delta G(c)$ = excess free enthalpy vs. concentration c, and $S_{cc}(q)$ at the concentration with strong AB interaction $c_{Li}=0.8$. Consider the pronounced peak of $S_{cc}(q)$ around q=1.5 Å^{-1}, fig. 2c: Its position shows the wavelength for which there is strong concentration modulation (one can read from figs. 2c, 2d a value of about 4 Å as corresponding to the average size of "one $Li_{0.8}Pb_{0.2}$ unit");

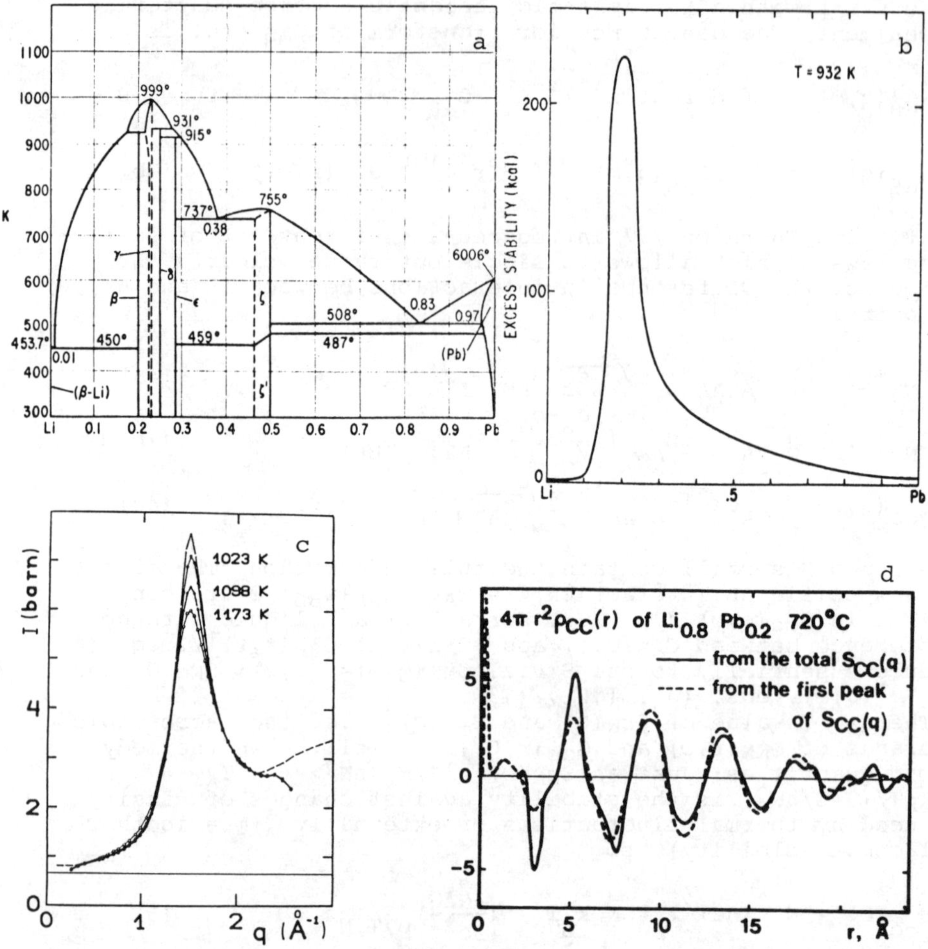

Figure 2: For Li-Pb the phase diagram and the excess free enthalpy of mixing ΔG (for T=932 K) from /9/, and the structure factor $S_{CC}(q)$ /10/ at the concentration of strongest AB interaction, c_{Li}=0.8, are shown. Fig. 2d gives the Fourier transform $g_{CC}(r)$ for c_{Li}=0.8 from /3/.

the peak height gives the average amplitude of the concentration modulation. This is, of course, only the projection of the true average concentration ordering onto plane waves with $q=|\bar{q}|$. The peak width reflects the range of spatial coherence.

A Scattering Experiment on the Dynamics of Heterocoordination

Inelastic neutron scattering is the most clearcut experiment known to measure G(r,t) via $S_{AB}(q,\omega)$, but very little information exists so far. Here we shall review an experiment by Soltwisch et al. on liquid $Li_{0.8}Pb_{0.20}$ which measured the quasielastic part of $S_{CC}(q,\omega)$ /10/. The coherent neutron scattering intensity of a binary system, $I(q,\omega)$, stems from the amplitudes scattered from AA-, BB- and AB- pairs with scattering strengths proportional to b_A^2, b_B^2 and $b_A b_B$ ($b_{A,B}$= coherent scattering length of element A,B) and with scattering laws S_{AA}, S_{BB}, S_{AB} respectively. Reformulating this in terms of S_{NN}, S_{Nc}, S_{cc} one has

$$I(q,\omega) \sim \left[\langle b\rangle^2 S_{NN}(q,\omega) + 2\langle b\rangle \Delta b S_{Nc}(q,\omega) + \Delta b^2 S_{cc}(q,\omega)\right] \quad (7)$$

with $\langle b\rangle = c_A b_A + c_B b_B$, $\Delta b = b_A - b_B$. This expression is anschaulich in the sense that the density scattering law (neglecting A - B differences) S_{NN} has the average scattering length squared $\langle b\rangle^2$ as its strength factor, the scattering law for deviations from the average concentration (viz. S_{cc}) has the contrast in scattering amplitudes Δb^2 as its strength factor, and the cross term S_{Nc} goes with $\langle b\rangle\Delta b$. In order to study AB-ordering in a liquid alloy, Ruppersberg and coworkers have made extensive use of so called zero alloys, i.e. combinations of isotopes of A,B and concentrations c chosen such that $\langle b\rangle=0$ (with $\Delta b \neq 0$) /3/. If in addition, there is strong heterocoordination one has the favourable conditions of $I(q,\omega)$ being essentially $S_{cc}(q,\omega)$ and deviating markedly from the more or less trivial $S_{cc}^{ideal}(q,\omega)$. This is the case for $^7Li_{0.80}{}^{nat}Pb_{0.20}$.

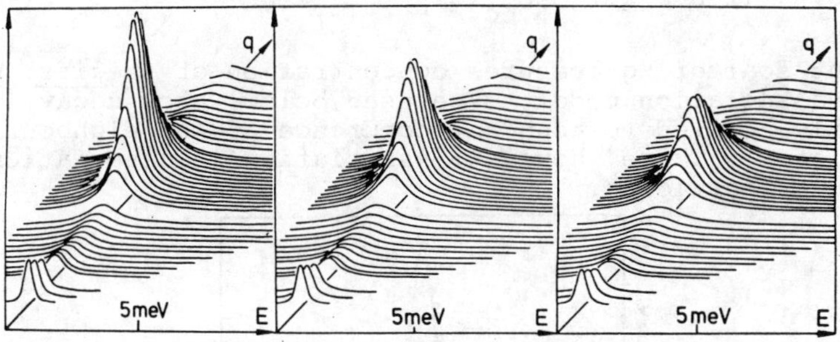

Figure 3: Scattered neutron intensity for $^7Li_{0.80}Pb_{0.20}$ from /10/. From left to right, temperatures are 1023 K, 1098 K, 1173 K. The S_{cc}-peak occurs at $q=1.5$ Å^{-1}. Note the strong temperature dependence of $S_{cc}(q_{peak},\omega=0)$. The small peak at $q \to 0$ is the incoherent scattering from ^7Li.

The experiment was performed at the ILL high flux reactor using a time of flight spectrometer: monoenergetic neutron pulses impinge on the sample, are quasielastically or inelastically scattered, and have therefore different times of flight until they reach the detectors. For details see /10/. Fig. 3 shows the total observed intensities

$$I(q,\omega) \sim 4\pi\Delta b^2 \, S_{cc}(q,\omega) + c_{Li}\sigma_{incLi} \, S_{sLi}(q,\omega) \quad . \quad (8)$$

The maximum developping at $q \to 0$ is due to the second term (incoherent scattering); it corresponds to self diffusion of Li and will not concern us here (see /11/). The discussion of the dynamics will proceed in three steps.

Lifetime of AB-order. An average lifetime of concentration modulations at wavevector q may be defined via the intermediate scattering function

$$I_{cc}(q,t) = \int_{-\infty}^{\infty} e^{-iEt/\hbar} S_{cc}(q,E)\, dE/\hbar \qquad (9)$$

by

$$\tau_{cc}(q) = \frac{1}{I_{cc}(q,0)} \int_{0}^{\infty} I_{cc}(q,t)\,dt = \pi\hbar \frac{S_{cc}(q,\omega=0)}{S_{cc}(q)} . \qquad (10)$$

Fig. 4 gives the experimental $S_{cc}(q,0)$ as well as the derived $\tau_{cc}(q)$. It is seen that the concentration modulation waves with $\lambda = 2\pi/q_{peak} \approx 4$ Å which have largest average amplitudes, i.e. are most abundant in the liquid, have also the longest lifetime $\tau_{cc}(q_{peak}) = 1.7 \ldots 1.1 \cdot 10^{-12}$sec, between 1023K and 1173K. Within the simplified diffusional picture which will be discussed presently $\tau_{cc}(q)$ is increased or decreased w.r.t. the ideal value $1/(\pi\bar{D}q^2\hbar)$ just by the factor $S_{cc}(q)$

$$\tau_{cc}(q) \approx (\pi\hbar\bar{D}q^2)^{-1} S_{cc}(q). \qquad (11)$$

Coherent scattering measures concentration of species. Within the association model to be used below, slow decay is similarly coupled to abundant occurence of AB-neighbour pairs (associates). There, however, association is association of individual particles.

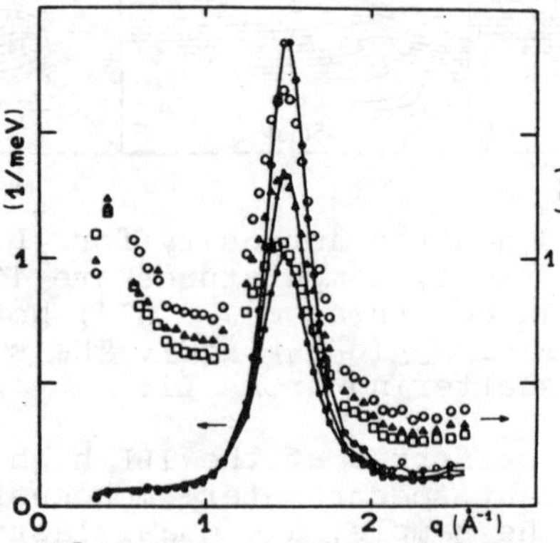

Figure 4: Lifetime of concentration modulation waves $\tau_{cc}(q)$ in $Li_{.8}Pb_{.2}$, unconnected data and r.h. scale. "Elastic" scattering from concentration fluctuations, $S_{cc}(q,0)$, left hand scale. 1023 K, 1098, 1173 K from top to bottom. From /10/.

Interdiffusion constant. In the hydrodynamic limit ($q\to 0, \omega\to 0$) the concentration fluctuations obey a diffusion equation /12/

$$S_{cc}(q,E) = S_{cc}(q) \frac{\Gamma^+/2\pi}{\omega^2 + (\Gamma^+/2)^2} \qquad (12)$$

$$\Gamma^+ = 2\hbar q^2 D^+ .$$

The width Γ^+ is given by the interdiffusion constant D^+ (see e.g. /13/) which in an ideal system (w=0) is

$$D^{+ideal} = \bar{D} = c_A D_{sB} + c_B D_{sA} \quad . \qquad (13)$$

Any concentration modulation, any "pattern" existing at t=0 decays by independent diffusion of the two species. Darken made an important observation in alloys /14/ viz. that concentration fluctuations in many cases decay with a value D^+ very much different from $\bar{D}$, as a consequence of the non-ideal stability ($\partial^2 G/\partial c^2$) of the system: $D^+ = (c_A D_{sB} + c_B D_{sA}) S_{cc}^{id}/S_{cc}(0)$, an expression which follows from the fact that in general the fluctuations are subject to the restoring force $\nabla\mu$ (μ = chemical potential), not simply to ∇c. In the spirit of generalized hydrodynamics, D^+ and Darken's equation may now be generalized to finite q by just introducing $S_{cc}(q)$,

$$D^+(q) = \bar{D} \cdot c_A c_B / S_{cc}(q) \qquad (14)$$

and describing the quasielastic peak as a Lorentzian also at intermediate q. Then the width $\Gamma^+(q)$, fig. 5, or exponential decay rate $1/\tau_{cc}(q)$ deviate from $1/q^2\bar{D}$ just as prescribed by $S_{cc}(q)$. It may be useful to stress that this approximation was found to be very good for $S_{cc}(q,0)$ in $Li_{.8}Pb_{.2}$. - While in heterocoordinated liquids fluctuations with large extension ($q \ll q_{peak}$) decay faster and the modulations on atomic scale ($q \approx q_{peak}$) slower, the case of demixing fluids has $S_{cc}(q)$ and $\tau_{cc}(q)$ increased at $q\to 0$ for $T\to T_c$. One should also note in this context a certain restriction to the possible shapes of $S_{cc}(q)$: $\int [S_{cc}(q) - c_A c_B]\, d^3q = 0$ which requires that if deviations $S_{cc}(q) > S_{cc}^{ideal}$ occur at some q, a compensating effect has to occur at some other q. In addition, because of the absence of long range order in the liquid, oscillations of $S_{cc}(q)$ vs. q cannot be very narrow i.e. spaced very closely. At large q, $S_{cc}(q)\to S_{cc}^{ideal}$ anyhow. - As a thermodynamic model, $S_{cc}(0)$ is often expressed by the ordering potential w, which then is generalized to w(q) yielding $S_{cc}(q) = S_{cc}^{ideal}/[1-2w(q)S_{cc}^{ideal}/k_B T]$ see e.g. /1/. Darken's equation can in the same spirit be derived by starting from this equation by just replacing S_{cc}^{ideal} with $S_{cc}^{ideal}(q,E)$ from equa. (12) and (13) /10/.

Second moment rule. Quite analogous to the monatomic liquid, de Gennes' second moment rule applies to $S_{CC}(q,0)$ /12/:

$$\hbar^2 \omega_o^2 = \frac{1}{S_{cc}(q)} \int E^2 \, S_{cc}(q,E) \, dE = \hbar^2 q^2 k_B T \, m_{cc}^{-1} \, c_A c_B / S_{cc}(q) \tag{15}$$

with $m_{cc}^{-1} = (c_B m_A^{-1} + c_A m_B^{-1})$. Since (15) is a monotonous function of q, large values of S_{CC} at some q enforce small widths there (width $\sim (k_B T/S_{CC}(q))^{-1/2}$ for gaussian). In the experiment referred here, the energy range covered was not sufficient to exhaust the E-integral (15). Note that although

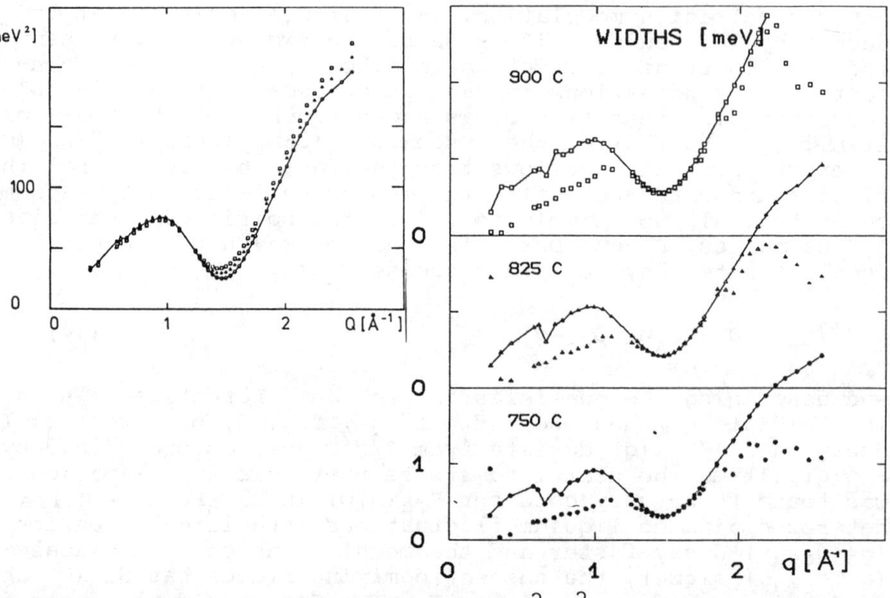

Figure 5: Square of second moment, $2\hbar^2 \omega_o^2$, for $Li_{.80}Pb_{.20}$ (left) and width Γ/2 of the quasielastic line if it is fitted simply by one Lorentzian (full curves, right part). Unconnected points show width of main Lorentzian, if a second (broader) Lorentzian is also allowed.

$k_B T \approx 100$ meV, the narrowing is clearly observed also in the quasielastic part of the scattering law (1 meV-region), see fig. 5. - We conclude the discussion of the Li-Pb experiment by asking: "Are there equivalents of the optical modes"? They show up clearly in analytic /15/ and molecular dynamics calculations /16/ for molten salts i.e. true Coulomb systems. A neutron scattering experiment on RbBr had however not given a clearcut demonstration of them /17/ and the experiment on Li-Pb gave only an indirect indication from fourth moment /10/. After the structure factor $S_{CC}(q)$ had allowed Copestake et al. to derive a rather detailed ordering potential /18/, a molecular dynamics calculation for Li_4Pb was performed by Jaccucci et al. . It shows beautifully the expected high frequency mode, with $\hbar\omega \approx 30$ meV, see /19/. Also in liquid alkali chlorides, this excitation has very recently been definitely identified in inelastic neutron scattering /20/. - We conclude this section by pointing out that in a metallic alloy like Li-Pb with partial electron localization, the $S(q,\omega)$ of the electron density is important, while in molten salts one has true two-component systems.

Quadrupolar Nuclear Spin Relaxation in Alloys with Strong AB-Attraction

Quite a different approach at first sight to the study of processes in liquid alloys uses hyperfine interactions. The Knight shift is a standard indicator of metallicity in liquid alloys and the magnetic nuclear spin relaxation for incipient electron localization. Less attention has been paid to the quadrupolar relaxation rate $R_Q = 1/T_{1Q}$ probably because there are fewer nuclei where it can be measured conveniently by NMR techniques (Cu, Ga, Hg, Rb, Sb .. , Bi, Na, ...). However, large increases of R_Q are known to occur in alloy systems (e. g. Ga-Te /21/; In-Sb /22/; Cs-Sb /23/). Studying many alloys - using the TDPAD method - it has been found that large increases are indeed encountered systematically in alloys which show strong AB-attraction /24/. Since the electric hyperfine interaction is of short range in metallic liquids, it is best discussed in r-space rather than in q-space. For a detailed recent discussion of the theory of R_Q, we refer to /25/.

Theory. Imagine a probe atom, the nucleus of which carries spin $I > 1/2$ and electric quadrupole moment $Q \neq 0$. The neighbour i is at a distance $\bar{r}_i$ and it produces effectively an electric field gradient efg $v_i(\bar{r}_i)$, taking as the reference axis for θ e.g. the direction of the external magnetic field B_o; in a pure Coulomb system, $v_i(\bar{r}_i) \propto eZ_i r_i^{-3} (3\cos^2\theta_i - 1)$. Each distance vector $\bar{r}_i$ changes with time. The energy levels of the probe nucleus are perturbed by the total efg

$$V(t) = \sum_i v_i\ (\bar{r}_i(t)) \qquad (16)$$

and one may think of an instantaneous perturbation frequency Ω which, apart from spin factors, is $\Omega(t) \propto e/\hbar\ V(t)\cdot Q$ with typical values for Ω of the order 10^8/sec. Due to the average spherical symmetry in the liquid around each atom, $\langle V(t)\rangle = 0$. But due to the fluctuations in the arrangement of atoms around the probe, $\langle V^2\rangle$ is nonzero, and so is the efg-correlation function $\langle V(0)\cdot V(t)\rangle \neq 0$ for reasonably short times. In as much as the correlation time τ of the efg, defined by $\int dt\ \langle V(0)V(t)\rangle = \tau\cdot\langle V(0)^2\rangle$ is nonzero, the perturbation Ω is not fully averaged out, the NMR line is not completely "motionally narrowed" /26/, the orientation of the nuclear spin relaxes at a rate R_Q which in the cases of interest here is proportional to τ /27/. It is this effect which allows conclusions about τ and thus about rearrangements in the liquid. The exact expression for R_Q is

$$R_Q = \frac{3}{40} k(k+1) \frac{4I(I-1) - k(k+1)-1}{2I^2 (2I-1)^2} (\frac{eQ}{\hbar})^2 \int dt \langle V(0)V(t)\rangle . \quad (17)$$

Here k is the multipole order observed (k=1 for NMR, k=2,4 for γ-ray anisotropy). It is seen that the product of expressions (16) for two times has the consequence that the correlation of two distance vectors $\bar{r}_i(0)$, $\bar{r}_j(t)$ enters and determines the time average in (17). Forgetting for a moment the angular factors (see /28/) one has

$$\langle V(0)\ V(t)\rangle \propto \iint d^3r_o\ d^3r \sum_{i,j} v_i(r_o)W_{ij}(r_o;r,t)\ v_j(r), \quad (18)$$

where $W_{ij}(r_o;r,t)$ is the probability density to have distance $|\bar{r}_o|$ between the probe atom and neighbour i for t=0, and distance $|\bar{r}|$ between probe and j for t. W is similar to a product of functions G(r,t) but different in basic meaning as well as in its $(r_o;r,t)$-dependence. $\langle r_i^2\rangle$ from fig. 1 is the self part $\iint d^3r_o\ d^3r\ W_{ii}(r_o;\ r,t)r^2$. For a discussion of two particle distances in normal liquids, see e.g. /29/.

Experiment and results. A description of the experiment may be found in /30/ for the generalities, and in /31/ for the cases discussed presently. Suffice it to say that one produces nuclear isomers (lifetimes $10^{-6}...10^{-2}$ sec) by a nuclear reaction (like $^{115}In(\alpha,2n)^{117m}Sb$) using pulsed particle beam from an accelerator. The isomers are produced as an aligned ensemble (population of states with |m| = 0 or 1/2 preferred). By the nuclear spin relaxation processes, e.g. by R_Q, the degree of alignment relaxes, and the relaxation rate can be measured during the lifetime of the isomeric state by the rate of decay of the anisotropy of the γ-radiation which depopulates the isomeric state (TDPAD-technique). While this nuclear method suffers no decrease of sensitivity at high temperature and opens up the possibility for quite a number of elements, to study the quadrupolar interaction (e.g. Ag, Cd, Pb, Sn, Te have no NMR isotope with $Q \neq 0$, but they have suitable isomers), the price paid is that in most cases the probe atom is now a chemical impurity. - In the cases considered below, the rate R_Q is by far dominant, or the magnetic relaxation rate R_M has been subtracted.

In fig. 6 examples are shown for the dependence of R_Q on partner element, on concentration c, and on temperature. The interpretation sketched below for R_Q is based on a systematical study of liquid alloys of s-p elements by the TDPAD technique /24/, and is in agreement with the much rarer NMR data /32/. From the systematics one can conclude three observations already suggested by fig.6: (1) R_Q increases w.r.t. the average of the pure liquid metals /22/; (2) the stronger the increase is, the larger is the magnitude of dR_Q/dT, which is always negative; (3) large increases of R_Q correlate with marked AB-attraction. Of the three major ingredients to R_Q -

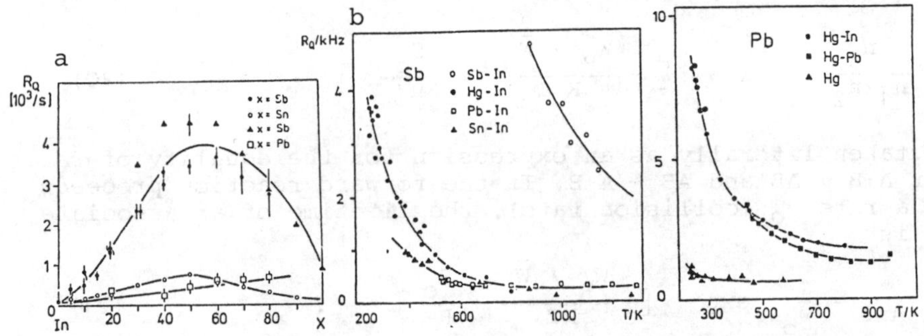

Figure 6: Quadrupolar nuclear relaxation rate R_Q of Sb in various In-X alloys vs. concentration, fig. 6a: Triangles are NMR data /22/, normalized to the TDPAD data /38/ for ^{117m}Sb (τ=340 μs) in In-Sb, Sn, Pb. (b) shows for c=0,5 dependence on temperature and alloy partner, for ^{117m}Sb and ^{206m}Pb (τ=120μs)

efg function v, structure g, and correlation time τ - the latter appears to be responsible for the cases of strong increases /21/ in liquid metallic alloys. To the extent that the other factors (v,g) change in an alloy, they seem to enhance the effect of τ.

Model of A-B associates. The increase in efg correlation time τ is imagined to occur by prolonged clustering of AB-neighbours. For R_Q, the relevant association is essentially that of individual atoms A-B, not that of species. The association is modelled in thermodynamic treatments /33/ by postulating, that A and B neighbours can be in essentially two energetic states ("associated" AB or "free" A and B) which are connected by a kind of chemical reaction

$$A + B \leftrightarrow AB \quad . \tag{19}$$

We discuss here the simple case of a 1 : 1 stoichiometry. It should be stressed that there are many different formulations of the basic model of associates, see e.g. /34/, /35/. From a discussion of thermodynamics /33/ one may take the appropriate parameters for the equilibrium (19) viz. bonding energy W_R for AB, entropy difference S_o between pure A, B and hypothetical pure AB in the liquid state, interaction energy of free A and B, of A with AB, and of B with AB; the latter three appear below in an average $\bar{W}(c)$. The parameters can be fitted to excess enthalpy (and entropy) of mixing if these exist, or have to be estimated. Connection with R_Q is supposedly as follows /36/: The law of mass action for the number of associates AB, free A and free B (n_3, n_1, n_2 respectively)

$$\frac{n_3}{n_1 \cdot n_2} = \exp \left[- \frac{W_R + \bar{W}}{R\,T} + \frac{S_o}{R} \right] \qquad (20)$$

is taken literally as an expression for the equality of rates for A+B → AB and AB → A+B. If the forward reaction proceeds at a rate ν_o (collision rate), the lifetime of an associate AB is

$$\tau_{Ass} = \frac{1}{\nu_o} \exp \left[- \frac{W_R + \bar{W}}{RT} + \frac{S_o}{R} \right] \quad . \qquad (21)$$

τ_{Ass} (rather $\nu_o \cdot \tau_{Ass}$) and n_3 are then known as functions of concentration and temperature. The contributions to R_Q when the probe atom is free was assumed to be proportional to $n_{free}=n_1$ (if A is the probe) and when it is associated, to $n_{Ass}=n_3$, by Elwenspoek et al. Then $R_Q = \text{const}\ [\langle V^2\rangle_{Ass}\ \tau_{Ass}\ n_{Ass} + \langle V^2\rangle_{free}\ \tau_{free}\ (1-n_{Ass})]$. Since the second term is small compared to the first, in alloys with marked attractive AB interaction, the relative changes of R_Q are

$$\frac{R_Q(c,T)}{R_Q(1,T)} \approx \frac{\langle V^2\rangle_{Ass} \cdot \tau_{Ass}(c,T)\ n_{Ass}(c,T)}{\langle V^2\rangle_{free} \cdot \tau_{free} \cdot 1} \quad . \qquad (22)$$

It was further assumed that $\langle V^2\rangle$ does not change between the two states (to the extent that electronic properties don't change). One can then fit (22) to $R_Q(c,T)$. Two examples are given in fig. 7 and the agreement is satisfactory in view of the early stage of refinement of the analysis. - One may further use τ_{free} from recent calculations for Rb and Ga /25/ /39/ as $\tau_{free} \approx 3 \ldots 5 \cdot 10^{-13}$sec to obtain estimates of τ_{Ass} from the experimental increase in R_Q, which lead to values of the order of 10^{-11}sec for both cases in fig. 7 at c=0.5, T near T_M. To the extent that both, electric field gradient and AB attraction, are essentially short range interactions in these highly metallic systems, one may consider the time τ_{Ass} (10^{-11}sec) as an experimentally derived estimate of the correlation time of the AB-bonding.

This analysis is certainly only a first step. The generalization of the interpretative concept, going from pure metals to alloys with $w/k_BT \ll 1$ to strongly AB-bonding alloys, can be based on a discussion of the functions W_{ij}/32/, /39/. - For very stable associates, rotation may become important /36/.- The weak points seem to be at present the general reconciliation of chemical short range order as derived from $S_{cc}(q)$ with models of association, and the efg function.

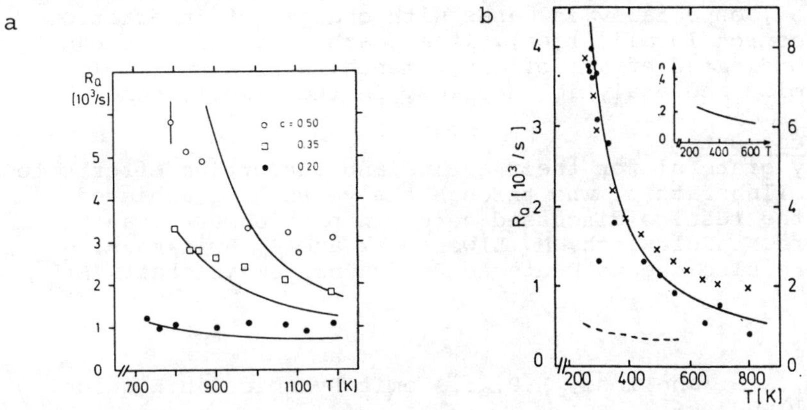

Figure 7: (a) Fit of $R_Q(c,T)$ of ^{117m}Sb in liquid $In_{1-c}Sb_c$ using thermodynamic model described in the text. Normalization of set of curves is at c=0.5, T=1000 K. From /36/.
(b) $R_Q(0.5, T)$ of ^{206m}Pb -crosses, rhs - and ^{117m}Sb - circles, lhs - in liquid $In_{0.5}Hg_{0.5}$ /37/. Curve is thermodynamic model. Dashed curve is for ^{206}Pb in pure liquid Hg. Inset shows $n_{Ass}(T)$. From thermodynamics of the In-Hg, In-Pb and Hg-Pb systems it may be assumed that the Pb probe atom behaves in the In-Hg alloy rather similar to the Hg atoms.

Conclusion

We conclude by summarizing results from experiments discussed here, for dynamics of A-B relative motion in liquid metallic alloys, by a simplified schematic, see below. Columns of the table are placed below the corresponding r-regions of the ordering potential. Only order-of-magnitude times are quoted.

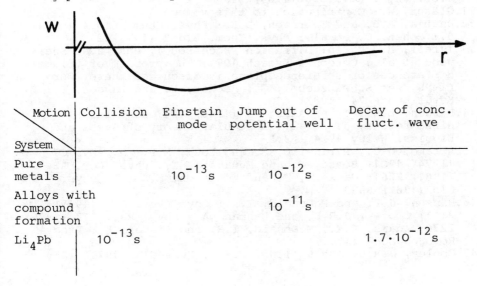

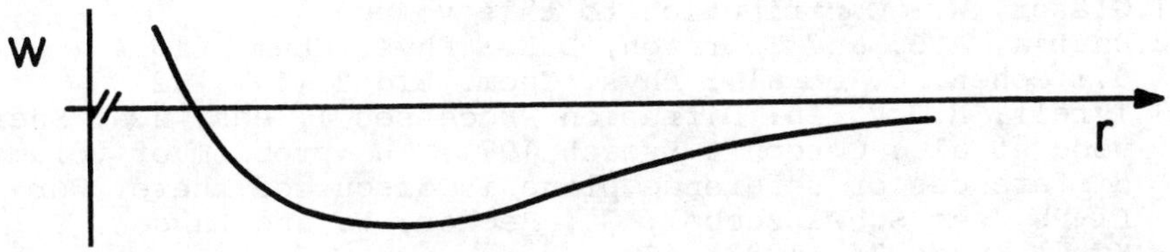

Motion / System	Collision	Einstein mode	Jump out of potential well	Decay of conc. fluct. wave
Pure metals		10^{-13} s	10^{-12} s	
Alloys with compound formation			10^{-11} s	
Li_4Pb	10^{-13} s			$1.7 \cdot 10^{-12}$ s

At present, atomic dynamics in AB-alloys is far from being understood, especially in cases with changes of interaction (e.g. electron localization). It appears as an ideal subject for a coordinated effort of experimental approaches, molecular dynamics and analytic theories, in the time to come.

Acknowledgement
I am very grateful for their active and supporting efforts to all my collaborators, who through the years have achieved many of the results discussed here. In particular I want to thank here M. Soltwisch, M. Elwenspoek and R. Brinkmann. - Thanks are also due to Deutsche Forschungsgemeinschaft (Sfb 161).

REFERENCES

1. March, N.H. and Tosi, M.P. : Atomic Dynamics in Liquids, Macmillan 1976
2. Quitmann, D.: J. Noncryst. Sol. 61+62 (1984) 249
3. Chieux, P. and Ruppersberg, H.: J. de Physique 41 (1980) C8-145; Boos, A. et al.: Z. Naturforschung 32a (1977) 1222
4. Jaccucci, G., McDonald, I.R. and Rahman, A.: Phys. Rev. A13 (1976) 1581;
Dixon, M. and Sangster, M.J.L.: J. Phys. C9 (1976) 909;
Dixon, M.: Phil. Mag. B47 (1983) 531;
Haan, S.W., Mountain, R.D., Hsu, C.S. and Rahman, A.: Phys. Rev. A22 (1980) 767;
Gonzales, J.M. & Torra, V.: J. Phys. F13 (1983) 281
5. Goetze, W.: Contribution to this volume
6. Tosi, M.: Contribution to this volume
7. Bhatia, A.B. and Thornton, D.E.: Phys. Rev. B2 (1970) 3004
8. Hafner, J.: Contribution to this Volume
9. Saboungi, M.L, Marr, J., Blander, M.: J. Chem Phys. 68 (1978) 1375
10. Soltwisch, M., Quitmann, D., Ruppersberg, H., and Suck, J.B.: Phys. Rev. B28 (1983) 5583
11. Gläser, W.: Contribution to this volume
12. Bhatia, A.B. and Thornton, D.E.: Phys. Chem. Liq 4 (1974) 97; Cohen, C. et al.: Phys. Chem. Liq 2 (1971) 213
13. Tyrell, H.J.V. in: Diffusion Processes I, eds, J.N. Sherwood et al., Gordon & Breach 1971. The problem of volume differences or reference plane is discussed there. For Li-Pb, see Schwitzgebel, G., Becker, W. and Langen, G.: Z. Metallkunde 74 (1983) 430
14. Darken, L.S.: Trans AIME 175 (1948) 184; see also Hirschfelder, J. et al.: Molecular Theory of Gases and Liquids, Wiley 1954, §9.2
15. Abramo, M.C., Parinello, M. and Tosi, M.P.: J. Phys C7 (1974) 4201; Bosse, J. and Munakata, T.: Phys. Rev. A24 (1981) 2261; Gaskell, T. and Woolfson, M.S.: J. Phys. C15 (1982) 6339
16. Hansen, J.P. and McDonald, I.R.: Phys. Rev. A11 (1975) 2111; Copley, J.R.D. and Rahman, A.: Phys. Rev. A13 (1976) 2276; Adams, E.M., McDonald, I.R. and Singer, K.: Proc. Roy Soc A357 (1977) 37
17. Copley, J.R.D. and Dolling, D.: J. Phys C11 (1978) 1259

18. Copestake, A.P., Evans, R., Ruppersberg, H. and Schirmacher, W.: J. Phys. F13 (1983) 1993
19. Jaccucci, G., Ronchetti, M. and Schirmacher, W.: in Condensed Matter Research Using Neutrons, eds. S.W. Lovesey and W. Scherm, Plenum 1984, p. 139 and Contribution to this volume
20. McGreevy, R.L., Mitchell, E.WJ., Margaca, F.M.A.: J Phys. C17 (1984) 775; 4725
21. Warren, W.: Phys. Rev. B3 (1971) 3708
22. Claridge, E., Moore, D.S., Seymour, E.F.W. and Sholl, C. A.: J. Phys. F2(1972) 1162
23. Dupree, R., Kirby, D.J. and Freyland, W.: Phil. Mag. B46 (1982) 595
24. Weihreter, E., von Hartrott, M., Höhne, J., Quitmann, D., Rossbach, J. and Willeke, F.: Phys. Lett. 67A (1978) 394; Liquid and Amorphous Metals, eds. E. Lüscher and H. Coufal Sijthoff 1980 p. 623; v. Hartrott, M., Höhne, J., Quitmann, D., Rossbach, J., Weihreter, E., and Willeke, F. Phys. Rev. B19 (1979) 3449
25. Bosse, J., Quitmann, D., and Wetzel, Ch.: Phys. Rev. A28 (1983) 2459
26. Abragam, A.: Nuclear Magnetism, Clarendon 1961
27. Quitmann, D.: in "Site Characterization etc", eds. A. Perez and R. Coussement, Plenum 1980, p. 139
28. Brinkmann, R., Elwenspoek, M., Maxim, P., and Quitmann, D.: Hyperfine Int. 15+16 (1983) 581
29. Haan, S.W.: Phys. Rev. A20 (1979) 2516; Posch, H.A, Balucani, V and Vallauri, R.: Physica 123A (1984) 516
30. Ackermann, H., et al.: Topics in Current Physics 31 (1983) 291
31. von Hartrott, M., Hadijuana, J., Nishiyama, K., Quitmann, D., Riegel, D., and Schweickert, H.: Z. Physik A278 (1976) 303
32. Maxim, P., Brinkmann, R., Paulick, C.A., Elwenspoek, M., v. Hartrott, M., Kiehl, M. and Quitmann, D.: to be published
33. Sommer, F.: Z. Metallkunde 73 (1982) 72, 77; Guggenheim, E.A.: Mixtures, Clarendon 1952
34. Bhatia, A.B and Singh, R.N: Phys. Chem. Liq 11 (1982) 285, 343
35. van der Lugt, W.: Contribution to this volume
36. Elwenspoek, M., Brinkmann, R., Maxim, P., Paulick, C.A. and Quitmann, D.: Ber. Bunsenges. 87 (1983) 823; Phys. Lett. 96A (1983) 435
37. Elwenspoek, M., Brinkmann, R., v. Hartrott, M., Kiehl, M. Maxim, P., Paulick, C.A., Willeke, F., and Quitmann, D.: J. Noncryst. Sol 61+62 (1984) 153
38. v. Hartrott, M., Quitmann, D., Roßbach, J., Weihreter, E., and Willeke, F.: Hyperfine Int. 10 (1981), 1027
39. Brinkmann, R., Elwenspoek, M., Maxim, P., Paulick, C.A. and Quitmann, D.: submitted to J. de Physique

TRANSITION FROM GAUSSIAN TO DISPERSIVE ATOMIC TRANSPORT IN AMORPHOUS MATERIALS

W. SCHIRMACHER

Institut Max von Laue/Paul Langevin, 156X,
F 38042 Grenoble, France*

1. INTRODUCTION

A.c. ionic conductivity data in disordered solids show a frequency dependence according to $\sigma'(\omega) \propto \omega^{1-\alpha}$ with $0<\alpha\lesssim 0.2$ over many decades of frequency (1-3).

Since the velocity fluctuation spectrum (4) of charged particles is proportional to $\sigma'(\omega)$ this implies that the mean square distance covered by a particle does not increase linearly with time t but rather as t^{α}. Such anomalous diffusion which is associated with non-exponential relaxation (1) has been termed "non-Gaussian" or "dispersive" transport in connection with a random walk description of transient carrier motion in amorphous semiconductors (5-7).At low frequencies on the other hand one observes a crossover from a frequency dependent $\sigma'(\omega)$ to a frequency independent one leading to the d.c. conductivity $\sigma'(0)$ which is proportional to the diffusion coefficient D (diffusivity) according to the Nernst-Einstein relation (4).

In glasses and other disordered solids D varies strongly with temperature according to an Arrhenius law $D \propto \exp\{-E_A/k_BT\}$, where E_A is of the order of an electron volt (1,8). Therefore the transition from Gaussian (frequency independent σ') to dispersive (frequency dependent σ') transport occurs also as a function of temperature if σ' is studied in a fixed frequency range or time scale (6,7). The details of this transition, which can be studied quite well experimentally, will be described below in terms of a barrier hopping model developed recently (9) for diffusion in disordered solids. It will be shown that the transition is accompanied by a peak in the polarizability and involves an interesting scaling property.

2. GENERAL FORMALISM

Performing an investigation of single-particle motion one is usually interested in the incoherent density density correlation function ("single particle propagator") $G(\underline{r},t)$ which gives the conditional probability for finding a particle at $\underline{r}$ at time t if it started initially at the origin. In terms of the time dependent radius vector of the particle $\underline{r}_o(t)$ this quantity is defined as follows (4)

$$G(\underline{r},t) = \frac{1}{\rho} < \rho_o(\underline{r},t)\,\rho_o(\underline{0},0)>, \tag{1}$$

where $\rho_o(\underline{r},t) = \delta[\underline{r} - \underline{r}_o(t)]$ and <...> denotes an ensemble and configurational average. ρ is the number of particles per volume.

Permanent Address: Phys. Dept. E13, Technische Universität München
D 8046 Garching / FRG

Its Fourier and Laplace transform

$$G(\underline{k},\omega) = \int_{-\infty}^{+\infty} d^3\underline{r} \int_0^{\infty} dt \exp\{i\, \underline{k}.\underline{r}-\omega t\}\, G(\underline{r},t) \tag{2}$$

is related to the generalized frequency dependent diffusion coefficient $D(\omega)$ by the Green-Kubo identity

$$\lim_{\underline{k}\to 0} G(k,\omega) = [i\omega + D(\omega)k^2]^{-1} . \tag{3}$$

The real part of $D(\omega)$ gives the velocity fluctuation spectrum of the particles (4). If the particles are charged with charge q, $D(\omega)$ is related to the complex dynamic conductivity $\sigma(\omega)$ by the Nernst-Einstein relation

$$\sigma(\omega) = (\rho^* q^2/k_B T) D(\omega) , \tag{4}$$

where ρ^* is the carrier density times a correlation factor of the order of 1.

Studying stochastic random walk models it is useful to parametrize $D(\omega)$ as (10-13)

$$D(\omega) = \frac{1}{6} R^2 \nu(\omega) , \tag{5}$$

where $\nu(\omega)$ is a generalized complex jump frequency and R^2 a mean square jump distance.

Another quantity of interest is the time dependent mean square distance covered by the particle $\langle r^2(t)\rangle = \langle(\underline{r}(t) - \underline{r}(o))^2\rangle$. It is related to the other quantities by

$$\langle r^2(t)\rangle = -\frac{1}{2}(\nabla_{\underline{k}})^2_x\, G(\underline{k},t) \Big|_{\underline{k}\to\underline{o}} = L^{-1}\{\frac{D(\omega)}{(i\omega)^2}\} , \tag{6}$$

where L^{-1} denotes an inverse Laplace transform.

It is now useful to make a distinction between three qualitative different types of behaviour of $\langle r^2(t)\rangle$ at long times (which corresponds to the low frequency behaviour of $D(\omega)$) (14):

a. Gaussian Transport

If we have $\langle r^2(t)\rangle = Dt$ with diffusivity D the velocity fluctuation spectrum does not depend on frequency (white noise spectrum).

Since the probability cloud $G(\underline{r},t)$ of a particle which was initially at $\underline{r} = \underline{o}$ in this case has the form (4)

$$G(r,t) = [4\pi Dt]^{-3/2} \exp\{-r^2/4Dt\} , \tag{7}$$

which broadens in a Gaussian fashion this type of transport has been called Gaussian.

The function $G(k,t)$ has the form

$$G(\underline{k},t) = \exp\{-Dk^2 t\} . \tag{8}$$

We see that Gaussian transport corresponds to exponential density relaxation.

b. Localization

If $<r^2(t)>$ has an upper bound, i.e. if for large times $<r^2(t)>$ = const. $= \gamma$ the particle is localized in a volume of diameter $\gamma^{1/2}$.

In this case $D(\omega) = i\omega\gamma$, and the density fluctuations are manifestly non-ergodic: $G(\underline{k},t)$ does not decay at all (14):

$$G(\underline{k},t) = [1 + \gamma k^2]^{-1} . \tag{9}$$

A transition from ergodic to non-ergodic behaviour has been shown to be the underlying feature of the Anderson transition (14) as well as the glass transition (14-16).

c. Dispersive Transport

If $<r^2(t)>$ increases sublinearly with t, e.g. $<r^2(t)> \propto t^{\alpha}$ $(0<\alpha<1)$ we have $D(\omega)=D_1(i\omega)^{1-\alpha}$. Clearly $G(\underline{k},t)$ is neither constant nor decays exponentially (14).

In the special case $\alpha = \frac{1}{2}$ $G(k,t)$ can be obtained analytically (14,17)

$$G(k,t) = \exp\{\omega_k t\}\mathrm{erfc}\{[\omega_k t]^{1/2}\} \tag{10}$$

with $\omega_k = D_1^2 k^4$. This function decays asymptotically as $t^{-1/2}$, which is a much slower decay than the exponential one. In other words, if the particles perform dispersive transport they disappear much more slowly from a given region than they would in the case of Gaussian diffusion. The experimentally observed transition from dispersive to Gaussian transport has therefore some similarities to the glass transition.

3. HOPPING MODEL

Let us consider a single particle which can hop among sites "i" and "j" separated by distances r_{ij} and energy barriers E_{ij} with transition frequencies

$$W_{ij} = W(r_{ij}, E_{ij}) = \nu_o(r_{ij})\exp\{-E_{ij}/k_B T\} . \tag{11}$$

The change in the occupation n_i of the sites is then given by the following set of Markovian master equations:

$$\frac{d}{dt} n_i = \sum_j W_{ij} (n_j - n_i) . \tag{12}$$

r_{ij} and E_{ij} are assumed to be positive definite random variables with distributions p(r) and P(E). Eqs. (14) are mathematically equivalent to Kirchhoff's equations of a random impedance network (19) as well as the equation of motion of electrons in the tight-binding approximation with fluctuating site energies and overlap integrals (20).

The formal solution of eqs. (14) is given by

$$G(\underline{k},\omega) = <\sum_{ij} [i\omega\underline{\underline{1}}-\underline{\underline{K}}]^{-1}_{ij} \exp\{i\underline{k}\underline{r}_{ij}\}> , \tag{13}$$

where $\underline{\underline{1}}$ is the unit matrix and $\underline{\underline{K}}$ is a matrix with diagonal elements $K_{ii} = -\sum_j W_{ij}$ and off-diagonal elements $K_{ij} = W_{ij}$.

4. THE EFFECTIVE MEDIUM APPROXIMATION

A widely used mean field theory of disorder for obtaining the averaged propagator of eq. (13) is the effective medium (EMA) or coherent potential (CPA) approximation. The idea of the EMA is to replace the actual disordered system by a uniform one with equal but frequency dependent hopping rates.

The uniform but frequency dependent hopping rate of the effective medium is then calculated from a consistency requirement. In the random inpedance network the consistency requirement is that there should be on the average no extra voltages induced by replacing the true system by the effective medium. In the tight binding treatment of electrons in disordered solids one requires that the averaged t-matrix $\langle t \rangle$ corresponding to the difference between the true and the effective system be zero. These two procedures are equivalent and are a generalization of the EMA used by Bruggeman (21) to calculate the dielectric constant of inhomogeneous media.

The EMA has been used to study nearest-neighbour hopping on disordered lattices (22-24) as well as hopping in topologically and energetically disordered systems (11-13).

The former EMA versions have been obtained by the standard $\langle t \rangle = 0$ procedure, whereas the latter are derived using diagrammatical techniques (11) and the renormalized perturbation expansion (13), resp. These derivations will not be repeated here.

Instead it will be shown in the following paragraph how the EMA for a topologically disordered system (11-13) can be derived from the lattice EMA (22-24).

5. LATTICE EMA AND TOPOLOGICALLY DISORDERED SYSTEM

Let us consider a particle hopping on a d-dimensional hypercubic lattice with random transition rates W_{ij}. "i" and "j" denote now lattice sites and the transition is allowed to nearest neighbours only. The corresponding effective medium consists of the same lattice but with a frequency dependent uniform transition rate W_m. The EMA equation is (22-24)

$$\langle t \rangle = \int P(W)\,dW \, \frac{W_m - W}{\frac{1}{2}Z_m - (W_m - W)(1-\varepsilon G_m(\varepsilon))} = 0 \, . \qquad (14)$$

Here G_m is the single-site Green's function of the medium, Z_m the coordination number and $\varepsilon = i\omega/W_m$. $P(W)$ is the distribution of the random variables W_{ij} subject to the normalization condition $\int P(W)\,dW = 1$. Eq. (14) can be manipublated to take the form

$$W_m = \int P(W)\,dW \, \frac{W}{\frac{2}{Z_m}\left(\frac{W}{W_m} - 1\right)(1-\varepsilon G_m(\varepsilon))} \, . \qquad (15)$$

Let us now consider a disordered system with no topological order and no nearest neighbour hopping restriction. This system ("real system") is now mapped to the d-dimensional hypercubic lattice with nearest neighbour hopping ("fictitious system"). Since in the real system each site can in principle be reached from a given one the number of sites in the real system N_S is set equal to the coordination number Z_m of the fictitious system. The site distance r_{ij} of the real system has no topological significance in the fictitious system but is simply a random number. As above the hop rates W_{ij} are assumed to depend on r_{ij} and another independent random variable E_{ij} only. In this case we have $P(W)dW=(1/V)p(|\underline{r}|)P(E)d^3\underline{r}dE$ where V is the volume of the real system. Defining $\nu(\omega)=(Z_m/2)\bar{W}_m(\omega)$ eq. (15) takes the form

$$\nu(\omega) = \frac{N_S}{2V}\int p(r)d^3\vec{r}\int P(E)dE \frac{W}{1+(\frac{W}{\nu}-\frac{2}{N_S})(1+\varepsilon G)} . \tag{16}$$

The Green's function of the medium is given by (23)

$$G_m(\varepsilon) = \int_0^\infty \exp\{-(\tfrac{1}{2}Z_m+\varepsilon)x\}[I_o(x)]^d dx , \tag{17}$$

where I_o is the zero order modified Bessel function.

In the limit $Z_m = N_S \to \infty$ with $n_s = N_S/V$ kept constant one has $G_m(\varepsilon) \to [\frac{Z_m}{2}+\varepsilon]^{-1}$, and one obtains

$$\nu(\omega) = a_p n_s \int_{-\infty}^{+\infty} d^3\underline{r}\, p(r) \int_0^\infty P(E) \frac{1}{\frac{1}{i\omega+\nu(\omega)}+\frac{1}{W(r,E)}} \tag{18}$$

(with $a_p = 1/2$), which is identical to the EMA equation of Gochanour et al. (11). The corresponding equation of Movaghar et al. (12,13) contains a density renormalization factor $a_p = \exp\{-1\}$ which comes from a correction for double counting. Eq. (18) has also been derived from the random impedance network analogy by Summerfield and Butcher (25). They take the prefactor to be the inverse of the threedimensional percolation number (19). Calculations of various transport properties of electrons in disordered semiconductors based on eq. (18) and generalizations of it have yielded quite accurate results for $D(\omega)$ compared with the averaged numerical solution (simulation) of eqs. (12) for electronic hopping (12), as well as with experiment.

Recently it has been shown (9) that ionic diffusion in glasses can be discribed quite successfully in terms of the EMA using a fixed-range hopping model with a constant distribution of energy barriers. This includes the crossover from the Arrhenius behaviour of the conductivity in the dc limit to the $\omega^{1-\alpha}$ law at high frequencies. These aspects will be discussed in the following sections.

6. EMA RESULTS FOR FIXED RANGE HOPPING

For fixed range hopping let us put

$$p(r) = (Z_S/4\pi R^2 n_S)\delta(r-R) , \tag{19}$$

where Z_S is the number of nearest neighbour sites. Eq. (18) now takes the form (9)

$$\nu(\omega) = Z_s a_p \int_0^\infty dE\, P(E) \frac{1}{G(\omega)+\frac{1}{W(E)}} . \tag{20}$$

Here $G(\omega) = [i\omega+\nu(\omega)]^{-1}$ and $W(E) = W(R,E)$.

At low temperatures one obtains from (20) in the dc limit:

$$1 = Z_s\, a_p \int_0^{\mu} dE\, P(E) \qquad \mu = -k_B T\, \ell n(\nu(o)/\nu_o) \tag{21}$$

showing that the EMA together with a fixed range barrier hopping model always predicts an Arhenius law $\nu(o) = \nu_o \exp\{-\mu/k_B T\}$ independent of the shape of P(E) for the temperature dependence of D(o) and $\sigma'(o)$ at low temperatures. The experimental data in fact show such a behaviour as stated in the beginning.

7. THE GAUSSIAN-DISPERSIVE TRANSITION IN THE CONSTANT BARRIER DENSITY MODEL

The salient features of $\nu(\omega)$ and especially the transition from Gaussian to dispersive transport can be studied quite successfully (9) already with the simplest possible model for P(E), namely

$$P(E) = \text{const.} = \bar{P} k_B / Z_s a_p \,. \tag{22}$$

This is quite analogous to the assumption of a constant density of energy separations in the two-level system (TLS) model for the low temperature tunneling motion of glasses (26).

Inserting (22) into (20) we obtain

$$\nu(\omega) = (\bar{P}T/G(\omega))\,\ell n[1 + \nu_o G(\omega)] \,. \tag{23}$$

In the dc limit this implies

$$\nu(o) = [\exp\{1/\bar{P}T\} - 1]^{-1} \,. \tag{24}$$

For PT << 1 this becomes an Arrhenius law with activation energy $E_A = 1/\bar{P}$ in agreement with the general result of eq. (21).

In Fig. 1 the real part of $\nu(\omega)$, $\nu'(\omega)$ is plotted against frequency for five different values of $\bar{P}T$. It is seen that at low temperatures ($\bar{P}T$ << 1) a large dispersive region with $\nu'(\omega) \propto \omega^{1-\alpha}$ (α << 1) exists. For higher temperatures this region becomes successively smaller until it disappears altogether at $\bar{P}T^{-1}$.

It is also interesting to look at the imaginary part of ν. $D''(\omega)/\omega$ is the frequency dependent polarizability of the system (14). It is found that $\nu''(\omega)$ as calculated from eqn. (23) exhibits a pronounced change in its temperature and frequency dependence at the transition from Gaussian to dispersive transport. This leads to a cusp-like peak in its temperature dependence as can be seen from Fig. 2.

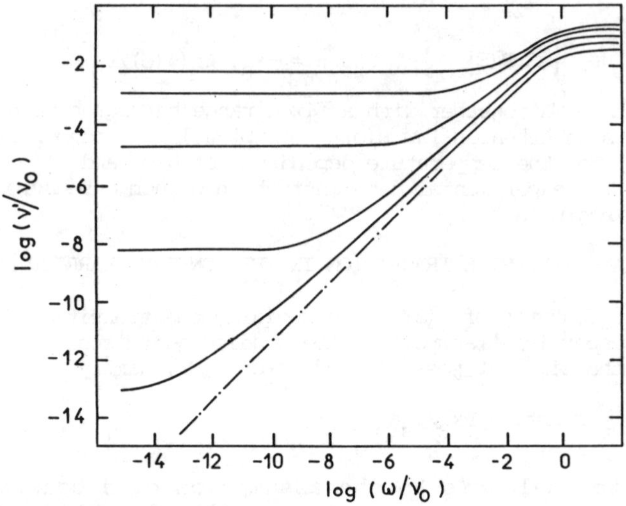

Fig. 1: Normalized a.c. conductivity $\nu'(\omega)/\nu_o$ against ω for 5 different temperatures $\bar{P}T$ = 0.1, 0.0768, 0.0456, 0.0263 and 0.0165 as given by eq. (1). The dash-dotted line indicates the slope 1.

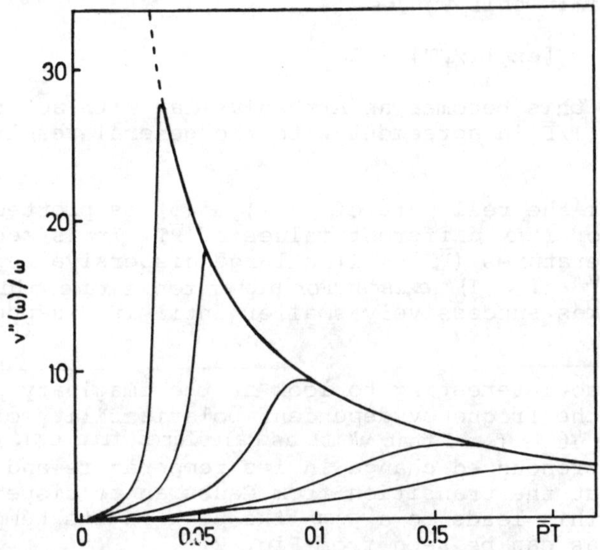

Fig. 2: Normalized polarizability $\nu''(\omega)/\omega$ against $\bar{P}T$ for 5 different frequencies $\omega/\nu_o = 10^{-15}$, 10^{-10}, 5.10^{-7}, 10^{-4}, 8.10^{-4}. The dashed line corresponds to $\nu''/\omega = (1/\bar{P}T)-1$.

At low temperatures $\bar{P}T \ll 1$ ν' and ν'' have interesting scaling properties as far as the frequency and temeprature dependence is concerned.

Let us define the following dimensionless quantities:

$$y = \omega/\nu(o)\bar{P}T = (\omega/\nu_o\bar{P}T)\ \exp\{1/\bar{P}T\},$$
$$z = \nu(o)G(\omega) = \nu_o\ \exp\{-1/\bar{P}T\}\ G(\omega). \tag{25}$$

Then eq. (22) in the limit PT << 1 takes the form

$$iyz = -\ell nz, \tag{26}$$

which means that z is a universal function of y. From this it follows that $\nu'/\nu(o)$ is a universal function of y as well as $\bar{P}T[(\nu''/\omega) + 1]$. Such universal behaviour has in fact been found in a lot of conductivity and permittivity data in disordered systems (1-3).

The form of the universal function $\nu'/\nu(o)$ is identical to that of the low frequency part of the two bottom curves of Fig. 1. A plot of the universal function $F \equiv \bar{P}T[(\nu''/\omega)+1]$ is given in Fig. 3. The peak in the temperature dependence of ν'' corresponds to the edge of the plateau and occurs at $y \approx 1$. This means that in the frequency-temperature plane the curve y=1 (i.e. $\omega = (\nu_o/\bar{P}T)\ \exp\{-1/\bar{P}T\}$) divides the region between dispersive (y > 1) and Gaussian (y < 1) transport.

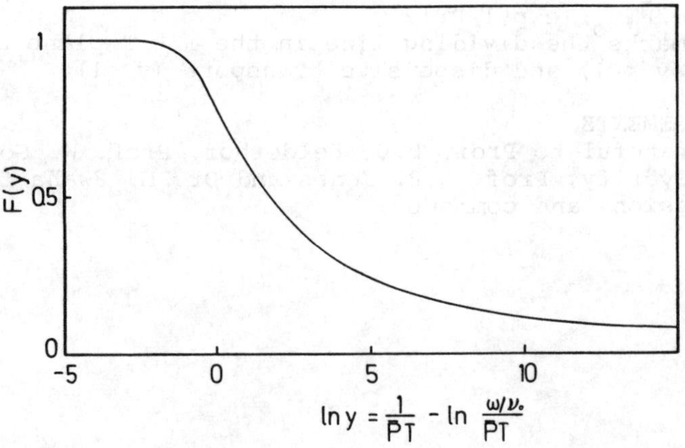

Fig. 3: Universal plot of $F = \bar{P}T[(\nu''/\omega) + 1]$ against y. The plateau corresponds to the region of Gaussian transport.

8. SUMMARY

In this lecture arguments have been put forward showing that the single particle motion in glasses and other diordered solids exhibits a transition from Gaussian to dispersive transport both as a function of frequency and temperature. The transition consists in a change of the velocity fluctuation spectrum from a white noise to a strongly frequency dependent one. In the regime of dispersive transport the mean square distance covered by a particle increases sublinearly with time which means that the particles are almost localized. Therefore the transition from normal diffusion (Gaussian transport) to anomalous diffusion (dispersive transport) has some features in common with a glass transition.

The main difference to the glass transition (14-16) lies in the fact that only the single particle motion is involved. The consistence of the "medium" does not change at all. Correspondingly it is possible to describe the Gaussian-dispersive transition by a model in which non-interacting partiles perform a random barrier hopping process in a static medium characterized by a uniform energy barrier distribution. The model has been solved using the effective medium approximation. In agreement with most of the experimental data the model predicts a $D'(\omega)$ which in the d.c. limit exhibits an Arrhenius-type temperature variation with apparent activation energy $\bar{P}^{-1}$, whereas at high frequencies/low temperatures $D'(\omega)$ becomes weakly temperature dependent but strongly frequency dependent according to an $\omega^{1-\alpha}$ law with very small α.

The transition from Gaussian to dispersive transport studies as a function of temperature is accompanied by a cusp-like peak in the polarizability $D''(\omega)/\omega$.

At low temperatures $\bar{P}T \ll 1$ the normalized decay function $Z(\omega) = \nu(o)\underline{G}(\omega)$ becomes a universal function of the parameter $y = (\omega/\bar{P}T\nu_o)\exp\{1/\bar{P}T\}$.

$y = 1$ marks the dividing line in the $\omega - T$ plane between Gaussian $(y < 1)$ and dispersive transport $(y > 1)$.

ACKNOWLEDGEMENTS

I am grateful to Prof. B.U. Felderhof, Prof. W. Götze, Prof. B. Györffy, Prof. R.B. Jones and Dr. M. Uwaha for helpful discussions and comments.

REFERENCES

1. Wong J. and Angell CA: Glass, Structure by Spectroscopy, Marcel Dekker, N.Y. (1976)
2. Jonscher AK: Nature 267, 673 (1977)
3. Ngai KL: Comm. Sol. State Phys. 9, 127 (1979)
4. Hansen JP and McDonald IR: Theory of Simple Liquids, Academic Press, N.Y. and London 1976
5. Scher H and Montroll EW, Phys. Rev. B12, 2455 (1975)
6. Pfister G. and Scher H, Adv. Phys. 27, 747 (1978)
7. Schirmacher W, Sol. St. Comm. 39, 893 (1981)
8. Tuller HL, Button DP and Uhlmann DR: J. Noncryst. Sol. 40, 93 (1980)
9. Schirmacher W, Phys. Rev. B, to appear
10. Scher H and Lax M: Phys. Rev. B7, 4491 and 4502 (1973)
11. Gochanour CR, Anderson HC and Fayer DM, J. Chem. Phys. 70, 4254 (1979)
12. Movaghar B, Pohlmann B and Sauer GW: Phys. Stat. Sol. (b) 97, 553 (1980)
13. Movaghar B and Schirmacher W: J. Phys. C14, 859 (1980); see also Movaghar B, Grünewald M, Pohlmann B, Würz D and Schirmacher W, J. Statist. Phys. 30, 315 (1983)
14. Götze W, Sol. Comm. 27, 1393 (1978); Phil. Mag. B43, 219 (1981); see also the review articles in "Recent Developments in Condensed Matter Physics", Ed. T.T. Devreese, Plenum Press, New York, Vol. I, 1984, p. 133 and this volume
15. Leutheusser E: Phys. Rev. A29, 2765 (1984), see also Leutheusser E and Yip S, preprint, 1984
16. Bengtzelius U, Götze W and Sjölander A: J. Phys. C17, 5915 (1984)
17. Erdelyi A, Magnus W, Oberhettinger F and Tricomi FG: Tables of Integral Transforms, Vol. 1, Mc Graw-Hill, New York 1954
18. Abramowitz M and Stegun A: Handbook of Mathematical Functions, NBS Appl. Math. Ser. 55, Washington, D.C.
19. Kirkpatrick S, Rev. Mod. Phys. 45, 574 (1973)
20. Elliott RJ, Krumhansl JA and Leath PL: Rev. Mod. Phys. 46, 465 (1974)
21. Bruggeman DAG, Ann. Phys. (Leipz.) 24, 636 (1935)
22. Odagaki T and Lax M: Phys. Rev. B24, 5284 (1981)
23. Webman I: Phys. Rev. Lett. 47, 1496 (1981)
24. Summerfield S: Sol. St. Comm. 39, 401 (1981)
25. Summerfield S and Butcher PN, J. Phys. C15, 7003 (1982)
26. Anderson PW, Halperin BI and Varma CM, Phil. Mag. 25, 1 (1971); Phillips WA: J. Low Temp. Phys. 7, 351 (1972); for more recent reviews see e.g. P.W. Anderson in 'ILL Condensed Matter", Eds. Balian R, Maynard R, Toulouse G, North Holland, Amsterdam 1979 or Black JL in: Glassy Metals I, Eds. H.-J. Güntherodt, Beck H, Topics in Applied Physics, Vol. 46, Springer, Heidelberg 1981, p. 167

INTERACTION EFFECTS IN LIQUIDS WITH LOW ELECTRON DENSITIES

W. W. WARREN, JR., AT&T Bell Laboratories

1. INTRODUCTION

It is a familiar fact that electrons in most elemental liquid metals and many liquid alloys behave as if they are nearly free. This is a consequence of the high conduction electron densities in these systems and resultant screening of the electron-ion interaction. The effects of electron-electron interactions are also relatively small in ordinary metals except for corrections to a few physical properties such as the magnetic susceptibility. The situation is quite different, however, if we consider systems with electron densities reduced substantially below those of ordinary metals. These conditions may be found in a number of liquids such as expanded elemental liquid metals, solutions of metals in non-metals, or strongly interacting liquid alloys such as Cs-Au. Structural disorder becomes much more important as the full unscreened ionic potentials are exposed and may lead to electron localization by the Anderson transition or other means such as self-trapping. Alternatively, electron-electron interactions drive the Mott transition and cause localization at low densities in systems with one electron per atom. In some systems, of course, both effects may be at work.

In these lectures I will discuss two complementary classes of systems in which strong electron-electron or electron-ion interactions appear at low electron densities. The first are the expanded liquid alkali metals in which electron correlation effects have a profound effect on the magnetic properties on the *metallic* side of the metal-nonmetal transition. The second group are molten alkali halides containing low densities of *localized* electrons introduced, say, by dissolution of small amounts of excess metal. We shall see that the electron ion interaction dominates in this case.

2. LECTURE I: EXPANDED ALKALI METALS

The liquid-gas critical points of the heavier alkali metals (Cs and Rb) are sufficiently low that these metals may be studied experimentally up to static supercritical conditions (see lectures of F. Hensel). Pure alkali metals contain exactly one valence electron per atom so it is reasonable to expect correlation effects to become important on approaching the metal-nonmetal transition. In fact, the expanded alkali metals may the closest realization of the fictitious expanded monovalent crystals considered by Mott[1] in his original discussion of the metal-nonmetal transition. The actual transition to a non-metallic state is continuous and occurs close to the critical point.

The susceptibility data obtained by Freyland[2] for expanded cesium (Fig. 1) show that the low density metal does indeed exhibit unusual behavior. As the density decreases, the susceptibility initially decreases due, in part, to the effects of d-bands,[3] then rises to a peak at about twice the critical density. With further expansion, the

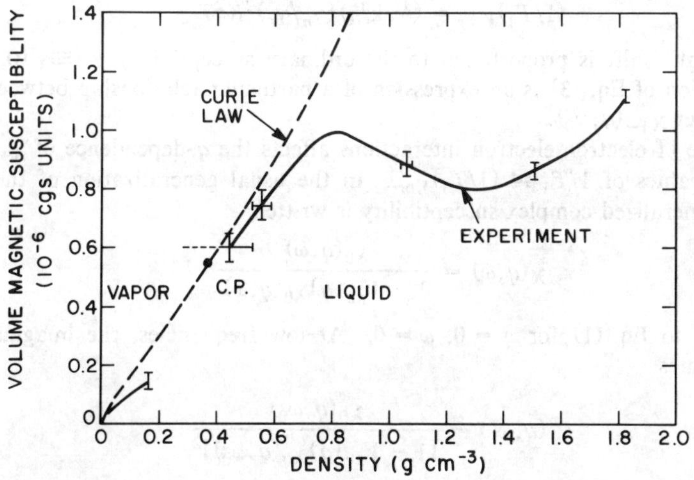

FIGURE 1. Electronic volume susceptibility versus density for liquid cesium along coexistence curve. Solid line: experiment (Ref. 2); broken line: calculated Curie susceptibility.

susceptibility roughly follows the Curie law to nearly the critical point. Similar enhancement effects are observed in expanded sodium[4] and rubidium.[5]

It is natural to ask whether the susceptibility at low density can be explained by a straightforward extension of the Stoner enhancement model. This model is widely used to account for the effects of electron-electron interactions on the magnetic behavior of alkali metals at ordinary densities. For this purpose one usually writes the susceptibility χ in the form

$$\chi = \chi_0/(1-\alpha) , \tag{1}$$

where the enhancement parameter $\alpha = V_{xc}\chi_0$ is the product of the susceptibility χ_0 of the non-interacting system and an exchange-correlation potential V_{xc}. For cesium at ordinary densities, $1/(1-\alpha) \simeq 1.8$.[6] Within this model, the low density enhancement could be attributed either to an increase in V_{xc} or χ_0, although band calculations[3] for low density structures suggest that χ_0 actually *decreases* at low density. In fact, as we discuss next, nuclear spin relaxation data from NMR are clearly inconsistent with an increase in α for any reason.

The nuclear spin relaxation rate in a liquid metal may be expressed as an integral over the imaginary part of the generalized transverse susceptibility $\chi''_{+-}(q,\omega)$:

$$1/T_1 = (64/9) <|\psi(0)|^2>_F^2 \gamma_n^2 \, \mathrm{kT} \, \Omega^2 \omega_0^{-1} \int dq \, q^2 \chi''_{+-}(q, \omega_0) \, , \tag{2}$$

where $<|\psi(0)|^2>_F$ is the average probability density of Fermi surface electrons at the nucleus, γ_n is the nuclear gyromagnetic ratio, Ω is the volume per electron and ω_0 is the NMR frequency ($\omega_0 \sim 10^8$ rad/s). Now, for non-interacting electrons, Eq. (2) reduces to the Korringa relaxation rate which may be expressed in terms of the Knight shift K:

$$(1/T_1)_{\mathrm{Korr}} = (4\pi k/\hbar)(\gamma_n/\gamma_e)^2 K^2 T \,. \tag{3}$$

Since the Knight shift is proportional to the ordinary susceptibility χ ($\equiv \chi'(0,0)$), the Korringa relation of Eq. (3) is an expression of a particular relationship between χ and the integral over $\chi''_{+-}(q,\omega_0)$.

The presence of electron-electron interactions affects the q-dependence of $\chi''_{+-}(q,\omega_0)$ and leads to values of $1/T_1 \neq (1/T_1)_{\mathrm{Korr}}$. In the usual generalization of the Stoner model,[7] the generalized complex susceptibility is written

$$\chi(q,\omega) = \frac{\chi_0(q,\omega)}{1 - V_{xc}(q)\chi_0(q,\omega)} \,, \tag{4}$$

which reduces to Eq. (1) for $q = 0$, $\omega = 0$. At low frequencies, the imaginary part $\chi''(q,\omega_0)$ becomes

$$\chi''(q,\omega_0) = \frac{\chi_0''(q,\omega_0)}{[1 - V_{xc}(q)\chi_0'(q,\omega_0)]^2} \,. \tag{5}$$

Now the essential point is that $V_{xc}(q)\chi_0'(q,\omega_0)$ decreases monotonically with q so that the average enhancement of $\chi''(q,\omega_0)$ in Eq. (2) is less than that of $\chi'(0,0)$ and K. Thus the model predicts that the ratio $\eta \equiv (1/T_1)/(1/T_1)_{\mathrm{Korr}} \leqslant 1$ and η *decreases* further for greater $q = 0$ enhancement $1/(1-\alpha)$. This behavior is a manifestation of the well known bias of the Stoner model toward ferromagnetism whereby the maximum exchange-correlation enhancement occurs for the uniform ($q = 0$) susceptibility.

The foregoing discussion of the Stoner model suggests that if the increase in χ at low density were due to larger values of α, we should expect η to decrease at low density. In fact experimental results for cesium[6] shown in Fig. 2 and for sodium[4] show the

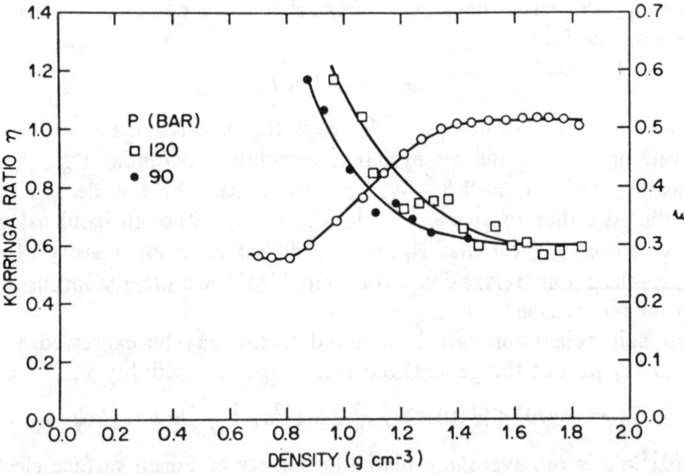

FIGURE 2. Korringa ratio η at two pressures (left-hand scale) and normalized charge density at nucleus ξ on coexistence curve (right-hand scale) versus density for liquid cesium (Ref. 6, 10).

opposite behavior. These results have been interpreted[4,6] to show that the Stoner picture breaks down at low electron density because of a qualitative change in the q-dependence of the susceptibility, namely, development of antiferromagnetic (finite q) enhancement character.

An alternative to the Stoner model for χ is the enhancement predicted by Brinkman and Rice[8] for a highly correlated metal. Near the metal-nonmetal transition where only a small fraction f of the ions are occupied instantaneously by two electrons, these authors predicted an enhancement of the effective mass (or density of states) $m^*/m_0 = 1/2f$. The correlation effect lowers the effective degeneracy temperature of the electrons by the same factor ($T_d = 2fT_F$) and leads to a limiting Curie susceptibility when $T \gtrsim T_d$. There is no corresponding restriction on the Stoner-enhanced susceptibility which becomes arbitrarily large as $\alpha \rightarrow 1$. The observation that the susceptibility is limited by the Curie law at high temperatures (Fig. 1) is strong evidence that the enhancement in cesium is of this type.[9]

The Knight shift data in cesium[6,10] and sodium[4] reveal interesting changes in the form of the conduction electron wave functions at low density. Using the standard expression for the Knight shift

$$K = (8\pi/3) < |\psi(0)|^2 >_F \Omega \chi_V \tag{6}$$

together with the measured volume susceptibility χ_V, we can obtain $<|\psi(0)|^2>_F$ from experiment. The result for cesium, expressed as the ratio ξ of $<|\psi(0)|^2>_F$ to the atomic value $|\psi(0)|^2_{\text{atom}}$ is shown in Fig. 2. The surprising observation is that ξ tends to decrease strongly at low density, only bending toward the atomic value $\xi = 1$ at the lowest densities studied.

I would like to conclude this lecture by suggesting how the above experimental observations can be understood in terms of structural changes occuring in the liquid at low density. Diffraction experiments[11] show that the density reduction develops mainly from a decrease in the average number of near-neighbors rather than from uniform expansion to larger interatomic distances. In this important respect, the expanded liquid differs from the expanded crystals considered in most theoretical treatments. As the liquid expands, electrons can lower their kinetic energy by spreading their wave functions into neighboring vacancies (thus lowering $<|\psi(0)|^2>_F$). But because the states are primarily s-like and spherically symmetric, they can only expand into vacancies by increasing their overlap with the remaining near-neighbors. Correlation will then favor antiparallel ("antiferromagnetic") spin alignment on neighboring atoms and introduce enhancement of $\chi(q, \omega_0)$ at finite q. The essential point here is that the changes in electronic properties at low density are not, according to this picture, simply consequences of the change in electron density, as in an electron gas, but they explicitly reflect the evolution of the underlying structure of the expanded liquid.

3. LECTURE II: ELECTRONS IN MOLTEN SALTS

Excess electrons may be introduced into molten salts by several methods. The most common is by addition of metal to form a metal-molten salt solution. Many metals and their salts form such solutions[12,13] and in some cases solubility is complete so that homogeneous solutions exist over the full range of composition from pure salt to pure metal. More frequently, a region of liquid-liquid immiscibility intervenes and complete solubility exists only above the critical (consolute) temperature. Low concentrations of excess electrons can be achieved without addition of metal by cathodic injection or ultraviolet irradiation. Because a large number of cation sites is available to a relatively small number of electrons, we do not expect correlation effects to be strong at low electron densities. On the other hand, the electrons see the full coulomb potentials of the ions so that the electron-ion interaction can play a decisive role.

This lecture is concerned mainly with the dynamics of low concentrations of excess electrons in alkali halide melts. We begin with the central experimental fact that electrons in these melts are localized on the time scale of diffusive ionic motions. This may be seen clearly in the behavior of the nuclear spin relaxation rates in alkali metal-halide solutions. The relaxation rate may be expressed in the simple general form:

$$1/T_1 = \omega_i^2 \tau_e \,, \tag{7}$$

where ω_i is a frequency characterizing the strength of the magnetic hyperfine coupling and τ_e is a correlation time for the fluctuating local field. For mobile electrons, τ_e is essentially the association time for a particular electron-ion pair. The variation of τ_e with electron density n_e is shown in Fig. 3 for the full range of concentrations in Cs-CsI.[14] Starting from a value ($\sim 10^{-15}$ s) typical of electron-ion interactions in a metal, the correlation time lengthens rapidly with decreasing electron density until a limiting

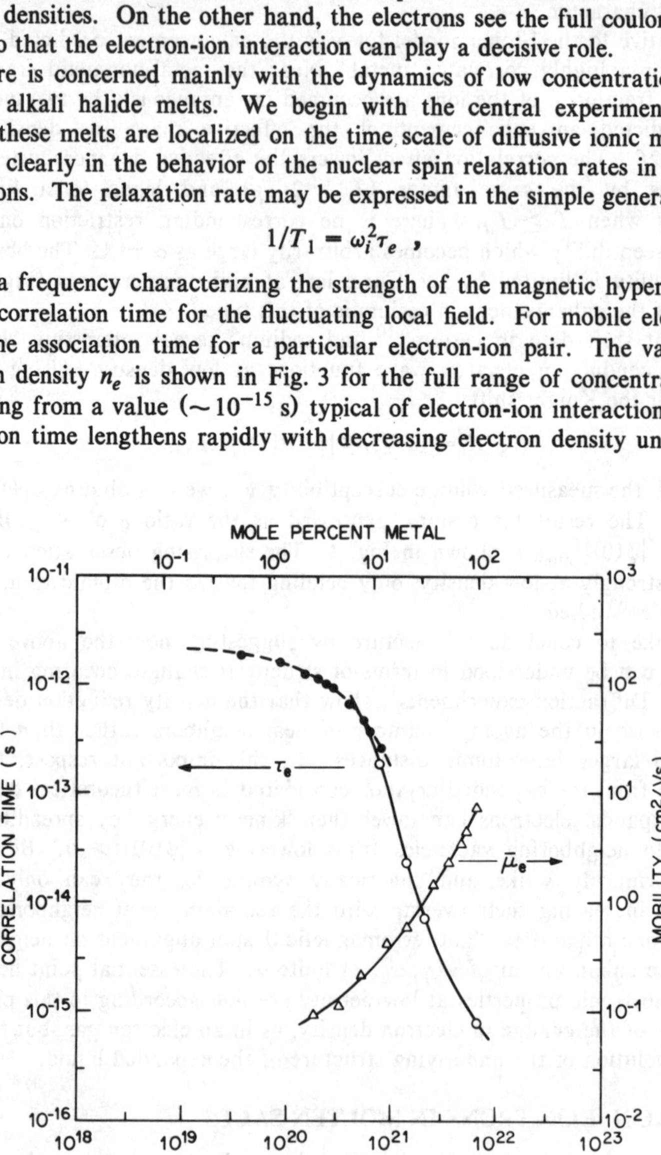

FIGURE 3. Hyperfine correlation time τ_e (left-hand scale) and average mobility $\bar{\mu}_e$ (right-hand scale) versus electron density for Cs-CsI (Ref. 14, 15). Upper scale gives composition in mole % Cs.

value ($\sim 10^{-12}$ s) is reached in the range of low metal concentration ($n_e \lesssim 10^{20}$ cm^{-3}). Thus, electrons are localized in this limit as long, but only as long, as is permitted by fluctuations of the underlying liquid structure.

A second indication of the development of localization follows from the average electronic mobility $\bar{\mu}_e$ derived from the DC electrical conductivity σ:

$$\bar{\mu}_e = \sigma / n_e e \ . \tag{8}$$

As may be seen from the data[15] for Cs-CsI in Fig. 3, $\bar{\mu}_e$ decreases strongly with decreasing electron density and reaches a value $\sim 10^{-1}$ cm^2/Vs in the low concentration range. This mobility is typical for electrons in molten alkali halides. It is important to note, however, that the mobility at low densities, while greatly reduced relative to the metal, nevertheless still exceeds the mobilities of the ions by roughly two orders of magnitude. In this respect the mobilities of the localized states are surprisingly high.

There is now considerable evidence that the structure of the localized states can be described as F-center analogues. In this model, originally suggested by Pitzer,[16] the electrons replace anions in the highly charge-ordered structure of the molten salt. That is, they are solvated by a shell of cations. Recent simulation studies by Parrinello and Rahman[17] indicate relatively strong self-trapping effects in which the neighboring cation shell contracts, deepening the potential well in which the electron is bound. While involving the electron-ion interaction, this type of localization does not require disorder (Anderson transition) except in the sense of a point defect.

The localized electrons are responsible for strong optical absorption bands in the red or near infrared ($\Delta E_{opt} \simeq 1$ eV). These bands closely resemble the F-center absorption in the corresponding crystals except for a shift to somewhat lower energy. Using a simple spherical approximation for the F-center potential, Senatore et. al.[18] calculated the trend of the optical absorption energy through the family of molten alkali halides and found good agreement with experiment. The hyperfine fields on neighboring nuclei, measured by NMR[14] are also in excellent agreement with the F-center model.

Recent optical transient absorption studies[19] yield an additional perspective on the dynamics of the localized electrons. These experiments utilize a pulsed YAG laser which produces pulses of 30-50 ps width at the 1064 nm fundamental wavelength. A fourth harmonic (266 nm) pulse creates an initial population of electrons and holes in a pure molten salt (KCl) and the subsequent F-center absorption is probed with the 1064 nm fundamental. Initial experiments show an induced absorption that develops in less than 100 ps and persists with no detectable decay for times exceeding 1 ns. The recombination time thus exceeds by at least an order of magnitude the mean time between electron-hole encounters estimated from the number of electron-hole pairs created ($\sim 10^{17}$ cm^{-3}) and the average mobility $\bar{\mu}_e$. The most likely explanation is that recombination kinetics are inhibited by the need to dissociate hole-center polarons such as the molecular ion I_2^-.

It is evident that electron-ion interactions in the form of polaron effects strongly influence the dynamics of localized electrons and holes in molten alkali halides. In fact, these effects offer a clue to the resolution of an apparent puzzle related to the high electron mobilities. The optical energies (~ 1 eV) indicate that localized electrons sit in potential wells whose depths greatly exceed thermal energies. The thermally

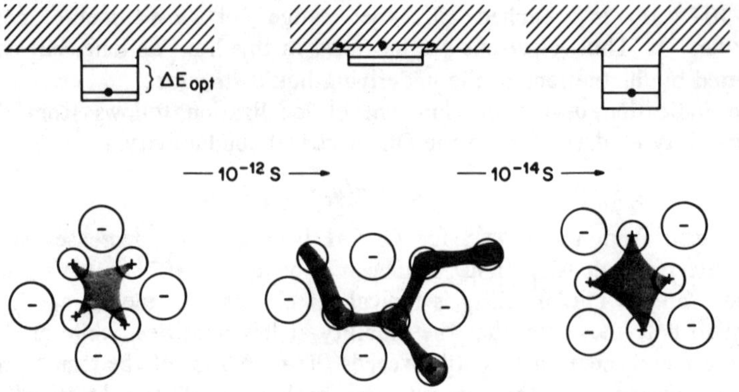

FIGURE 4. Proposed diffusion controlled electron transport process in molten alkali halides.

activated hopping rate from such a well cannot exceed $\sim 10^9$ s.$^{-1}$ Such a low rate is inconsistent with the NMR correlation times τ_e as well as the average mobility $\bar{\mu}_e$. The importance of the lattice polarization is that its relaxation provides a large change in coulomb energy which tends to offset the energy needed to eject an electron from the *static* potential well. Thus the energy of the final state (electron removed, structure relaxed) is much closer to that of the initial configuration (electron bound, structure polarized) than implied by the static potential well.

The foregoing considerations suggest a transport mechanism along the lines indicated in Fig. 4. A localized electron is initially bound in a well formed by polarized cations. Now a diffusing anion penetrating this configuration will repel the electron, pushing it onto the cations. In terms of the potential well, the presence of the anion (or, equivalently, departure of a cation) weakens the potential allowing the wavefunction to spread.[20] This further weakens the self-trapping potential until the bound state is pushed up into the continuum. Electrons can be delocalized by thermal fluctuations in this way on the time scale of a few picoseconds. Once in continuum states, electrons move rapidly to new sites for localization. The values of τ_e and $\bar{\mu}_e$ suggest that this takes $\sim 10^{-14}$ s and the electrons move an average distance of 10-20 Å.[21] Thus, in summary, it is the strong unscreened electron-ion interaction which both drives localization of electrons in the molten salt and provides the necessary coupling for non-activated transport of electrons from one localized state to another.

REFERENCES

1. N. F. Mott, Proc. Phys. Soc. London, Sect. A *62*, 416 (1949).
2. W. Freyland, Phys. Rev. B *20*, 5104 (1979).
3. W. W. Warren, Jr. and L. F. Mattheiss, Phys. Rev. B *30*, 3103 (1984).
4. L. Bottyan, R. Dupree, and W. Freyland, J. Phys. F *13*, L173 (1983).
5. W. Freyland, J. Phys. (Paris) Colloq. *41*, C8-74 (1980).

6. U. El-Hanany, G. F. Brennert, and W. W. Warren, Jr., Phys. Rev. Lett. *50*, 540 (1983).
7. See, for example, R. W. Shaw, Jr. and W. W. Warren, Jr., Phys. Rev. B *3*, 1562 (1971).
8. W. F. Brinkman and T. M. Rice, Phys. Rev. B *2*, 4302 (1970).
9. W. W. Warren, Jr., Phys. Rev. B *29*, 7012 (1984).
10. W. W. Warren, Jr., U. El-Hanany, and G. F. Brennert, J. Non-Crystalline Solids *61-62*, 23 (1984).
11. G. Franz, W. Freyland, W. Gläser, F. Hensel, and E. Schneider, J. Phys. (Paris) Colloq. *41*, C8-194 (1980).
12. M. A. Bredig in *Molten Salt Chemistry,* edited by M. Blander (New York, Interscience, 1964), p. 367.
13. W. W. Warren, Jr. in *The Metallic and Nonmetallic States of Matter,* edited by P. P. Edwards and C. N. R. Rao (Taylor and Francis, London, 1985), Chap. 6.
14. W. W. Warren, Jr., S. Sotier, and G. F. Brennert, Phys. Rev. B *30*, 65 (1984).
15. S. Sotier, H. Ehm, and F. Maidl, J. Non-Crystalline Solids *61-62*, 95 (1984).
16. K. S. Pitzer, J. Am. Chem. Soc. *84*, 2025 (1962).
17. M. Parrinello and A. Rahman, J. Chem. Phys. *80*, 860 (1984).
18. G. Senatore, M. Parrinello, and M. P. Tosi, Phil. Mag. *41*, 595 (1980).
19. W. W. Warren, Jr., B. Campbell, and G. F. Brennert (unpublished).
20. The spreading effect in clearly evident in the excited state charge distribution presented in Ref. 17.
21. W. W. Warren, Jr., S. Sotier, and G. F. Brennert, Phys. Rev. Lett. *50*, 1505 (1983).

NUCLEAR QUADRUPOLAR RELAXATION AND THERMODYNAMICS IN LIQUID TERNARY ALLOYS

K. OTT, M. KIEHL, M. HAGHANI, M. v. HARTROTT*, P. MAXIM, C.A. PAULICK, and D. QUITMANN

Institut für Atom- und Festkörperphysik, Freie Universität Berlin, Arnimallee 14, D-1000 Berlin 33, Fed. Rep. Germany

In liquid alloys with strong AB bonding (decrease of electrical conductivity etc. at stoichiometric compositions; for a review see /1/) indications for long living associates have been derived from nuclear spin relaxation /2/. In such systems experimental results show high increase of quadrupolar rate R_Q in comparison with the pure metals.

In ternary systems A-B-S (S is a third element of low concentration, e.g. an impurity produced by nuclear reaction) there will be competition if two (or more) kinds of associates are energetically favorable. To study this question the nuclear quadrupolar relaxation rate R_Q was measured in In-Na alloys at a ^{117m}Sb probe. The experiment was performed essentially in the way described earlier /3/ for other liquid In alloys.

We describe the concentration dependence of nuclear spin relaxation rate R_Q in liquid In-Na system with an Sb impurity as the probe atom by a thermodynamical model, assuming the existence of two associates /2/, /4/.

In the ternary In-Na-Sb system, phase diagrams, conductivity measurements etc. suggest that the associates In_8Na_5 and Na_3Sb are energetically favorable. InSb has small enthalpy of mixing ΔH in comparison with the other associates and is therefore neglected in our calculation. In the model calculation we assign InNa and NaSb 1 : 1 stoichiometry for both associates to simplify computation. Thus there are five species characterized by the indices:
1 free Na, 2 free In, 3 free Sb, 4 InNa, 5 NaSb. One has $n_1=x_a(1+n_4+n_5)-n_4-n_5$, $n_2=x_b(1+n_4+n_5)-n_4$, $n_3=x_c(1+n_4+n_5)-n_5$ and the laws of mass action

$$\frac{n_4}{n_1\ n_2} = \frac{\gamma_1\gamma_2}{\gamma_4} \exp(-\Delta G_{12}/RT) = 1/K_4 ,$$

$$\frac{n_5}{n_1\ n_3} = \frac{\gamma_1\gamma_3}{\gamma_5} \exp(-\Delta G_{13}/RT) = 1/K_5 .$$

The ΔG's are the standard Gibbs functions of the pure associates. The γ-factors contain weighted averages of the interaction parameters W_{13}, W_{23}, W_{12} analogous to the two component

*Guest of Sfb 161

alloys /4/. We get the five concentrations and K_4, K_5, using the method of Brebrick /5/ to solve the system of coupled equations iteratively.

For the quadrupolar relaxation rate the ansatz is used /2/:

$$R_Q = f(k,I) \langle \Omega_Q^2 \rangle \tau_Q .$$

The constant k=1 for NMR, k=2 for TDPAD, and

$$f(k,I) = 3/80 \frac{k(k+1)(4I(I+1)-k(k+1)-1)}{I^2(2I-1)^2} .$$

Actually we measure the sum $R=R_Q+R_{magn}$:

$$R_{magn} = k(k+1)/2 \langle \Omega_M^2 \rangle \tau_M .$$

Ω^2 represents square of the mean fluctuation of the electric field gradient or magnetic field. For the isomer ^{117m}Sb, R_{magn} is 0.15 kHz, thus 1% of R and therefore we set $R=R_Q$.

τ_Q is the fastest correlation time of the efg. This time is the lifetime of the associates (about 10^{-11}sec) multiplied with the probability P that the probe is part of the associate $P=n_5/(n_3+n_5)$. Seeing the mass action law as an equation for equality of rates we can write:

$$R_Q = \text{const} \langle \Omega_Q^2 \rangle P \, 1/K_5 .$$

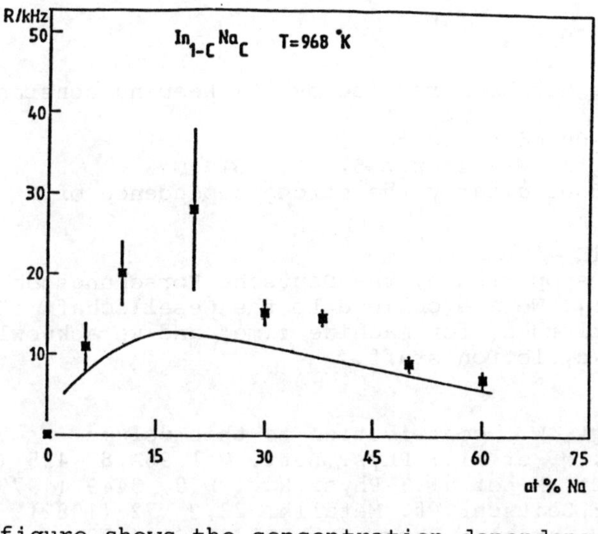

The first figure shows the concentration dependence of R_Q of ^{117m}Sb (squares). Curve is model calculation #1 from fig. 2.

Compared to other liquid metallic alloys, and to the electronic anomalies in In-Na, the maximum of R_Q is strongly shifted to small x. In the thermodynamic model this requires large ΔG_{13}.

The following figure shows how the concentration dependence of R_Q changes with variation of the interaction parameters W_{ij} and with the standard Gibbs functions of the associates.

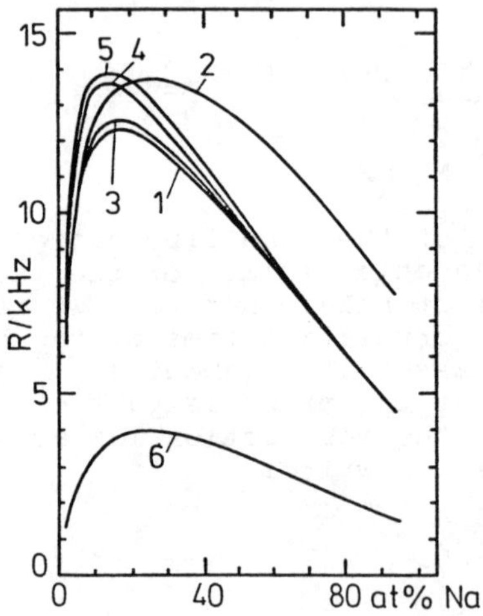

Each parameter has been reduced by 20% keeping constant the others.
Curves with reduced parameters:
1) none 2) W_{13} 3) W_{23} 4) W_{12} 5) ΔG_{12} 6) ΔG_{13}
The diagram shows clearly the strong dependence of R_Q on W_{13} and ΔG_{13}.

ACKNOWLEDGMENTS

This work was supported by the Deutsche Forschungsgemeinschaft through Sfb 161. We are obliged to the Gesellschaft für Kernforschung, Karlsruhe, for machine time, and we acknowledge the help from the cyclotron staff.

REFERENCES

1. van der Lugt, W., contribution to this volume
2. Elwenspoek, M. et al.: Phys. Lett. Vol 96A,8 435 (1983)
3. von Hartrott, M. et al.: Phys. Rev. B19, 3449 (1979)
4. Sommer, F.: Zeitschr. f. Metallk. 73,2 72 (1982)
5. Brebrick, R. et al.: Phys. Rev. B2, 3003 (1983)

SINGLE PARTICLE MOTION IN LIQUID SODIUM

C. MORKEL, W. GLÄSER
PHYSIK-DEPARTMENT, TU MÜNCHEN
8046 GARCHING, GERMANY

1. INTRODUCTION

In classical diffusion experiments (NMR, tracer methods) one can observe single particle motion only in the Brownian limit, which means long times compared with characteristic times of 10^{-13} sec and long distances in comparison with typical path lengths of 10^{-11} m in a liquid. Slow neutron scattering experiments however allow us to reveal the entire time and space range of atomic motion, beginning at short times (10^{-13} sec) and distances (10^{-11} m) and extending up to the Brownian limit of diffusion (10^{-10} sec resp. 10^{-9} m).

In neutron scattering experiments with the instruments IN4 and IN6 (ILL, Grenoble) the scattering law $S(Q,\omega)$ of liquid sodium was measured in three states in the temperature range from 400 K to 800 K at saturated vapour pressure.

The incoherent scattering law $S_s(Q,\omega)$, giving information about the single particle motion, was separated from the total scattering intensity using a model for the coherent scattering law $S(q,\omega)$. Thus $S_s(Q,\omega)$ could be analysed in terms of a collision rate $\alpha(Q)$ and several parameters of $S_s(Q,\omega)$ such as the reduced half width $\gamma(Q) = \omega_{1/2}(Q)/(DQ^2)$ (D: diffusion constant).

The observed step in the Q-dependent collision rate $\alpha(Q)$ as well as the resonance like behaviour of the reduced half width $\gamma(Q)$ indicates a transition from diffusive single particle motion at low Q ($Q < 1Å^{-1}$) and "hard sphere gas" like behaviour of the fluid particles at high Q ($Q > 3Å^{-1}$). The characteristic Q-value of the change-over shifts to lower Q with increasing temperature. These features are well described by a kinetic theory, with which the experimental results are compared, showing good qualitative agreement.

2. EXPERIMENT

The scattering law $S(Q,\omega)$ has been measured for scattering vectors Q between 0.3 $Å^{-1}$ and 4.6 $Å^{-1}$, according to the kinematic region of the two tof-instruments IN4 (E_o = 12.55 meV, full curve, Fig. 1a) and IN6 (E_o = 2.35 meV, dashed curve, Fig. 1a). In Fig. 1b the measured resolution function of both the instruments are shown (IN4: ΔE = 0.57 meV, IN6: ΔE = 0.053 meV).

For further evaluation of the experiment the structure factor S(Q) was deduced for the temperatures 403K, 602K and 803K (Fig. 2). For comparison x-ray data/1/ for T = 373K are included in Fig. 2 (solid curve).

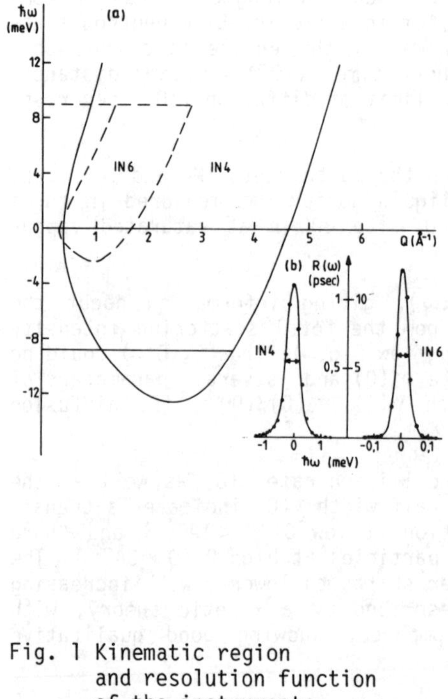

Fig. 1 Kinematic region and resolution function of the instruments.

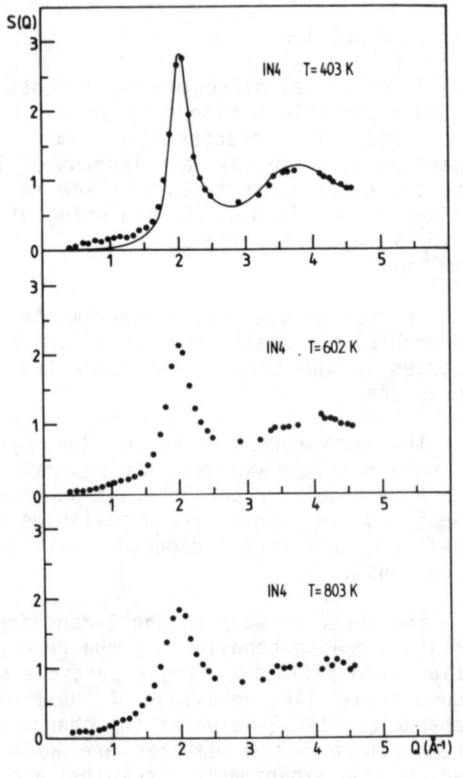

Fig. 2 Structure factor of liquid sodium at three temperatures.

The experimental S(Q)-data were used to separate the incoherent scattering-law $S_s(Q,\omega)$ from the coherent part $S(Q,\omega)$, for which the visco-elastic model/2/ was used. After subtraction of the coherent scattering the remaining incoherent scattering law $S_s(Q,\omega)$ was fitted with the Nelkin-Ghatak-model/3/, which describes the single particle motion in terms of a collision rate.

3. RESULTS

Taking the collision rate α of the Nelkin-Ghatak-model as fit parameter, a change from diffusive behaviour to nearly free single particle motion is indicated by a strong decrease of the collision rate $\alpha(Q)$ at a certain Q-value. This Q-value shifts to lower Q with increasing temperature (Fig. 3).

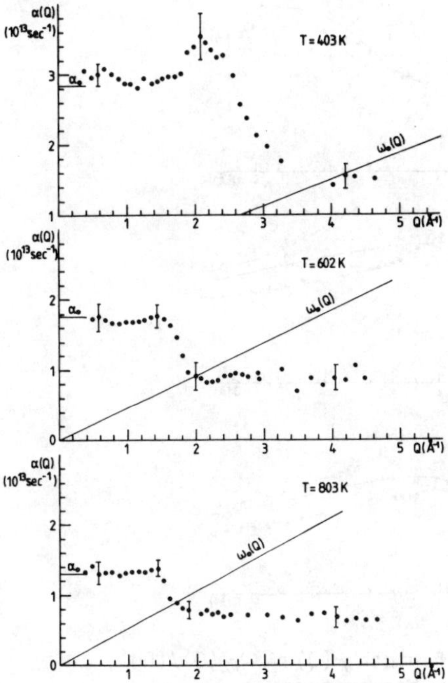

Fig. 3 Collision rate $\alpha(Q)$ compared with the kinetic frequency $\omega(Q) = (KT/M)^{1/2} \cdot Q$.

T(K)	$\bar{l}$ (Å)	$\bar{l}_E$(Å)	σ_{hs}(Å)
403	0.15	0.20	3.26
602	0.26	0.32	3.12
803	0.37	0.45	3.00

Table I. Particle mean free path $\bar{l}$ of liquid sodium at different temperatures.

For comparision with theories the reduced half width $\gamma(Q) = \omega_{1/2}(Q)/(DQ^2)$ is best suited. In Fig. 4 $\gamma(Q)$ is compared with the classical value $\gamma(Q) = 1$ (········) at low Q and with the perfect gas behaviour $\gamma(Q) \propto 1/Q$ (for 602K and 803K) at high Q (—·—·—).

In the high Q region a comparison with the Enskog theory of a hard sphere gas/4/ is possible (— — — —) and yields mean free path lengths $\bar{l}$ of the fluid particles, which are in good agreement with the calculated values $\bar{l}_E$ of the Enskog theory. (Table I).

The experimental results (•••••)/5/ are further compared with the kinetic theory of Götze and Zippelius (————)/6/.

This theory describes diffusion retardation ($\gamma(Q)<1$) at small Q via a Q-and ω-dependent many particle potential created by the motion of the tagged particle, which polarizes the surrounding liquid/6/. At a characteristic Q_c-value, which depends on temperature, this additional effective potential breaks down. Thus the particle is allowed to move more freely, which is reflected in a sudden increase of the measured reduced half width $\gamma(Q)$ and the approach to the ideal gas behaviour (—·—·—) at high Q.

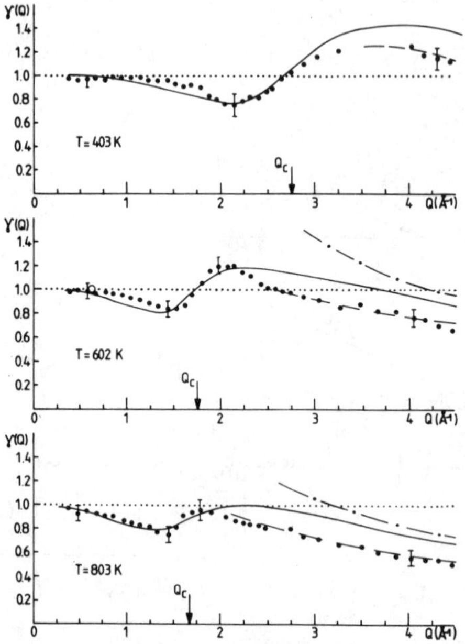

Fig. 4 Reduced half width $\gamma(Q) = \omega_{1/2}(Q)/(DQ^2)$ at different temperatures.

REFERENCES

1. Greenfield, A., Wellendorf, J., Wiser, N., Phys. Rev. A4, 1607 (1971)

2. Lovesey, S., J. Phys. C4, 3057 (1971)

3. Nelkin, M., Ghatak, A., Phys. Rev. 135, A4 (1964)

4. Sears, V., Phys. Rev. A5, 452 (1972)

5. Morkel, C., Ph.D., Techn. Univ. München (1984)

6. Götze, W., Zippelius A., Phys. Rev. A14, 1842 (1976)

5. ELECTRONIC PROPERTIES OF LIQUID AND AMORPHOUS MATERIALS

ELECTRONIC STRUCTURE AND TRANSPORT IN AMORPHOUS AND LIQUID METALS

H. Beck and R. Frésard
Institut de Physique
CH-2000 Neuchâtel (Switzerland)

We review some recent theoretical work concerning the electronic density of states and electrical transport in disordered metals, namely tight-binding calculations, self-consistent transport theory and a multiple scattering approach.

1. INTRODUCTION

The quantitative description of electronic properties in non-crystalline matter is still a difficult problem. This contribution reviews some work performed in this field since the 1979 NATO school on the same subject. It is impossible to present an exhaustive overview on 6 years of progress, thus the choice of subjects is purely personal. We shall omit, in particular, specific low temperature effects, covered in another contribution.

The basis of the calculations presented below is a one-electron Hamiltonian describing an electron in the field of a random arrangement of ions. In Schrödinger's representation H reads

$$H = \sum_i \frac{\vec{p}_i^{\,2}}{2m} + \sum_i \sum_{n=1}^{N} V(\vec{r}_i - \vec{R}_n), \tag{1}$$

$\vec{R}_n$ being the (possibly time dependent) ionic positions. In a tight-binding basis

$$H = \sum_{n\lambda} E_{n\lambda} |\varphi_{n\lambda}\rangle\langle\varphi_{n\lambda}| + \sum_{nm}{}' \sum_{\lambda} V_{nm,\lambda} |\varphi_{n\lambda}\rangle\langle\varphi_{m\lambda}|, \tag{2}$$

λ denoting the quantum numbers of the local orbitals at site n.

For the calculation of the resistivity ρ two different approaches can be used [1] :

(a) the conductivity $\sigma(=\rho^{-1})$ is given by the velocity correlation function (Kubo's formula) :

$$\sigma = \frac{e^2}{3\Omega} \int_0^\infty dt\, e^{-\varepsilon t} \int_0^\beta d\lambda \langle \vec{v}\cdot\vec{v}(t+i\hbar\lambda)\rangle \tag{3}$$

Here $\vec{v}$ is the velocity operator and Ω the volume of the system.

(b) the resistivity ρ is expressed by force correlation functions : $\rho = \rho_1/(1+D)$ with

$$\rho_1 = \frac{1}{3\Omega n^2 e^2} \int_0^\infty dt\, e^{-\varepsilon t} \int_0^\beta d\lambda \langle \vec{F} \cdot \vec{F}(t + i\hbar\lambda) \rangle \qquad (4)$$

$$D = \frac{1}{3n\Omega} \int_0^\infty dt\, e^{-\varepsilon t} \int_0^\beta d\lambda \langle \vec{F} \cdot \vec{v}(t + i\hbar\lambda) \rangle . \qquad (5)$$

Here n is the electron density and $F_\alpha = [p_\alpha, H]$ the force acting on an electron.

These expressions (more precisely the corresponding generalizations for the tensors $\rho_{\alpha\beta}$ and $\sigma_{\alpha\beta}$) are valid in presence of a magnetic field (replacing $\vec{p}$ by $\vec{p} - e\vec{A}$ in H) and can be used for evaluating Hall coefficient and magnetoresistance. The equivalence between (3) and (4,5) has been examined in [1]. There, it is shown that when $\varepsilon \to 0$ both ρ_1 and (1+D) vanish, their ratio tending to $1/\sigma$, as required. It seems, however, that this equivalence proof neglects boundary terms in partial integrations and is therefore not conclusive. Moreover, ρ_1 has been transformed into [2] :

$$\rho_1 = \frac{3\hbar}{2^7 \pi \Omega e^2 k_F^4} \int d\Omega_k \int d\Omega_{k'} (\vec{k} - \vec{k}')^2 \left| T(\vec{k}, \vec{k}'; E_F) \right|^2 , \qquad (6)$$

which is a positive (!) quantity, involving the full T-matrix for the ionic potential in (1), integrated over the directions of incoming and outgoing wave vectors $\vec{k}$ and $\vec{k}'$ ($k^2 = k'^2 = 2mE_F$) on the Fermi surface. Thus the precise relationship between (3) and (4,5) still needs to be clarified.

II. TIGHT-BINDING CALCULATIONS

The tight-binding description of s and d bands in liquid or amorphous transition metals (TM), associated with recursion methods or momentum expansions and the continued fraction technique, has been used by many authors to calculate the resolvent $(z-H)^{-1}$ and the density of states D(E), see, e.g., Refs. [3]-[5] and further work quoted therein. Many of these calculations are done for a finite cluster of ions, the positions $\vec{R}_n$ being taken from a suitable structural model. The hopping elements V in (2) are calculated for each pair of ions as a function of their distance. The "d-band" density of states shows usually a two-peak structure (bonding and antibonding LCAO states) and has a width similar to that for the corresponding crystal. Projecting $(z-H)^{-1}$ onto running-wave states one can find k-dependent spectral functions [3]. The "s-band" shows, in general, a crystal-like dispersion with typical s-d hybridization effects. However the negative slopes proposed to explain the positive Hall constant [6] are not confirmed by Ref. [3]. In general it seems preferable to evaluate transport coefficients directly by expressions like (3)-(5), without relying on some effective dispersion relation.

Using the same recursion techniques, eq. (3) for σ has been evaluated [3,4] for a tight-binding system. The calculated resistivities for Fe [3] and La [4] compare rather favorably with experiments. By splitting the eigenstates into their s and d parts, the s and d contributions to σ can be sorted out (since eq. (3) involves a two-point Green function there is also an s-d cross term which is, however, numerically small). Thus the conduction process is represented by two parallel resistors, carrying the s and d current, respectively : $\sigma = \sigma_s + \sigma_d$, with $\sigma_i = \mu_i D_i(E_F)$. The d part is dominant ($\sim$ 80%), owing to the high density of states D_d, although the mobility μ_d is much lower that μ_s. We shall come back to a similar decomposition of ρ in section IV.

These LCAO-type calculations for D and ρ have the advantage of using the full structural information of a given model, rather than making some approximations for higher order ionic correlation functions. Moreover, they do not make use of dispersion curves, while still retaining information on the s and d contributions to the physical quantities. On the other hand the numerical effort seems still too high to make it an "every day tool" for evaluating transport coefficients in disordered metals.

III. SELF-CONSISTENT TRANSPORT THEORY

The velocity correlation function in (3) can be expressed by the two-point Green function

$$\Phi_{\vec{k}\vec{k}'}(\vec{q}\vec{q}',t) = (-i)\,\theta(t)\left\langle\left[f^{+}_{\vec{k}}(\vec{q}t),\, f_{\vec{k}'}(\vec{q}'0)\right]\right\rangle \tag{7}$$

with $f_{\vec{k}}(\vec{q}) = c^{+}_{\vec{k}-\vec{q}/2}\, c_{\vec{k}+\vec{q}/2}$. Since we have no electron-electron interaction ϕ factorizes into a product of one-point functions G, but the average over a probability distribution $p(\{\vec{R}_n\})$ of the ionic positions (denoted by a bar) introduces "correlations", since $\overline{GG} \neq \overline{G}\cdot\overline{G}$. In order to obtain useful approximations for the Laplace transform

$$\overline{\Phi}_{\vec{k}\vec{k}'}(\vec{q}\vec{q}';z) \equiv \delta_{\vec{q}\vec{q}'}\,\varphi_{\vec{k}\vec{k}'}(\vec{q},z) \tag{8}$$

we start from the equation of motion

$$z\,\Phi_{\vec{k}\vec{k}'}(\vec{q}\vec{q}';z) = \delta_{\vec{k}\vec{k}'}\,\delta_{\vec{q}\vec{q}'}\left[N(\vec{k}+\tfrac{\vec{q}}{2}) - N(\vec{k}-\tfrac{\vec{q}}{2})\right]$$

$$+\,\vec{v}\cdot\vec{q}\,\Phi_{\vec{k}\vec{k}'}(\vec{q}\vec{q}';z) + \sum_{\vec{k}_1\vec{q}_1} R_{\vec{k}\vec{k}_1}(\vec{q}\vec{q}_1)\,\Phi_{\vec{k}_1\vec{k}'}(\vec{q}_1\vec{q}';z) \tag{9}$$

with $N(\vec{Q}) = \langle c^{+}_{\vec{Q}}\, c_{\vec{Q}}\rangle$, $\vec{v} = \hbar\vec{k}/m$,

$$R_{\vec{k}\vec{k}'}(\vec{q}\vec{q}') = U(\vec{q}-\vec{q}')\left[\delta_{\vec{k}',\vec{k}-\frac{\vec{q}'-\vec{q}}{2}} - \delta_{\vec{k}',\vec{k}+\frac{\vec{q}'-\vec{q}}{2}}\right]$$

$$\equiv M_{\vec{k}\vec{k}'}(\vec{q}\vec{q}')\,A(\vec{q}-\vec{q}') \tag{10}$$

$$\mathcal{U}(\vec{q}) = A(\vec{q})\hat{V}(\vec{q}) \tag{11}$$

$$\hat{V}(\vec{q}) = \int d\vec{r}^{\,3} e^{i\vec{q}\cdot\vec{r}} V(\vec{r})$$

$$A(\vec{q}) = \sum_n e^{i\vec{q}\cdot\vec{R}_n} \tag{12}$$

The problem is to express the average $\overline{R\phi}$ in (9) by $\overline{\phi}$. This can be achieved in an iterative way by introducing an external field Λ, acting on the ionic positions, and using functional derivative techniques similar to those applied for interacting systems [7]. Write the average of any quantity x, depending on the $\vec{R}_n$:

$$\overline{x}_\Lambda \equiv \overline{(x\, e^{i\Lambda A})} / \overline{(e^{i\Lambda A})} \tag{13}$$

with $\Lambda A = \sum_{\vec{q}} \Lambda(\vec{q})\, A(\vec{q})$. The quantity Λ is set equal to zero at the end. The average of $\overline{A\phi}$, needed in equation (9) can be obtained by

$$\overline{\left(A(\vec{q})\Phi\right)}_\Lambda = -i\, \delta\overline{\Phi}_\Lambda / \delta\Lambda(\vec{q}) + \overline{A(\vec{q})}_\Lambda \cdot \overline{\Phi}_\Lambda \,. \tag{14}$$

Introducing a self-energy C by

$$\sum_{\vec{k}_1\vec{q}_1} \overline{R_{\vec{k}\vec{k}_1}(\vec{q}\vec{q}_1)\,\Phi_{\vec{k}_1\vec{k}'}(\vec{q}_1\vec{q}';z)} \equiv \sum_{\vec{k}_1} C_{\vec{k}\vec{k}_1}(\vec{q}z)\,\varphi_{\vec{k}_1\vec{k}'}(\vec{q}z) \tag{15}$$

it is possible to obtain a functional derivative equation by techniques described in [7] :

$$D_{\vec{k}\vec{k}_1}(\vec{q}\vec{q}_1,z) = M_{\vec{k}\vec{k}_1}(\vec{q}\vec{q}_1)\overline{(A(\vec{q}-\vec{q}_1))}_\Lambda + i\sum_{\vec{k}_2\vec{k}_3}\sum_{\vec{q}_2\vec{q}_3} M_{\vec{k}\vec{k}_2}(\vec{q}\vec{q}_2)\cdot$$

$$\cdot H_{\vec{k}_2\vec{k}_3}(\vec{q}_2\vec{q}_3,z)\,\frac{\delta D_{\vec{k}_3\vec{k}_1}(\vec{q}_3\vec{q}_1,z)}{\delta\Lambda(\vec{q}-\vec{q}_2)} \,. \tag{16}$$

Here H obeys the same equation (9) as ϕ, but with the inhomogeneous term, involving N, replaced by unity. D reduces to C when $\Lambda \to 0$. Eq. (16) is valid when the Λ-dependence of N can be neglected (see the end of this section). Using (15), one obtains the "generalized" transport equation"

$$(z-\vec{v}\cdot\vec{q})\,\varphi_{\vec{k}\vec{k}'}(\vec{q}z) - \sum_{\vec{k}_1} C_{\vec{k}\vec{k}_1}(\vec{q}z)\,\varphi_{\vec{k}_1\vec{k}'}(\vec{q}z) = \delta_{\vec{k}\vec{k}'}\, I_{\vec{k}}(\vec{q}), \tag{17}$$

I being the averaged inhomogeneity of (9) and C playing the role of the "collision kernel". The latter depends self-consistently on φ (or its counterpart H), as the first order iteration of (16)

$$C^{(1)}_{\vec{k}\vec{k}'}(\vec{q}z) = N \sum_{\vec{k}_1 \vec{k}_2 \vec{q}_1} M_{\vec{k}\vec{k}_1}(\vec{q}\vec{q}_1) S(\vec{q}-\vec{q}_1) M_{\vec{k}_2\vec{k}'}(\vec{q}_1\vec{q}) H_{\vec{k}_1\vec{k}_2}(\vec{q}_1 z) \qquad (18)$$

indeed shows.

A transport equation of this form, obtained by a mode decoupling procedure in a Mori approach, is the starting point of Götze's self-consistent theory of localization in disordered metals [8]. He solves (17) by introducing a basis $a_\alpha(\vec{k})$ in the space of functions of $\vec{k}$. Truncating the expansion of φ after $\alpha = 0$ (a_0 = const) and $\alpha = 1$ ($a_1 \propto \vec{k}.\vec{q}$) allows to find density and velocity correlation functions. The static resistivity ρ is finally given by the q = 0, z = 0 value of the current relaxation kernel

$$m(\vec{q}z) = \sum_{\vec{k}\vec{k}'} a_1^*(\vec{k},\vec{q}) C_{\vec{k}\vec{k}'}(\vec{q}z) a_1(\vec{k},\vec{q}). \qquad (19)$$

Rather than giving details that can be found in [8] and subsequent publications (see e.g. [9]) we summarize some results of this self-consistent transport theory and its generalizations to dynamic electron-ion interactions :

(i) The solution of (17), (18) allows to discuss the influence of increasing electron-ion coupling V on ρ: when V is weak ρ is close to ρ_0, the "Ziman-result" [10]. When V increases, ρ gets larger than ρ_0, and finally ρ diverges at a critical coupling strength. At this point the Fermi energy crosses the mobility edge. The corresponding ρ_0, evaluated with a pseudopotential appropriate for a simple metal, would be around 300 to 400 μΩcm. Thus the theory predicts non-conducting behavior when the mean free path calculated in Born's approximation is somewhat shorter than the mean interionic distance d. While the details of this Anderson transition may not be of immediate interest for amorphous metals (which show metallic behavior), the theory offers the possibility for evaluating ρ, for a given metal, beyond the usual golden rule level, with a moderate numerical effort. For a quantitative improvement over the Ziman approximation higher order contributions to C may be important, however, since the calculations in section IV show that the corrections to ρ_0 may have both signs, while (18),(19) always yield $\rho \geqslant \rho_0$.

(ii) In order to discuss the T-dependence of ρ an additional electron-ion coupling has been added to (1), see [9] :

$$H_2 = \sum_{\vec{k}\vec{q}} \frac{k^2}{2m} c^+_{\vec{k}-\vec{q}/2} c_{\vec{k}+\vec{q}/2} W(\vec{q}) \left[A(\vec{q}) - \overline{A(\vec{q})}\right]. \qquad (20)$$

Here the (time dependent) ionic density $A(\vec{q})$ couples to the electronic kinetic energy — a coupling which has been used in the theory of superconductivity. The procedure sketched above to find an equation for φ can be repeated when H_2 is added to H in

(1). The structure factor showing up in C (eq. (18)) is replaced by an integral over the dynamic ionic density correlation function and C contains an additional term involving $W(\vec{q})$. The longitudinal current now consists of two parts (the "new" one involving W and A) and thus two relaxation kernels $m(\vec{q},z)$ and $\ell(\vec{q},z)$ are introduced. The conductivity is given by

$$\sigma = m(\vec{o},o)^{-1} + \ell(\vec{o},o). \qquad (21)$$

This is again a parallel resistor picture, the interesting fact being that an increasing coupling strength can now improve the conductivity by increasing ℓ. The calculations in [9] yield the following results : for small coupling ℓ is negligible and the usual variational solution of Boltzmann's equation is retrieved. In particular the T-coefficient $\alpha \equiv d\rho/dT$, due to ionic vibrations, is in general positive. When ρ is larger than some critical value ρ_c the second term in (21) leads to $\alpha < 0$. This corresponds to what is now called "Mooji's rule" [11] : α is typically negative for a disordered metal (particularly transition metal) when $\rho > \rho_c$, the latter being of the order of 150 $\mu\Omega$cm.
Physically the result derived from (21) is understood as a "prelocalization effect" : whereas potential scattering alone (based on eq. (1)) would lead to a mobility break-down, the fluctuations in kinetic energy described by (20) allows the electron to move on by "random tunnelling", even in a regime where the wave functions are already rather "scrambled" by the disordered medium [12].

Schirmacher [13] has also investigated the influence of electron-electron interaction on these facts (at least in the random phase approximation) : it does not modify the second term in (21), and hence the occurence of $\alpha < 0$, but it may suppress the diffusion pole in the electronic density propagator, thus altering the analytic behavior of $m(\vec{q},z)$ at small z.

(iii) It seems interesting to generalize this scheme in the following ways :

- Equation (16) can be iterated to higher orders, using some approximation for the higher ionic structure factors. It is interesting to note that the approximation (18) is equivalent to the (Markovian) factorization approximation for the Feynman propagator ρ of an electron subject to Hamiltonian (1). Beck and Nettel [14] have used the double path integral representation of ρ, averaged over Gaussian ionic disorder, to derive a self-consistent equation for the mobility of electrons in a non-degenerate (high T) semiconductor. A mobility break-down (at a critical V_c or T_c), similar to the one found by Götze [8] came out of this approach, which has already been generalized for the non-Markovian situation [15].

- For TM alloys the equations have to be renormalized, such that the ionic t-matrix appears, rather than V. Moreover the "true" equilibrium electronic structure should be used in eq's (7) to

(21), which amounts to considering the influence of V on the occupation number $N(\vec{Q})$ in eq's (9) and (16). This approach should then allow for a non-trivial discussion of the Hall coefficient for which a simple Boltzmann equation yields the free electron value for an isotropic one-band system, unless "life time" and "band structure" effects are taken into account.

IV. MULTIPLE SCATTERING CALCULATIONS

Eq. (6) for the resistivity is particularly useful when V, in eq. (1), consists of non-overlapping muffin-tins. Then the total (energy-shell) T-matrix in angular momentum representation can be expressed by the single-ion scattering matrix $\tau_\ell = -\mathcal{K}^{-1}\cdot\exp(i\delta_\ell)\cdot\sin\delta_\ell$ and its phase-shifts δ_ℓ ($L \equiv \ell,m$) :

$$T(\vec{k},\vec{k}') = (4\pi)^2 \sum_{\substack{LL' \\ nm}} Y_L(\vec{k})\, \mathcal{T}^{LL'}_{nm}(\mathcal{K})\, Y_{L'}(\vec{k}')\, e^{i\vec{k}\cdot\vec{R}_n - i\vec{k}'\cdot\vec{R}_m} \tag{22}$$

with

$$\mathcal{T}^{LL'}_{nm}(\mathcal{K}) = \left(\left[\tau^{-1}(\mathcal{K}) - B(\mathcal{K})\right]^{-1}\right)^{LL'}_{nm}, \quad \mathcal{K}^2 = k^2 = k'^2 = 2mE . \tag{23}$$

Expressions for B and other details are given in [16]. Neglecting B yields the usual single-site approximation ρ_{SS} for ρ [17] :

$$\rho_{SS} = \frac{3\hbar\pi^3}{e^2\Omega_0 k_F^6} \int_0^{2k_F} dq\, q^3 S(q) \left| \sum_\ell (2\ell+1)\, \tau_\ell(k_F)\, P_\ell(\cos\theta) \right|^2 , \tag{24}$$

which has been used in numerous calculations for liquid and amorphous TM and RE (rare earth) alloys. It yields $\alpha < 0$ when $2k_F \approx k_p$, the latter wave number giving the position of the first peak in S(q). Expression (24) has also been criticized : for d-scatterers it typically yields much too high ρ-values when the relation $k_F^2 = 2mE_F$ is respected [18]. Thus for liquid RE's an ad hoc renormalization procedure of the ionic potential was used [19]. Moreover, the multiple scattering (MS) contributions are probably not small when the mean free path is short. Some calculations going beyond (24) and using approximations for higher order ionic structure factors [20], [21] have shown that ρ_l can be larger or smaller than ρ_{SS}. Some recent calculations for small clusters for which (22), (23) can be evaluated exactly by matrix inversion have revealed some interesting aspects of multiple scattering, pointing to the limits in the use of (24) for strong potentials [22]. The same formalism allows to express the correction δD to D(E), due to electron-ion scattering, by the so-called "cluster phase-shifts" η_λ [23] :

$$\delta D(E) = \frac{2}{N\pi} \sum_\lambda \frac{d\eta_\lambda}{dE} . \tag{25}$$

The quantities ctg η_λ are the eigenvalues of a matrix related to $\mathcal{T}$.

The results of these calculations are in principle valid for clusters immersed in a quasi-free-electron environment, but they can also be used to judge the role of MS in a large system, made up of many such cluster domains. This is an "incoherent single cluster approximation", in which ρ and δD are taken to be the sums of individual cluster contributions, neglecting the scattering processes between clusters. The first calculations were performed for clusters of up to 19 identical s-scatterers (the s phase-shifts being those for a square well of depth $-V_0$ and radius a) of pure or distorted fcc-like symmetry. Here are some results :

(i) Density of states D(E)

While the details, particularly at low E, are specific to the value of V_0, there is a common feature, namely a minimum in D(E) at some energy E_m (at large E D(E) approaches free-electron behavior), see Fig. 1. This effect of the ionic structure has also been found by an effective medium approach [24] and in self-consistent perturbation theory [25]. Photoelectron spectroscopy has revealed such minima in amorphous AuSn alloys [26]. It also plays a role in a stability argument for metallic glasses [27].

The value of E_m varies considerably with V_0. However, when E_F for a given valence Z is evaluated by the true D(E), the value of Z for which $E_F = E_m$ is rather independent of V_0 : $1.3 \lesssim Z \lesssim 1.8$, in agreement with Z = 1.6 in [24]. It is tempting to interpret this minimum as the "pseudogap" of an average crystal band structure. Indeed an effective dispersion $\varepsilon(k)$, defined such as to yield the correct D(E), shows the expected deviation from $k^2/2m$ near $k_m \approx k_p/2$, corresponding to an average zone boundary. Systematic differences between $2k_m$ and k_p can be understood in terms of attractive and repulsive pseudopotentials (yielding the same phase-shifts as V but no bound states).

(ii) Resistivity

As in [20], [21] ρ_l can be larger or smaller than ρ_{ss}. The ratio $\beta \equiv \rho_l/\rho_{ss}$ is found to be typically > 1 for k_p values lying in the ascending slope of S(q) ($k_F \lesssim k_p/2$), where ρ_{ss} underestimates the scattering power through the relatively low values of S. On the other hand, specially for large V_0, $\beta < 1$ for $k_F \approx k_p/2$. Thus MS somewhat washes out the usual marked k_F-dependence in ρ_{ss}, which comes from S(q). The de Broglie wave lengths λ, for which β has a maximum, are $\lambda \approx 2d$,d,etc. Thus the sign of the MS contributions to ρ_l is an interference effect : when λ and d are "commensurate", the scattering is enhanced over the single-site approximation.

By comparing ρ_l for clusters with the same number of ions but with "more or less ordered" structures one can obtain information about the T-dependence of ρ, the structure factor of more (less) ordered clusters looking like the low (high) T forms of the measured S(q). The quantity $\mathscr{T} \equiv \rho_{DIS}/\rho_{ORD} - 1$ thus represents the T-coefficient α of ρ. For weak potentials $\mathscr{T}$ has the expected "Ziman" behavior : $\mathscr{T} < 0$ for k_F near the peaks of S(q) and

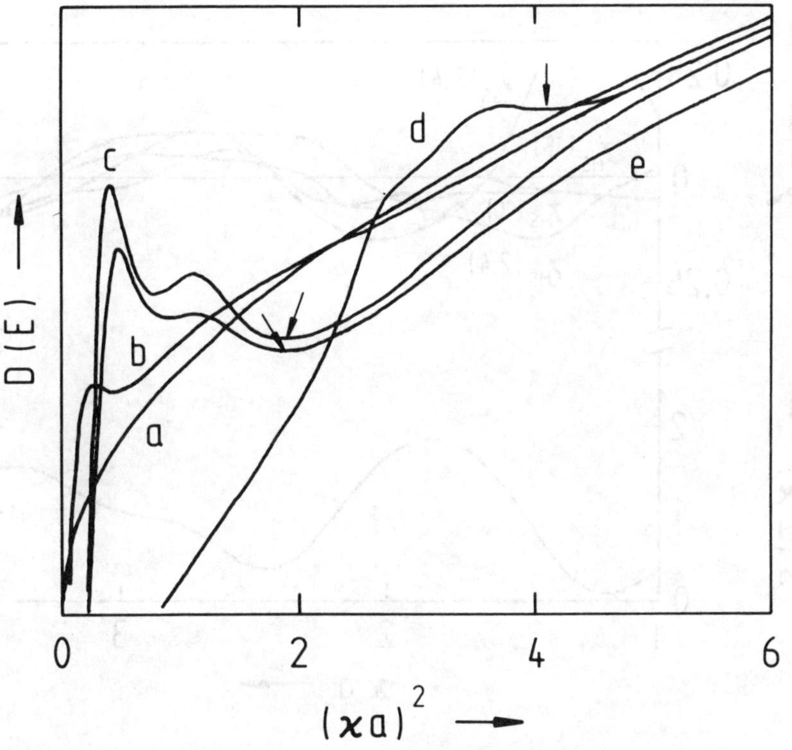

Fig. 1 : Electronic densities of states (free electron part plus correction δD) of a 13-site cluster of distorted fcc structure for various well depths: V_0a^2 = 0(a), 0.9(b), 2.4(c), 15(d), 22(e). The arrows indicate the position of the structure induced minima at $E = E_m$.

> 0 inbetween. The situation changes when V_0 is increased. First the region of negative α is pushed to lower energies (curve b of Fig. 2) : $k_F \lesssim k_p/2$ is now the condition for $\alpha < 0$. It is interesting to note, however, that the domain of negative α roughly corresponds to the same valence as in the weak scattering (Ziman) limit, namely $Z \approx 1.8$, since E_F is also lowered (curve b is for an "attractive" pseudopotential). Thus, if $2k_F \approx k_p$ is no more the right condition for $\alpha < 0$, the valence criterion $Z \approx 1.8$ still holds. For even larger V_0 (curve c) α is slightly negative for all $k_F < 1.3(k_p/2)$. The Ziman criterion has been replaced by a Mooji behavior [11] : $\alpha < 0$ as long as $\rho_l \gtrsim 120$ μΩcm, whereas for larger values of k_F, $\rho_l < 120$ μΩcm and α again follows the oscillations of $S(q)$.

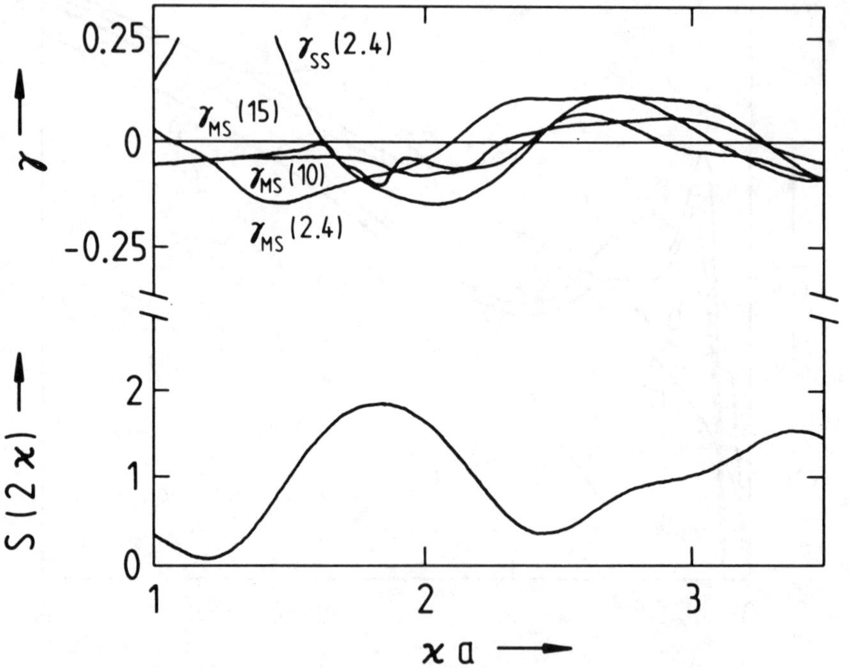

Fig. 2 : The quantity γ (defined in the text) for a 19-site cluster, in single-site approximation (γ_{SS}) and from the full multiple scattering calculation (γ_{MS}). The values of V_0a^2 (potential strength) are give in parentheses. The structure factor S(q), shown below, allows to locate the domains of $\gamma < 0$ with respects to the peaks of S(q).

The influence of increasing disorder on the scattering strength of a cluster can be understood in terms of the phase-shifts $\eta_\lambda(E)$. The scattering is strong when $\eta_\lambda \approx \pi/2$ (resonance). When the cluster is relatively symmetric there are (quasi) degeneracies : several η_λ go through $\pi/2$ at the same energy E_R which yields a high value for ρ at $k_F^2 \approx 2mE_R$. When the cluster is more disordered the degeneracies are lifted : each η_λ resonates at a somewhat different E_R and ρ is lower than before. If, in the language of wave functions, disorder is said to destroy phase coherence [12], one may say, in the scattering approach, that disorder destroys "local degeneracies" of the scattering modes : there is a "fanning" of the η_λ-curves.

Preliminary calculations for d-wave scattering show that here this effect is even more important. In ρ_{SS} (eq. (24)) all the phase shifts are equal and ρ_{SS} is very large when $k_F^2 \approx 2mE_R$. This is responsable for the quantitative failure of ρ_{SS} in TM or RE systems [18], [19]. The cluster phase-shifts η_λ also show resonant behavior, but passing from the symmetry of a single ion to that of the cluster has removed the high degeneracy and ρ_l is considerably lower than ρ_{ss}. Again MS diminishes the effective scattering power of the cluster. In many cases two "bundles" of resonance curves are found, corresponding to the two peaks seen in D(E) in section II. Again additional disorder lowers ρ_l even more, in agreement with Mooji's rule [22].

Taking s and d scattering into account one may (neglecting again cross terms showing up, for example, in (24)) represent ρ as a sum, $\rho \approx \rho_s + \rho_d$. However, in contrast to section II the two resistors are now in series, carrying the same current but contributing differently to the total potential difference. Again, in a TM, the d-contribution is dominant, owing to the strong d-scattering. The relation to the end of section II is established by observing that a relation like $\rho \propto D_d(E_F)$, postulated by Mott [28] for TM's, indeed holds for the total scattering cross-section $\bar{\sigma}$. The latter is expressed by the η_λ as usual,

$$\bar{\sigma} = \varkappa^{-2} \sum_{\lambda} \sin^2 \eta_\lambda \, . \tag{26}$$

When the η_λ form a bunch of resonance curves, σ (eq. (26)) and δD (eq. (25)) indeed have the same energy dependence in the vicinity of the resonant energies, thus $\bar{\sigma} \propto D_d$.

These multiple scattering calculations build a bridge between the usual ρ_{ss} and a more adequate treatment of strong scattering systems. Like the self-consistent transport theory, they point to a Mooji behavior, but, unlike the static version of the latter, it allows for $\rho_l < \rho_{ss}$. It is obvious that the calculations on relatively small clusters are not yet suited for the evaluation of the resistivity of a real metal. Future work should aim at developping sufficiently simple, but adequate analytic approximation schemes into which MS is incorporated. This should finally provide us with a mean to obtain good quantitative results for ρ in TM's and RE's without making inconsistent or ad hoc assumptions on k_F and E_F [18], [19] or on the number on conduction electrons (see the criticism put forward in [4]), and to understand the negative α of alloys like Gd Co [29] which do not seem to conform with the rule $2k_F \approx k_p$.

Part of this work was supported by the Swiss National Science Foundation.

[1] Huberman M. and G.V. Chester, Adv. in Physics 24, 489 (1975)
[2] Rousseau J.S. et al., 1973 Proc. Conf. on Properties of Liquid Metals, ed. S. Takeuchi, Taylor and Francis, London, p. 249
[3] Bose S.K. et al., J. Physics F 13, 2089 (1983)
[4] Ballentine L.E. and J.E. Hammerberg, Can. J. Phys. 62, 692 (1984)
[5] Cyrot-Lackmann F., J. Non-Cryst. Solids 62, 1027 (1984)
[6] Morgan G.J. and G.F. Weir, Phil. Mag. 47, 177 (1983)
[7] Kadanoff L.P. and G. Baym, Quantum statistical mechanics, Benjamin, N.Y. 1962, ch.5.
[8] Götze W., Phil. Mag. B 43, 219 (1981)

[9] Belitz D. and W. Schuhmacher, J. Phys. C 16, 913 (1983)
[10] Ziman J.M., J. Phys. C 1, 1532 and C 2, 1230 (1969)
[11] Mooji J.H., Phys. Stat. Solidi (a)17, 521 (1973)
[12] Imry Y., Phys. Rev. Lett. 44, 469 (1980)
[13] Schirmacher W., Sol. State Comm. 53, 1015 (1985)
[14] Beck H. and S. Nettel, Phys. Lett. A 105, 319 (1984)
[15] Nettel S. and H. Beck, Phys. Rev. B 28, 4535 (1983)
[16] Ehrenreich H. and L.M. Schwartz, Sol. State Phys. 31, 149 (1976)
[17] Dreirach O. et al., J. Phys. F 2, 709 (1972)
[18] Esposito E. et al., Phys. Rev. B 18, 3913 (1978)
[19] Delley B. and H. Beck, J. Phys. F 9, 517 (1979)
[20] Dunleavy H.N. and W. Jones, J. Phys. F 8, 1477 (1978)
[21] Gorecki J. and J. Popielawski, J. Phys. F 13, 1197 and 2107 (1983)
[22] Frésard R. and H. Beck, submitted to J. Phys. F
[23] John W. and P. Ziesche, Phys. Stat. Solidi (b)47, 555 (1971)
[24] Nicholson D. and L. Schwartz, Phys. Rev. Lett. 49, 1050 (1982)
[25] Ballentine L.E., Adv. Chem. Phys. 31, 263 (1975)
[26] Häussler P. et al., Phys. Rev. Lett. 51, 714 (1983)
[27] Nagel S.R. and J. Tauc, Phys. Rev. Lett. 35, 380 (1975)
[28] Mott N.F., Phil. Mag. 26, 1249 (1972)
[29] Güntherodt H.J. et al., Journal de Physique, Colloque, C 8, 381 (1980)

ELECTRONIC STRUCTURE OF METALLIC GLASSES - BULK AND SURFACE PROPERTIES

P. OELHAFEN
Institut für Physik der Universität Basel
Klingelbergstrasse 82, CH-4056 Basel, Switzerland

ABSTRACT

The bulk electronic structure of a great variety of metallic glasses containing transition metals, normal metals, metalloids or actinide metals has been investigated by electron spectroscopy (UPS, XPS and AES), X-ray emission and band structure calculations. The general features of the valence band structure and their relation to other physical properties is reviewed.

The chemisorption of CO on glassy transition metal alloy surfaces has been studied. A close connection between the surface activity and the bulk electronic structure has been found.

1. INTRODUCTION

The knowledge of the valence band structure of a solid is significant for the understanding of many physical properties. The adequate description in the crystalline state consists of the dispersion relation E(k) along the major symmetry directions and the density of electronic states D(E). In an amorphous isotropic solid or in a liquid, E(k) is no longer defined, whereas the concept of the density of states (DOS) is still meaningful.

Photoelectron spectroscopy is perhaps the most widely used experimental technique in the study of the electronic structure of amorphous alloys. Ultraviolet photoelectron spectroscopy (UPS) with excitation energies of about 20...40 eV yields information about the valence electrons whilst the excitation energies in X-ray photoelectron spectroscopy (XPS) are high enough to ionize inner shell electron states as well and, therefore, additional information on the electronic structure can be obtained. The core level binding energy shifts which are related to the charge rearrangement on alloying and the changes of the core line asymmetries, which can yield at least qualitative information about the local density of states, are of particular interest here. XPS core level spectroscopy also provides important information in the form of a chemical analysis of the sample surface within the probing depth of photoelectron spectroscopy which varies between about 0.5 and 2 nm depending on the kinetic energy of the excited electrons[1].

An important question in connection with UPS valence band spectroscopy on amorphous alloys concerns the problem of how far the spectra reflect the density of initial states i.e. the DOS below the Fermi level E_F. In the crystalline case it is well known that UPS valence band spectra can deviate appreceably from the density of states for several reasons: (i) the final state of the excited electron is influenced by the DOS above E_F and (ii) due to k-conservation the direct optical transitions lead to structures in the energy distribution of the emitted photoelectrons which are not related to DOS effects. The first point becomes irrelevant as soon as the excitation energy is high enough and the final state is located in a free electron like continuum well above E_F. The second point

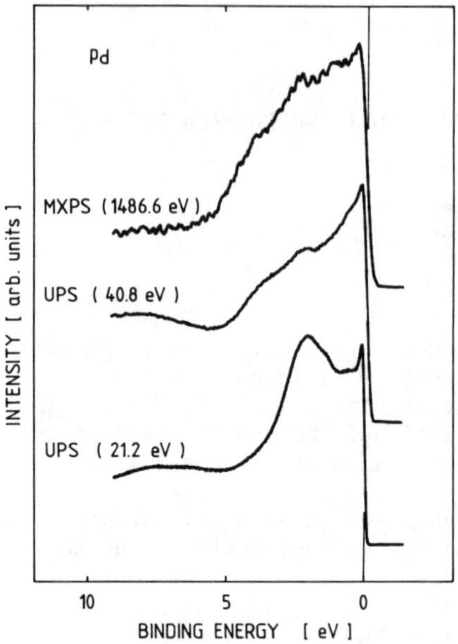

Fig.1: Photoelectron valence band spectra of polycrystalline Pd measured with HeI (21.1 eV), HeII (40.8 eV) and monochromatized Al k_α (1486.6 eV) excitation.

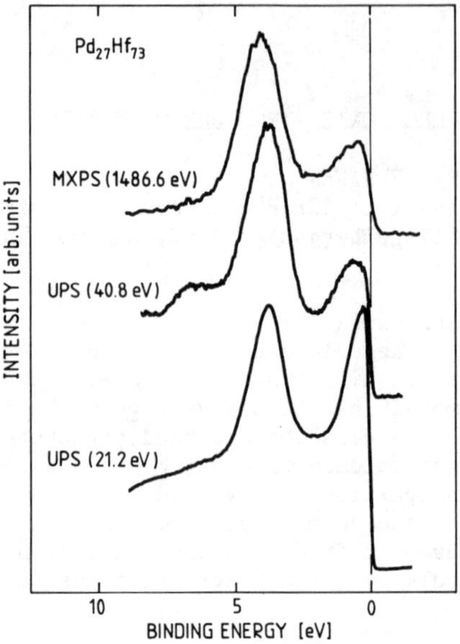

Fig.2: Photoelectron valence band spectra of glassy $Pd_{27}Hf_{73}$ measured with different excitation energies[3].

only holds for the crystalline case since k-conservation is no longer an important selection rule in the amorphous or liquid state.

In practice the photon energy should be varied to test how far a valence band photoelectron spectrum reflects the density of initial states. Figures 1 and 2 are shown in order to illustrate this point. In Fig.1 the valence band spectra of pure polycrystalline Pd measured with different excitation energies are shown. As has been shown by Höchst et al.[2] the XPS valence band spectrum of Pd can essentially be explained by the density of initial states if minor corrections for instrumental resolution, electron hole interaction, matrix element modulation, lifetime of the photohole and inelastic electron-electron scattering are taken into account. It is obvious that the spectrum measured with a photon energy of 21.2 eV deviates appreceably from the XPS valence band spectrum which is believed to represent quite closely the initial density of states. The Pd 4d-electron states near the Fermi level and the peak at about 2 eV are strongly emphasized compared with the Pd 4d bonding states in the lower part of the d-band. In addition, it is evident from Fig.1 that the spectrum measured with 40.8 eV photons reflects more closely the DOS than the spectrum measured with 21.2 eV.

A completely different behaviour can be observed in amorphous alloys or liquid metals. Figure 2 shows a comparison of valence band photoelectron spectra of glassy $Pd_{27}Hf_{73}$[3] measured with three different excitation energies: 21.2, 40.8 and 1486.6 eV. In contrast to the polycrystalline Pd, the general shape of the spectra of the amorphous alloy is very similar. The main difference in the spectra is the relative intensity of the two pronounced peaks. As will be discussed later on, the peak near the Fermi edge contains mainly Hf 5d-electron states whereas the Pd 4d-electron states are essentially located in the peak at a binding energy of about 4 eV. Since the photoelectron excitation cross section for the Hf 5d and the Pd 4d-electron states depend in a different way on the excitation energy the relative peak intensity is changing as a function of energy as well. The peak at 6 eV in the valence band spectrum measured with 40.8 eV is due to a minor oxygen contamination of the sample surface and is not visible in the other two spectra since the surface sensitivity is most pronounced in the the case of an excitation energy of 40.8 eV.

Moreover the similarity of the three spectra shown in Fig.2 indicate that the data reflect more bulk than surface properties of the alloys. The change in the escape depth of the photoelectrons of spectra measured with 40.8 and 1486.6 eV is of the order of a factor of two. Since the biggest change in the electronic structure is expected to occur within the the first two atomic layers at the surface, we can conclude that even the spectrum measured with 40.8 eV, where we are close to the minimum in the escape depth, reflects essentially a bulk DOS if we leave out of consideration the peak at about 6 eV which is due to oxygen contamination. An additional argument for the relevance of UPS valence band spectra to obtain bulk information will be given by comparing photoelectron spectra from clean glassy alloy surfaces with band structure calculations.

The stability model by Nagel and Tauc[4] is an interesting aspect in the context of the valence band density of states of amorphous alloys. Within this model the valence electrons are considered to be nearly free and an increased stability against crystallization has been found when the Fermi level E_F is located at a minimum in the DOS. The size and location of this minimum is in turn determined by the structure factor of the system. For an amorphous system with a spherically symmetric structure factor the system will be in a metastable state. Threfore, many photoemission experiments have been performed on amorphous alloys in order to investigate the DOS behaviour near E_F[5]. However, this kind of study is restricted to nearly free electron-like systems since the electronic structure in transition metal alloys is dominated by d-electrons and therefore a nearly free electron model is no longer applicable. A minimum in the DOS at E_F, which is obviously related to the model of Nagel and Tauc, has been observed in different amorphous alloys such as quench condensed films of noble metal-tin alloys[6], glassy Ca-Al[7] and Mg-Zn[8]. However, we are not going to discuss these systems in more detail. We will mainly focus on glassy transition metal alloys for which we will review the general common valence band properties in order to understand their relation to the chemisorption behaviour we have studied on these alloys.

2. VALENCE BAND STRUCTURE OF GLASSY TRANSITION METAL ALLOYS

The valence band spectra of a typical glassy transition metal alloy $Rh_{25}Zr_{75}$ is shown in Fig.3 along with the spectra of the pure polycrystalline alloy constituents[5,9]. The alloy spectrum is dominated by a distinct two peak structure which does not exhibit any similarities with the spectra of Rh and Zr. Moreover it is obvious from Fig.3 that the alloy spectrum cannot be obtained by a superposition of the pure element valence band spectra. From this we can already conclude that Rh-Zr is a strongly interacting system for which no simple picture such as a rigid band model can be applied. This behaviour is in contrast with weakly interacting systems such as the random compositionally disordered binary alloys with alloy constituents close to each other in the periodic table

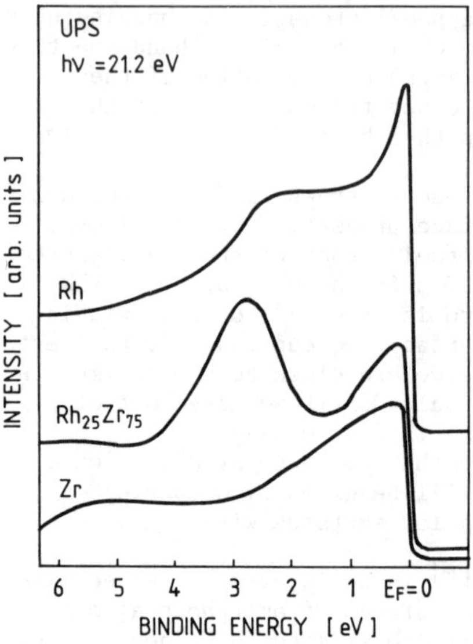

Fig.3: UPS valence band spectra of polycrystalline Rh, Zr and glassy $Rh_{25}Zr_{75}$[5,9].

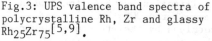

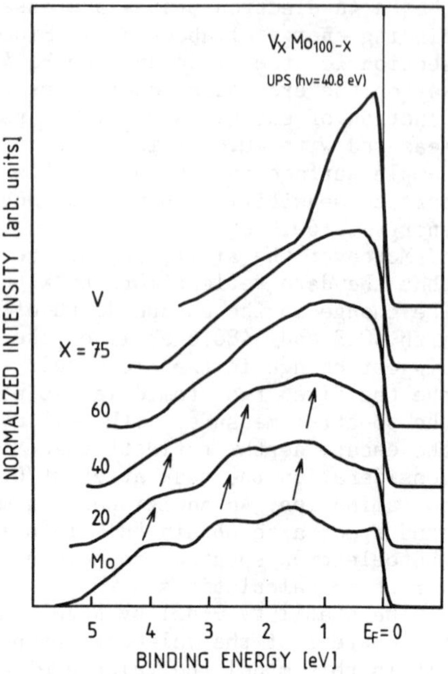

Fig.4: UPS (HeII) valence band spectra of Mo and crystalline V_xMo_{100-x} solid solutions[10].

and with only small electronegativity differences. An example is shown in Fig.4 with valence band spectra measured on bcc V-Mo alloys[10]. The rigid band behaviour in the alloys at the Mo rich side is clearly seen since three distinct structures in the spectra are shifting rigidly towards lower binding energies if we increase the V content and, as a consequence, are decreasing the average number of valence electrons per atom.

An important question in connection with the alloy valence band spectrum in Fig.3 concerns the local electronic structure or, in other terms, what is the contribution from the Rh 4d and the Zr 4d-electron states to the dominating peaks at E_F and at a binding energy of about 3 eV? A first qualitative information to this problem can be obtained by looking at a difference spectrum obtained from measurements on samples with two different alloy concentrations. From this it can clearly be concluded that an increase in the Rh content leads to a increase in the intensity of the peak at 3 eV and correspondingly a decrease in the intensity for the peak at E_F.

This qualitative picture can be confirmed by analizing other experimental data with more local character such as X-ray emission spectroscopy (XES), Auger elctron spectroscopy (AES) and XPS core level line shape analysis. Figures 5a and 5b show the X-ray emission spectra of the Rh and Zr emission bands from the alloy $Rh_{25}Zr_{75}$ and the pure alloy constituents[11]. Since the recombination of the electron from the valence band to the L_{III} shell is subject to dipole selection rules ($\Delta\ell = \pm 1$) the emission spectra are determined essentially by transitions of Rh and Zr 4d-electrons. In order to interpret the energy shifts of the emission bands with respect to E_F we need to know the XPS core level

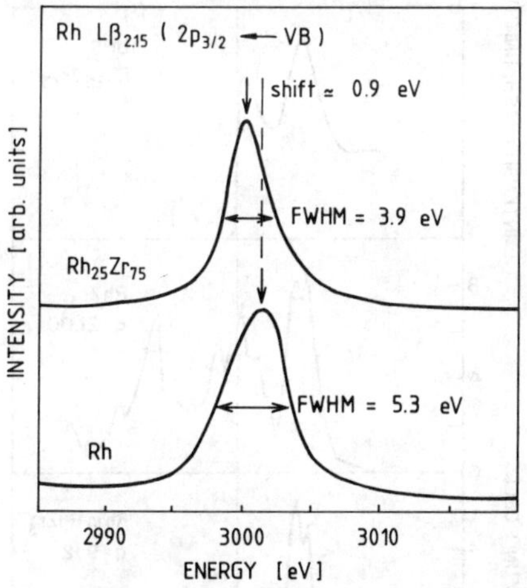

Fig.5a: XES Rh emission bands from crystalline Rh and glassy $Rh_{25}Zr_{75}$[11].

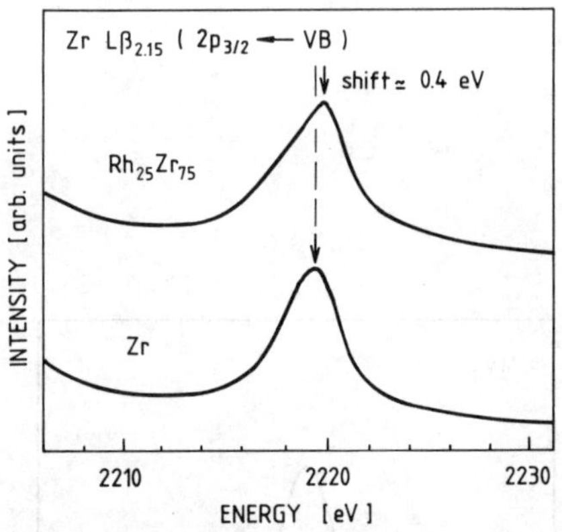

Fig.5b: XES Zr emission bands from crystalline Zr and glassy $Rh_{25}Zr_{75}$[11].

binding energy shifts on alloying for which we take the values of the Rh 3d and Zr 3d binding energy shifts which amount +0.2 and +0.1 eV, respectively. Since the Rh and Zr emission bands exhibit shifts of -0.9 and +0.4 eV a net shift of the d-bands of -1.1 and +0.3 eV with respect to the Fermi level is obtained for the Rh and Zr 4d-band, respectively[12]. We note that the Rh 4d-band shift of -1.1 eV is in reasonable agreement with the shift of the centroid position of the Rh d-band on alloying. Finally we observe besides the shifts of the emission bands a distinct decrease of the FWHM in the case of Rh which is due to the marked dilution in the Zr rich alloy and an increase of the FWHM for the Zr emission band due to a shoulder at the low energy side which indicates hybridization effects with the Rh 4d-electron states.

Similar information can be obtained from CVV Auger electron spectra i.e. transitions in which the valence electrons are involved. The necessary condition in order to have a band-like transition concerns the relative magnitude of the band width W and the effective coulomb interaction U_{eff}: the transition is band-like if $W > U_{eff}$ and is atomic-like if $W < U_{eff}$[13]. An example for a band-like transition which can yield information about the local electronic structure is the $M_5N_{4,5}N_{4,5}$ transition of Rh which is shown in Fig.6[14]. Again, a shift to lower kinetic energy and a narrowing is observed. If we assume the same effective coulomb interaction for the two hole final state of the Auger transition for pure Rh and the alloy we are able to explain the shift of the Auger peak by the core level binding energy shift and the Rh d-band shift by using the simple formula

$$\Delta E_{AES} = \Delta E_B^{core} - 2\Delta E_B^{d\text{-}band},$$

where ΔE_{AES} is the Auger line shift (kinetic energy) and ΔE_B^{core} and $\Delta E_B^{d\text{-}band}$ the core level and the d-band binding energy shifts, respectively. Since the Auger line shift is 2.1 eV (Fig.6) and the core level shift measured with XPS is 0.2 eV a d-band shift of 1.15 eV is obtained, which is in very good agreement with the shift obtained from XES.

The observed shift of the Rh 4d-states to higher binding energies on

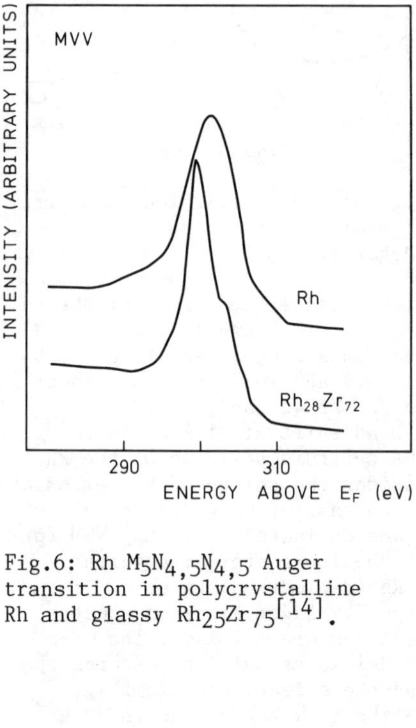

Fig.6: Rh $M_5N_{4,5}N_{4,5}$ Auger transition in polycrystalline Rh and glassy $Rh_{25}Zr_{75}$[14].

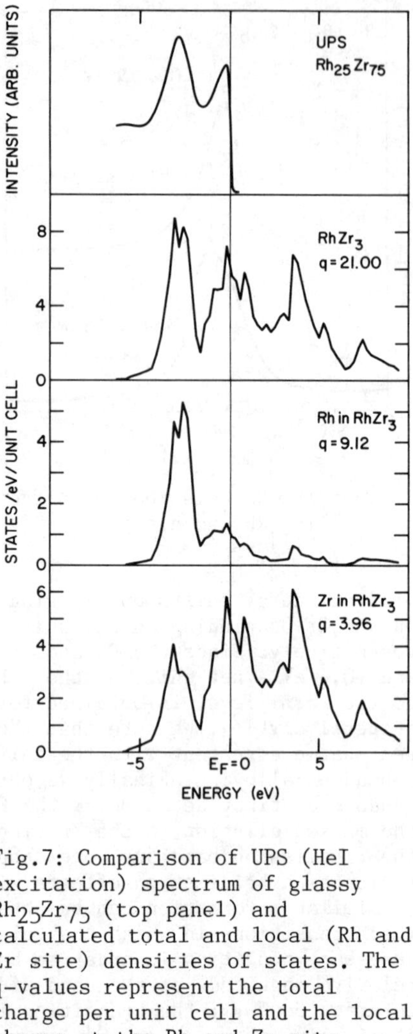

Fig.7: Comparison of UPS (HeI excitation) spectrum of glassy $Rh_{25}Zr_{75}$ (top panel) and calculated total and local (Rh and Zr site) densities of states. The q-values represent the total charge per unit cell and the local charge at the Rh and Zr site, respectively[9].

alloying is accompanied by a distinct decrease in the density of states at the Rh site at E_F. This effect can experimentally be observed by a marked change in the XPS core line asymmetry which is due to electron-hole excitations near E_F. Therefore, a decrease of the local density of states leads to a decrease in the probability for the excitation of electrons at E_F and in turn to a deacrease in the core line asymmetry.

Figure 7 shows a comparison of the UPS valence band spectrum of glassy $Rh_{25}Zr_{75}$ with calculated density of states[9]. The calculation is performed by the self-consistent ASW (augmented spherical wave) method for the fcc-like $AuCu_3$ type symmetry. The calculated state densities shown here refer to the theoretical equilibrium lattice separation obtained by energy minimization. The only input to these calculations are

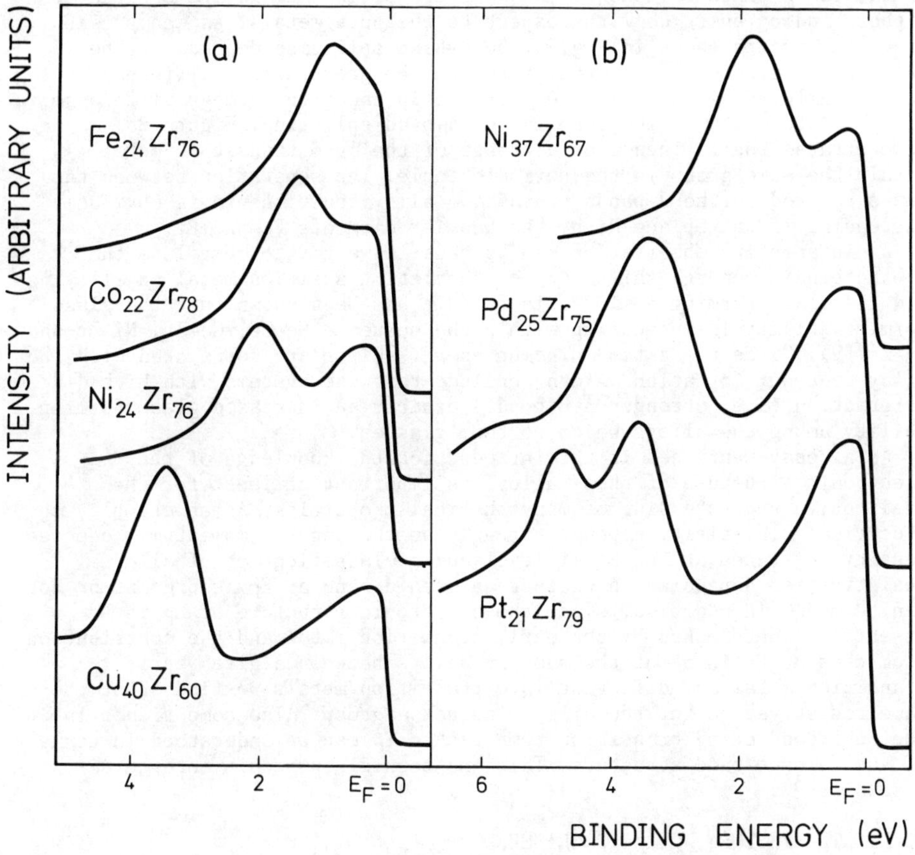

Fig.8: UPS (HeI excitation) spectra of glassy Zr alloys with late transition metals (a) from the same series and (b) from the same group of the periodic table[15].

the atomic numbers of the constituents and the crystal structure. The site decomposed densities of state are included in addition to the total density of states. The comparison of the experimental and theoretical DOS shows a close agreement between corresponding d-band positions. In addition, the calculations clearly confirm the important features obtained experimentally: the dominance of the d-states of the early transition metal at E_F and the position of the Rh 4d-states at a binding energy of about 3 eV.

The d-band behaviour of glassy Rh-Zr alloys as discussed above has basically been observed in all glassy alloys with late and early transition metals (i.e. transition metals with more than half and less than half filled d-bands, respectively). However, distinct trends in the d-band shifts on alloying with respect to the relative position of the two alloy constituents in the periodic table and the alloy concentration has been found. Figure 8 shows a comparison of different valence band spectra of glassy Zr-alloys measured with 21.2 eV photons[15]. An analysis similar to that described above for Rh-Zr revealed for all the alloys shown in Fig.8 a dominance of the Zr 4d-electron states at E_F and, again, the d-bands related to the late transition metals are shifted to higher binding energies with respect to the pure metals. As can be seen from the spectra shown in Fig.8a the d-band splitting depends on the valence difference (or in other terms on the group number difference) of the two alloy constituents: the increase in the group number difference is correlated with an increase in the d-band splitting. Figure 8b demonstrates the influence of a change of the late transition metal within the same group in the periodic table. The separation between the peaks related to the d-bands of the two alloy constituents is obviously increasing if we replace Ni by the heavier elements Pd and Pt.

An interesting observation can be made if we try to correlate the d-band binding energy shifts ΔE_B of the late transition metal on alloying and the glass forming ability. In fact it has been shown that the glass forming ability is increasing within the sequence Fe-Zr, Co-Zr, Ni-Zr and Cu-Zr[16]. Since the d-band binding energy shifts are correlated with the alloy heats of formation we can conclude that the systems with high d-d interaction (i.e. stronger A-B bonds) exhibit an increased glass forming ability among the alloys which do form glasses.

As already mentioned in the introduction the knowledge of the electronic structure of these alloys is important at least for the qualitative understanding of many physical properties. As an example, the electrical resistivity depends strongly on the angular momentum projected density of states at the Fermi level and a classification of alloy resistivities depending on whether an open d-band at E_F is present or not can be made. In addition, since the electronic structure at E_F is essentially determined by the early transition metal and the contribution from this at E_F is about the same in alloys between a given early transition metal and different late transition metals we find about the same resistivities for the alloys in such a group. Also some trends in the superconducting transition temperature T_c can be understood in terms of the valence band structure. This has been discussed in more detail elsewhere[5].

3. DILUTE SYSTEMS: ACTINIDE GLASSES

The possibility to change the alloy concentration continuously over a fairly large range is an important advantage in studies using metallic glasses. This has been applied in the context of uranium glasses in order to increase the mean U-U distance in an amorphous matrix and to examine the electronic configuration of uranium. A study of U-rich binary glasses with Fe, Co, and Ni did reveal a U $5f^3$ configuration[17]. Therefore, no change in the valence between these alloys and pure alpha-U has been found. However, the situation can be different in a dilute U-glass as we can see in Fig.9[18]. The UPS valence band spectra of the glassy alloys $(Pd_{25}U_{75})_{82}Si_{18}$, $(Pd_{95}U_5)_{83}Si_{17}$ and $Pd_{82}Si_{18}$ and pure uranium is shown in Fig.9a. The U-rich alloy is dominated by a U 5f peak at E_F and shows only a weak Pd 4d peak at a binding energy of about 4 eV. The corresponding U 4f core level spectrum is shown in Fig.9b. The close similarity to the U 4f spectrum of pure U is evident: the core level binding energy shift is quite small (0.2 eV) and the shape of the core lines exhibit almost no change on alloying which indicates that no change in the electronic configuration takes place. A drastic change, however, can be observed if we lower the U content. The U peak at E_F is no longer discernible in the valence band spectrum (even at very high magnification) and the U 4f core lvels exhibit a marked shift of 1.4 eV and a distinct change in the shape of the core line accompanied with the occurence of a satellite. An almost identical behaviour has been found in

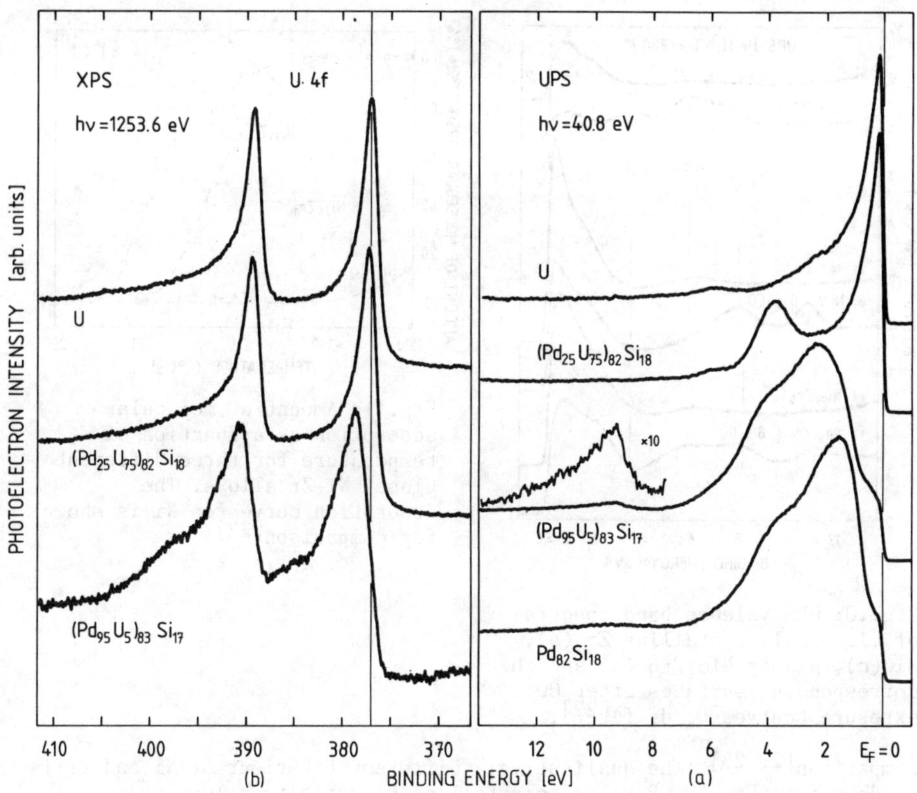

Fig.9: (a) UPS (HeII excitation) and (b) XPS U 4f core level spectra of alpha-U and glassy Pd-U-Si and Pd-Si alloys[18].

crystalline Pd_3U in which the U is considered to be in the tetravalent U $5f^2$ configuration[19]. Therefore, uranium exhibits a change in the valence as a function of alloy concentration. This observation has been confirmed by magnetic measurements which indicate a large increase in the magnetic moment per U atom in the dilute U alloy compared with the U-rich system[20]. Measurements of U glasses in other amorphous matrices such as Ni-Y-U or Ni-Zr-U behave differently and no change in the valence could be detected[18].

4. SURFACE PROPERTIES OF GLASSY TRANSITION METAL ALLOYS: CHEMISORPTION OF CARBON MONOXIDE

Some metallic glasses exhibit unusual surface properties such as high corrosion resistivity or high catalytic activity[21]. Despite these facts only few investigations using surface sensitive analytical methods have been used so far in order to understand the surface reactions and the role of substrate involved in corrosion and catalysis. As a first model reaction the chemisorption behaviour of CO on glassy transition metal alloys has been examined and an interesting correlation between the dissociation behaviour and the local electronic structure has been found.

A typical chemisorption experiment with CO on glassy $Ni_{91}Zr_9$ is shown in Fig.10 and the reaction of pure Zr and Ni with CO is included for

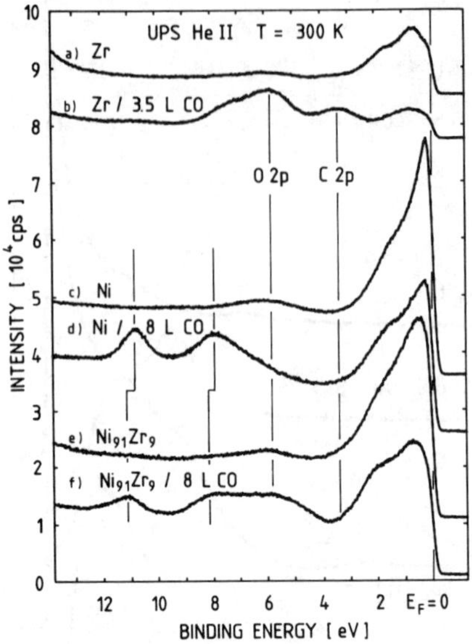

Fig.10: UPS valence band spectra of clean polycrystalline Zr (a), Ni (c), glassy $Ni_{91}Zr_9$ (e) and the corresponding surfaces after CO exposure (curves b, d, f)[22].

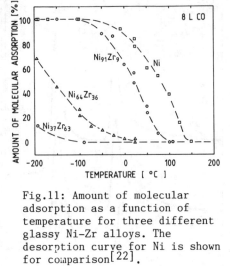

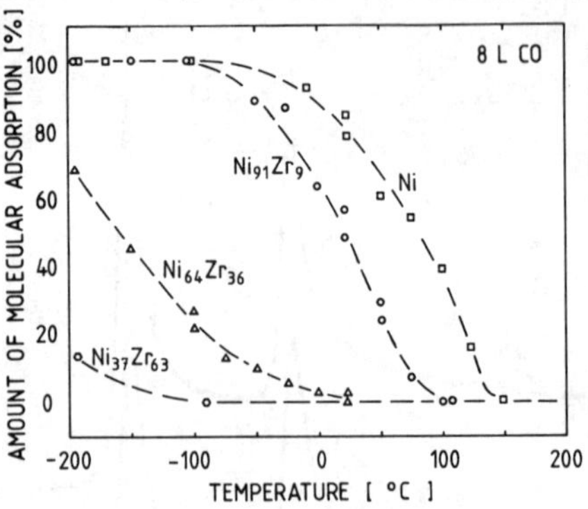

Fig.11: Amount of molecular adsorption as a function of temperature for three different glassy Ni-Zr alloys. The desorption curve for Ni is shown for comparison[22].

comparison[22,23]. The qualitatively different behaviour of Ni and Zr is evident from Fig.10. Pure polycrystalline Ni exhibits molecular adsorption at room temperature which is typical for the late transition metals whereas Zr, like most other early transition metals, dissociate the adsorbed CO molecule. This can be seen unambiguously from the peaks related to the 4-sigma and 5-sigma/1-pi molecular orbitals at about 8 and 11 eV in the case of Ni and the O 2p and C 2p peak at about 3.5 and 6 eV in the case of Zr. The spectrum of the CO exposed $Ni_{91}Zr_9$ alloy shows clearly a mixture of peaks which can be attributed to both molecular and dissociative adsorption. A quantitative analysis yields about an equal amount of dissociative and molecular adsorption at room temperature for this alloy.

The temperature and concentration dependence of the dissociation properties are of particular interest. The amount of molecular adsorption for three different glassy Ni-Zr alloys is shown as a function of temperature in Fig.11[22]. The complement to 100% correspond to the amount of dissociative adsorption on these alloys. The desorption curve of CO on pure Ni is shown for comparison. Two observations should be emphasized: (i) there is a distinct shift of the boundary between molecular and dissociative adsorption to lower temperatures with increasing Zr content revealing the strong influence of Zr present in Ni, and (ii) the fact that e.g. $Ni_{91}Zr_9$ shows at -100°C 100% molecular adsorption and at +100°C 100% dissociative adsorption. From this it is obvious that the adsorption behaviour of these alloys cannot be understood in terms of Ni-like and Zr-like adsorption sites but the local electronic structure, which is very different from that in the corresponding pure metals, has to be taken into account.

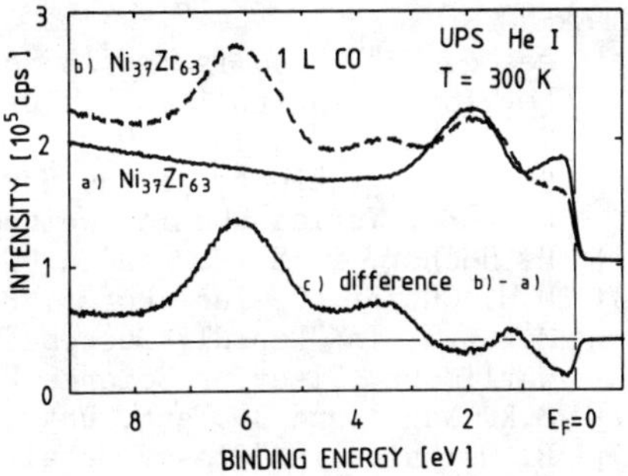

Fig.12: UPS valence band spectra of clean glassy $Ni_{37}Zr_{63}$ (b) and the same surface after chemisorption of 1 Langmuir CO[22] (1L = $1\cdot10^{-6}$ Torr·sec).

The split d-bands of the transition metal alloys under investigation with well separated contributions from the two alloy constituents makes it possible to identify the electron states which are responsible for dissociation and molecular adsorption. In the Ni-Zr alloys the Ni d-band reveals an increasing binding energy with decreasing Ni content with respect to pure Ni. Since the same binding energy shifts have been observed for the 4-sigma and 5-sigma/1-pi molecular orbitals we conclude that the molecular adsorption is essentially coupled with the antibonding Ni 3d-electron states. On the other hand the relevance of the bonding Zr 4d-electron states at E_F for the dissociation of CO can be seen from Fig.12. The $Ni_{37}Zr_{63}$ alloy has been exposed to CO at room temperature and 100% dissociative adsorption has been found[22]. Comparing the broken curve (after dissociation) with the spectrum of the clean alloy surface shows a distinct decrease in intensity at E_F, while the intensity of the Ni 3d peak at about 2 eV remains practically the same in both spectra.

The above observations are not specific to Ni-Zr alloys but are also found in other glassy transition metal alloys. In all the cases studied so far the bonding d-states at E_F of an early transition metal seems to be crucial for the dissociation of the CO molecule. As soon as these electron states are missing at E_F, dissociation is no longer observed e.g. in Pd-Si, Fe-Ni-B or similar alloys. However, the d-band shift of the late transition metal seems to be responsible for the decrease in the CO-substrate bond, which can be observed at a reduced temperature for the desorption of the molecules.

5. CONCLUSIONS

Glassy transition metal alloys exhibit strong d-d interactions which are manifested by a distinct d-band splitting. The degree of the splitting depends on the relative position of the alloy constituents in the periodic table and the alloy concentration. The strong d-d interaction is also responsible for the distinct change in the local electronic structure on alloying. The chemisorption behaviour of the glassy transition metal alloys is strongly determined by these local properties and the distinct separation of the d-states arising from the individual alloy constituents makes it possible to discriminate the role of the different d-electron states for molecular and dissociative adsorption.

ACKNOWLEDGEMENT

I would like to acknowledge fruitful discussions and collaborations in the field presented here with Y. Baer, K.H. Bennemann, U. Gubler, H.-J. Güntherodt, C.F. Hague, R. Hauert, P. Häussler, G. Indlekofer, V.L. Moruzzi, R. Schlögl, D. Tomanek and A.R. Williams. This work has been supported by the Swiss National Science Foundation and the "Kommission zur Förderung der wissenschaftlichen Forschung".

REFERENCES

[1] see e.g. "Photoemission in Solids I", ed. by M. Cardona and L. Ley,Topics in Applied Physics, Vol.26, Springer-Verlag, Berlin, Heidelberg, New York 1978
"Low Energy Electrons and Surface Chemistry", by G. Ertl and J. Küppers, Verlag Chemie, Weinheim 1974
[2] H. Höchst, S. Hüfner and A. Goldmann, Phys. Lett. 57A, 265 (1976)
[3] U.M. Gubler, G. Indlekofer, P. Oelhafen, H.-J. Güntherodt and V.L. Moruzzi, in "Rapidly Quenched Metals", ed. by S. Steeb and H. Warlimont, Elsevier Science Publishers B.V., 1985, p. 971
[4] S.R. Nagel and J. Tauc, Phys. Rev. Lett. 35, 380 (1975)
[5] P. Oelhafen in "Glassy Metals II" ed. by H. Beck and H.-J. Güntherodt, Topics in Applied Physics, Vol. 53, Springer-Verlag, Berlin Heidelberg, 1983, p.283
[6] P. Häussler, F. Baumann, J. Krieg, G. Indlekofer, P. Oelhafen and H.-J. Güntherodt, Phys. Rev. Lett. 51, 714 (1983)
P. Häussler, F.Baumann, U. Gubler, P. Oelhafen and H.-J. Güntherodt, in "Rapidly Quenched Metals" ed. by S. Steeb and H. Warlimont, Elsevier Science Publishers B.V., 1985, p.1007
[7] S.R. Nagel, U.M. Gubler, C.F. Hague, J. Krieg, R. Lapka, P. Oelhafen, H.-J. Güntherodt, J. Evers, A. Weiss, V.L. Moruzzi and A.R. Williams, Phys. Rev. Lett. 49, 575 (1982)
[8] P. Oelhafen, J. Krieg and H.-J. Güntherodt, unpublished results
[9] V.L. Moruzzi, P. Oelhafen, A.R. Williams, R. Lapka, H.-J. Güntherodt and J. Kübler, Phys. Rev. B27, 2049 (1983)
[10] G. Indlekofer, P. Oelhafen, J. Krieg, R. Lapka, U.M. Gubler, H.-J. Güntherodt,V.L. Moruzzi, H.R. Khan and K. Lüders, Z. Phys. B 58, 293 (1985)
[11] J.-M. Mariot, C.F. Hague, P. Oelhafen and H.-J. Güntherodt, to be published
[12] P. Oelhafen, E. Hauser and H.-J. Güntherodt in "Inner Shell and X-ray Physics of Atoms and Solids" ed. by D.J. Fabian, H. Kleinpoppen and L.M. Watson, Plenum Press, New York 1981, p.575
[13] J.C. Fuggle in "Electron Spectroscopy: Theory, Techniques and Applications", Vol.4, ed. by C.R. Brundle and A.D. Baker, Academic Press, London 1981, p.85
[14] P. Oelhafen, unpublished results
[15] P. Oelhafen, E. Hauser and H.-J. Güntherodt, Solid State Commun. 35, 1017 (1980)
[16] Y. Nishi, T. Morohoshi, M. Kawakami, K. Suzuki and T. Masumoto in "Proc. 4th Intern. Conf. on Rapidly Quenched Metals", ed. by T. Masumoto and K. Suzuki", 1982, p.111
[17] P. Oelhafen, G. Indlekofer, J. Krieg, R. Lapka, U.M. Gubler, H.-J. Güntherodt, C.F. Hague and J.M. Mariot, J. Noncryst. Sol. 61&62, 1067 (1984)
[18] G. Indlekofer, P. Oelhafen and H.-J. Güntherodt, in "Rapidly Quenched Metals", ed. by S. Steeb and H. Warlimont, Elsevier Science Publishers B.V., 1985, p.1011
[19] Y. Baer, H.R. Ott and K. Andres, Solid State Commun. 36, 387 (1980)
[20] J. Durand et al., to be published
[21] R. Schlögl in "Rapidly Quenched Metals", ed. by S. Steeb and H. Warlimont, Elsevier Science Publishers B.V., 1985, p.1723
[22] R. Hauert, P. Oelhafen, R. Schlögl and H.-J. Güntherodt, Solid State Commun. 55, 583 (1985)
[23] D. Tomanek, R. Hauert, P. Oelhafen, R. Schlögl and H.-J. Güntherodt, Surf. Sci. 160, L493 (1985)

FLUID METALS AT HIGH TEMPERATURES

F.HENSEL

Physical Chemistry, University of Marburg, D-3550 Marburg,FRG

1.INTRODUCTION

Fluid metals, because of their high latent heats of vaporization and heat-transfer coefficients, are prominent candidates for high temperature working fluids in turbine power converters and for heat-transfer media in nuclear reactors. For the selection of a particular metal and for the design and efficient operation of technical processes the precise knowledge of the physical and thermodynamic properties of fluid metals is essential. The necessity for safety analysis and risk assessment requires to extend our knowledge of those properties up to conditions far above the proposed operating conditions, preferably up to and beyond the liquid-vapour critical point temperatures of metals. Inspite of this technological interest, however, many fundamental properties of fluid metals - including the critical point phase transition- are poorly understood. No rigorous theoretical approach is available that describes the thermophysical behaviour of fluid metals over the entire liquid-vapour range.

In the absence of detailed experimental or theoretical information, engineers traditionally have relied heavily upon empirical methods, based on equations of state and the principle of corresponding states for correlating and predicting such properties. These methods are well established for certain nonmetallic fluids (1), but this does not guarantee the validity for liquid metals. As is well known the fundamental difference of dealing with the thermophysical behaviour of metals is the absence of an adequate interatomic-potential function for the entire liquid-vapour range. Regardless of the way in which the intermolecular interaction in a metal is described the description must change with density. Liquid metals near their melting points are reasonably well described by the nearly-free-electron approximation (2), whereas at sufficiently low densities, in the vapour phase, the valence electrons should occupy spatially localized atomic orbitals. This implies that the interatomic cohesion should change qualitatively on going from liquid-like to vapour-like densities. By contrast, for most insulating molecular substances the interatomic interaction may be considered independent of density to a good approximation. Several attempts have been made to study this contrast theoretically (3) and experimentally (4), but the subject has remained elusive.

The fundamental theoretical difficulty in dealing with the thermophysical properties of fluid metals is the presence of competing interactions in the system. In particular, the rapid variation in the electronic properties in course of the metal-nonmetal transition near the critical point of a fluid metal manifests itself in a correspondingly strong density- and temperature-dependence of the effective particle interaction. Corresponding to the theorist's intractable many-body effects is the experimental difficulty that the high cohesive energies of metals place the critical region at temperatures and pressures too high for easy experimental investigation. The resulting problems of temperature and pressure measurement and control together with the highly reactive nature of fluid metals have chiefly limited the accuracy with which properties have been measured in the past. The main problem was that the analysis of electrical, magnetic and thermophysical data close to the critical point of metals was hampered by the presence of spurious effects due to temperature gradients.

This situation is now changing. Improvements and extensions of experimental technique make it possible to measure the equation of state, the electrical transport properties and partly optical properties close to the critical points of Hg, Cs and Rb with quite high precision and comparatively accurate temperature control and optimal elimination of temperature gradients. This new experimental information seems to be accurate enough to allow one to determine the exact location of the critical points of Hg, Cs and Rb (table 1) and to study the asymptotic behaviour of the physical properties of metals near the vapour-liquid critical point.

TABLE 1.

Metal	T_c (K)	p_c (bar)	ρ_c (g/cm^3)	Ref.
Hg	1750	1670	5.8	(5)
Cs	1924	92.5	0.38	(6)
Rb	2017	124.5	0.29	(6)

A comprehensive review of the existing experimental and theoretical results of fluid metals is unnecessary because of the large number of surveys (3,4) that exists and in the following I will restrict myself to those novel developments which have not been adequately reviewed elsewhere, e.g. critical phenomena in fluid metals and the interrelation between the vapour-liquid critical point phase transition and the metal-nonmetal transition.

2. THE CRITICAL POINT PHASE TRANSITION OF METALS

2.1. The metal-nonmetal transition

For most metals the critical region lies at higher pressures and temperatures than are accessible to conventional experimental techniques. Consequently, a considerable effort has been carried out to estimate their critical conditions (7). The estimation methods are based on different versions of the relation between the critical data and other thermophysical properties. The most frequently used methods are the principle of corresponding states and the law of rectilinear diameters (8) which are quite well obeyed for nonmetallic molecular liquids. However, on theoretical grounds, one should not place too much faith in their validity for liquid metals. In the case of metallic liquids the effective particle interactions in the vapour state are very different from the metallic binding forces in the liquid state. This difference is a consequence of the metal-nonmetal transition which leads to a rapid variation in electronic properties near the critical point of a metal which manifests itself in a strong state-dependence of the effective particle interaction. The magnitude of this state-dependence distinguishes the particle interactions in metals from those in insulators.

From the experimental standpoint it is unquestionable that the gross changes in the electronic structure at the metal-nonmetal transition occurs in the vicinity of the critical point. This is convincingly demonstrated by a comparison of fig.1 and fig.2 which present a selection of the most recent and most accurate density ρ (9) and electrical conductivity σ (9) results for fluid cesium in form of isotherms as a function of pressure at sub- and supercritical conditions. The conductivities of the coexisting liquid phases are given by

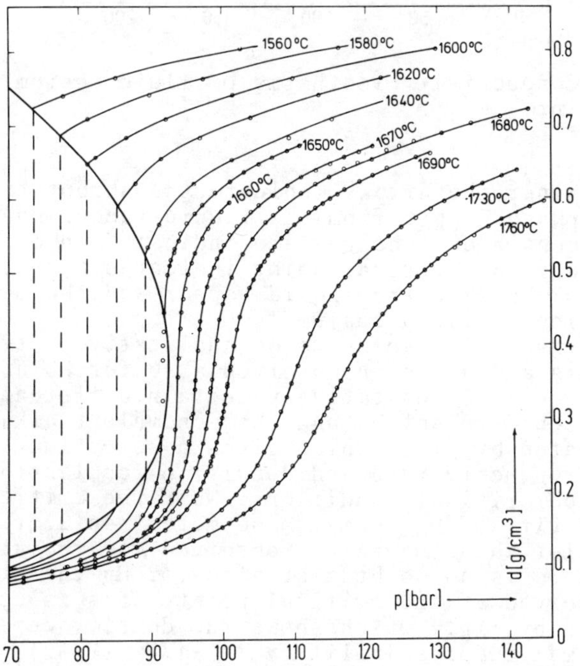

FIGURE 1. Equation of state data of fluid cesium near the critical point.

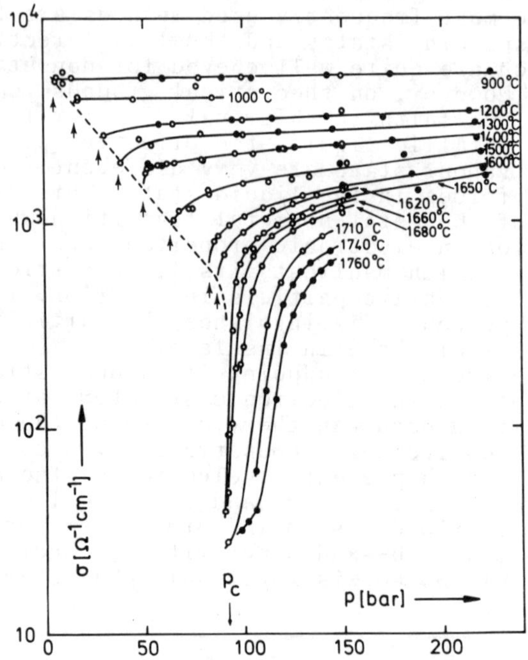

FIGURE 2. Conductivity isotherms of fluid cesium near the critical point.

the dashed line; the arrows indicate the abrupt transition to the vapour phase. Apart from the liquid-vapour phase transition no discontinuous changes are indicated in σ. The steep fall in σ near the critical point indicates that there exists a strong link between the liquid-vapour and the metal-nonmetal transition in fluid cesium.

This behaviour of σ seems to be qualitatively typical for liquid metals and has been observed also for Rb and K and for divalent mercury. Quantitatively certain differences are observed between divalent Hg and the monovalent alkali metals as demonstrated by fig.3 which gives a selection of recent electrical conductivity σ and density ρ results (5) for Hg at sub- and supercritical conditions. Near the critical point the conductivity of Hg seems to be more than 2 orders of magnitude smaller than the values observed for the alkali metals. However, there is no doubt that also for Hg the steepest fall in σ is observed at the critical point. This is convincingly demonstrated by fig.4 which shows the density dependence of the isothermal compressibility $\chi_T = -1/V \times (\partial V/\partial p)_T$ together with the pressure coefficient $1/\sigma \times (\partial\sigma/\partial p)_T$ of the conductivity of fluid mercury at temperatures T close to the critical temperature $T_c = 1478^oC$. There is a close correlation between the increase in χ_T, which diverges as the critical point is

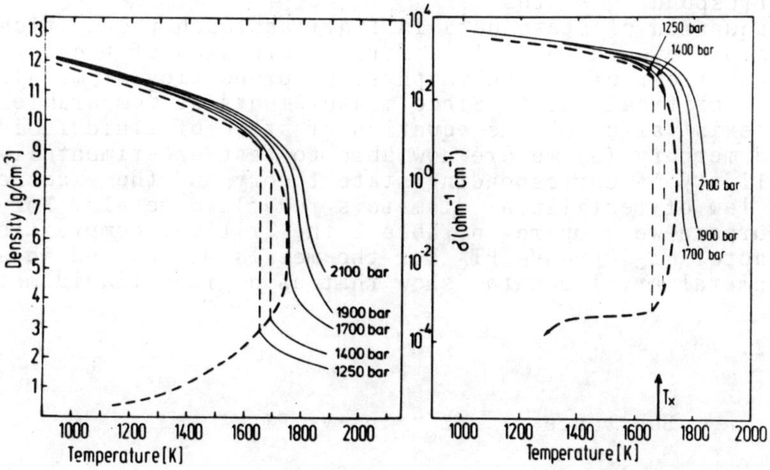

FIGURE 3. Density ρ and electrical conductivity σ of fluid mercury. The dotted lines indicate ρ and σ of the coexisting liquid and vapour phases.

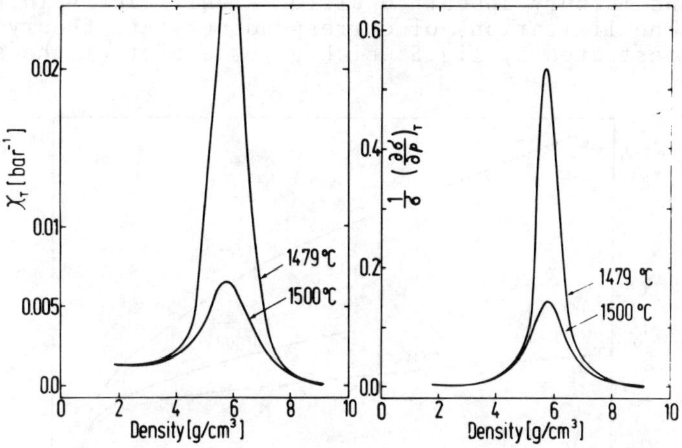

FIGURE 4. Isothermal compressibility, $\chi_T = -1/V\times(\partial V/\partial p)_T$, and the isothermal pressure coefficient of the d.c. conductivity, $1/\sigma\times(\partial\sigma/\partial p)_T$, of fluid Hg for densities and temperatures near the critical point.

approached, and the increase in the pressure coefficient of σ. Both quantities seem to diverge at the critical point.

2.2 Corresponding states

The equation of state data in fig.1 approach the critical point close enough to yield a first impression of the asymptotic behaviour of the thermophysical properties of metals near the critical point. Since measurements of comparable accuracy exist also for the equation of state of fluid rubidium (6) and mercury (5) we are now able to test experimentally the validity of corresponding state theory and the exactness of the law of rectilinear diameters for fluid metals. For that purpose we compare in table 2 the critical compressibility factors $Z_C = p_C \times V_C / RT_C$ for the metals Hg, Cs and Rb and the nonmetal Ar. The data show that as a group liquid metals

TABLE 2.

Substance	$Z_c = p_c \times V_c / RT_c$
Ar	0.291
Hg	0.385
Cs	0.22
Rb	0.22

do not appear to have very similar values of Z_C. The alkali metals and mercury appear to differ significantly in their values. The limitations of corresponding state theory are also demonstrated by fig.5 which gives a plot of the satu-

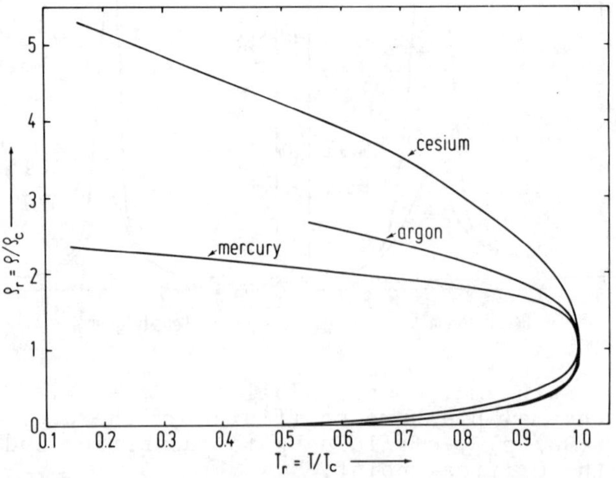

FIGURE 5. Density variation along the liquid-vapour coexistence curve for metals in comparison with argon: plotted are reduced densities ρ/ρ_c versus reduced temperature T/T_c.

rated liquid and vapour densities of Cs, Hg and Ar in reduced variables. Poor reduced correlation between Cs and Hg is observed and the metal curves are clearly distinguishable from that for the group of nonmetallic liquids (e.g. Ar) which obey a principle of corresponding states. Thus the experimental evidence shows that metals and nonmetals cannot be included together in a group obeying a principle of corresponding states, and furthermore that reduced correlations are unlikely to hold for liquid metals as a group. On the other hand, it has been shown that the behaviours of Cs and Rb are very similar for many properties (6,10) (cp. e.g. table 2). Hence it can be expected that the p-V-T data of Cs and Rb can be reduced in corresponding regions of the phase diagram, i.e. that reduced correlations may be found only within groupings of metals. This expectation is supported by fig.6 which shows a reduced plot of the coexisting liquid (ρ_L) and vapour (ρ_V) densities together with the diameter $\rho_d = 1/2\times(\rho_L + \rho_V)$ versus temperature. A law of corresponding states seems to be valid for Cs and Rb over the whole liquid vapour range. This can best be demonstrated when the p-V-T data are analyzed in terms of the thermodynamic equation of state

$$p = T\left(\frac{\partial S}{\partial V}\right)_T - \left(\frac{\partial U}{\partial V}\right)_T = \gamma_V T + p_i \qquad (1)$$

in which $T\times(\partial S/\partial V)_T = (\partial p/\partial T)_V\times T = \gamma_V\times T$ depends only on the repulsive core of the particles. The separation in eq.(1) permits immediate evaluation of γ_V and p_i from an isochoric plot of experimental p-V-T data. Fig.7 shows as an example the p-T-diagram for Cs. The isochores are straight lines within the accuracy of the measurements (6,9). In practice,

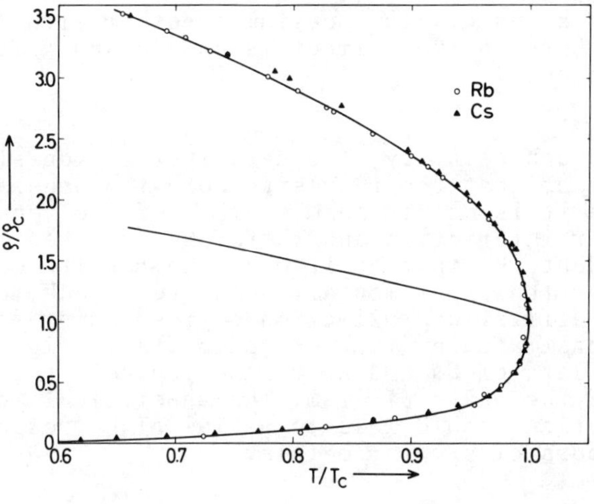

FIGURE 6. Reduced plot of the liquid-vapour coexistence curve of Cs and Rb.

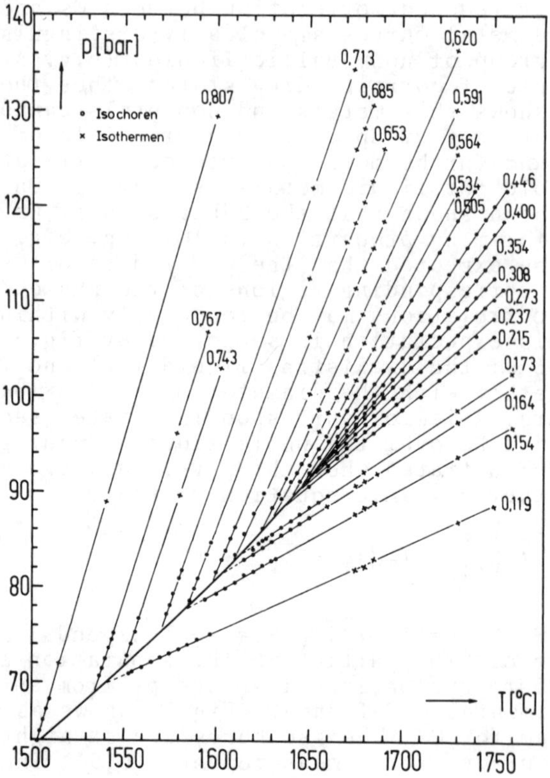

FIGURE 7. Isochores of fluid cesium together with the vapour pressure curve. o from direct isochoric and x from isothermal measurements.

i.e. at high enough accuracy, the data must lie on slight curves. That means that the repulsive core is temperature dependent because it is mainly some average of the repulsive part of the pair interaction and therefore expected to be temperature dependent. Be that as it may, the results of p_i and γ_V obtained from the experimentally observed isochores of Cs and Rb show to a first approximation corresponding state behaviour as demonstrated by the reduced plots in fig.8 and 9, i.e. the p-V-T data of Cs and Rb can be reduced in corresponding regions of the phase diagram. The theoretical reasoning of this observation should lead to a "law of corresponding states" for groups of similar metals.

2.3 Critical phenomena

The coexistence curves of Cs and Rb are remarkably asymmetric compared to those of simple nonmetallic fluids (11). The asymmetry, however, is very similar to that observed for other

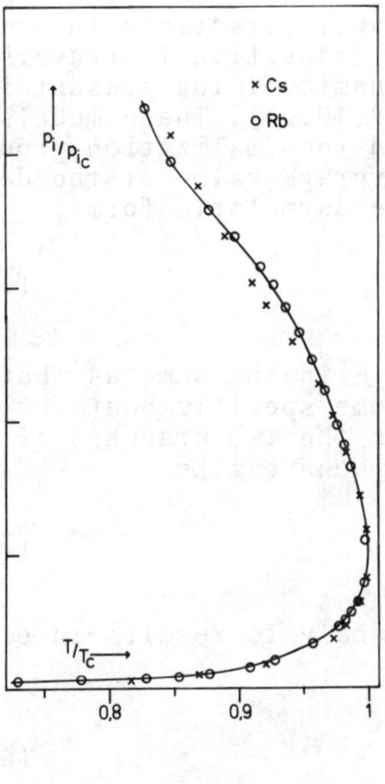

FIGURE 8. Reduced diagram of the inner pressure of Cs and Rb along the liquid-vapour coexistence curve.

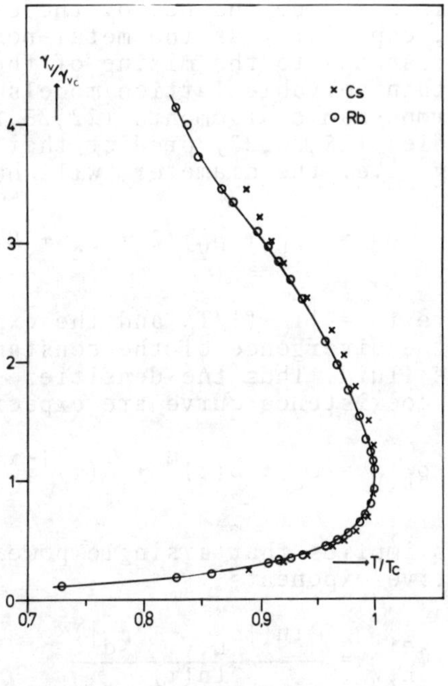

FIGURE 9. Reduced diagram of the isochoric pressure coefficient $\gamma_V = (\partial p/\partial T)_V$ of Cs and Rb along the liquid-vapour coexistence curve.

metallic systems in which the interactions vary strongly with state. In particular, the metal-ammonia (12) and the electron-hole (13) liquid phase diagrams have asymmetries (14) which have been attributed to the change in the range of the screened interactions near the critical points, due to the proximity of metal-nonmetal transitions. It is obvious from fig.6 that Cs and Rb violate the law of rectilinear diameter over a surprisingly large temperature range. By contrast, this law is experimentally valid for the coexistence curves of nearly all simple nonmetallic one-component fluids to within the capacity of present-day experimentation. Thus far the only other one-component system to exhibit an appreciable deviation from rectilinear diameter behaviour is SF_6, as studied by Weiner et al.(15). However, Moldover and Gammon (16) have recently suggested that wetting phenomena may have contributed to this experimental observation.

Recently, it has been suggested by Goldstein and Ashcroft (28) that the contrast between the diameter data of Cs and Rb, and the apparent experimental linearity of the diameters of essentially all nonmetallic one-component fluids arises from many-body effects whose magnitudes distinguish the particle interactions in metallic fluids from those in nonmetallic fluids. In particular, it is argued that the strong thermodynamic state dependence of the effective interactions in a metal, especially as the metal-nonmetal transition is traversed, corresponds to the mixing of thermodynamic fields present in certain solvable lattice models (17,18,19,21). These models, thermodynamic arguments (22,23,24) and renormalization-group studies (25,26,27) predict that the average value of the density, i.e. the diameter, will have the asymptotic form

$$\rho_d = (\rho_L + \rho_V)/2 = \rho_c + D(\tau)^{1-\alpha}, \tag{2}$$

where $\tau = |T_c - T|/T_c$ and the exponent α is the same as that of the divergence of the constant volume specific heat for a pure fluid. Thus the densities $\rho_{L,V}$ in the two branches of the coexistence curve are expected to behave like

$$\rho_{L,V} = \rho_c \pm B(\tau)^{\beta} + A(\tau)^{1-\alpha} + \ldots . \tag{3}$$

This implies that a single power law analysis results in effective exponents

$$\beta^{eff}_{L,V} = \frac{\partial \ln(|\rho_{L,V} - \rho_d|)}{\partial \ln(\tau)} \tag{4}$$

in the two branches of the coexistence curve which may differ (9,4) because of the different relative importance of the $A\times(\tau)^{1-\alpha}$ term in eq.(3) and which may strongly deviate from the values calculated with the renormalization-group approach (4,9). However, the asymptotic exponents

$$\beta^{asym}_{L,V} = \lim_{\tau\to 0} \beta^{eff}_{L,V} \tag{5}$$

must be the same. It is an empirical fact that the higher order terms in eq.(3) cancel to a large extent when the difference $\rho_L - \rho_V$ is formed. For this difference a power law with the same exponent $\beta = \beta_L = \beta_V$ is found to hold, and the asymptotic range of validity is normally quite large in τ. The latter fact is of great help in the analysis of coexistence curves. In fig.10 we have plotted $\log(\rho_L - \rho_V/2\rho_c)$ versus $\log|\tau|$ (6) for Cs and Rb. Fitting to the equation

$$\frac{(\rho_L - \rho_V)}{2\rho_c} = B|\tau|^{\beta}, \tag{6}$$

we find B = 2.25 and β = 0.355 for Cs and B = 2.45 and β =

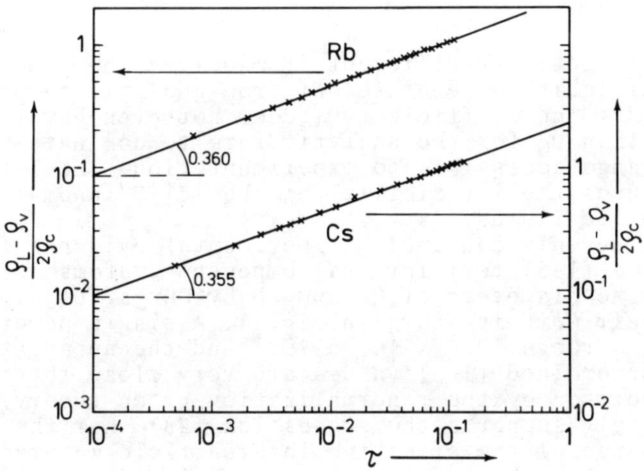

FIGURE 10. Single power law analysis of the coexisting liquid and vapour densities of Cs and Rb (ref.6).

0.36 for Rb. A single power law applies over a range $10^{-3} < |\tau| < 10^{-1}$ and the apparent experimentally determined ß-values are very close to those observed for normal nonmetallic fluids which belong to the same static universality class as an uniaxial ferromagnet represented by the three-dimensional Ising model or the Landau-Ginzburg-Wilson model with an one-component order parameter. The main difference between the coexistence curves of nonmetallic and metallic fluids seems to be the different magnitude of the $A|\tau|^{1-\alpha}$ term in eq.(3) which manifests itself in large amplitudes of the singularities in the coexistence curve diameters (28).

The predicted singularity in the diameter (see eq.(2)) is difficult to verify experimentally for several reasons. Firstly, it has been shown (29) that if one particular function, e.g. ρ, has a $|\tau|^{1-\alpha}$ singularity, then any less symmetric function ρ, where ρ' is an analytic function of (e.g. $\rho = V^{-1}$), behaves as $|\tau|^{2ß}$. Thus the sought-for effect will be missed unless the correct function is chosen. The mass density has long been known empirically to give a more symmetric coexistence curve than volume (30). However, this is only a strong, but not a conclusive argument for supposing that ρ is the appropriate function for the order parameter. Secondly, the size of the asymptotic range (or, equivalently, the amplitudes of the correction terms to eq.(3) depends on the choice of the order parameter. Wegner (31) has used renormalization group theory to produce series expansions for representing data over a wide range of thermodynamic space. The expansions provide the following correction terms to eq.(3) for the diameter

$$\rho_d = \frac{\rho_L + \rho_V}{2\rho_c} = 1 + D_0|\tau|^{1-\alpha} + D_1|\tau| + \ldots \quad (7)$$

Since (1-α) is not very different from unity, the true singularity is difficult to separate out from analytic temperature term of eq.(7). The coefficient D_1 does not even have to be much larger than D_0 for the analytic term to dominate over the entire range accessible to experimentation. The latter certainly causes the invisibility of the $|\tau|^{1-\alpha}$ anomaly in most nonmetallic fluids.

Up to now the only convincing experimental evidence for the existence of a (1-α) term for one-component systems is the analysis of the diameters of Cs and Rb by Jüngst et al.(6). A plot from their work is shown in fig.11. A single power law applies over a range $10^{-3} < |\tau| < 10^{-1}$ and the apparent experimentally determined (1-α) values are very close to the value 0.89 predicted by the renormalization-group theory. This finding strongly supports the suggestion (28) that the strong state-dependence of the effective interparticle interactions, and especially the changes in such forces in course of the metal-nonmetal transition, lead to very large amplitudes of the (1-α) anomaly in the diameters of liquid-vapour coexistence curves of metals.

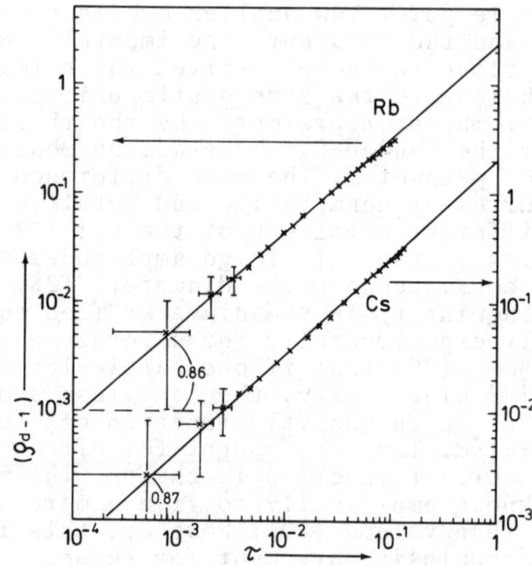

FIGURE 11. Single power law analysis of the diameter of Cs and Rb (ref.6).

2.4 The effect of impurities near the critical point of fluid metals

The thermodynamic properties of dilute binary solutions near the critical points of pure solvents are of significant theorectical and practical importance for the investigation of pure fluid critical points at which the presence of trace amounts of impurity may lead to dramatic effects. The impurities do affect the value of the critical temperature T_c rather strongly, and properties like the activity coefficient or the partial molar volume of solvent and solute undergo anomalies near the critical point of the solvent which reflect the singularities in both the isothermal compressibility, χ_T, and the constant volume specific heat, c_V, of the pure solvent.

A knowledge of this unusual phenomenon is of great importance for any solution (31), but for metallic systems it is of special interest. This is because of the fundamental connection of the critical point phase transition and the metal-nonmetal transition in metallic fluids. The latter implies that the above mentioned anomalies in thermodynamic properties must be accompanied by correspondingly drastic effects in the electrical properties. It was first shown by Zillgitt et al.(32) that a small percentage of indium has a remarkable effect on the conductivity σ of mercury in the metal-nonmetal transition range close to the critical point for temperatures $T > T_x$. Subsequent experiments with indium and other solutes (33,34) confirmed this and connected this behaviour with large negative excess volumes od mixing (35,36). This is shown in fig.12 in which the density and the conductivity of the amalgams is plotted at the same constant pressure and temperature

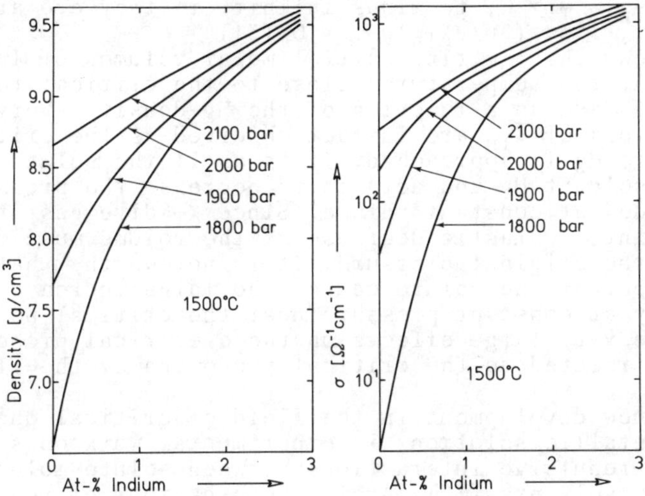

FIGURE 12. Density and conductivity of Hg-In-alloys at constant pressure and T = 1500°C as a function of In-concentration.

as a function of the In-concentration. It is obvious that the increase in σ is primarily caused by a corresponding increase in the density.

This effect can be well understood in terms of the existing models for the critical behaviour of the partial molar volume of the solute In $\overline{V}_{In}$ defined as

$$\overline{V}_{In} = V + (1 - X_{In}) \left(\frac{\partial V}{\partial X_{In}}\right)_{p,T} \tag{8}$$

in dilute solutions, as mole fraction of In $X_{In} \rightarrow 0$. Using the relation

$$\left(\frac{\partial V}{\partial X_{In}}\right)_{p,T} = - \frac{\left(\frac{\partial p}{\partial X_{In}}\right)_{V,T}}{\left(\frac{\partial p}{\partial V}\right)_{T,X_{In}}} . \tag{9}$$

Eq.(8) is rewritten for the limiting case $X_{In} \rightarrow 0$ in the form:

$$\frac{1}{V} \overline{V}_{In} = 1 + \chi_T \left(\frac{\partial p}{\partial X_{In}}\right)_{V,T} . \tag{10}$$

The isothermal compressibility χ_T diverges at the critical point of pure solvent. Consequently the limiting value of $\overline{V}_{In}$ tends at the solvent critical point to plus infinity if the forces between solute and solvent molecules are repulsive, i.e. $(\partial p/\partial X_{In})_{V,T} > 0$, to minus infinity if they are attractive enough , i.e. $(\partial p/\partial X_{In})_{V,T} < 0$ (31).

Fig.13 shows the limiting partial molar volumes of In $\overline{V}_{In}$, as $X_{In} \rightarrow 0$, at two temperatures close to the critical temperature $T_c = 1478^oC$ as a function of the Hg-density. Very high negative values of $\overline{V}_{In}$ are in fact observed as the critical point of pure Hg is approached. It is as if the solute In attracts the solvent Hg and acts so to decrease the pressure, as In is added at constant volume. Since χ_T diverges at the critical point, a drastic decrease of the volume must occur to restore the original pressure. It is noteworthy that mainly this effect of the solute causes the large increase of the conductivity at constant pressure near the critical point of mercury. The very large effects on the electrical properties are closely related to the critical phenomena of the fluid solvent Hg.

A rather new development in the field of critical phenomena in dilute metallic solutions is experimental work on systems with highly repulsive interactions between solute molecules and metallic solvents (37). Again the behaviour of the partial molar volume for the limiting case $x \rightarrow 0$ near the solvent critical point is physically sensible. This is demonstrated by fig.14 which contrasts the behaviours of the partial molar volumes for $x \rightarrow 0$ for In-Hg- and Xe-Hg-solutions at a temperature close to the critical temperature of pure mercury. The

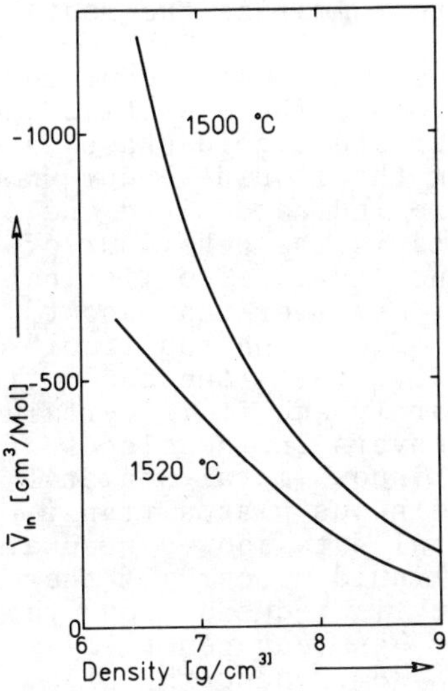

FIGURE 13. Partial molar volume $\overline{V}_{In}$ for $x_{In} \to 0$ for In-amalgams as a function of the Hg-density at temperatures close to the critical temperature of pure Hg.

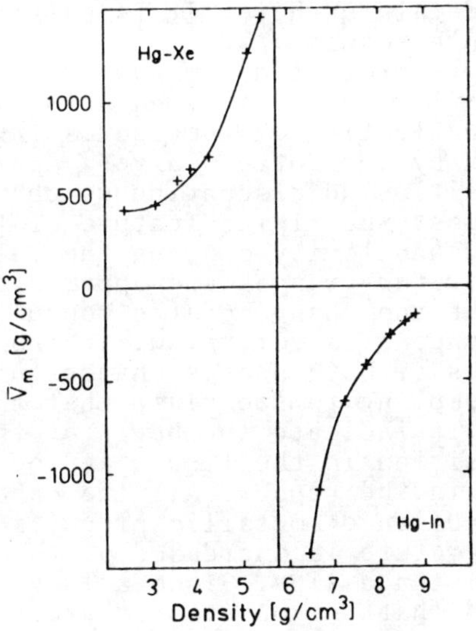

FIGURE 14. Comparison of the density dependence of the partial molar volumes $\overline{V}$ for $x \to 0$ for In-Hg- and Xe-Hg mixtures near the critical point of pure mercury.

partial molar volumes of Xe in Hg (a similar behaviour has been observed for He (37)) diverge to plus infinity at the critical point of mercury. Addition of Xe or He to Hg at constant volume causes an increase in pressure due to the repulsive interaction between the solute molecules and the solvent mercury. Both systems - the Xe-Hg- and the He-Hg-mixtures -, for which the mutual miscibility is extremely low, and whose critical curves extend directly from the critical point of the less volatile mercury component with a positive slope to higher temperatures and pressures, exhibit the so-called gas-gas immiscibility of the first kind (38). The most important aim of such work is a better understanding of the interaction between neutral inert gas atoms with itinerant electronic states.

3. ELECTRONIC PROPERTIES OF FLUID METALS NEAR THE CRITICAL POINT

3.1 Divalent mercury

Mercury has the lowest known critical temperature for any fluid metal. Consequently, extensive experimental work has been devoted to study properties like conductivity, thermoelectric power, Hall effect, Knight shift, optical reflectivity and -absorption, low frequency dielectric constant, and thermodynamic properties at sub- and supercritical consitions.

Most of these results have been fully reviewed (4) and need not detain us here. It is sufficient to summarize the most notable results.

Fig.3 presented already a selection of the most recent conductivity and density data for mercury (5). The densities and conductivities of the coexisting vapour and liquid phases are given by the dotted curves. Apart from the liquid-vapour phase transition no discontinuous changes are indicated in σ and ρ. The most surprising feature of the data is the behaviour of the conductivity σ along the coexistence curve. Two distinct temperature ranges are apparent: For T_x smaller than about 1680 K the change of the conductivity of both phases liquid and vapour is very slow; in the range $T_x < T < T_c$ the conductivities of both phases change quite sharply and it is certainly tempting to speculate that the turnovers in the slopes of σ at T_x indicate the begin of the continuous metal-nonmetal transition in the liquid and of a continuous plasma transition in the vapour. All the experimental data show a gradual diminution of metallic properties in liquid mercury if the temperature is increased or the density is reduced along the coexistence line. Since a long time it has been generally assumed that the onset of characteristic semiconducting electrical transport properties (39) and the relatively sharp drop in the Hg Knight shift (40) indicates that somewhere around T_x = 1680 K, which corresponds to a liquid density under saturation condition of about 9 g/cm^3, mercury changes macroscopically to a nonmetallic, i.e. effectively "semiconducting" state. This assumption is consistent with the analysis of the optical conductivity (41,42), as determined from optical reflectivity measurements, which indicates that a real gap opens for densities smaller than 9 g/cm^3. From accurate d.c. conductivity data close to the critical point (5) it can be concluded that the highly ionized "semiconducting" regime terminates in the critical point. Thus for $T > T_x$ (i.e. $\rho < 9$ g/cm^3) the liquid-vapour change seems to be from a nonmetallic, effectively "semiconducting" (highly ionized) liquid to an essentially dielectric (only slightly ionized) less conducting vapour.

The approach of the transition from the insulating vapour side has recently been extensively studied by measurements of the density- and temperature dependence of the optical properties of the vapour (42,43,45,46). At very low densities a line-spectrum is observed with the main absorption lines at 4.89 eV and 6.7 eV corresponding to transitions between the 6s ground state and the 6p triplet and singlet state of the Hg-atom. As the density is increased the sharp lines broaden due to interactions with neighbouring atoms resulting in a relatively steep absorption edge which moves rapidly to lower energies with increasing density. Fig.15 gives a few selected data (46) for the density dependence of the edge at a constant temperature 1481°C, i.e. near the critical temperature. Bhatt and Rice (47) and Uchtmann et al.(45) have shown that a uniform density increase is insufficient to explain a line broadening as large as the observed shift in fig.15 and that one must take density fluctuations into account. The absorp-

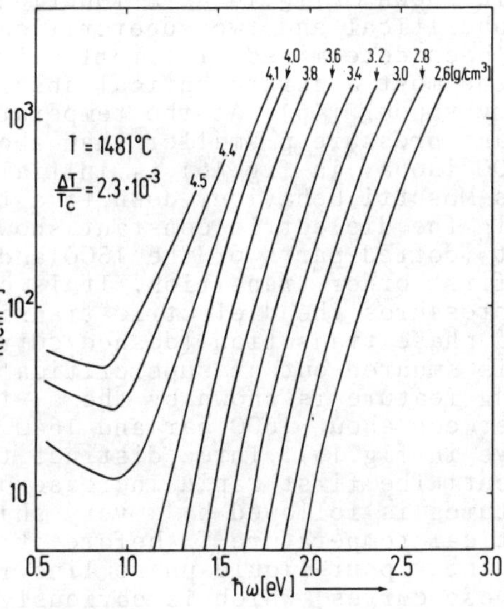

FIGURE 15. Optical absorption edge of dense Hg vapour at different densities for a constant supercritical temperature of 1481°C; the absorption constant is plotted against photon energy (46).

tion edge is then lowered by the environment of the atom being excited, and the edge is thus explained in terms of absorption by excitonic states of large randomly distributed clusters. From the large values of the absorption coefficient it can be concluded that the singlet exciton (6^1p_1) with large oscillator strength broadens faster than the triplet exciton (6^3p_1) with small oscillator strength. A detailed analysis (45) of the density dependence of the absorption edge shows that the singlet contribution dominates for densities larger than 1 g/cm^3 whereas for $\rho < 1$ g/cm^3 the shape of the edge is dominated by the triplet transition.

If one approaches the critical region, i.e. for densities larger than about 4 g/cm^3 in fig.15, the optical absorption spectrum qualitatively changes. A third absorption regime appears. At low frequency a foot extends in the infrared spectral range at least down to 0.5 eV. The intensity of this absorption foot is strongly density- and temperature dependent (46). The appearance of the infrared absorption foot seems to be intimately related to the liquid vapour critical point (46).

Recent experimental works on the near-infrared (48,49) and low frequency (50) dielectric constant in mercury vapour indicate that the liquid-vapour critical point phase transition is preceded by a "dielectric transition" which is signaled by a relatively steep anomaly in the real part of the dielectric constant ε_1. A few selected experimental results for the real part of the dielectric constant ε_1 at the constant photon energy 1.27 eV are shown in fig.16 as a function of temperature for three subcritical and two supercritical pressures. The ε_1 data have been determined from reflectivity measurements employing the most accurate optical absorption data of compressed mercury vapour (46). As the temperature is decreased at constant pressure p smaller than about 1600 bar (see e.g. the 1500 isobar in fig.16) ε_1 initially follows close to Clausius-Mosotti behaviour down to a transition temperature T_D. At T_D the dielectric constant shows an extremely sharp enhancement (dotted parts of the 1500 and 1600 isobars); it looks like a first order transition. It is obvious that for subcritical pressures the dielectric transition preceds the vapour-liquid phase transition (dashed curve). The enhancement of ε_1 is smeared out at supercritical temperatures. A very interesting feature is shown by the ε_1-isobars in the pressure range between about 1600 bar and 1640 bar (see e.g. the 1633 bar curve in fig.16). Three distinct temperature ranges are apparent: The first rapid increase of ε_1 at supercritical temperatures is followed by a very sharp decrease at about the critical temperature T_c before it starts again to increase when the vapour-liquid phase line is approached. The pattern of these curves, which is obviously caused by two

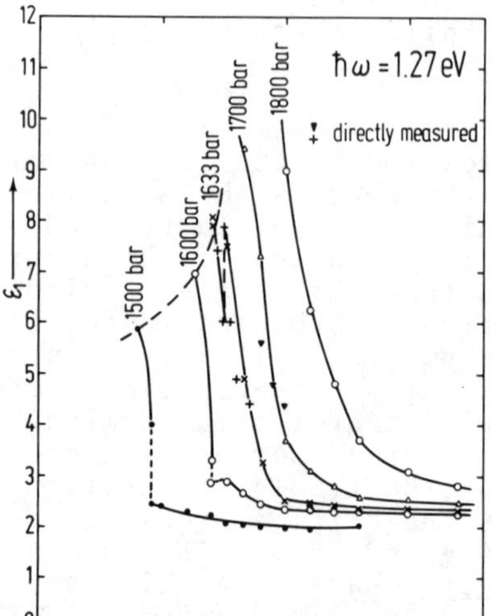

FIGURE 16. The real part of the dielectric constant ε_1 of Hg at the constant photon energy 1.27 eV as a function of temperature T for different sub- and supercritical pressures. The dotted lines indicate the sharp dielectric transitions at subcritical pressures. The dashed line indicates the vapour-liquid phase line.

competing effects, immediately suggests that the dielectric anomaly in mercury is intimately related to the vapour-liquid critical point.

More physical insight is obtained when ε_1 is plotted in the vicinity of the critical point as a function of density at constant temperature T, as shown in fig.17 for a number of selected isotherms. Values of ρ for each set of p-T coordinates were obtained from ref.(5).

It is obvious that the form of the ε_1-curves is caused by two competing effects. Far away from the critical point (for $T/T_c \gtrsim 1.05$ and $\rho/\rho_c \lesssim 0.7$) ε_1 is only slightly density dependent at constant temperature and shows only slight positive deviations from the simple Clausius-Mosotti value. This effect changes close to the phase line where the dielectric transition which precedes the metal-nonmetal transition is signaled by a steep enhancement of ε_1. The pattern of these curves is especially interesting close to the usual critical point where ε_1 shows a strongly temperature dependent part $\Delta\varepsilon_1$, i.e. the different between the experimental value and the Clausius-Mosotti value. The amplitude of $\Delta\varepsilon_1$ becomes very large close to the critical point. By contrast, the anomalous critical point contribution in the dielectric constant of normal nonpolar fluids is immeasurably small (51).

Hensel and Hefner (48) have speculated that the enhanced dielectric anomaly in mercury is due to a transition from a normal dielectric gas phase to an equilibrium disperse system containing charged dense droplets. The underlying idea of

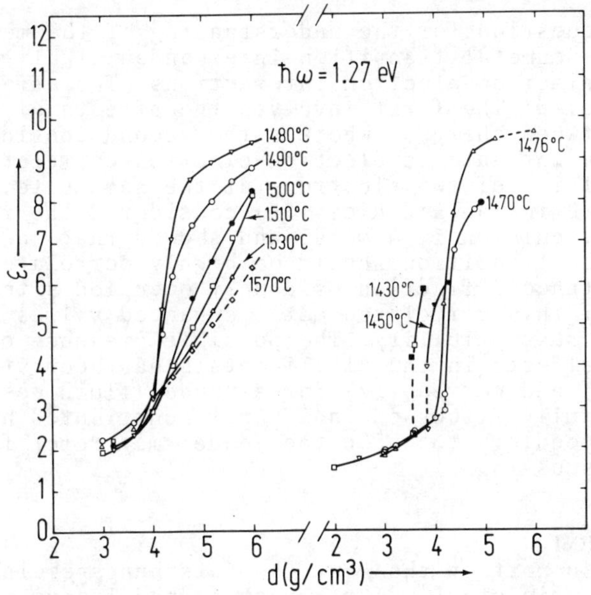

FIGURE 17. Real part of the dielectric constant ε_1 of Hg at constant temperature versus density. The critical temperature $T_c = 1477^oC$, the critical density is 5.77 g/cm^3.

this proposal is that a charge present in a vapour of neutral atoms can cause a local density increase which can lead in the vicinity of the vapour-liquid critical temperature to an analogue of a vapour-liquid transition in the compressed region, i.e. dense charged clusters are formed which approach metallic conditions while the macroscopic system is still in the nonmetallic region (52). The picture of charged dense droplets in mercury vapour has been questioned by Turkevich and Cohen (53). Alternatively, they make the interesting proposal that mercury for densities smaller than the M-NM transition density and larger than the dielectric transition density constitutes a disordered, inhomogeneous excitonic insulator phase. The underlying suggestion is that improved coordination in the fluid, i.e. clustering, can reduce the bottom of the Frenkel exciton band, leading ultimately to an exciton condensation instability. The permanent dipole moments of the condensed excitons are estimated to order as a ferroelectric phase (54).

3.2 Monovalent alkali metals

For the monovalent alkali metals the experimental situation is more complicated than in the case of divalent mercury. Because of the highly reactive nature of these metals only conductivity measurements of limited accuracy have been possible which extend up to conditions close to the critical point of cesium (see fig.1 (2)). The steep fall in σ as the critical point is traversed suggests a close link between the metal-nonmetal transition and the liquid-vapour critical point phase transition.

A central question for the understanding of the mechanism of the metal-nonmetal transition in expanded alkali metals is the role of electron-electron interactions. Two effects have to be considered. The first involves the effects of long range screening between charges, whereas the second considers the effect of the intraatomic electron-electron correlation, the Coulomb repulsion of two electrons at the same site, the Hubbard energy. Brinkman and Rice (55) considered the role of the intraatomic repulsion in a metal and showed that the metallic state near the transition should be highly correlated, having a low instantaneous fraction of doubly occupied sites. They predicted for this correlated metal enhanced values for the paramagnetic susceptibility. The possible presence of large correlation effects in the alkali metals has been studied by Freyland (56) and Warren (57) for expanded fluid cesium. These results are fully discussed in a paper contributed by Warren to these proceedings to which the reader may refer for more extensive discussing.

ACKNOWLEDGEMENT

Financial support by the Deutsche Forschungsgemeinschaft and the Bundesministerium für Forschung und Technologie is gratefully acknowledged.

REFERENCES

1. Guggenheim E.A.: J.Chem.Phys. 13, 253 (1945).
2. Shimoji M.: Liquid Metals. New York: Academic Press, 1977.
3. see e.g. Mott N.F.: Phil.Mag. 37, 377 (1978); Nara S., Ogawa T. and Matsubara T.: Prog.Theo.Phys. 57, 1474 (1977); Ebeling W. and Sändig R.: Ann.Phys.(Leipzig) 28, 289 (1973); Ebeling W., Kraeft W.D. and Kremp D.: Theory of Bound States in Plasmas and Solids. Berlin: Akademie-Verlag, 1976; Krumhansl J.A.: in Physics of Solids at High Pressures, Tomizuka C.T. and Emrick R.M.(eds): New York: Academic Press, 1965; Landau L. and Zeldovitch G.: Acta Phys.Chim.USSR 18, 194 (1943); Yonezawa F. and Ogawa T.: Prog.Theo.Phys.(Japan) Suppl. 72, 1 (1982).
4. see e.g. Freyland W. and Hensel F.: in The Metallic and the Nonmetallic States of Matter: An Important Facet of Chemistry and Physics of Condensed Matter, Edwards P.P. and Rao C.N.R. (eds.): London: Taylor and Francis, 1985; Hensel F.: Angew.Chem. 92, 598 (1980); Angew.Chem.Int. Ed.Engl. 19, 593 (1980); Hensel F.: Proc.8th Symp.Thermophysical Properties, Sengers J.V. (ed.): New York: ASME, 1982, p.151; Cusack N.E.: in "Metal Non-Metal Transitions in Disordered Systems, Friedman L.R. and Tunstall D.P. (eds.): Edinburgh, 1978; Endo H.: Prog.Theo.Phys.(Japan) Suppl. 72, 100 (1982); Alekseev V.A. and Iakubov I.T.: Phys.Reports 96, 1 (1983).
5. Götzlaff W.: Diploma Thesis, University of Marburg (1983).
6. Jüngst S., Knuth B. and Hensel F.: Phys.Rev.Lett. 55, 2160 (1985).
7. for a review see Horvath A.L.: J.Chem.Education 50, 335 (1973).
8. Ross R.G. and Greenwood D.A.: Prog.Mat.Sci. 14, 173 (1969).
9. Hensel F., Jüngst S., Noll F. and Winter R.: in Localization and Metal-Insulator Transition, Adler D. and Fritzsche H. (eds.): New York: Plenum Press, 1985, p.109.
10. Knuth B.: Diploma Thesis, University of Marburg (1985).
11. see e.g. Sengers J.V. and Levelt-Sengers J.M.H.: in Progress in Liquid Physics, Croxton C.A. (ed.): Chichester, U.K.: Wiley, 1978.
12. Chieux P., Damay P., Dupuy J. and Jal J.F.: J.Phys.Chem. 84, 1211 (1980).
13. Thomas G.A.: Nuovo Cimento 39, 561 (1977).
14. Thomas G.A.: J.Phys.Chem. 88, 3749 (1984).
15. Weiner J., Langley K.H. and Ford, Jr. N.C.: Phys.Rev. Lett. 32, 879 (1974).
16. Moldover M.R. and Gammon R.W.: J.Chem.Phys. 80, 528 (1984).
17. Rowlinson J.S.: Adv.Chem.Phys. 41, 1 (1980).
18. Widom B. and Rowlinson J.S.: J.Chem.Phys. 52, 1670 (1970).
19. Mermin N.D.: Phys.Rev.Lett. 26, 957 (1971).
20. Mermin N.D.: Phys.Rev.Lett. 26, 169 (1971).
21. Hemmer P.C. and Stell G.: Phys.Rev.Lett. 24, 1284 (1970).
22. Green M.S., Cooper M.J. and Levelt-Sengers J.M.H.: Phys. Rev.Lett. 26, 492 (1971).

23. Mermin N.D. and Rehr J.J.: Phys.Rev.Lett. 26, 1155 (1971).
24. Rehr J.J. and Mermin N.D.: Phys.Rev.A 8, 472 (1973).
25. Ley-Koo M. and Green M.S.: Phys.Rev.A 16, 2483 (1977).
26. Nicoll J.F.: Phys.Rev.A 24, 2203 (1981).
27. Nicoll J.F. and Albright P.C.: Proc.8th Symp.Thermophysical Properties, Sengers J.V. (ed.): New York: ASME, 1982, p.377.
28. Goldstein R.E. and Ashcroft N.W.: Phys.Rev.Lett. 55, 2164 (1985).
29. Buckingham M.J.: in Phase Transitions and Critical Phenomena, Domb C. and Green M.S. (eds.): London: Academic Press, 1972, Vol.2.
30. Levelt-Sengers J.M.H.: Physica 73, 73 (1974).
31. Wheeler J.C.: Ber.Bunsenges.Phys.Chem. 76, 308 (1972).
32. Zillgitt M., Schmutzler R.W. and Hensel F.: Phys.Lett. 5, 419 (1972).
33. Neale F.E., Cusack N.E. and Johnson R.D.: J.Phys.F: Metal Phys. 9, 113 (1979).
34. Tsuji K., Yao M. and Endo H.: J.Phys.Soc.Jpn. 42, 1594 (1977).
35. Even U. and Jortner J.: Phil.Mag. 30, 325 (1974).
36. Schönherr G. and Hensel F.: Ber.Bunsenges.Phys.Chem. 85, 361 (1981).
37. Seyer H.P.: Doctoral Thesis, University of Marburg, (1984).
38. Rowlinson J.S. and Swinton F.L.: Liquids and Liquid Mixtures, 3rd edition. London: Butterworth Scientific,1981.
39. Schönherr G., Schmutzler R.W. and Hensel F.: Phil.Mag.B 40, 411 (1979).
40. Warren W.W. and Hensel F.: Phys.Rev.B 26, 5980 (1982).
41. Hefner W., Schmutzler R.W. and Hensel F.: J.Phys.(Paris) Colloq. 41, C8-62 (1980).
42. Ikezi H, Schwarzenegger K., Simons A.L., Passner A.L. and McCall S.L.: Phys.Rev.B 18, 2494 (1978).
43. Hensel F.: Phys.Lett. 31A, 88 (1970).
44. Overhof H., Uchtmann H. and Hensel F.: J.Phys.F 6, 523 (1976).
45. Uchtmann H, Popielawski J. and Hensel F.: Ber.Bunsenges. Phys.Chem. 85, 555 (1981).
46. Brusius U.: Doctoral Thesis, University of Marburg (1986).
47. Bhatt R.N. and Rice T.M.: Phys.Rev.B 20, 466 (1979).
48. Hefner W. and Hensel F.: Phys.Rev.Lett. 48, 1026 (1982).
49. Yao M., Uchtmann H. and Hensel F.: Phil.Mag.B 52, 499 (1985).
50. Schönherr G.: in Nato ASI Series C, Acrivos J.V., Mott N.F. and Yoffe A.D.(eds.): Dordrecht: Reidel, 1984, Vol.130.
51. Sengers J.V., Bedeaux D., Mazur P. and Greer S.C.: Physica 104A, 573 (1980).
52. Hefner W., Sonneborn-Schmick B. and Hensel F.: Ber.Bunsenges.Phys.Chem. 86, 844 (1982).
53. Turkevich L.A. and Cohen M.H.: Ber.Bunsenges.Phys.Chem. 88, 297 (1984).
54. Turkevich L.A. and Cohen M.H.: Phys.Rev.Lett. 53, 2323 (1984).

55. Brinkman W.F. and Rice T.M.: Phys.Rev.B 2, 4302 (1970).
56. Freyland W.: Phys.Rev.B 20, 5104 (1979).
57. El-Hanany U., Brennert G.F. and Warren W.W.: Phys.Rev. Lett. 50, 540 (1983).

LOW TEMPERATURE TRANSPORT PROPERTIES: THE ELECTRICAL RESISTIVITY OF SOME AMORPHOUS ALLOYS

G. FRITSCH*, A. SCHULTE, E. LÜSCHER

Physik Department, TU München, D-8046 Garching, *Inst. f. Physik, Fak. f. BauV/I 1, Universität der Bundeswehr München, D-8014 Neubiberg

We report on contributions to the resistivity due to quantum corrections. These effects should show up at temperatures $T \lesssim 30$ K. The temperature- and the magnetic-field dependence of the resistivity is discussed in detail. Superconducting fluctuations are also considered. Finally, theoretical results are compared to experimental data at low magnetic fields for the amorphous alloys CuTi and CuZr. The degree of agreement obtained leads to the conclusion that quantum corrections are also important in three dimensional systems.

1. INTRODUCTORY REMARKS

In the electronic transport properties of the amorphous state new phenomena show up, which are absent or not easily to detect in crystalline solids. Fig. 1 indicates schematically on which level these effects will manifest in resistivity, Hall-effect and thermopower. With the exception of the superconductivity transition the basic transport can be described by straight lines. The new phenomena to be discussed are small corrections, being in the range of 10^{-1} to 10^{-3}. They are indicated by the dashed lines.

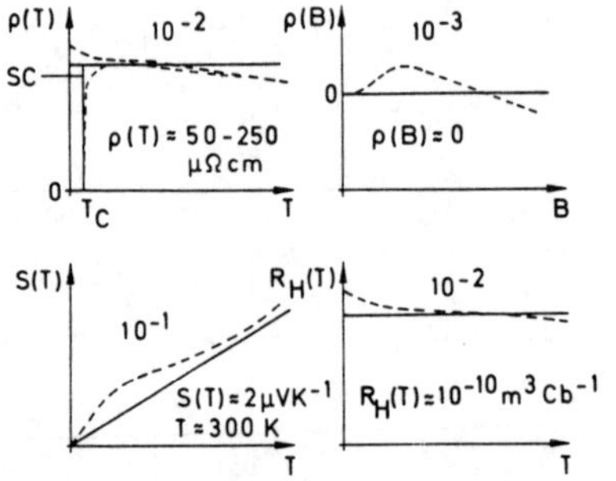

FIGURE 1. Rough sketch of the transport properties $\rho(T)$, $\rho(B)$, $S(T)$ and $R_H(T)$ for amorphous alloys. The dashed lines indicate the kind of effects discussed in this paper (Scale exaggerated). The full lines correspond to the predictions of the nearly free electron model.

In the following only the resistivity will be analysed. We will show that phenomena such as quantum corrections to the resistivity and superconductivity fluctuations are of importance. The behaviour of the Hall-constant will be discussed elsewhere in this volume /1/. Also, we only mention the effects occurring in the thermopower, which are interpreted as being caused by electron-phonon enhancement /4/. In order to illustrate the situation for amorphous alloys, some explicit experimental data are reproduced in Fig. 2. The temperature dependence of the resistivity for several alloys including ones showing superconductivity ($Pd_{30}Zr_{70}$,

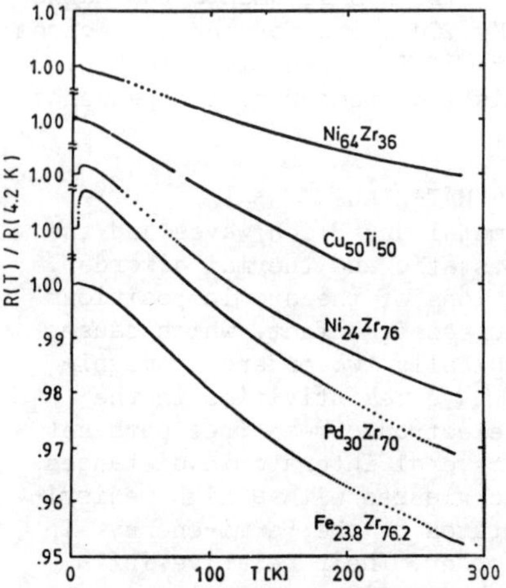

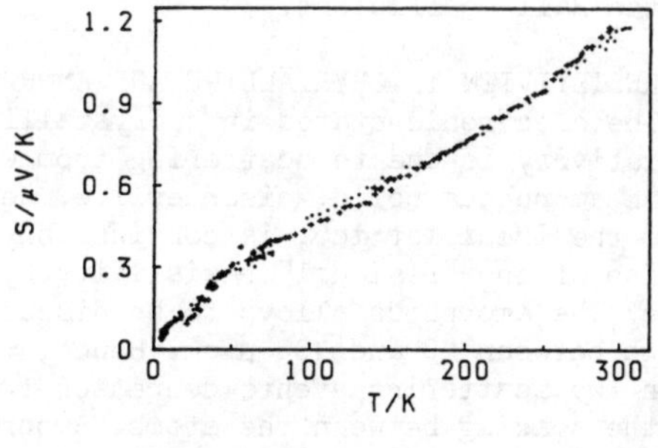

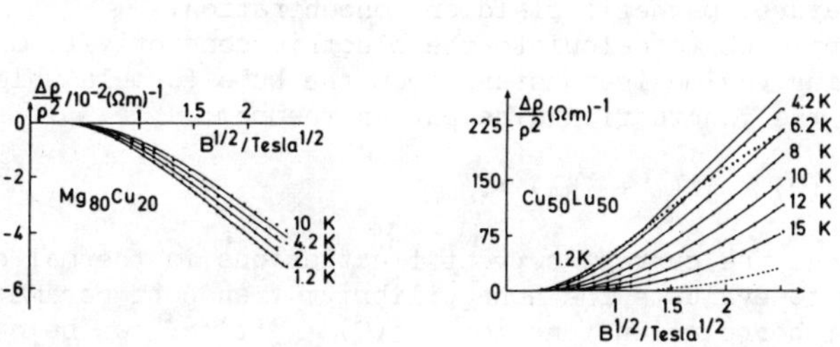

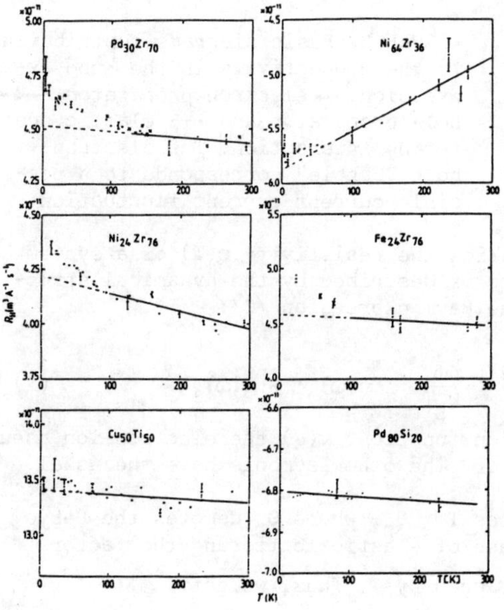

FIGURE 2. Some experimental data for amorphous alloys. a) resistance R(T), b) magnetoresistivity $\Delta\rho/\rho^2$ after Bieri et al /34/, c) Thermopower S /4/ and d) Hall-constant R_H/5/.

$Ni_{24}Zr_{76}$) is indicated in Fig. 2 a, whereas Fig. 2 b documents the magnetoresistivity for the free electron alloy $Mg_{80}Cu_{20}$ and for the d-electron alloy $Cu_{50}Lu_{50}$. Data for the thermopower of $Cu_{50}Ti_{50}$ are given in Fig. 2 c and finally, Fig. 2 d demonstrates a distinct temperature dependence of the Hall-coefficient.

2. RESISTIVITY IN CRYSTALLINE AND AMORPHOUS MATERIALS

The electronic states in a crystalline metal are Bloch waves and the resistivity is due to scattering from weak static and thermal disorder. In an amorphous solid, disorder i.e. deviations of the atomic positions from the ideal lattice, is considerably increased, a fact, which causes a rise of the "residual" resistivity by typically two orders of magnitude. The amorphous alloys to be discussed have resistivities in the range between 40 and 400 µΩcm. Hence, the electronic mean free path between two scattering events decreases from several interatomic distances to the spacing between the atoms. Among the glasses with a high resistivity are a large number of alloys with d-states at the Fermi-energy. In general, we observe both, d-states and disorder. Their relative influence on the transport may be distinguished by varying external parameters such as temperature, magnetic field or concentration.

The usual approach to calculate the electric conductivity σ in the strong scattering regime is to start from the Kubo-formula which is a consequence of the fluctuation-dissipation theorem:

$$\sigma \sim \iint d\lambda dt < j(0) \cdot j\,(t + i\hbar\lambda) >. \qquad (1)$$

Hence, we need the current-current fluctuations in thermal equilibrium in order to evaluate the nonequilibrium transport parameter σ. The current-current correlation function $< j(0) \cdot j(t) >$ can be calculated by using propagator (Greens-functions) techniques. The diagrammatic representation of a perturbation series in order to determine this function helps to make the situation physically plausible (Fig. 3).

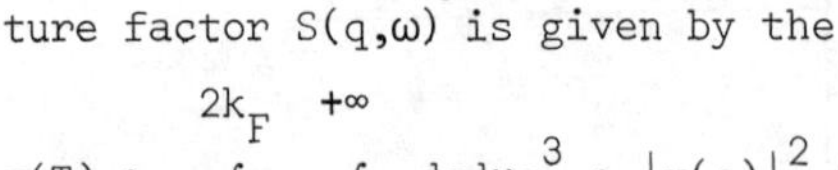

FIGURE 3. Basic diagrams contributing to the conductivity in the Kubo expression. ⟶ electron propagator, ⟵ hole propagator and --- elastic scattering interaction. The electron-hole "bubble" corresponds to a possible current-current fluctuation.

In the weak scattering formulation the resistivity $\rho(T)$ of a system whose static and dynamic structure is described by the dynamical structure factor $S(q,\omega)$ is given by the Baym expression /2/:

$$\rho(T) \sim \int_0^{2k_F} \int_0^{+\infty} dq d\omega q^3 \cdot |v(q)|^2 \cdot \frac{\hbar\omega}{k_B T} \cdot n(\omega) \cdot S(q,\omega), \qquad (2)$$

where $n(\omega)$ is the Bose distribution function, $v(q)$ the electron-ion pseudopotential and k_F the Fermi-vector. The other symbols have the usual meaning.

In the case of high temperatures $T \gtrsim \theta_D$, where θ_D denotes the Debye temperature, as well as in the case of elastic scattering the factor $\frac{\hbar\omega}{k_B T} \cdot n(\omega)$ in equ. (2) may be replaced by 1. Thus, we get:

$$\rho(T) \sim \int_0^{2k_F} q^3 \, |v(q)|^2 \, S_\rho(q) \, dq. \tag{3}$$

Here, $S_\rho(q)$ denotes the static resistivity structure factor which will be discussed below. Equ. (3) is the Ziman formula. It should be applicable for crystalline and amorphous as well as liquid elements*.

In a crystalline solid the function $S_\rho(q)$ can be decomposed generally as follows (schematically):

$$S_\rho(q) = S^{(c)}_{elast}(q) + S^{(c)}_{inelast.}(q) \tag{4}$$

with

$$S^{(c)}_{elast}(q) = (\text{Elastic Bragg}) \cdot DW(T) \cdot DW(\text{Defect}) + (\text{Elastic Defect}) \tag{5}$$

and

$$S^{(c)}_{inelast}(q) = (\text{1-Phonon}) \cdot DW(T) + (\text{2-Phonon}) \cdot DW(T) + \ldots + (\text{n-Phonon}) \cdot DW(T) \approx (\text{1-Phonon}). \tag{6}$$

This latter approximation stems from a compensation between the multiphonon-terms ($n > 2$) and the dynamic Debye-Waller-factor DW(T) in the 1-phonon-term (DW(Defect): elastic Debye-Waller factor). However, the first term in equ. (5):"(Elastic Bragg) · DW(T) · DW(Defect)" has to be omitted in $S_\rho(q)$ since the symmetry of the electronic wave functions is the one of the lattice. Hence, elastic Bragg-scattering doesn't contribute to the resistivity and we get:

$$S^{(c)}_\rho(T) = (\text{Elastic Defect}) + (\text{1-Phonon}), \tag{7}$$

yielding a constant rest-resistivity ($T \to 0$) and a linear T-dependence at higher temperatures.

In an amorphous solid, the remaining short range order (SRO) generates a kind of smeared out "Bragg-peak", which dominates the function $S^{(a)}(q)$ at the position q_p of the first peak. This elastic contribution is T-dependent, because the SRO is decreased by the thermal motion. Therefore, the height of the first peak should be diminished with rising temperature. In other q-regions inelastic scattering dominates $S^{(a)}(q)$, its magnitudes will increase with rising temperature.

Since the symmetry of the electronic wave functions is no longer adapted to the structure, the full $S^{(a)}(q)$ will contribute to the resistivity adding a huge amount due to elastic scattering. From this discussion it is clear that the resulting T-dependence is a superposition of the one caused by inelastic and the other one produced by the elastic part. If the elastic part dominates (i.e. in d-electron alloys with $q_p \approx 2k_F$/7/), then the Debye-Waller factor will dictate the T-dependence to a first approximation. This assumption yields a negative linear T-coefficient of the resistivity at high temperatures and a quadratic behaviour at lower temperatures /8/. The latter may be approximated by a constant at temperatures below about 30 K.

---

*Within certain approximations this is also true for an amorphous alloy /6/.

However, the arguments, given above are based on weak scattering theory. It is by no means clear if this model will hold when the resistivities are as high as 200 μΩcm. For purpose of comparison we mention that the resistivity of many crystalline elements amount to about 1 μΩcm. The problem can be restated by observing, that $\rho \sim (k_F \cdot l)^{-1}$, where the Boltzmann theory holds as long as $k_F l >> 1$. Here, l denotes the mean free path of the electrons. If $k_F l \lesssim 1$, we are in the strong scattering regime and corrections $\sim (k_F l)^{-2}$ and higher orders have to be considered.

An estimate of the quantity $k_F l$ for $\rho \approx 200$ μΩcm can be performed as follows /9/:

$$\sigma = n\, e^2 \tau / m = (3\pi^2)^{-1} \cdot (k_F l) \cdot k_F \cdot e^2/\hbar. \tag{8}$$

With $k_F = 1.4$ Å^{-1}, we deduce $k_F l \approx 4.5$.

This number is larger than 1 but not large against this value. Hence one should worry whether the weak scattering approach is still correct or if higher order corrections in $(k_F l)$ must be taken into account.

Goetze and collaborators /10/ have based their formulation of the conductivity σ on equ. (1). They obtained corrections proportional to $(k_F l)^{-2}$. However, it seems that for resistivities up to 200 μΩcm in d-element alloys the Ziman formulation is still valid, at least effectively. This statement is true especially if the resistivity is dominated by elastic scattering, which is the case when d-elements are involved /7/. Scattering events of this type do not destroy the phase correlation between the incoming and the outgoing electronic wave functions. Therefore, the phase correlation length is much larger than the mean free path l. We may introduce an effective mean free path $l_{eff} > l$. Such a conclusion is also supported by results from effective medium theories /11/. In the following, we will focus on the T- and B-dependent parts of σ at low temperatures and fields. It will be assumed that the conductivity can be successfully described by a modified- in the above sense - free electron-model. We would like to discuss corrections proportional to $(k_F l)^{-1}$: the quantum corrections.

3. T-DEPENDENCE OF THE QUANTUM CORRECTIONS

In order to obtain these additional contributions to the conductivity elastic scattering should dominate the inelastic one. Such a situation may occur especially in case of d-electron amorphous alloys /7/. The consequences of the elastic scattering from disorder are twofold. Firstly, it will give rise to quantum interference (QI) and secondly it will modify the Coulomb-interaction between the electrons (I). We should emphasize that these corrections as presented in the following are evaluated assuming a free electron-like density of states at the Fermi-surface. Since the d-electrons are hybridized with the s-electrons at the Fermi-surface this means that such processes are more sensitive to disorder than to "bandstructure".

i) Quantum Interference Effect (QI)

To understand this phenomenon we need to consider the fact that the phase correlation of the electronic wave function is not destroyed by elastic scattering. Since there are many paths for an electron to diffuse between two points $\vec{r}_1$ and $\vec{r}_2$ the phase is averaged out. However, the starting and the final points are the same for selfcrossing paths depicted in Fig. 4/12/. The electronic wave function may go either way around the loop. Thus, there is an interference at the cross point if only elastic scattering occurs in between. It can be shown that the wave functions interfere constructively /13/. Sometimes this is termed weak localization, as there is an increased probability for the electron to return to its starting position. The relevant diagram for the conductivity is shown in Fig. 5 left: the maximally crossed diagram. In the fluctuation

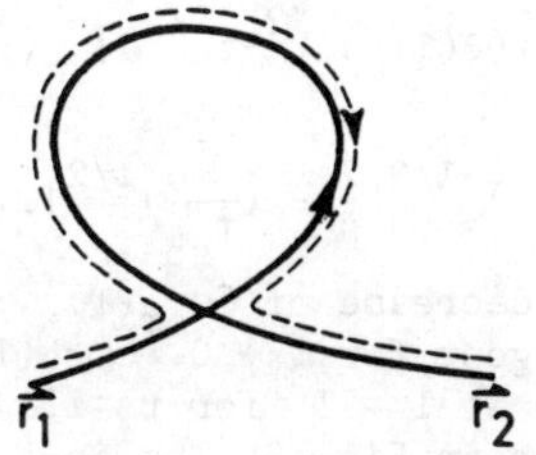

FIGURE 4. The self-crossing path in $\vec{r}$-space. The two interfering wave functions are shown by the full and by the dashed line. If only elastic scattering is present between $\vec{r}_1$ and $\vec{r}_2$, the wave functions interfere constructively.

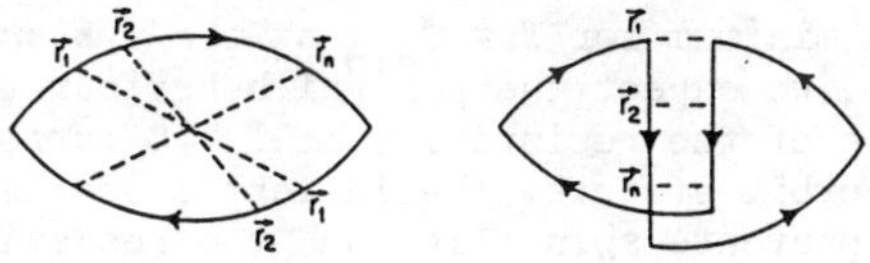

FIGURE 5. Diagram, contributing to the quantum interference effect (Langer-Neal-Diagram). For further explanation see Fig. 3. Left: normal diagram showing the electron-hole propagator and maximal crossed scattering lines, right: same diagram, redrawn to indicate the equivalence to the particle-particle propagator with elastic "ladder" interactions: the "Cooperon".

picture the two interfering wave functions ("Quantum Interference") are represented by an electron-hole pair. This diagram can be redrawn as indicated in Fig. 5 right. Here, we see, that repeated elastic scattering interactions (---) occur between the particle-particle propagator. Due to the similarity to superconductivity the quasi-particle 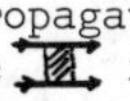 is called a Cooperon. Since weakly localized electrons contribute less to the conductivity than free ones, we expect a rise of the resistivity at T = 0. At higher temperatures the inelastic scattering destroys the phase coherence of the wave function more and more, the quantum interference effect disappears gradually. The inelastic scattering time τ_i is approximately proportional to T^{-p}, where the exponent $p \approx 1.5$ for electron-electron and $p \approx 2$ to 4 for electron-phonon scattering /13/. Therefore, one gets as a correction to the conductivity σ /25/:

$$\Delta\sigma_{QI}(T) = -\frac{e^2}{2\pi^2\hbar}\,(D\cdot\tau_o)^{-1/2}\cdot[2/\pi - (\tau_o/\tau_i)^{1/2}] \qquad (9)$$

with τ_o representing the elastic scattering time ($\tau_o \neq \tau_o(T)$), D the electronic diffusion constant ($D = \sigma/(e^2 N(E_F))$) and the other symbols have the usual meaning. $N(E_F)$ indicates the bare density of states. $\tau_o \ll \tau_i$ should be obeyed throughout. From equ. (9) we expect a decay of the QI-contribution proportional to $T^{p/2}$.

However, there is an additional scattering process which modifies the QI-effect: We have to consider spin-orbit scattering. Physically speaking the spin-orbit scattering preserves time reversal but it changes the constructive interference into a destructive one via spin-flip scattering. The characteristic time involved is τ_{so} which is assumed to be independent of T. General arguments let us expect a shorter τ_{so} for higher nuclear charge Z of the alloy constituents /14/. Fukuyama et al /15/ obtained for the quantum interference with spin-orbit scattering:

$$\Delta\sigma_{QI}(T) = -\frac{e^2}{2\pi^2\hbar}(D\cdot\tau_o)^{-1/2}\cdot\left[\frac{2}{\pi}-\left(\frac{\tau_o}{3\tau_{so}}\right)^{1/2}\cdot\left(3(1+t)^{1/2}-t^{1/2}\right)\right] \quad (10)$$

with $t = 3\,\tau_{so}/[4\tau_i(T)]$ and $\Delta\sigma_{QI}(0) = -\frac{e^2}{2\pi^2\hbar}(D\cdot\tau_o)^{-1/2}\left[\frac{2}{\pi}-\left(\frac{3\tau_o}{\tau_{so}}\right)^{1/2}\right]$.

The T-dependence* of this expression gives a decrease of $\delta\sigma_{QI}(T) \sim -T^{p/2}$ for small T which yields a horizontal tangent for $T \rightarrow 0$. $\delta\sigma_{QI}(T)$ has a minimum for $T = T_{min}$ at $t = 1/8$ and a zero at $T = T_o$ for $t = 1.175$. Hence, we expect the general behaviour reproduced in Fig. 6. The increase of the resistivity at T = 0 turns out to be smaller than without spin-orbit effects. The inelastic scattering time $\tau_i(T)$ on the other hand prevents spin-flips and the resistivity increases towards the QI-limit. However, afterwards, the loss of phase coherence in the wave function due to $\tau_i(T)$ takes over, the QI-effect disappears at higher temperatures.

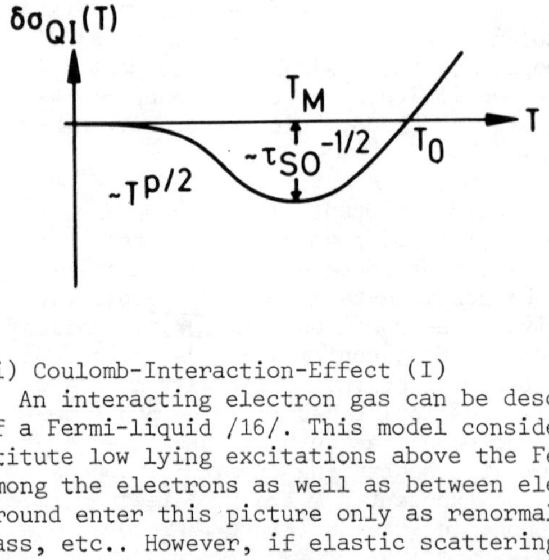

FIGURE 6. The T-dependence of the quantum-interference effect with spin-orbit scattering. Only the T-dependent part is shown. $T_M = [\tau_{io}/(6\cdot\tau_{so})]^{1/p}$ and $T_o = [0.881\,\tau_{io}/\tau_{so}]^{1/p}$, where $\tau_i(T) = \tau_{io}\cdot T^{-p}$.

ii) Coulomb-Interaction-Effect (I)

An interacting electron gas can be described via the Landau-theory of a Fermi-liquid /16/. This model considers quasiparticle which constitute low lying excitations above the Fermi energy. The interactions among the electrons as well as between electrons and the positive background enter this picture only as renormalizations such as effective mass, etc.. However, if elastic scattering is present, the dynamic screening of the electrons is no longer perfect. The result is an electron-electron interaction producing an additional resistivity. The combination of elastic scattering and Coulomb-interaction is essential because otherwise the latter would not contribute to the resistivity (momentum conservation). Fig. 7 shows some of the relevant diagrams. They are evaluated to first order in perturbation theory. The diagrams in Fig. 7 left des-

*The T-dependence part of $\Delta\sigma$ is called $\delta\sigma$.

cribe the exchange-corrections in the particle-hole channel, the ones in Fig. 7 right the Hartree- and exchange-contributions in the particle-particle-channel. The outcome is (Altshuler et al /17/):

$$\Delta\sigma_I(T) = - C + 0.65 \cdot \frac{e^2}{2\pi^2\hbar} \left(\frac{4}{3} - \frac{3}{2} \tilde{F}\right) \cdot \left(\frac{k_B T}{2\hbar D}\right)^{1/2} , \quad (11)$$

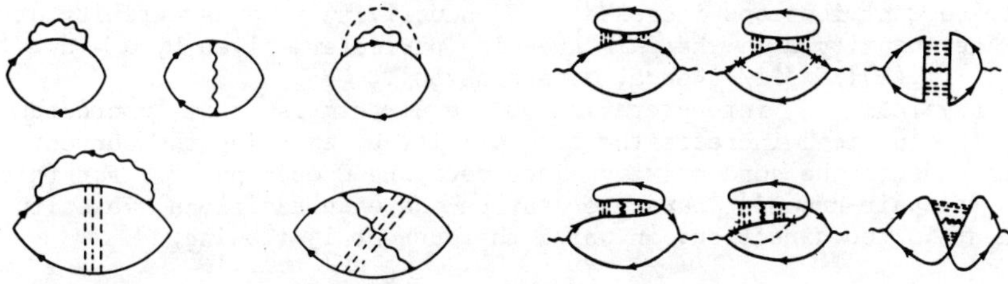

FIGURE 7. Diagrams contributing to the interaction correction. For further information see Fig. 3. ~~: Coulomb interaction, ▲: Vertex dressed with elastic scattering. Left: Particle-Hole-Channel (Diffuson), right: Particle-Particle channel (Cooperon), upper part: Hartree correction, lower part: exchange correction.

where C denotes a constant. It should be mentioned that the term 4/3 is the leading term in the exchange correction, whereas the term $-(3/2)\cdot\tilde{F}$ contains the second order term of exchange- and the leading one in the Hartree-interaction.

The new quantity appearing in equ. (11) is the function $\tilde{F}(F)$, which depends on the screening parameter F. This function reads /18/:

$$\tilde{F}(F) = (-32/3)\cdot[1 - 3F/4 - (1-F/2)^{3/2}]/F, \quad (12)$$

with $\lim_{F \to 0} \tilde{F}(F) = F$ and $0 \leq F \leq 1$ /26/.

This contribution reduces the conductivity at T = 0. As long as $\tilde{F}(F) \leq 8/9$, its T-dependence is positive with respect to conductivity and proportional to $\sqrt{T}$.

iii) Superconductivity Fluctuations (SC)

There exists a further mechanism modifying the conductivity at low temperatures: the paraconductivity caused by SC-fluctuations. In the case of disordered ("dirty") superconductors the coherence-length $\xi_d(0)$ of Cooper pairs is reduced to /19/ (weak coupling limit):

$$\xi_d(0) = 0.85 \cdot [\xi_o(0) \cdot l]^{1/2}, \quad (13)$$

yielding values around 100 - 200 Å at T = 0. $\xi_o(0)$ denotes the coherence length in the "clean" crystalline case. This fact allows the system to produce Cooper pairs as fluctuations with only minor free energy consumption, since the superconducting volume which must be generated is in the order of $\xi_d^3(0)$. Actually, two different contributions to the conductivity are expected which are termed $\sigma_{Al}(T)$ according to Aslamasov and Larkin /20/ and $\sigma_{MT}(T)$ according to Maki and Thompson /21/:

$$\sigma_{AL}(T) = \frac{e^2}{32\hbar} \cdot \xi_d^{-1}(0) \cdot [\frac{T_c}{T-T_c}]^{1/2} \quad (14)$$

and

$$\sigma_{MT}(T) = 4\,\alpha \cdot \frac{e^2}{32\hbar} \cdot \xi_d^{-1}(0) \cdot [\frac{T_c}{T-T_c}]^{1/2}, \quad (15)$$

where T_c denotes the superconductivity transition temperature. Both these contributions increase the conductivity towards infinity at $T = T_c$. The quantity α is the pair breaking parameter, given by $\alpha = \hbar \cdot [8 \cdot k_B T \cdot \tau_i(T)]^{-1}$ /22/. Hence, α depends on T!

The physical interpretation of the Al-term is, that superconductivity fluctuations increase the conductivity by carrying the current. The MT-term adds to the conductivity since decaying Cooper pairs - for this reason the pair-breaking parameter enters- have an additional velocity component in the direction, in which the current is flowing.

4. TOTAL T-DEPENDENCE OF THE CONDUCTIVITY

The total conductivity can be written as the sum:

$$\sigma_{tot}(T) = \sigma_o(T) + \Delta\sigma_{QI}(T) + \Delta\sigma_I(T) + \sigma_{SL}(T). \quad (16)$$

Since $\Delta\sigma_{QI}, \Delta\sigma_I << \sigma_o(T)$, we may calculate $\rho_{tot}(T) = \sigma_{tot}^{-1}(T)$ to first order as:

$$\rho_{tot}(T) = \frac{\rho_o(0)}{A(T)} [1 - \frac{\rho_o(0)}{A(T)} [\delta\sigma_{QI}(T) + \delta\sigma_I(T)] \quad (17)$$

with $A(T) \approx 1 + \rho_o(0) \cdot \sigma_{SL}(T)$ and $\rho_o(0) = \sigma_o(0)^{-1}$.

Here $\sigma_{SL}(T) = \sigma_{AL}(T) + \sigma_{MT}(T)$, the abbreviations $\delta\sigma_{QI}(T)$, $\delta\sigma_I(T)$ refer only to the T-dependent parts of $\Delta\sigma_{QI}(T)$, $\Delta\sigma_I(T)$, respectively. Potential scattering as well as other contributions, not discussed in this paper such as scattering from two level systems /23/ and any magnetic effects are lumbed together in $\rho_o(T)$. Here, we will assume that these additional effects are much smaller or absent as compared to the quantum corrections.

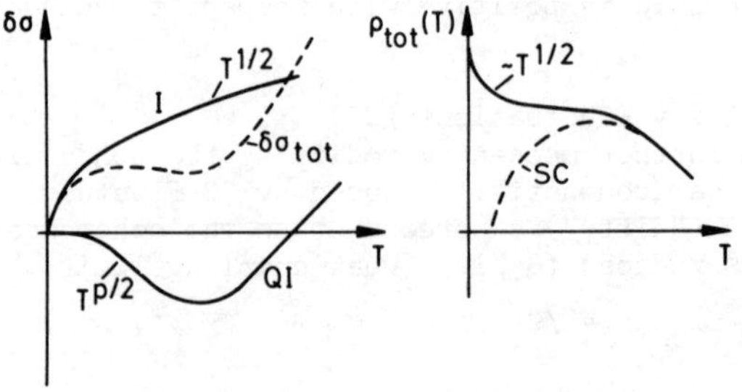

FIGURE 8. The temperature dependence of the conductivity. Left: Contributions of the interaction effect $\delta\sigma_I$ and the quantum interference effect $\delta\sigma_{QI}$ as a function of T and total conductivity. Right: Total resistivity with (dashed line) and without paraconductivity (full line).

The potential scattering should be independent from temperature at low temperatures. Hence, we will use $\rho_o(T) \approx \rho_o(0)$ for $T \leq 30$ K.

Equ. (17) contains some characteristic features which are shown in Fig. 8. The effect of $\delta\sigma_{QI}(T)$ and $\delta\sigma_I(T)$ produces the typical "wavy" behaviour of $\rho_{tot}(T)$. If superconductivity is present, this structure is suppressed, a peak in $\rho_{tot}(T)$ occurs. For purpose of comparison some experimental results are reproduced in Fig. 9. At higher temperatures the T-dependence of $\rho_o(T)$ changes our predictions.

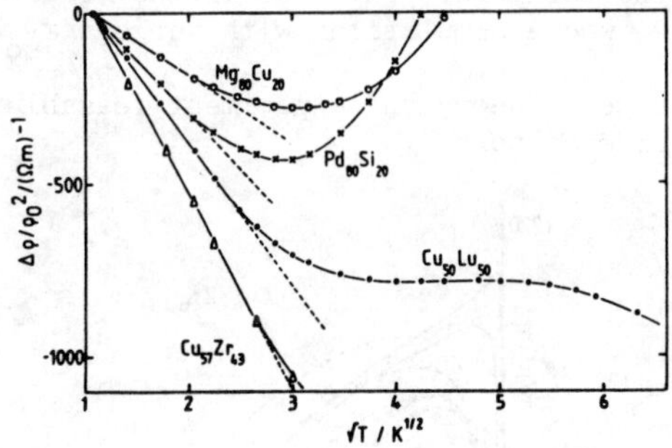

FIGURE 9. Resistivity of some alloys at low temperatures versus $T^{1/2}$ in order to show the interaction effect. After Bieri et al /3/.

TABLE I. Parameters governing the T-dependence of the resistivity

Contribution	Parameters
QI	$D,\ \tau_{so},\ \tau_i(T)\ \{^{\tau_{io}}_{p}$
I	$D,\ \tilde{F}$
SC	$T_c,\ \xi_d(0),\ \tau_i(T)\{^{\tau_{io}}_{p}$

The parameters listed in Table I must be known in order to explain the resistivity $\rho_{tot}(T)$ quantitatively according to equ. (17). T_c can be taken from experimental data. The electronic diffusivity D is defined by $D = v_F^2\tau/3$, where $\tau \approx \tau_o$ in our case. One may use either

$$\sigma = e^2 N(E_F) \cdot D \tag{18}$$

or

$$B'_{C2}(T_c) = \frac{dB_{C2}(T)}{dT}\Bigg/_{T_c} = (4/\pi)\cdot(k_B/e)\cdot D^{-1} \tag{19}$$

to obtain D from measured data. Equ. (19) is especially suited for superconductors. The coherence length $\xi_d(0)$ is known from the mean free path and from the coherence length in the pure case $\xi_o(0)$. This quantity in turn is given by /19/:

$$\xi_o(0) = 0.18\,\hbar v_F/(k_B T_c), \tag{20}$$

so we finally get with the help of the definition of the electronic diffusivity D and equ. (13):

$$\xi_d(0) = 0.625 \cdot [\hbar D/(k_B T_c)]^{1/2}. \tag{21}$$

Therefore, the remaining parameters are τ_{so}, $\tilde{F}$ and $\tau_i(T)$, which actually contains two parameters: $\tau_i(T) = \tau_{io} T^{-p}$.
It is for this reason that further information is needed. This is taken from determinations of the magnetoresistivity and of the Hall-constant. The latter quantity allows to deduce $\tilde{F}$, if electron-electron interaction is important for its T-dependence /17/:

$$\delta R_{H,I}/R_H = 2 \cdot \delta\sigma_I/\sigma. \tag{22}$$

A feeling for the magnitude of the parameters involved in the resistivity can be obtained from Fig. 10, where simulations with various τ_{so}, $\tilde{F}$ and T_c are reproduced.

In order to make a comparison between theory and experiment feasable, the magnetoresistivity will be considered next.

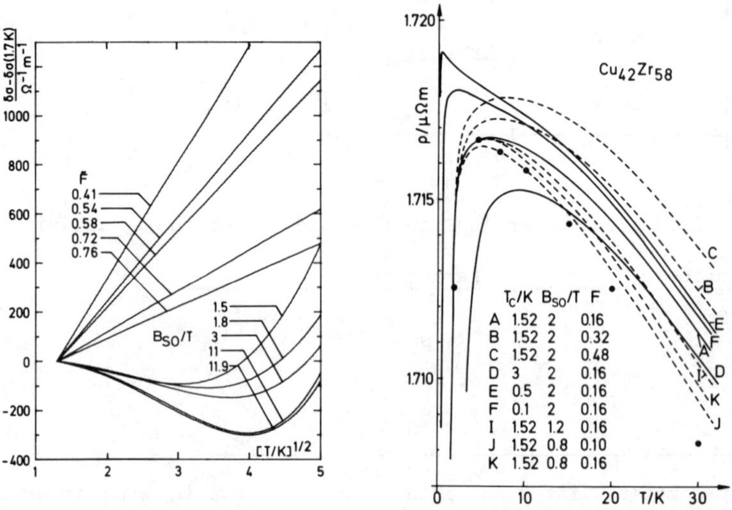

FIGURE 10. Corrections to the resistivity as calculated for various parameters. Left: quantum interference part for certain B_{so} (lower curves) as well as interaction effect for certain $\tilde{F}$ (upper curves). Right: resistivity with parameters for $Cu_{42}Zr_{58}$ for a variation of T_c, B_{so} and F. The experimental data are indicated by dots.

5. MAGNETORESISTIVITY (MR)

Amorphous alloys are isotropic systems. Therefore, the Fermi surface has spherical symmetry and the magnetoresistivity should disappear to first order. The higher order terms are /24/:

$$\Delta\sigma/\sigma = (\pi^2/12)\,(\sigma \cdot R_H)^2 \left(\frac{k_B T}{E_F}\right)^2 \cdot \frac{B^2}{1+(\sigma R_H)^2 B^2}, \tag{23}$$

where B denotes the magnetic field and E_F the Fermi energy. Since $(\sigma R_H)^2 \approx 10^{-10}$ for typical amorphous alloys, this contribution is too small to be detected for fields up to 10 Tesla. Thus, we should look what a MR we could expect from the quantum corrections!

i) Quantum Interference (QI)

Spin-orbit-scattering is supressed by a magnetic field B and therefore, we expect a rise in the resistivity (positve MR) for small fields. The magnetic field B also shifts the phase of the electronic wave function and destroys the constructive interference. At large fields this effect takes over. A decrease in resistivity (negative MR) occurs. In between we expect a maximum. If the spin orbit contribution to the resistivity is negligible, only a negative MR will show up. According to Kawabata /25/ we have for the magnetoresistivity:

$$\delta\sigma_{QI}(B,T) = \frac{e^2}{2\pi^2\hbar}\left(\frac{eB}{\hbar}\right)^{1/2} \cdot \frac{1}{2}\left[3 \cdot f_3\left(\frac{B}{B_i(T)+\frac{4}{3}B_{so}}\right) - f_3\left(\frac{B}{B_i(T)}\right)\right]. \quad (24)$$

In this expression the scattering times τ_n ($n = i,so$) have been replaced by characteristic fields B_n:

$$B_n = \hbar \cdot (4eD\tau_n)^{-1}. \quad (25)$$

Here, $f_3(x)$ denotes a complicated function, which can be approximated by 0.605 for $x \gg 1$ and $x^{3/2}/48$ for $x \ll 1$. This is shown in Fig. 11.

For small fields $B < B_i(T)$, we derive from equ. (24):

$$\delta\sigma_{QI}(B,T) = -5 \cdot \frac{B^2}{B_i(T)^{3/2}} \cdot \left[1 - 3 \cdot \left(1 + \frac{4B_{so}}{3B_i(T)}\right)^{-3/2}\right] [(\Omega m)^{-1}], \quad (26)$$

where the fields have to be inserted in Tesla.

We recognize the proportionality to B^2. One should also bear in mind that $B_i(T) \sim T^p$ according to equ. (25).

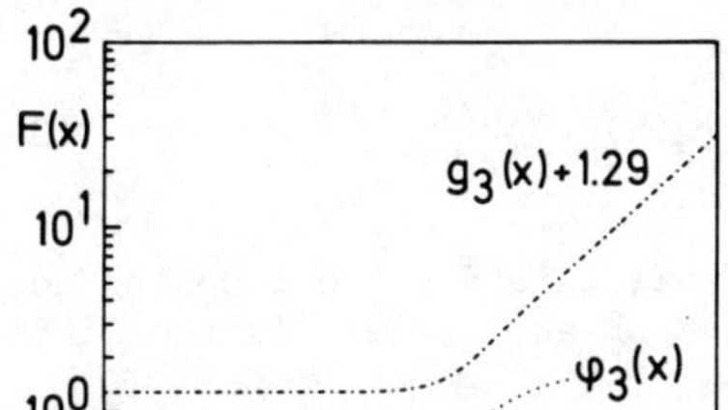

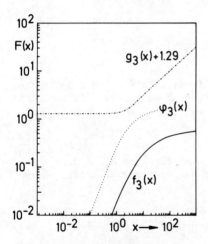

FIGURE 11. The function $f_3(x)$ according to Kawabata /25/: $x \gg 1$; $f_3(x) \approx 0.605$, $x \ll 1$: $f_3(x) \approx x^{3/2}/48$, the function $g_3(x)$ according to Lee et al /26/: $x \gg 1$: $g_3(x) \approx x^{1/2} - 1.29$, $x \ll 1$: $g_3(x) \approx 0.053 \cdot x^2$ and the function $\varphi_3(x)$ according to Fukuyama /27/: $x \gg 1$: $\varphi_3(x) \approx 1.90$, $x \ll 1$: $\varphi_3(x) \approx 0.33 \cdot x^{3/2}$.

ii) Coulomb-Interaction (I)

The magnetic field B may influence the electron-electron interaction (I-effect, see equ. (11)) via the spin- and via the orbital-magnetic field couplings. In principle, both could change the correlations between the electrons. However, it can be shown, that the spin-effect mainly affects the particle-hole channel. Hence, only diagrams of the type given in Fig. 7 left are relevant. The dominating contribution comes from the Hartree-diagrams (Fig. 7 right, upper diagrams with the arrow reversed in the bubble). The diagrams of Fig. 7 right turn out to be insensitive to magnetic fields. These facts yield a MR as follows (Lee et al /26/):

$$\delta\sigma_{IPH}(B,T) = - \frac{e^2}{2\pi^2\hbar} \cdot \frac{\tilde{F}}{2} \cdot \left(\frac{k_B T}{2D\hbar} \right)^{1/2} \cdot g_3(h), \tag{27}$$

where $g_3(h)$ is a function of the variable $h = g\mu_B B/(k_B \cdot T)$. The quantity μ_B denotes the Bohr magneton in $A \cdot m^2$ and g is the spin degeneracy (g = 2 for electrons). Borderline cases are $g_3 = h^{1/2} - 1.294$ for h>>1 and $0.053 \cdot h^2$ for h << 1. We illustrate such a behaviour in Fig. 11. We add, that this contribution disappears if the spin-orbit scattering is too strong, i.e. $\tau_{so} \ll \hbar/(g\mu_B B)$.

This Zeman splitting always yields a positive MR. For small fields, i.e. $B \ll k_B T/(g\mu_B)$ we get:

$$\delta\sigma_{IPH}(B,T) = - 6.39 \frac{B^2}{T^{3/2}} \cdot \left(\frac{\tilde{F}}{0.3} \right) \left(\frac{5\cdot 10^{-5}}{D} \right)^{1/2} [(\Omega m)^{-1}]. \tag{28}$$

Again, we derive a proportionality to B^2. The fields have to be inserted in Tesla, the diffusivity D in $m^2 s^{-1}$.

The orbital effect is dominated by the particle-particle channel. Only the diagrams, shown in Fig. 7 b, will contribute to the MR. Since the particle-particle propagator is called the Cooperon, the following expression describes the field dependence of these quasi-particles. The calculations were performed by Fukuyama /27/ and Altshuler et al /28/. The result is:

$$\delta\sigma_{IPP}(B,T) = \frac{e^2}{2\pi^2\hbar} (e/\hbar)^{1/2} \cdot [-g(B,T)] \cdot \varphi_3(h) \cdot B^{1/2} \tag{29}$$

with $h = 2\, eDB/(\pi k_B T)$. The quantity g(B,T) is an effective coupling constant, which has a complicated dependence on B and T /29/. In the limit of small fields, we get:

$$-g(B,T) = [\ln[T/T_c(B)]\}^{-1}. \tag{30}$$

The function $\varphi_3(h)$ possesses the limits 1.90 for h >> 1 and $0.33 \cdot h^{3/2}$ for h << 1. From these approximations we derive for small fields $B \ll \pi k_B T/(2eD)$:

$$\delta\sigma_{IPP}(B,T) = 35.9 \frac{B^2}{T^{3/2}} \cdot \ln^{-1}(T/T_{co}) \cdot \left(\frac{D}{5\cdot 10^{-5}} \right)^{3/2} [(\Omega m)^{-1}]. \tag{31}$$

Again the units are B/Tesla and $D/m^2 s^{-1}$. This contribution is thus also proportional to B^2 in the case of small fields. It gives a negative MR. The behaviour of the function $\varphi_3(x)$ is shown in Fig. 11 ($T_{co} = T_c(0)$).

vi) Paraconductivity

Superconductivity is supressed by a magnetic field since the field tends to favour parallel spins and therefore to destroy the Cooper pairs. Hence, we expect a positive MR in case of the paraconductivity. With respect to the Al-term a complicated behaviour (Usadel /30/, Mikeska et al /31/) shows up which is not reproduced here. In addition, inelastic scattering is not included. We will only report the low field result (See Appendix A):

$$\delta\sigma_{AL}(B,T) = -\ 185.3 \cdot \frac{B^2}{T^{3/2}} \cdot (\ln T/T_{co})^{-5/2} \cdot \left(\frac{D}{5\cdot 10^{-5}}\right)^{3/2} [(\Omega m)^{-1}]. \quad (32)$$

This contribution is quadratic in B.
The field dependence of the MT-term was discussed by Larkin /32/. He founds:

$$\delta\sigma_{MT}(B,T) = -\frac{e^2}{2\pi\hbar^2}(e/\hbar)^{1/2} \cdot B^{1/2} \cdot \beta(-g(T,B))\cdot f_3\left(\frac{B}{B_i(T)}\right), \quad (33)$$

where $\beta(x)$ is a function tabulated by this author (see Fig. 12) and $f_3(x)$ has been given above. For temperatures above 8 K the function $\beta(B)$ is almost a constant, independent of B. Therefore, we get:

$$\delta\sigma_{MT}(B) = -\ 10 \cdot \frac{B^2}{B_i(T)^{3/2}} \cdot \beta\ [(\Omega m)^{-1}]. \quad (34)$$

Since $B_i \sim T^p$, this effect decreases with rising temperature.

From this discussion we conclude that the magnetoresistivity is proportional to B^2 in a certain B- and T-range. It also decreases with rising temperature.

The relevant parameters entering the theoretical predictions are summarized in Table II. Since in principle D, $\tilde{F}$ and T_c are known from other data, we have a second set of measurements for the determination of the quantities τ_{so} and $\tau_i(T)$.

In the following section the results of some experiments will be discussed with respect to the corrections outlined above.

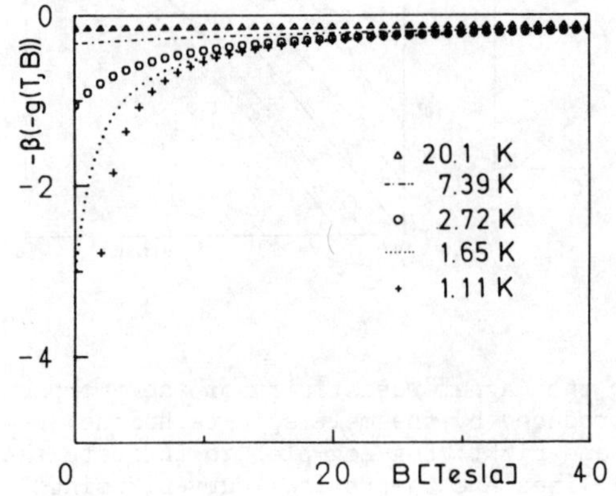

FIGURE 12. The function $\beta(B,T)$ as given by Larkin /32/, calculated with $-g(T,B) = [\ln(T/T_{co}) + \Psi(1/2 + B \cdot 2eD/(\pi k_B T)) - \Psi(1/2)]^{-1}$ /29/. $\Psi(x)$ denotes the diagramma-function and $\pi k_B/(2eD) = 4$ (SI-units).

TABLE II. Parameters governing the magnetoresistivity at low fields

Contribution	Parameters
QI	$\tau_{so},\tau_i(T)\ \{^{\tau_{io}}_{p}$
IPH	$D,\ \tilde{F}$
IPP	$D,\ T_c$
SC_{AL}	D, T_c
SC_{MT}	$\tau_i(T)\ \{^{\tau_{io}}_{p}$

6. COMPARISON WITH EXPERIMENT

The experimental data as shown in Figs. 2 a and b have been interpreted before in terms of weak localization or/and electron-electron interaction /33-39/. However, no consistent quantitative comparison between theory and experiment has yet been presented in order to clarify the relative importance of the various contributions to the resistivity $\rho(B,T)$. In addition, some of the work has been performed on superconductors with transition temperatures well above 0.1 K /34,40/. Hence, the paraconductivity dominates the behaviour (see Fig. 10 right) and no conclusion can be drawn regarding the other effects. In this paper we attempt to describe consistently the magnetoresistivity and the T-dependence of the resistivity using similar parameters.

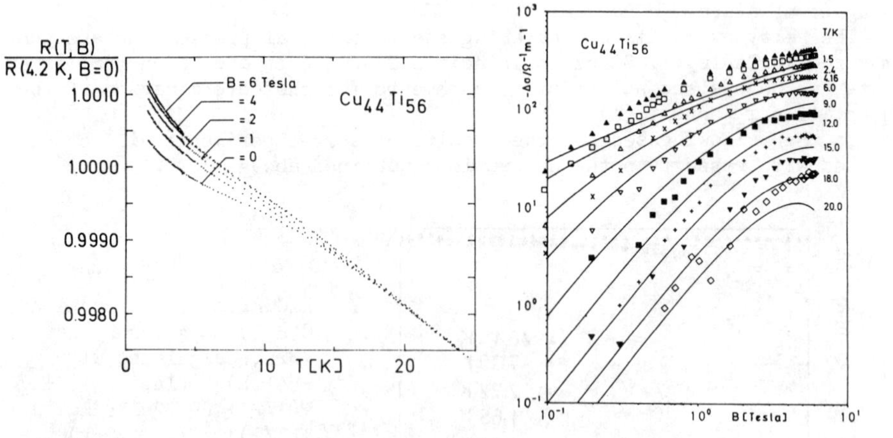

FIGURE 13. Experimental data for the magnetoresistivity of the amorphous alloy $Cu_{44}Ti_{56}$. This alloy was produced by the melt-spin-technique. Left: MR as a function of temperature and right: Log-Log-plot to indicate the quadratic B-dependence. The full lines are theoretical curves, using the full field-dependence.

i) Magnetoresistivity (MR)

At first, we would like to examine the MR alone, a procedure, which yields the parameter τ_{so} and $\tau_i(T)$ or B_{so} and $B_i(T)$, respectively. For each value of B_{so} a function $B_i(T)$ can be derived. The screening parameter $\tilde{F}$ is unimportant, since the interaction contribution turns out to be negligible in our case.

Some data referring to the amorphous alloys $Cu_{44}Ti_{56}$ and $Cu_{42}Zr_{58}$ are reproduced in Figs. 13 and 14. It can be recognized that a MR is detected within the limits or error: $\Delta\rho/\rho \gtrsim 10^{-5}$ up to 20 K. We have observed a quadratic field dependence in these systems. It disappears for low temperatures, since $B \ll B_i(T)$ is no longer guaranteed for the lowest fields attainable.

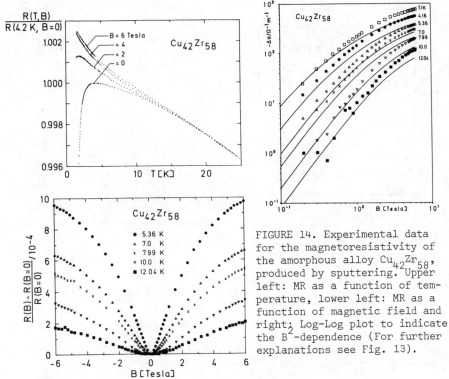

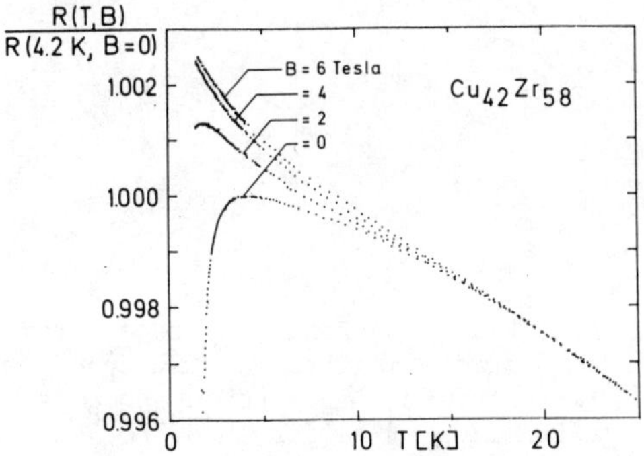

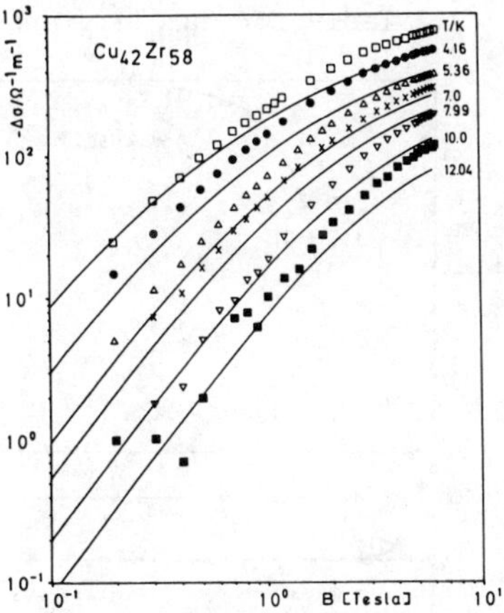

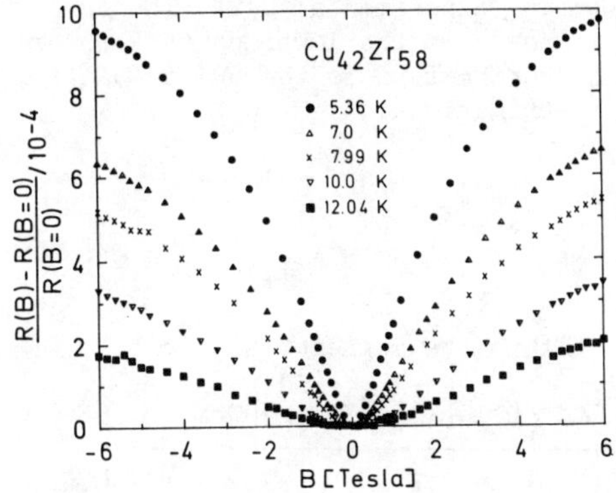

FIGURE 14. Experimental data for the magnetoresistivity of the amorphous alloy $Cu_{42}Zr_{58}$, produced by sputtering. Upper left: MR as a function of temperature, lower left: MR as a function of magnetic field and right: Log-Log plot to indicate the B^2-dependence (For further explanations see Fig. 13).

In combining the various results of Section IV we may write the total magneto conductivity $\delta\sigma(B,T)$ as:

$$\delta\sigma(B,T) = \delta\sigma_{QI}(B,T) + \delta\sigma_{IPH}(B,T) + \delta\sigma_{IPP}(B,T) + \delta\sigma_{AL}(B,T) + \delta\sigma_{MT}(B,T). \quad (35)$$

With the help of the equs. (26,28,31,32 and 34), we get from equ. (35):

$$\delta\sigma(B,T) = \{- \frac{5}{B_i(T)^{3/2}} \; [1-3(1+ \frac{4B_{so}}{3B_i(T)})^{-3/2}] - \frac{6.39}{T^{3/2}} \cdot (\frac{\tilde{F}}{0.3}) \cdot (\frac{5 \cdot 10^{-5}}{D})^{1/2}$$

$$+ \frac{35.9}{T^{3/2} \ln T/T_{co}} \cdot \left(\frac{D}{5\cdot 10^{-5}}\right)^{3/2} - \frac{185.3}{T^{3/2}(\ln T/T_{co})^{5/2}} \cdot \left(\frac{D}{5\cdot 10^{-5}}\right)^{3/2} -$$

$$- \frac{10}{B_i(T)^{3/2}} \cdot \beta(T)\} \cdot B^2 = a_2(T) \cdot B^2 \; [(\Omega m)^{-1}]. \qquad (36)$$

The fields have to be inserted in Tesla, the diffusivities in m^2s^{-1}.

The parameters used for the fits of the data to equ. (36) are summarized in Table III. The quantity $a_2(T)$, as defined in equ. (36) is shown in Fig. 15 for both alloys. These results refer to special choices of B_{so}: 1.8 Tesla for $Cu_{44}Ti_{56}$ and 2 Tesla for $Cu_{42}Zr_{58}$. In case of $Cu_{44}Ti_{56}$

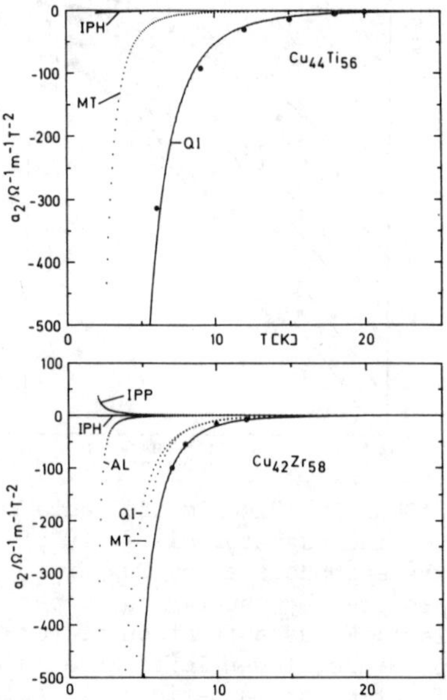

FIGURE 15. Contributions to a_2 from the various magnetoresistivity effects for $Cu_{44}Ti_{56}$ and $Cu_{42}Zr_{58}$ as a function of temperature. For definition of a_2 see text. The dots are experimental data. The full curve in the lower part of the figure refers to the sum of all contributions.

TABLE III. Parameters used for the fit to the magnetoresistivity

alloy	T_c/K	$-B'_{c2}$/Tesla K^{-1}	$D\cdot 10^5/m^2s^{-1}$	$\rho/\mu\Omega cm$	$\tilde{F}$
$Cu_{44}Ti_{56}$	< 0.05	-	3.1	192	0.5
$Cu_{42}Zr_{58}$	1.52	2.6	5	172	0.2

the magnetoresistivity is dominated by the quantum-interference contribution. The paraconductivity derived from the MT-term is small against the QI-effect. All the others, as listed in equ. (35), yield also only minor corrections. To illustrate this point, the $\delta\sigma_{IPH}$-part is shown in Fig. 15. The situation is partly different for $Cu_{42}Zr_{58}$ showing a T_c of 1.52 K. Here, both the QI-and the MT-terms determine the magnetoconductivity. Clearly, again the interaction terms do not play any role. This conclusion is not changed when other values of B_{so} are considered in the fits.

The various functions $B_i(T)$ for a range of B_{so}-values ($1T \leq B_{so} \leq 5T$) are reproduced in Fig. 16. A power law $B_i \sim T^p$ is a good description of the data.

ii) T-dependence of the Resistivity

Next, we would like to discuss the temperature- and concentration-dependence of the resistivity $\rho_{tot}(T)$. In the case of the CuTi-alloys we have analysed both the resistivity and the MR consistently. Preliminary results for the CuZr-alloy are indicated in Fig. 10 right. Paraconductivity is important in this case. The full analysis will be presented elsewhere. The T-dependence of the resistivity is a complicated interplay between the quantum-interference effect including spin-orbit scattering and the Coulomb interaction among the electrons, as was outlined in Section III. We tried to fit our results for a series of Cu_xTi_{100-x}-

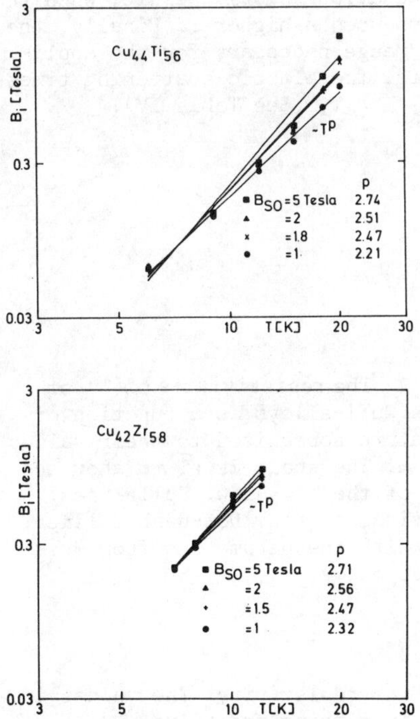

FIGURE 16. The temperature dependence of the inelastic scattering time τ_i (T) for various choices of B_{so} as derived from magnetoresistivity-data for the alloys $Cu_{44}Ti_{56}$ and $Cu_{42}Zr_{58}$.

alloys ($41 \leq x \leq 63$) to the theoretical expressions.
Since the $\overline{T}_c$-values for these alloys are smaller than 0.3 K /40/, the contributions σ_{SL} were put to zero. Hence, equ. (17) reads:

$$\rho_{tot}(T) \approx \rho_o(0) \cdot \{1-\rho_o(0)[\delta\sigma_{QI}+\delta\sigma_I]\}. \quad (37)$$

Defining $\delta\rho(T) = \rho_o(0) - \rho_{tot}(T)$ yields:

$$\delta\rho(T) = \rho_o^2(0) \cdot [\delta\sigma_{QI}+\delta\sigma_I] = \rho_o^2(0)\, \{2.35\cdot 10^3 \cdot \left(\frac{B_{so}}{2}\right)^{1/2} \cdot \left[\left(1+\frac{3B_i(T)}{4B_{so}}\right)^{1/2} - 0.2887 \cdot \left(\frac{B_i(T)}{B_{so}}\right)^{1/2} -1\right] + 2.88 \cdot 10^2 \cdot T^{1/2} \cdot \left(\frac{5\cdot 10^{-5}}{D}\right)^{1/2} \cdot \left(\frac{4}{3} - \frac{3}{2}\tilde{F}\right)\}\ [\Omega m]. \quad (38)$$

Here, B_{so} as well as $B_i(T)$ are in units Tesla and D is in units m^2s^{-1}. A term -1 was added to $\delta\sigma_{QI}$ in order to get the correct limit of $\delta\sigma_{QI}(0)=0$.

A fit of the data to equ. (38) involves the parameters B_{so}, $B_i(T) = a \cdot T^p$ as well as $\tilde{F}$. The results for the various alloy compositions are shown in Fig. 17. They were obtained with the parameter sets given in Table IV. It can be seen that good agreement is possible with parameters similar to the ones necessary to describe the MR. For x = 44 the curves for both sets are plotted in Fig. 17. Systematic trends with composition are observed for the spin-orbit field B_{so} and for the screening parameter $\tilde{F}$, whereas within the limits of error, the function $B_i(T)$ seems to stay the same. Theoretically, we expect an increase of B_{so} with rising Cu-content, as this element has a higher nuclear charge and therefore a stronger spin-orbit interaction. Since the Ti d-states are at the Fermi-level /41/, a tendency towards a more free electron behaviour should result, when the Ti-concentration is lowered. This fact produces a higher $\tilde{F}$. Finally, the characteristic times τ_i and τ_{so} are in a range necessary for the application of the theoretical models discussed. The elastic scattering times τ_o should be around 10^{-15} s, hence: τ_i, $\tau_{so} >> \tau_o$ (See Table IV).

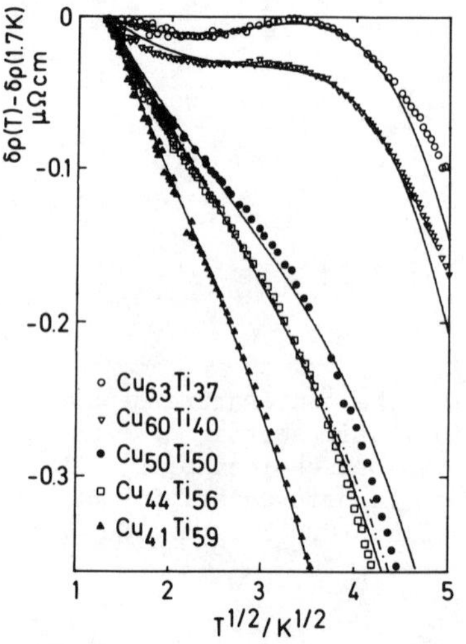

FIGURE 17. The resistivities $\delta\rho(T)$ of various CuTi-alloys as a function of temperature normalized to their values at 1.7 K. The scale is $\sqrt{T}$ to show the region of the $T^{1/2}$-law. Full lines: theoretical curves. Dot-dashed line: Result with the parameters from MR (see Table IV).

TABLE IV. Parameters used for the fit to the resistivity. The values in paranthesis are from the magnetoresistivity fitting procedure.

alloy	B_{so}/T	$\tilde{F}$	p*	$a^* \cdot 10^3/TK^{-p}$	$\tau_{so} \cdot 10^{12}$/s	$\tau_i(10K) \cdot 10^{11}$/s
$Cu_{41}Ti_{59}$	1.5	0.41	2.6	0.74	3.5	1.8
$Cu_{44}Ti_{56}$	1.8 (1.8)	0.55	2.6 (2.5)	0.47 (0.66)	2.9	2.5
$Cu_{50}Ti_{50}$	3	0.58	2.8	0.34	1.8	2.6
$Cu_{60}Ti_{40}$	11	0.72	3.0	0.47	0.48	1.1
$Cu_{63}Ti_{37}$	11.9	0.76	2.8	0.77	0.44	1.0

*From $B_i(T) = a \cdot T^p$

7. CONCLUSION

We may state that a $T^{1/2}$-law in the resistivity in the absence of a magnetic field is only present for T < 4 K, if superconducting fluctuations are not important. Otherwise it can't be seen at all. The deviations of the experimental from the theoretical curve above 20 K may be due to a nonconstant background resistivity $\rho_o(T)$. In the region 4 K < T < 20 K the resistivity is dictated by the interplay between quantum interference and electron interaction effects. The increase of B_{so} with rising Cu-concentration is consistent with the assumption, that the higher atomic number-material Cu shows the higher spin-orbit effect.

The magnetoresistance in the alloys considered is dominated by the quantum-interference effect as long as no paraconductivity is present. In the latter case both effects add up to give the total MR.

There is now strong evidence, that the low temperature transport properties in 3-dimensional amorphous alloys document the quantum corrections to the resistivity. In this respect the situation seems to be completely analogous to the 2-dimensional case /13,42/.

ACKNOWLEDGEMENT

The authors would like to thank A. Fleischmann and W. Dyckhoff for the help in producing the samples. Discussions with Y. Brynseraede are gratefully acknowledged. This work was supported in part by DFG, contract No. Lu 109/25-1.

REFERENCES

1. Schulte A, Fritsch G, Roithmayer A, Lüscher E: this volume.
2. Jäckle J: Einführung in die Transporttheorie, Vieweg Verlag Braunschweig 1978.
3. Bieri JB, Fert A, Creuzet G: LITVIM 94 (1984).
4. Fritsch G, Dyckhoff W, Pollich W, Zottmann W, Lüscher E: Z. f. Physik B59 (1985) 27.
5. Schulte A, Eckert A, Fritsch G, Lüscher E: J. Phys. F 14(1984)1877.
6. Wochner P, Jäckle J: Z. f. Physik B44(1981)293.
7. Fritsch G, Willer J, Lüscher E: Z. f. Naturforschung 37 a(1982)1235.
8. Nagel S: Phys. Rev. B16(1977)1694.
9. Altshuler BL, Aronov AG: Sov. Phys. JETP 50(1979)968.
10. Goetze W: J. Phys. C12 (1979)1297, Belitz D, Schirmacher W: J. Non. Cryst. Solids 61 & 62 (1984) 1073.
11. Nicholson D, Schwartz L: Phys. Rev. Letters 49(1982)1050
12. Altshuler, BL, Aronov AG, Khmel'nitskii ED, Larkin AI in: Quantum Theory of Solids, ed. Lifshitz IM, MIR Publishers Moscow 1983.
13. Bergmann G: Phys. Reports 107(1984)1.
14. Abrikosov AA, Gor'kov LP: Sov. Phys. JETP 15(1962)752
15. Fukuyama H, Hoshino K: J. Phys. Soc. Japan 50(1981)2131.
16. Pines D, Nozières P: The Theory of Quantum Liquids, Benjamin New York 1966.
17. Altshuler BL, Khmel'nitskii, Larkin AI, Lee PA: Phys. Rev. B22 (1980)5152.
18. Altshuler BL, Aronov AG: Solid State Comm. 46(1983)429, Finkelstein AM: Sov. Phys. JETP Letters 31(1980)219.
19. Ginzburg VL, Landau LD: Zh. Eksp. Teor. Fiz. 20(1950)1064.
20. Aslamasov LG, Larkin AI: Phys. Letters 26A(1968)238.
21. Maki K: Progr. Theor. Phys. 40(1968)193.
    Thompson RS: Phys. Rev. B1(1970)327.

22. Patton BR in: Proc. 13th Int. Conf. on Low Temperature Physics, ed. O'Sullivan WS: Plenum New York 1974.
23. Vladar K, Zawadowski A: Solid State Comm. 41(1982)649, Schirmacher W in: Rapidly Quenched Metals, eds. Steeb S, Warlimont H, Elsevier Science Publishers, B.V. Amsterdam 1985, p.995.
24. Busch G, Schade H: Vorlesungen über Festkörperphysik, Birkhäuser Basel 1973.
25. Kawabata A: J. Phys. Soc. Japan 49(1980)628.
26. Lee PA, Ramakrishnan TV: Phys. Rev. B26(1982)4009.
27. Fukuyama H: J. Phys. Soc. Japan 48(1980)2169.
28. Altshuler BL, Aronov AG, Larkin AI, Khmel'nitskii DE: Sov. Phys. JETP 54(1981)411.
29. McLean WL, Tsuzuki T: Phys. Rev. B29(1984)503.
30. Usadel KD: Z. f. Physik 227(1969)260.
31. Mikeska HJ, Schmidt H: Z. f. Physik 232(1970)443.
32. Larkin AI, JETP Letters 31(1980)219.
33. Rapp O, Bhagat SM, Gudmundsson H: Solid State Comm. 42(1982)741.
34. Bieri JB, Fert A, Creuzet G, Ousset JC: Solid State Comm. 49(1984) 849.
35. Howson MA, Greig D: J. Phys. F 13(1983) L 155.
36. Cochrane RW, Ström-Olsen JO: Phys. Rev. B29(1984)1088.
37. Tsai CL, Lu FC: J. Appl. Phys. 55(1984)1945.
38. Poon SJ, Wong KM, Drehman AJ: Phys. Rev. B31(1985)1668.
39. Saub K, Babić E, Ristić R: Solid State Comm. 53(1985)269.
40. Greig D, Howson MA in: Localization and Interaction Effects in Metallic Glasses, PTB-Bericht, Braunschweig 1984, p. 90.
41. Oelhafen P in: Glassy Metals II, eds. Güntherodt HJ, Beck H, Springer Verlag Berlin 1983, p. 295.
42. Lee PA,Ramakrishnan TV: Rev. Mod. Phys. 57(1985)287.

APPENDIX A

Calculation of the magnetoresistivity for the Al-term in the paraconductivity in the limit of small fields.

Usadel /30/ as well as Mikeska and Schmidt /31/ have presented the following expression for the full temperature and magnetic field dependence of the Al-term:

$$\delta\sigma_{AL}(B,T) + \sigma_{AL}(T) = \frac{e^2}{2\pi\hbar}\left(\frac{e}{\hbar}\right)^{1/2} \cdot B_T^{1/2} \cdot \sum_{n=0}^{\infty} U_n(\varepsilon,p) \qquad \text{(A1)}$$

with

$$U_n(\varepsilon,p)=(n+1)\{(\varepsilon+pn)^{-1/2}+(\varepsilon+p(n+1))^{-1/2}-2[\varepsilon+p(n+1/2)]^{-1/2}\}. \qquad \text{(A2)}$$

Here, the quantities p, ε and B_T are defined as:

$$p = (\pi^2/4) \cdot B/B_T,$$
$$\varepsilon = \ln(T/T_{co})+p/2 \quad \text{and} \qquad \text{(A3)}$$
$$B_T= \pi k_B T/(2eD),$$

where T_{co} denotes the superconducting transition temperature at zero field. The sum in equ. (A1) will be calculated under the assumption that the variation from one term in the sum to the next is small. Therefore, we replace the sum by the integral over n:

$$\sum_{n=0}^{\infty} U_n(\varepsilon,p) \approx \int_0^{\infty} U_n(\varepsilon,p)dn = S(\varepsilon,p). \qquad \text{(A4)}$$

The integration can be done easily, the result is:

$$S = \lim_{n \to \infty} \{- \tfrac{2}{3} [\varepsilon^{3/2} + (\varepsilon+p)^{3/2}-2(\varepsilon+p/2)^{3/2}]-2[(p-\varepsilon)\varepsilon^{1/2}-\varepsilon(\varepsilon+p)^{1/2}$$
$$- 2(p/2-\varepsilon)\cdot(\varepsilon+p/2)^{1/2}] +2/3 [(pn)^{3/2}(1+ \frac{\varepsilon}{pn})^{3/2}+(pn)^{3/2}\cdot(1+\frac{\varepsilon+p}{pn})^{3/2}$$
$$- 2(pn)^{3/2} (1+ \frac{\varepsilon+p/2}{pn})^{3/2}] + 2\cdot[(p-\varepsilon)(pn)^{1/2}(1+ \frac{\varepsilon}{pn})^{1/2}- \varepsilon(pn)^{1/2}$$
$$(1+ \frac{\varepsilon+p}{pn})^{1/2} - 2(p/2 - \varepsilon)(pn)^{1/2}(1+ \frac{\varepsilon+p/2}{pn})^{1/2}]\}. \qquad \text{(A5)}$$

In performing the limes $n \to \infty$ we observe, that the coefficients of the terms proportional to $(pn)^{3/2}$ and to $(pn)^{1/2}$ are tending towards zero in such a way that these terms disappear. Hence, we are left with:

$$S = - \frac{2}{3} \cdot [\varepsilon^{3/2}+(\varepsilon+p)^{3/2}-2(\varepsilon+p/2)^{3/2}]-2[(p-\varepsilon)\varepsilon^{1/2}-\varepsilon(\varepsilon+p)^{1/2}- (p-2\varepsilon)$$
$$(\varepsilon+p/2)^{1/2}]. \qquad \text{(A6)}$$

After some rearranging, expression (A6) reads):

$$S = \frac{2}{3}[(\varepsilon+p)^{1/2}(2\varepsilon-p)+(\varepsilon+p/2)^{1/2}\cdot 4(p-\varepsilon)+\varepsilon^{1/2}(2\varepsilon-3p)]. \qquad \text{(A7)}$$

Since $\varepsilon = \ln T/T_{co} + p/2$, we write $\varepsilon = \varepsilon_o + p/2$, with $\varepsilon_o = \ln T/T_{co}$. With the abbreviation $x \equiv p/\varepsilon_o$ equ. (A7) delivers:

$$S = \frac{4\varepsilon_o^{3/2}}{3}\left[(1+\tfrac{3}{2}x)^{1/2} - 2(1-x/2)(1+x)^{1/2} + (1-x)(1+x/2)\right]. \tag{A8}$$

We now consider the case of small fields, which is defined by $p \ll \varepsilon_o$. This is equivalent to: $B \ll (2/\pi)\cdot[k_B T/(eD)]\cdot \ln T/T_{co}$.
The result of a series expansion procedure with respect to x is:

$$S = \frac{4\varepsilon_o^{3/2}}{3}\left[\frac{3x^2}{16} - \frac{69}{1024}\cdot x^4 + O(x^6)\right] = \frac{1}{4\varepsilon_o^{1/2}} - \frac{23}{256}\cdot\frac{p^2}{\varepsilon_o^{5/2}}. \tag{A9}$$

Inserting this result into equ. (A1) and observing that $T \approx T_c$, it can be shown, that the first term just gives $\sigma_{AL}(T)$. The second term contains the field dependence. Putting all the abbreviations back, it yields:

$$\delta\sigma_{AL}(B,T) = -\frac{e^2}{2\pi\hbar}\left(\frac{e}{\hbar}\right)^{1/2}\cdot\frac{23\sqrt{2}\pi^{5/2}}{2048}\cdot\left(\frac{eD}{k_B T}\right)^{3/2}\cdot\frac{B^2}{(\ln T/T_{co})^{5/2}}. \tag{A10}$$

Finally, inserting the numerical values, we end up with equ. (32) as given in the main part of the text.

APPENDIX B

Summary of the T- and B-dependence of the conductivity of amorphous alloys

i) T-Dependence

Quantum interference	$\delta\sigma_{QI}(T) \sim + T^{p/2}$
Quantum interference plus spin-orbit	$\delta\sigma(T\to 0) \sim - T^{p/2}$
Interaction	$\delta\sigma_I(T) \sim + T^{1/2}$
Paraconductivity	$\sigma_{AL}(T) \sim + (T-T_c)^{-1/2}$
	$\sigma_{MT}(T) \sim T^{p-1}(T-T_c)^{-1/2}$

ii) B-Dependence (small fields)

Quantum interference plus spin-orbit	$\delta\sigma_{QI}(B,T) \sim B^2/T^{3p/2}$ (sign change without spin-orbit)
Interaction (Particle-Hole channel)	$\delta\sigma_{IPH}(B,T) \sim -B^2/T^{3/2}$
Interaction (Particle-Particle channel)	$\delta\sigma_{IPP}(B,T) \sim B^2/(T^{3/2}\ln T/T_c)$
Paraconductivity	$\delta\sigma_{AL}(B,T) \sim -B^2/[T^{3/2}(\ln T/T_c)^{5/2}]$
	$\delta\sigma_{MT}(B,T) \sim -B^2/T^{3p/2}$.

AMORPHOUS METALS AT HIGH PRESSURE

E. LÜSCHER, G. FRITSCH*

Physik-Department, Technische Universität München, D-8o46 Garching
*Fak.f.BAUV/I1, Physik, Universität d.Bundeswehr München, D-8o14 Neubiberg

1. INTRODUCTION

Formally we can describe a set of experiments in natural sciences as a determination of the function

$$F = f(q_i) \qquad i = 1,2.......n,$$

which represents an n-dimensional surface in a (n+1)-dimensional space. For simplification we should exclude chaotic behaviour. Examples of variables are: q_1: temperature, q_2: pressure, q_3: stress, q_4: volume, q_5: number density, q_6: current density etc. Very often time is also considered as a common variable like the other q_i, but discussions are still going on if the physical time is not a very special variable and that time itself is a quantity with more than one dimension. Examples of functions f_i e f are: f_1: enthalpy, f_2 = specific heat, f_3: thermal conductivity, f_4: electrical resistivity etc. In the most primitive experiment one measures and plots $f_i(q_i)$ using another q_1 ($1 \neq i$) as a parameter. The states of an alloy series $A_{1-x}B_x$ for example can be represented in a 3-dimensional plot, where f_i = enthalpy, q_1 = concentration of B and q_2 = temperature. The local minima in this topografic map represents quasiequilibrium states - e.g. amorphous states - and the absolute minimum for example corresponds to the crystalline ordering. In the following discussion we will limit ourself mainly to the function f_i: electrical resistivity and the variables q_1: chemical composition, q_2: temperature and q_3: quasihydrostatic pressure.

2. REMARKS ON HIGH PRESSURE TECHNIQUE

High pressure experiments can be classified in 3 categories:

1. Studies of phenomena which occur only or primarily at high pressure - for example : rockformation in the earth - interior.
2. The variable pressure - unaxial and isotropic - serves to gain a better understanding on characterization of states of matter or processes (for example: chemical reactions). One important goal is the test of physical models (theories) for properties and processes of matter.
3. Production of special material, which can be obtained only by applying high pressure - for example: artificial diamonds.

Despite the recommandations from the 1979 conference on weights and measures to use only the Pascal (Pa) as the pressure unit, there are still many other units in use (TABLE 1) - a physicist should be able to work with any units!

The known pressure range in nature is extended over more than 6o orders of magnitude, of which some examples are given in TABLE 2.
It is relatively easy to generate very high dynamic pressure for a few $1o^{-6}$ sec but to maintain high static (isotropic) pressure needs much more effort. This field of experimentation was opened in 19o5 by P.W. Bridgman (Nobelprice 1946), then a graduate student at the Harward University. He

TABLE 1 COMPARISON OF VARIOUS PRESSURE UNITS

1 Pascal	1 Nm^{-2} = 1 Pa
1 Atmosphere (standard)	101325 Pa
1 Atmosphere (technical)	98o66,5 Pa
1 bar	100 000 Pa
1 cm Hg (0^oC)	1 333,2 Pa
1 Torr	133,32 Pa
1 cm H_2O (4^oC)	98,064 Pa
1 psi	6 894,8 Pa

introduced the anvil-technique using refractory metalcarbides and surrounding the probe by soft material - AgCl at that time - in order to hydrostatise the axial pressure from the anvils. This is represented schematically in Fig.1/1/.
Already 19o5 Bridgman reached pressure up to 20 MPa. With todays diamond-anvils one can get up to 2oo MPa. In Fig.2 the thrust mechanism of the

TABLE 2 MAGNITUDES OF PRESSURES, OCCURRING NATURALLY

approx.pressure in Pa	Situation
10^{-27}	Nonequilibrim pressure of hydrogene in the intergalactic space
10^{-15}	"Radiation-pressure" of the 2.7 K radiation
10^{-11}	Best vacuum achieved in laboratory
10^{-5}	Pressure of sound wave at the threshold of hearing
10^{4}	Mean arterial blood overpressure in man
10^{5}	Athmospheric pressure at sea level
$1o^{9}$	Pressure at which Hg solidifies at $2o^oC$
10^{10}	Highest pressure achieved in laboratory with WC-anvils
10^{11}	Radiation pressure of a focused and pulsed laser beam
10^{12}	Pressure in the center of Saturn
10^{13}	Pressure at which He becomes metallic
10^{17}	Pressure at the center of the sun
10^{35}	Pressure at the center of a neutron star

diamond-anvil cell of the Geophysical Laboratory of the Carnegie Institution in Washington is shown.
The principle of this machine is similar to that of a nutcracker, where the force is applied through a screw device. The diamonds are mounted on half-cylindrical rockers assuring the parallel alignment of the two diamonds. With this machine, which is only 2o cm long and weighs about 3 kg, a pressure of about $1.8 \cdot 10^{11}$ Pa has been generated.
For low temperature at high pressure the whole device can be inserted in a cryostat.

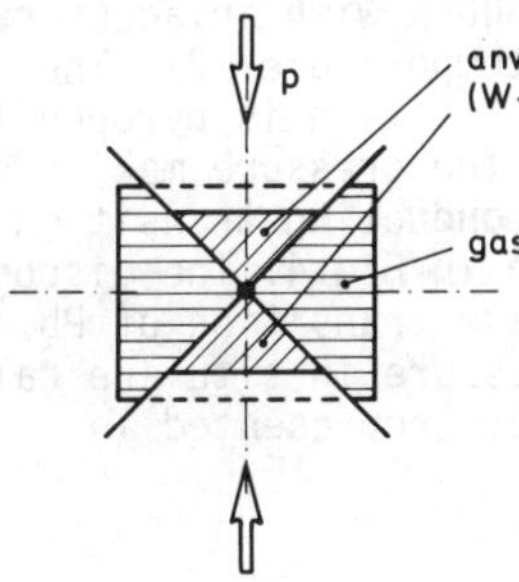

Fig.1: Schematics of a Bridgman-type pressure apparatus

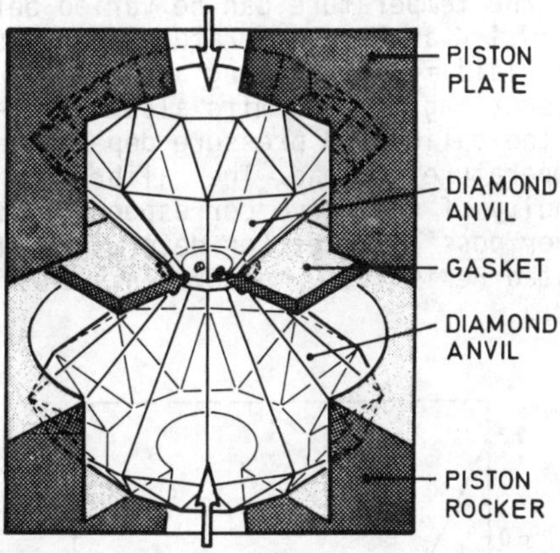

Fig.2: Diamond-anvil-cell. The gasket plays a very important role, because due to its softness it deforms and seals the cell. Since the pistons are transparent, optical as well as X-ray measurements can be done using the holes in the piston plates.

Unfortunately we do not have a diamond pressure machine, we are working with sintered aluminia anvils[2]; for electronic conductivity measurements tungsten-carbide anvils are not suitable. The pressure applied in a hydraulic press is conserved by a clamp-screw-device as indicated in Fig.3. This technique was also developed by Bridgman.

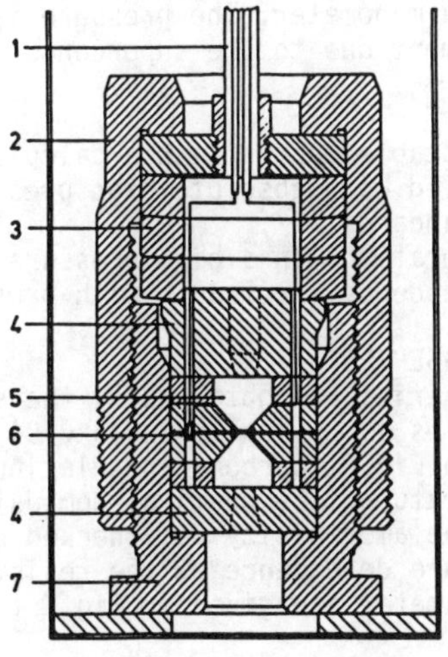

Fig.3: Screw technique for pressure fixation. The springs store the pressure. They also compensate length changes caused by thermal expansion due to temperature variation.

1: Supporting tube, 2: clamp-body,
3: springs, 4: piston-plates,
5: anvil, 6: gasket
7: clamp-screw.

With this clamp pressures up to 13 GPa with accuracy of 5% can be reached and the temperature can be varied between 1.5 and 300 K with an accuracy of ±0.1 K at low temperature and ±0.5 K at higher temperature /2/. The "soft material" for generating an isotropic pressure distribution is pyrophyllite. This arrangement permits also to determine in situ the pressure making use of the calibrated pressure dependence of the superconducting transition temperature of lead. The calibration curve is shown in Fig.4. The discontinuity of the curve corresponds to the fcc-hcp phase transition of Pb. Other possibilities for determining the ambient pressure in situ are calibrated resistivity measurements on Bi, Tl, Cs and Ba, represented in Fig.5.

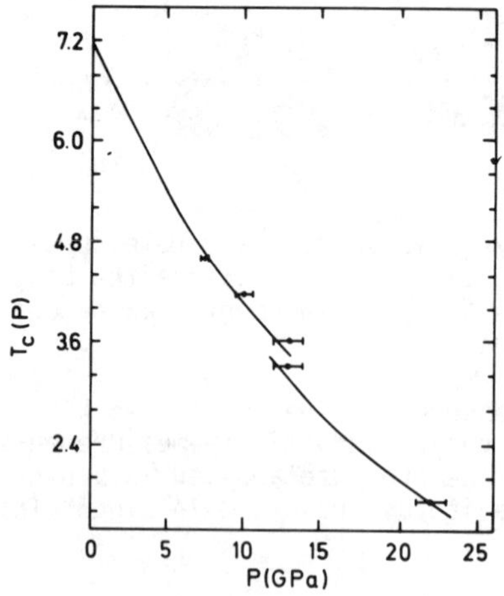

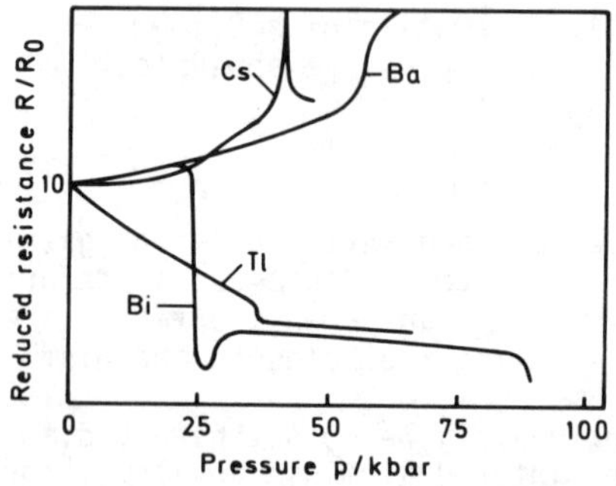

Fig.5: Pressure determination for other calibration metals.

Fig.4: Calibration curve for a lead-pressure manometer. The pressure is determined from a resistance measurement due to the superconducting transition.

In order to relate the variables in a predictable manner with a calculable uncertainty this measurement has to be coupled with absolute high pressure measurements from the National Bureau of Standards /3/.
Another example for secondary pressure calibrations in a high pressure optical cell is based on the shift of fluorescence wavelength with pressure.

3. EXPERIMENTAL RESULTS FOR THE Cu-Ti ALLOY-SERIES

Amorphous ribbons of the Cu-Ti-alloy series are obtained by the spin-melting technique. The geometry of the samples is typically 2x0.5x0.04 mm. A side and a top view of samples is given in Fig.6. Carbon- and Platinum resistors have been used to measure the temperature. The chemical composition was determined by plasma-ion analysis and the amorphicity was checked by X-ray scattering. Examples for the temperature dependence of the resistance of $Cu_{1-x}Ti_x$-alloys with the pressure as parameter are given in Fig.7 to 11 /4/.
The resistance value must be reduced in order to get the resistivity. The normalization is down with respect to the values at zero-pressure. In the geometrical factor corrections due to the compressibility and thermal ex-

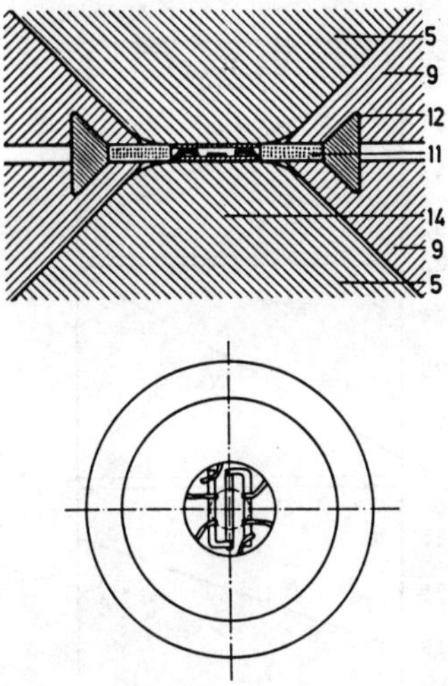

Fig.6: Sample arrangement for the measurement of ρ(p,T) for CuTi metallic glasses. The strip in the middle of the cell is a Pb-piece for pressure calibration. Two samples are arranged on both sides. The potential- and the current-leads are also shown.
5: anvil, 9: gasket, 12: stainless steel ring, 11: pyrophillite gasket and 14: anvil.

Figs.7-11: Results for CuTi in the T-range 4.2-300 K and the p-range 0-10 GPa /5/

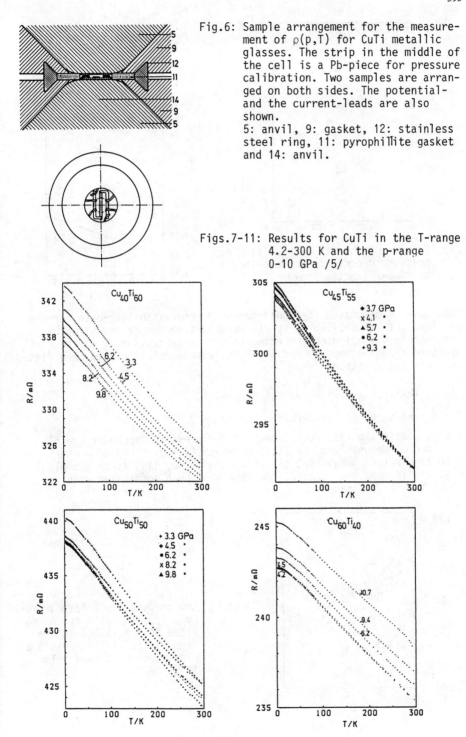

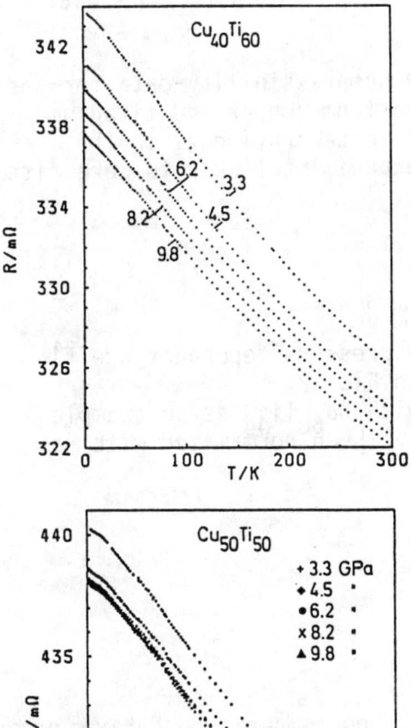

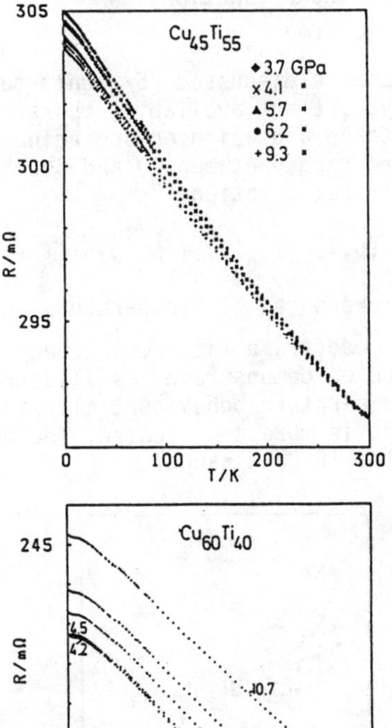

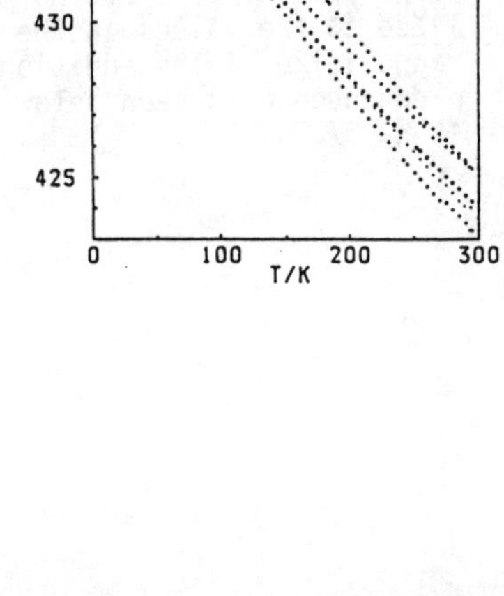

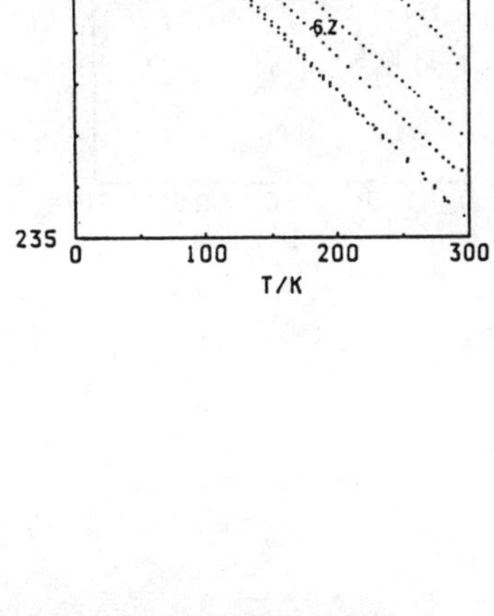

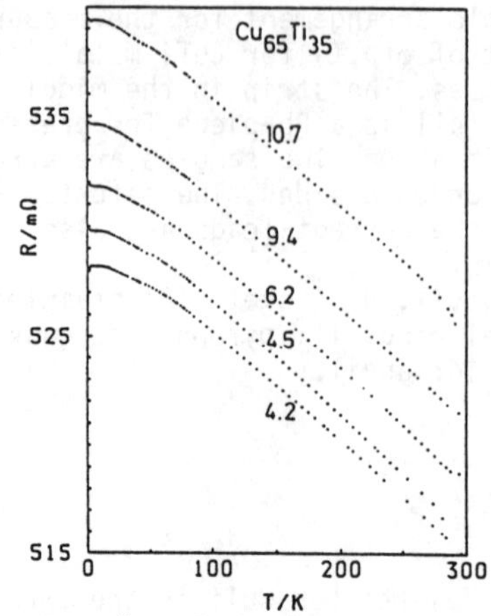

Fig.12: Example of the fit to equ.(1). The pressure dependence is described by linear functions. →

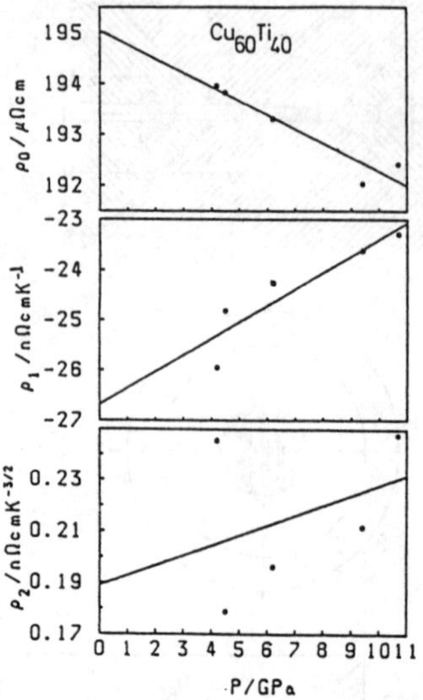

pansion must be discussed /5/. Unfortunately compressibility-data for the CuTi-alloys are not available, therefore data from copper and titanium had been taken and weighted according to the concentration x. In the temperature range between 30 and 300 K the experimental results were fitted to an empirical relation:

$$\rho(p,T) = \rho_0(p) + \rho_1(p) \cdot T + \rho_2(p)\, T^{\gamma} \qquad (1)$$

with : p pressure, T temperature and $\gamma \approx 2/3$.

With this pedestrian fit the tendency of the pressure dependent coefficent ρ_i can be demonstrated as illustrated in Fig.12.

The low temperature behaviour, illustrated with $Cu_{6o}Ti_{4o}$ as an example in Fig. 13, is more complicated. The data have been normalized with respect to the 15-2o K range.

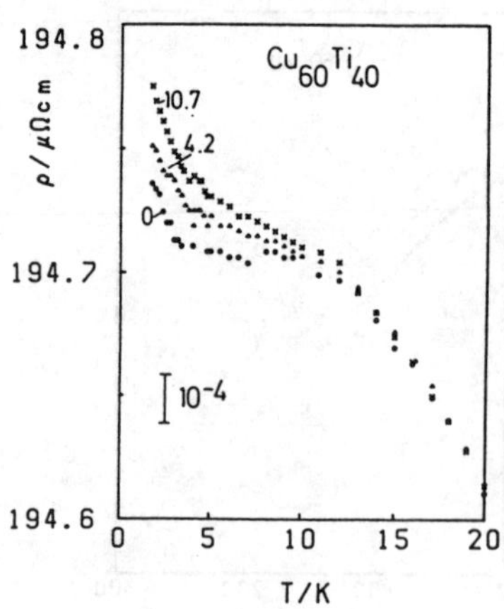

Fig.13: Low temperature behaviour for $Cu_{60}Ti_{40}$. The data are normalized to the values in the range 15-2o K. The additional p-dependence is seen below 15 K /5/.

4. DISCUSSION OF DATA FOR $Cu_{100-x}Ti_x$

Let us discuss first the low temperature behaviour of the resistivity. At zero pressure, its T-dependence can be explained by assuming a constant background resistivity and adding corrections due to quantum-effects /6/. The background describes the incoherent scattering of the electrons by the disordered structure. The quantum corrections are the sum of a quantum-interference and a Coulomb-interaction contribution. In addition the paraconductivity due to superconducting fluctuations plays a minor role. The pressure data shown in Fig.13 have not been analysed yet on the basis of the theoretical calculations, mentioned above.
In case of the high temperature regime we made an effort to explain the T- and p-dependence of the resistivity in terms of a modified Ziman description for an "amorphous element" /7,8/. This model is based on the weak scattering approximation. In view of the high resistivity of the CuTi-alloys of about 200 μΩcm its validity may be doubted. In addition d-electrons are present at the Fermi-surface /9/. However, this latter fact turns out to be very helpful, since d- and s-electrons may be hybridized /10/, and act as quasi-electrons with an effective negative dispersion ($dE/dk<0$). Their scattering behaviour is dominated by elastic backscattering, due to the d-electron-ion potential. Hence, inelastic scattering is not important, the T- and p-dependence of the resistivity should be governed by the elastic scattering contribution alone. The dominance of elastic scattering may also be the explanation for the ability of this model to describe the situation down to $T \cong \Theta_D/2$, where Θ_D denotes the Debye-temperature, despite of its conceptional shortcomings. Hence, we claim, that the resistivity (p- and T-dependence) may be analysed effectively by the modified Ziman-model mainly due to the dominance of elastic scattering. In such a case, the coherence length of the electronic wave function is much larger than the mean free path, since elastic scattering events preserve the phase of the wavefunction. From these assumptions we derive for the resistivity:

$$\rho(p,T) = \rho_o(p) + \rho_T(p,T)\cdot T, \qquad (2)$$

where explicit expressions can be given for $\rho_o(p)$ and $\rho_T(p,T)$. The relevant parameters, which determine these quantities are the Grüneisenparameter γ_G and the volume dependence of the scattering potential. Whereas the second one dictates the overall p-dependence, the first one governs the difference between $\rho_o(p)$ and $\rho_T(p,T)$. The γ_G-volumes necessary to explain the data are in good agreement with numbers determined otherwise /5/. Some information on the effective dispersion of the electrons at the Fermi surface can be obtained by considering the relation beween the volume (V) derivative of the resistivity $\partial \ln\rho/\partial \ln V$ and the thermopower S. Within the context of the model, considered above, this relation should read:

$$S/(S_oT) = -\gamma^{-1}\cdot|\partial \ln\rho/\partial \ln V + 1| \qquad (3)$$

with $S_o=-\pi^2 k_B{}^2/(3|e|E_F)$ and $\gamma=-(1/3)\partial \ln E_F/\partial \ln k_F$. Here k_B denotes the Boltzmann-constant, e the elementary charge and E_F the Fermi energy. Whereas the free electron value for γ is -2/3, the data yield numbers between 1.2 and 0.9 for CuTi, indicating a negative effective dispersion of the electrons at the Fermi-surface /5/.

5. FURTHER RESULTS

Many authors have contributed to the knowledge of the volume dependence of the electrical resistivities in amorphous alloys. Since only part of the measurements cover the full T-range from 4 K to room temperature (RT) it is convenient to compare the results at RT. This comparison is made by using the quantities $\partial \ln\rho/\partial \ln V$ or $\partial \ln\rho/\partial p$. The first one is more relevant for a theoretical analysis, however, it must be computed from the data with the help of the isothermal compressibility κ_T. κ_T is also needed for the correction of the sample geometry when converting the resistance to the resistivity. For this reason we give in TABLE 3 not only the numbers of $\partial \ln\rho/\partial \ln V$, but also the ones for $\partial \ln\rho/\partial p$ and for κ_T. The compressibility is not always known directly from a measurement. In such cases it is calculated from the Young modulus and the Poisson-ratio or from suitably averaging elemental values. It should also be mentioned that no corrections are applied to the data stemming from the p-dependence of κ_T and from the thermal expansion coefficient. These effects may be appreciable for non d-element alloys, such as MgZn.

The results generally show some scatter due to the inherent inaccuracy of the data of about 10 % . This fact may be caused by different sample production procedures, different pressure calibrations and possibly nonhydrostatic pressure conditions. In addition, also the compressibilities applied vary from one determination to the other. However, when the concentration dependence of an alloy series is considered good results are obtained. One should mention, that the data for $Fe_{80}B_{20}$ and $Fe_{40}Ni_{40}P_6B_{14}$ are especially inconsistent for yet unknown reasons.

The results for $Mg_{70}Zn_{30}$ are the only ones showning a negative value for $\partial \ln\rho/\partial \ln V$. They may be compared with theoretical calculations /11/. Qualitative agreement can be stated, however, the absolute numbers differ somewhat (Fig.14). The theoretical value yields:

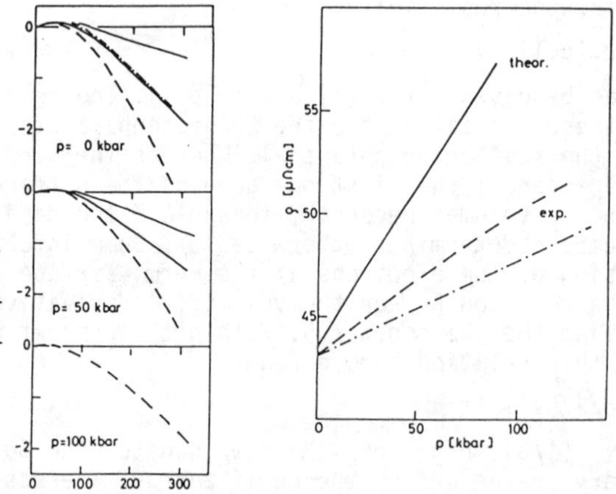

Fig. 14:
Pressure dependence of the resistivity for $Mg_{70}Zn_{30}$ as a function of temperature /11/.
Left: Full lines: Theory without (upper) and with (lower) corrections for thermal expansion. Broken lines: Experiment /16/ and dash-dot line: Experiment /31/. Right: dash-dot line: Experiment /16/ with κ_T=const. and dashed line: Experiment with $\kappa_T=\kappa_T(p)$.

TABLE 3 RESULTS FOR THE P-DEPENDENCE OF THE ELECTRICAL RESISTIVITY FOR AMORPHOUS ALLOYS

a) Alloy series:

x	reference	$\kappa_T \cdot 10^2 GPa^{-1}$	$\partial \ell n\rho/\partial p \cdot 10^3 GPa^{-1}$	$\partial \ell n\rho/\partial \ell nV$
$Cu_{100-x}Zr_x$				
27,5	a	1.01	-13.2	1.31
4o	b	1.0	-12.3	1.23
42	c	0.72	-10.9	1.52
45	a	0.81	- 9.9	1.22
5o	a	0.77	- 8.5	1.10
57	b	0.96	- 6.9	0.72
6o	a	0.70	- 6.5	0.93
$Ni_{100-x}Zr_x$				
80	d	1.36	-16.9	1.24
71	d	1.25	-15.1	1.21
66.7	d	1.21	-13.9	1.15
60	d	1.12	-11.6	1.04
55	d	1.10	-10.1	0.92
50	d	0.97	- 7.76	0.80
36.3	d	0.83	- 3.15	0.38
33	d	0.82	- 2.30	0.28
30	d	0.80	- 2.00	0.25
$Cu_{100-x}Ti_x$				
60	b	0.85	-3.67	0.43
55	b	0.85	-2.71	0.32
50	b	0.84	-1.93	0.23
40	b	0.82	-0.82	0.10
35	b	0.82	0	0

b) d-alloys:

$Pd_{30}Zr_{70}$	b	1.0	-9.2	0.92
$Ti_{50}Be_{40}Zr_{10}$	d	0.64	-7.7	1.21
$Ti_{50}Be_{40}Zr_{10}$	b	1.02	-8.3	0.81
$Ni_{65}Nb_{25}$	a	0.44	-5.3	1.19

TABLE 3 continued

Ni Y	d	0.64	-1.9	0.3
$Pd_{82-x}Si_{18}V_x$	c	0.6	-2.0	0.33
$Pd_{82}Si_{18}$	c	0.69	-3.5	0.51
$Pd_{80}Si_{20}$	b	0.61	-4.3	0.72
c) "free-electron"-alloys:				
$Mg_{70}Zn_{30}$	b	2.5	16.4	-0.66
d) magnetic alloys:				
$Fe_{80}B_{20}$	d	0.61	-5.8	0.95
$Fe_{80}B_{20}$	c	0.69	-7.0	1.02
$Fe_{80}B_{20}$	b	0.61	-10.5	1.73
$Fe_{75}B_{25}$	b	0.61	-8.4	1.39
$Fe_{78}Mo_2B_{20}$	d	0.64	-12.5	1.95
$Fe_{78}P_{22}$	d	?	-5.8*	?
NiP	d	?	-0.44*	?
$Co_{80}P_{20}$	b	0.614	-5.04	0.82
$Fe_{40}Ni_{40}P_6B_{14}$	f	0.57	-6.0	1.06
"	d	0.64	-6.4	1.0
"	b	0.64	-1.6	0.25
$Fe_{19}Ni_{61}P_{14}B_6$	c	0.64	-4.0	0.63
$Fe_4Ni_{63}C_{12}Si_8B_{13}$	e	?	-1*	?
$Fe_4Ni_{68}Cr_6Si_8B_{14}$	e	?	0*	?

* uncorrected for compressibility-effects.

a) Leeds group, see ref.14

b) Munich group, see ref.15, 19, 5, 7

c) Urbana group, see ref. 2o,21

d) Montreal group, see ref. 22, 23

e) P.J.Cote, L.V.Meisel, ref. 24

f) D.G.Ast,D.J.Krenitsky, see ref. 25

$\partial \ln\rho/\partial p \approx +48\ GPa^{-1}$. These results have to be compared with the experimental value of 16.4 GPa^{-1} (see TABLE 3). This discrepancy could be due to a pressure dependence of the compressibility and to the volume expansion coefficient! Both effects are not included in the data analysis.
The results for $Pd_{30}Zr_{70}$ as well as $Ti_{50}Be_{40}Zr_{10}$ and of some CuZr-alloys have been compared with the predictions of the simple model outlined above. The qualitative agreement is again good. In all other cases only plausible arguments have been presented in order to understand the ρ-dependence of

the resistivity. The correlation between $\partial\ell n\rho/\partial\ell nV$ and the thermopower parameter $\xi=S/(S_o \cdot T)$ as given with equ.(3) can be analysed for the alloy series $Cu_{100-x}Zr_x$, $Ni_{100-x}Zr$ and $Cu_{100-x}Ti_x$. The data are shown in Fig.15 together with the Hall-coefficient results and the calculated numbers for γ. Normally, the values of γ decrease with falling Zr or Ti concentration, however, in NiZr we observe an abrupt change $\gamma(x)$ at about 40 % Zr. The reason for such a behaviour is unknown.

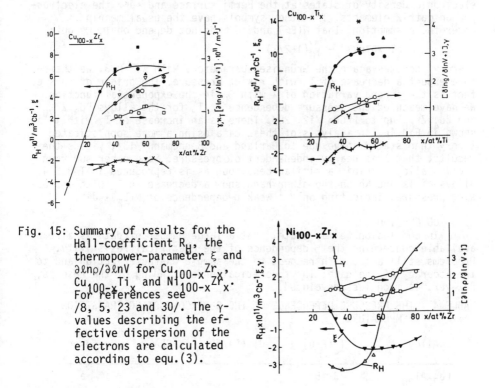

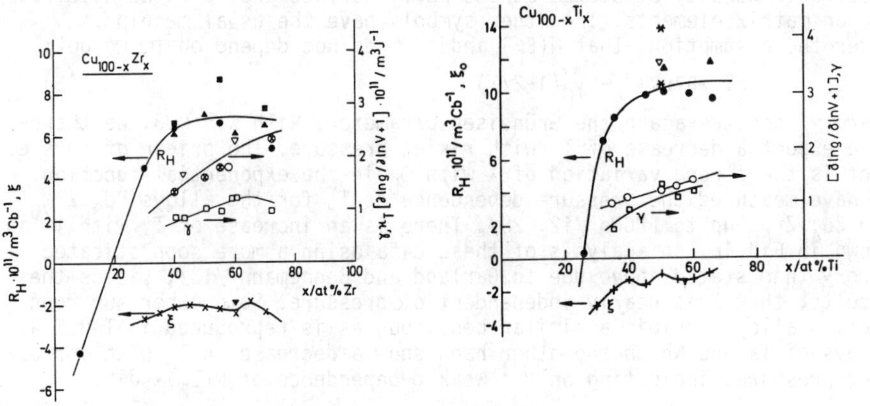

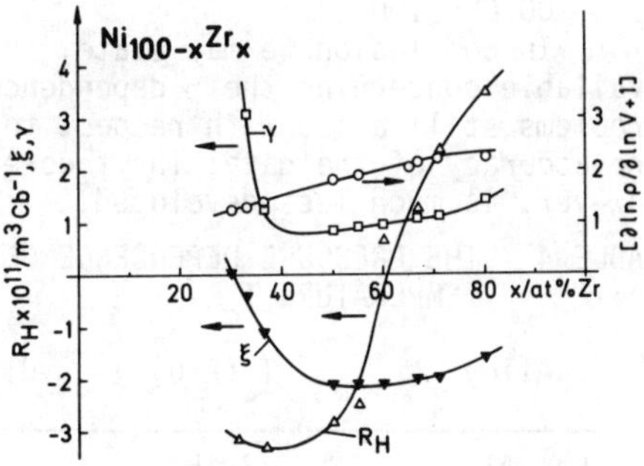

Fig. 15: Summary of results for the Hall-coefficient R_H, the thermopower-parameter ξ and $\partial\ell n\rho/\partial\ell nV$ for $Cu_{100-x}Zr_x$, $Cu_{100-x}Ti_x$ and $Ni_{100-x}Zr_x$. For references see /8, 5, 23 and 30/. The γ-values describing the effective dispersion of the electrons are calculated according to equ.(3).

6. SUPERCONDUCTING PROPERTIES

Most of the nonmagnetic transition metal alloys show superconductivity /12/. They are superconductors of type II with a very low B_{c1} (milli-Tesla) and a very high B_{c2} (several Tesla). In addition, a paraconductivity due to superconducting fluctuations, can be detected easily. All these effects are caused by the very short mean free path of the electrons and, hence, by the short coherence length of the Cooper-pairs. We have to deal with superconductors of weak and medium coupling strength in the dirty limit.

The pressure dependence of the paraconductivity and of the superconducting transition temperature T_c have been determined with resistivity measurements. Whereas it is too early to discuss the paraconductivity - the p-dependence of the background has to be subtracted -, the variation of T_c with p can be analysed /32/.*

* In this paper we proposed a linear background resistivity, an assumption which is only correct approximately.

In principle it is a simple task to predict the p-dependence from the BCS-equation ($\lambda \ll 1$, i.e. weak coupling).

$$T_c = 1.14 \cdot \Theta_D \exp |\lambda^{-1}| \quad (4)$$

with $$\lambda = N(E_F) \cdot \hbar^2 \langle J^2 \rangle / (m k_B^2 \Theta_D^2)$$

Here, λ denotes the electron-phonon coupling constant at $T=0$, $N(E_F)$ the electronic density of states at the Fermi surface and $\langle J^2 \rangle$ the electron-phonon matrix elements. The other symbols have the usual meaning. Under the assumption, that $N(E_F)$ and $\langle J^2 \rangle$ do not depend on p, we get:

$$\partial \ln T_c / \partial \ln V = -\gamma_G (1 - 2/\lambda), \quad (5)$$

where γ_G denotes again the Grüneisen-parameter. With $\lambda \approx 0.3$, we therefore expect a decrease of T_c with rising pressure. The origin of this effect is the strong variation of λ with Θ_D in the exponential function. We have measured the pressure dependence of T_c for the alloys $Pd_{30}Zr_{70}$ and $Cu_{40}Zr_{60}$ up to 11 GPa /12, 28/. There is an increase of T_c with p as shown in Fig.16. An analysis of these data using a more sophisticated theory than stated above due to Garland and Bennemann /13/, yields the results, that λ is nearly independent of pressure. Some other superconducting alloys exhibit a similar behaviour as is reproduced in TABLE 4. Alloys of Ta and Nb on the other hand show a decrease in T_c with increasing pressure, indicating only a weak p-dependence of $N(E_F) \cdot \langle J^2 \rangle$.

7. CONCLUSION

In conclusion we may state, that a large number of data is already available concerning the p-dependence of the electrical resistivity. Problems still arise with respect to the sample characterization and to the accuracy of the data. The theoretical understanding of the results, however, is much less developed.

TABLE 4 THE PRESSURE DEPENDENCE OF THE SUPERCONDUCTING TRANSITION TEMPERATURE T_c

Alloy	$T_c(P=0)$	$dT_c(P)/dP$ / $K \cdot GPa^{-1}$	reference
$La_{70}Al_{30}$	3.15	0.34	a
$La_{80}Al_{20}$	3.84	0.43	b
$La_{80}Ga_{20}$	3.97	0.39	b
$Pd_{30}Zr_{70}$	2.63	0.036	c
$Cu_{40}Zr_{60}$	1.69	0.046	c
$Nb_{55}Rh_{45}$	4.75	-0.024	d
$Ta_{55}Rh_{45}$	3.19	-0.031	d
$Nb_{55}Ir_{45}$	4.59	-0.027	d

a) Ref. 26, b) Ref. 27, c) Ref. 28, d) Ref. 29

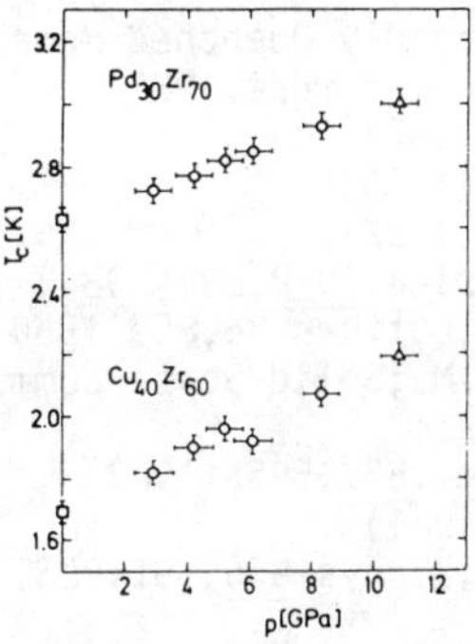

Fig. 16: Superconducting transition temperatures T_c for $Pd_{30}Zr_{70}$ and $Cu_{40}Zr_{60}$ as a function of pressure /32/.

There are some calculations indicating that simple alloys ($\rho \lesssim 50\mu\Omega cm$) may be explained within weak scattering theory. The superconducting transition temperature can be described successfully using the BCS-theory in the dirty limit.

8. ACKNOWLEDGEMENT:

The authors would like to thank Dr. J.S. Schilling for helpful discussions. The support of many individuals in the authors' laboratories is gratefully acknowledged. This work was supported in part by DFG.

REFERENCES

1 Bridgman,P.W.;The Physics of High-Pressure, Bell London 1958
2 Willer,J.,Moser,J.;J.Phys.E:Sci.Inst.12,886(1975)
3 Heydemann,P.L. in High Pressure Chemistry, ed.:H.Kelm,Reidel Publ. (1978)
4 Dyckhoff,W.;Thesis,TU Munich, in Preparation
5 Dyckhoff,W.,Fritsch,G.,Lüscher,E.; Proceedings of MRS Meeting, Strasbourg, 1985, to appear
6 Fritsch,G.,Schulte,A.,Lüscher,E.; this volume
7 Fritsch,G.,Willer,J.,Lüscher,E.;Z.f.Naturforschung,37a,1235(1982)
8 Fritsch,G.,Dyckhoff,W.,Pollich,W.,Zottmann,W.,Lüscher,E.;Z.f.Physik, B59,27(1985)
9 Oelhafen,P. in Glassy Metals II, eds.H.J.Güntherodt,H.Beck,Springer-Verlag,Berlin(1983)p.295
10 Weir,C.F.,Howson,M.A.,Gallgher,M.A.,Morgan,S.J.:Phil.Mag.47,163(1983)
11 Hafner,J.;Non Cryst.Solids,69,325(1985)
12 Lüscher,E.,Fritsch,G.;IEEE Trans-Magn.-MAG.20/5/Pt2,1651(1984)
13 Garland,J.W.,Bennemann,K.H. in AIP Conf.Proc.4,ed.D.H.Douglas,AIP, New York 1972
14 Greig,D.,Howson,M.A.;Solid State Comm.42,729(1982)
15 Willer,J.,Fritsch G.,Lüscher,E.; Non Cryst.Solids 46,321(1981)
16 Fritsch,G.Willer,J.,Wildermuth,A.,Lüscher,E.;J.Phys.F12,2965(1982)
17 Fritsch,G.,Willer,J.,Schink,H.,Wildermuth,A.,Lüscher,E.;J.Magn.Mat. 37,30(1983)
18 Lüscher,E.,Willer,J.,Fritsch,G.;J.Non Cryst.Solids,61 u. 62,1109(1984)
19 Fritsch,G.,Dyckhoff,W.,Pollich,W.,Lüscher,E.;J.Phys.F15,1537(1985)
20 Lazarus,D.;Solid State Comm.32,175(1979)
21 McNeil,L.E.,Lazarus,D.;Phys.Ref.B27,6007(1983)
22 Cochrane,R.W.,Strom-Olsen,J.O.,Rebouillat,J.P.,Blanchard,A.;Solid

State Comm.35,199(1980)
23 Cochrane,R.W.,Destry,J.,El Amrani,M.;in Rapidly Quenched Metals,eds.: S.Steeb,H.Warlimont,Elsevier Science Publ.Amsterdam,1985
24 Cote,P.J.,Meisel,L.V.; Phys.Rev.B25,2138(1982)
25 Ast,D.G.,Krenitsky,D.J.;Scr.Metall.10,247(1976)
26 Razavi,F.S.,Schilling,J.S.;Z.Phys.B48,123(1982)
27 Bindilatti,V.,Missell,F.P.,Tenover,M.;Physica,107B,391(1981)
28 Willer,J.,Fritsch,G.,Lüscher,E.;Appl.Phys.Letters 36,859(1980)
29 Müller,R.Shelton,R.N.,Koch,C.C.,Kroeger,D.M.;Solid State Comm.45,327 (1983)
30 Cochrane,R.W.,Destry,J.,Trudeau,M.;Phys.Rev.B27,5955(1983)
31 Mizutani,V.,Mizoguchi,T.;J.Phys.F11,1385(1981)
32 Willer,J.,Fritsch,G.,Rausch,W.,Lüscher,E.;Z.Phys.B50,39(1983)

ELECTRON LOCALIZATION IN THE LIQUID Tl-Te-SYSTEM AS SEEN BY Bi-IMPURITY

M.A. HAGHANI, R. BRINKMANN*, M. von HARTROTT**, P. MAXIM, K. OTT, C.A. PAULICK, and D. QUITMANN

Institut für Atom- und Festkörperphysik, Freie Universität Berlin, Arnimallee 14, D-1000 Berlin 33, Fed. Rep. Germany

Electron localization is observed in numerous liquid alloys of s-p-metals, e.g. in alloys of Tellurium, at stoichiometric compositions. There they show the property of strong decrease of the magnetic susceptibility χ, electric conductivity σ, and Knight shift K as well as strong increase of the viscosity, Hall effect and of the magnetic nuclear spin relaxation rate R_M /1, 2, 3, 5/.

In order to analyze the influence of local or band contribution on these effects we have measured K and R_M of an impurity atom with additional electrons, viz. Bi, in the liquid Tl-Te-system.

Our NMR-measurements of K for ^{209}Bi in the liquid Bi-Tl-system /4/ show that Bi forms, as expected, with Tl a good metallic system. Also Bi and Te form an almost metallic system /3/. Therefore, Bi appears as a good probe for studying local versus band contritution in the liquid Tl-Te system around the Tl_2Te-composition. Generally, little is so far known about impurities in liquid semiconductors (contrary to the solid).

Knight shift is given for simple metals by the s-electron conduction band value K_s: $K=K_s=8\pi/3\ \chi_p\langle|\psi_s(0)|^2\rangle_{FS}$. Here the Pauli paramagnetic susceptibility per atom is χ_p and $\langle|\psi_s(0)|^2\rangle_{FS}$ is the square of s-wave function at the nucleus averaged over these electrons at the Fermi surface. For the nucleus studied here, ^{206m}Bi, quadrupolar relaxation rate $R_Q=R-R_M$ is most probably neglegible /8/.

Usually the reduction of K below the free electron value K_{FE} and the increase of the magnetic spin relaxation rate R_M in the liquid semiconductors are ascribed to a reduction of electron density of states at the Fermi energy E_F

$$K = g\cdot K_{FE} \quad , \quad g = \frac{N(E_F)}{N_{FE}(E_F)} \leqslant 1 ,$$

(where g is Mott's g-factor) and to a concomitant slowing down of electron motion,

$$R_M = R_M^{Korr} \cdot \eta \quad , \quad \eta = \frac{\tau_M}{\tau_{FE}} \geqslant 1 ,$$

---

*Now at DESY, Hamburg.

**Guest of Sfb 161.

where η is Warren's enhancement factor /9/. The Korringa rate is $R_M^{Korr} = k(k+1) \cdot 2\pi \gamma_n^2/\gamma_e^2 \hbar \cdot K_S^2 T$. For a recent treatment see /10/.

The experimental technique was TDPAD (Time Differential Perturbed Angular Distribution /7/), where we produced the isomer ^{206m}Bi, lifetime τ_n=880 µs, by the nuclear reaction $^{205}Tl(\alpha,3n)^{206m}Bi$. In such experiments the concentration of the produced impurity ^{206m}Bi and decay products is very small indeed (of the order 10^{-10}%). Spin relaxation and Knight shift are detected through decay and precession frequency, respectively, of the asymmetry of the γ radiation depopulating the isomeric state (k=2):

$$I_\gamma(t) = I_o \exp(-t/\tau_n) \left[1+A_2 \exp(-t/\tau_R)(\tfrac{1}{4}+\tfrac{3}{4} \cos(4\pi\nu_L t+\phi))\right]$$

and so $R = \frac{1}{\tau_R}$; $K = (\nu_L-\nu_{Lref})/\nu_{Lref}$.

Here $I_\gamma(t)$, I_o are γ-intensity at the time t and t=0, A_2 is amplitude of the γ-anisotropy at t=0, ν_L is Larmor frequency and ν_{Lref} is the Larmor frequency for the nucleus in a reference substance; here we have used the equality of the Knight shift for ^{209}Bi in Tl_2Bi_3 (NMR) and for ^{206m}Bi in Tl_2Bi_3 (TDPAD). The measurements are performed at constant external magnetic field $B_{ext} \approx$ 800 gauss, both at T = 740 K.

We have measured K and R on ^{206m}Bi as a function of the concentration C and the temperature T at compositions close to Tl_2Te, viz. Tl-concentrations 67.6, 67.3, and 66.7 at % with an estimated uncertainty of ±0.1%. Results are presented in figure 1 which includes from the literature /5/ K for ^{205}Tl in Tl-concentrations 66.2, 67.4, and 68 at %. Figure 2 gives the concentration dependence of R_M for ^{206m}Bi in liquid Tl-Te-systems at the temperature T = 800 K.

The impurity-Knight shift shows a strong dip at the concentration corresponding to Tl_2Te which is comparable to NMR-results of ^{205}Tl in the liquid Tl-Te-system /5/. The difference against the pure Tl-Te-system is a strong temperature dependence of K. Also R_M of the Bi-impurity displays a very strong enhancement upon approaching the stoichiometric composition from the metal side. Further work is in progress.

We thank our colleagues M. Soltwisch, M. Kiehl and I. Senel for their help. We are obliged to Kernforschungszentrum Karlsruhe (KfK) for machine time at the cyclotron and to Deutsche Forschungsgemeinschaft, Sfb 161, for financial support.

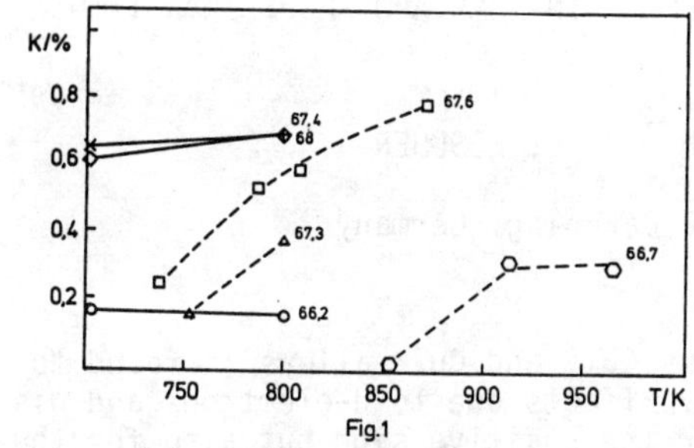

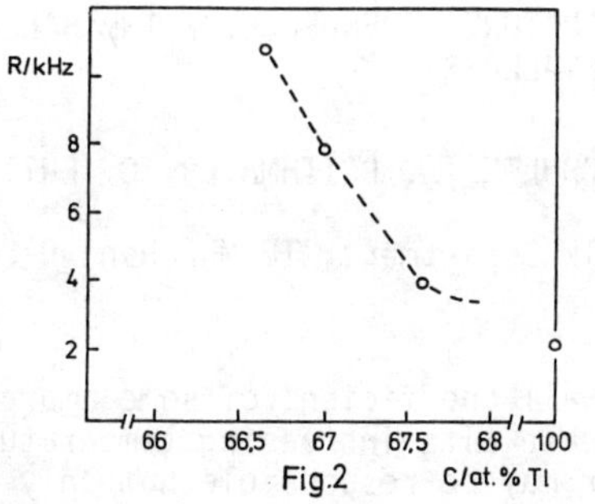

Figure 1: Temperature dependence of Knight shift K for ^{205}Tl-nucleus (with Tl concentration 66.2, 67.4 and 68 at %) and for ^{206m}Bi-impurity (with Tl-concentration 66.7, 67.3, 67.6 at %) in liquid Tl-Te-system.

Figure 2: Concentration dependence of the magnetic spin relaxation rate for ^{206m}Bi-impurity in liquid Tl-Te-system at the temperature T = 800 K.

REFERENCES

1. Carter, G.C., et al.: Metallic Shifts in NMR (Pergamon Press Oxford 1977)
2. Cutler, M.: Liquid Semiconductors (Academic Press, New York, San Francisco, London 1977)
3. Haghani, M.A.: Diplomarbeit, Freie Universität Berlin 1983 (unpublished)
4. Senel, I.: private communication
5. Warren, W.W.: Nuclear Magnetic Resonance in Semiconducting Liquid Alloys, J. Non Cryst. Sol. 8, 241 (1972)
6. Brown, D. et al.: Magnetic properties of Liquid Alloys of Te and Tl. Phil. Mag. 23, 1249, 1971
7. Ackermann et al.: Topics in current Physics 31 291 (1983)
8. Willeke, F., v. Hartrott, M., and Quitmann, D.: J. Phys. F: Metal Phys. 11, 275 (1981)
9. Warren, W.W.: Phys. Rev. B3, 3708 (1971)
10. Götze, W., and Ketterle, W.: Z. Phys. B-Condensed Matter 54, 49-57 (1983)

TEMPERATURE DEPENDENCE OF THE HALL-EFFECT IN SOME AMORPHOUS TRANSITION METAL ALLOYS

A. SCHULTE, A. ROITHMAYER, G. FRITSCH[+] and E. LÜSCHER

Physik Department, TU München, D-8046 Garching, Germany

The Hall coefficient of some amorphous CuTi and CuZr alloys is found to decrease with increasing temperature. Effects due to d-electrons and disorder may be responsible not only for the positive sign but also for the temperature dependence of this quantity. Below about 25 K the data can be described by a $T^{1/2}$-law. This fact gives support to the presence of electron-electron interaction effects in accordance with theoretical predictions.

1. INTRODUCTION

The electronic transport properties of amorphous alloys are determined by the disordered atomic structure and by the electronic density of states. The latter quantity behaves very similar for crystalline compounds and for liquid and amorphous alloys containing transition-metal elements /1/. Usually, the Fermi energy is located within the d-states of the early TM-metal. Consequently, these TM-TM alloys exhibit high resistivities ($\approx$ 200 μΩcm) and an (anomalous) positive sign of the Hall coefficients /2,3/. Such a behaviour differs strongly from the one observed in simple metallic glasses, which show low resistivities ($\approx$ 50 μΩcm) and a Hall coefficient roughly compatible with the nearly free electron model. Therefore, two mechanisms may be of importance in understanding this difference, namely strong scattering and the density of d-states at the Fermi level, i.e. "disorder"- and "band structure"-effects

2. EXPERIMENTAL DETAILS AND RESULTS

In order to lend further support to this point, we will present and discuss results for the temperature dependence of the Hall effect between 1.5 and 100 K. These data were determined in magnetic fields up to 6 Tesla. The measurements were performed on samples produced by rapid quenching from the melt as well as by sputtering. In detail, the $Cu_{44}Ti_{56}$ $Ni_{35}Ti_{65}$ and $Cu_{39}Zr_{61}$ samples were prepared from melt-spun ribbons the others by dc-magnetron sputtering. These data confirm the temperature dependence of the Hall coefficient R_H/2/, which was observed recently. Hence, we conclude that this property is independent from the details of the disorder present. The Hall coefficient R_H is determined from the Hall resistivity $\rho_H = R_H \cdot B$ with a standard dc-dc technique described in more detail elsewhere /2/. The Hall resistivity is related to the longitudinal (σ_{xx}) and transverse (σ_{xy}) conductivities via

$$\rho_H = \sigma_{xy}/\sigma_{xx}^2. \quad (1)$$

+ Inst. f. Physik, Fak. f. BauV/I 1, Universität der Bundeswehr München, D-8014 Neubiberg, Germany

Some results for the temperature dependence of the normalized electrical resistance R are reproduced in Fig. 1, in order to characterize the samples. The variation of R with temperature compares well with literature data and, thus, demonstrates the amorphous state of the samples.

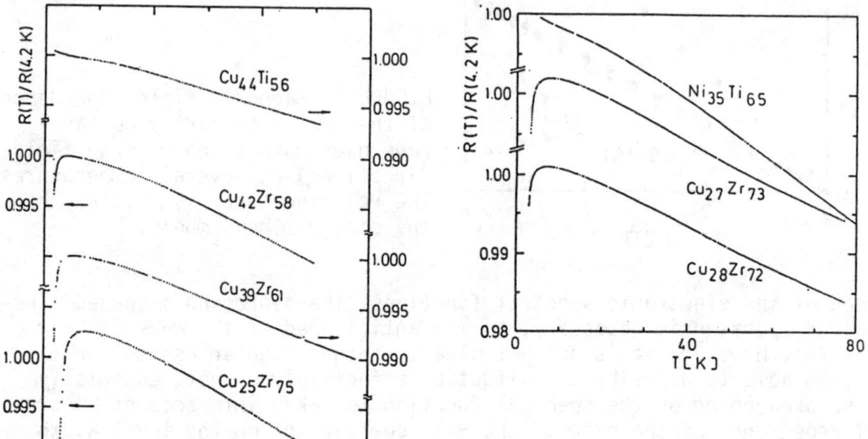

FIGURE 1. Temperature dependence of the electrical resistance normalized to the value at 4.2 K.

As an example, the magnetic field dependence of the Hall resistivities for $Cu_{39}Zr_{61}$ and $Cu_{44}Ti_{56}$ are shown in Fig. 2 at various temperatures. No sign fo nonlinearities is detected in the ρ_H (B) curves. Hence, contributions from magnetic impurities are ruled out altogether. Some experimental results for the Hall coefficient determined with a high resolution of about 10^{-3} are plotted as a function of temperature in Fig.3. The accuracy of the absolute value is limited by a 10% error in the measurement of the sample thickness. All the transition metal rich alloys which were investigated here exhibit a positive Hall coefficient R_H. Below about 25 K the increase in the magnitude of R_H with decreasing temperature can be described by a $T^{1/2}$-law.

3. DISCUSSION OF THE OVERALL T-DEPENDENCE

Only in metallic conductors containing d-electrons positive Hall-coefficients have been observed to our best knowledge. Therefore, the origin of the sign should be related to the "band structure". Recently, the effect of level broadening in a two band model has been examined by Hänsch /4/. This author studied one band Hamiltonians for a wide s- and a narrow d-band which were coupled by a hybridization interaction. Disorder is included by using spectral functions with finite widths. The diagonal and non-diagonal elements of the conductivity tensor are evaluated with the help of the Kubo formalism in the presence of a magnetic field. The free electron result is recovered in the quasiparticle limit, i.e. with vanishing level broadening. In Hänsch's very schematic model the appearance of the sign change is essentially determined by the finite

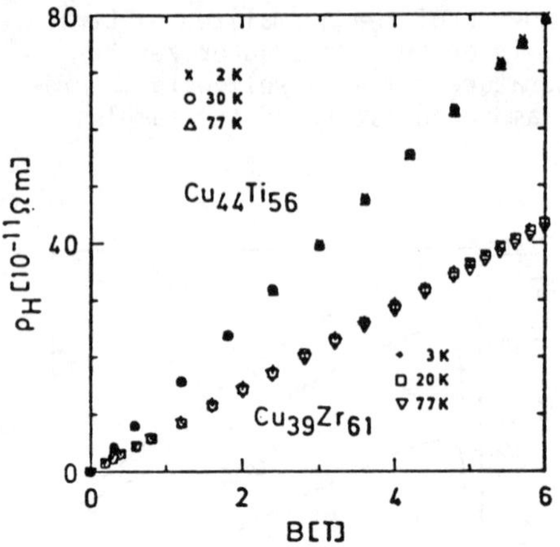

FIGURE 2. Magnetic field dependence of the Hall resistivity ρ_H (average over normal and reverse field directions) at several temperatures. The relative error is smaller than the size of the symbols.

spread of the electronic spectral functions. Therefore, no s-shaped electronic dispersion is necessary, as was anticipated in the work of Weir et al /5/. However, it is not yet clear, whether such an assumption may still be able to describe the situation effectively or not, especially if the broadening of the spectral function is taken into account. A crucial dependence of the sign of the Hall coefficient on the level width has been noted also by Movaghar /6/, using standard linear theory for extended states. The model outlined above can probably also account for the T-dependence observed. Since the electronic self energy is related to the deviations of the actual atomic arrangement from that of lattice, it determines the width of the spectral function. Thus, thermally induced variations of the disorder will change this quantity and consequently the Hall coefficient, too. Actually calculations of the self energy taking into account an electron-"phonon" coupling show that there is a considerable effect due to the temperature /4,7/.

4. DISCUSSION OF THE LOW TEMPERATURE BEHAVIOUR

At low temperatures precursors of localization may show up in the electronic transport properties of disordered systems /8,9/. Crudely speaking, interaction effects are enhanced since the electrons show a diffusive motion due to elastic scattering rather than propagating freely as plane waves. Altshuler et al /10/ have pointed out that electron-electron interaction in a weakly disordered system should produce a variation of R_H with temperature. The analytical expression for the fractional change of $\delta R_H(T)/R_H(0)$ obtained by these authors reads

$$\frac{\delta R_H(T)}{R_H(0)} = 1.3\,\frac{e^2}{2\pi^2\hbar}\,\rho(0)\,\left(\frac{4}{3} - \frac{3}{2}\tilde{F}\right)\cdot\left(\frac{k_B T}{2\hbar D}\right)^{1/2}, \qquad (2)$$

where $\tilde{F}$ is a screening parameter /11/ and D denotes the electronic diffusion constant. The other symbols have the usual meaning.

In Table I the parameters determined from the experimental results ($T \lesssim 25$ K) are summarized assuming

$$R_H(T) = R_H(0) + \delta R_H \cdot T^{1/2}. \qquad (3)$$

Here δR_H is defined by eqs. (2) and (3). This interpretation relies on the fact that any thermal effects have died out in the temperature range considered. Hence, the first term in equ. (3) is replaced by $R_H(0)$. With the help of the resistivity the quantity $\tilde{F}$ was found to vary between 0 and 1. The electronic diffusivity D is derived from the resistivity $\rho(0)$ and the bare density of states $N(E_F)$ by using $D = [e^2\rho(0)\, N(E_F)]^{-1}$. In case of $Cu_{44}Ti_{56}$, D deduced in this way is consistent with the magnetic field- and the temperature-dependence of the electrical resistivity /9/. According to Altshuler et al /10/, the effect of electron-electron interaction should be given by $\delta R_H(T)/R_H(0) = -2 \cdot \delta\sigma_{xx}(T)/\sigma_{xx}(0)$. Hence due to equ. (1) σ_{xy} must be constant in the T-range, where the $T^{1/2}$-behaviour is observed. It can be seen in Fig. 3 (inserts) that this prediction is verified for the non-superconducting samples CuTi and NiTi. The situation is more complicated for the CuZr-case. Here, superconducting fluctuations interfere. Quenching them with a magnetic field B introduces additional contributions from magnetoresistivity. However, it is clear from Fig. 3, that the curve labelled 0 and 6 Tesla embrace the anticipated behaviour: σ_{xy} = const.

This analysis invalidates an earlier interpretation given by Gallagher et al /12/. These authors argue that electron-electron interaction should lead to a significant temperature dependence of R_H even above 200 K. In our opinion such a contribution, which in addition is unable to account

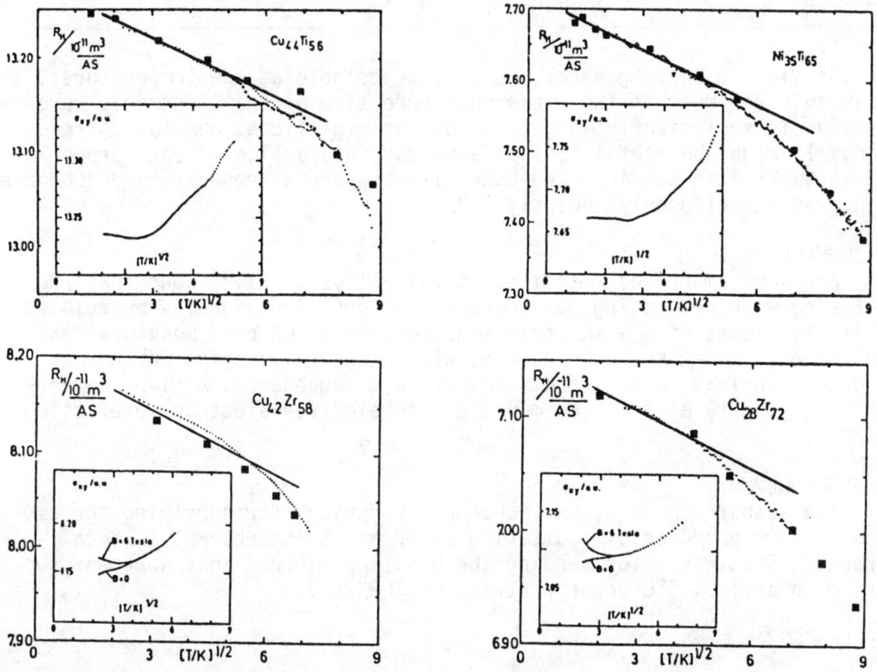

FIGURE 3. Hall coefficient R_H as a function of the square root of the temperature. Symbols (■) denote results from the magnetic field dependence of the Hall resistivity at constant temperature. The other data were taken by varying the temperature at constant magnetic field (...:6 Tesla), ●●●:3 Tesla). Insets show the transverse conductivity σ_{xy} obtained from R_H and the resistance. The $T^{1/2}$-dependence is indicated by

straight lines.

TABLE 1. Parameters determined from a comparison of eq. (2) and (3) to the data. The quantity D was derived from the conductivity as described in the text. In case of the NiTi-alloy the bare density of states $N(E_F)$ is not known, hence, no values for D and $\tilde{F}$ can be given.

alloy	R_H $10^{-11} m^3 A^{-1} s^{-1}$	δR_H $10^{-13} K^{-1/2} m^3 A^{-1} s^{-1}$	$\rho(0)$ $10^{-6} \Omega m$	D $10^{-5} m^2 s^{-1}$	$\tilde{F}$
$Cu_{44}Ti_{56}$	13.28	- 1.91	1.9	3.1	0.51
$Cu_{42}Zr_{58}$	8.2	- 1.72	1.7	1.3	0.16
$Cu_{28}Zr_{72}$	7.15	- 1.56	1.6	1.3	0.056
$Ni_{35}Ti_{65}$	7.72	- 2.65	>1.5	-	-

for the sign of R_H in general, is only detectable at low temperatures. It is in this regime that the phase coherence time of the electronic wave function is sufficiently long to allow for significant quantum corrections. It might be useful to look also for interaction effects proportional to $T^{1/2}$ in coevaporated CuZr films, where a temperature dependence of R_H was reported only recently /13/.

5. SUMMARY

For some amorphous transition metal alloys we have found that $/R_H/$ is rising with decreasing temperature. The positive sign may be related to finite widths of the spectral functions in a two band model (s- and d-electrons) as proposed by Hänsch. Within such a description thermal disorder can lead to an overall temperature dependence. At low temperatures ($T \lesssim 25$ K) at $T^{1/2}$-law may indicate electron-electron interaction effects.

ACKNOWLEDGEMENT

The authors thank A. Fleischmann, TU Munich, for supplying the amorphous alloys produced by sputtering and Prof. S. Steeb, MPI f. Metallforschung Stuttgart, for sending the $Ni_{35}Ti_{65}$ ribbon. This work was supported in part by DFG under contract Lu 109/25-1.

REFERENCES

1. Oelhafen, P. in: Glassy Metals II, Eds.: H.-J. Güntherodt, H. Beck, Topics in Applied Physics, Vol. 53, Springer, Heidelberg 1983, p. 283.
2. Schulte, A., Eckert, A., Fritsch, G., Lüscher, E., J. Phys. F 14 (1984) 1877.
3. Naugle, D.G., J. Phys. Chem. Solids 45 (1984) 367.
4. Hänsch, W., to be published.

5. Morgan, G.J., Weir, G.F., Phil. Mag. B47 (1983) 177.
6. Movaghar, B., to be published.
7. An-Ban Chen, Weisz, G., Sher, A., Phys. Rev. B 5 (1972) 2897.
8. Lee, P.A., Ramakrishnan, T.V., Rev. Mod. Phys. 57 (1985) 287.
9. Fritsch, G., Schulte, A., this volume
10. Altshuler, B.L., Khmelnitzkii, D., Larkin, A.I., Lee, P.A., Phys. Rev. B 22 (1980) 5142
11. Altshuler, B.L., Aronov, A.G., Sol. Stat. Comm. 46 (1983) 429.
12. Gallagher, B.L., Greig, D., Howson, M.A., J. Phys. F 14 (1984) L 225.
13. Minnigerode, v., G., Göttjer, H.G., European Conference Abstracts, Vol. 9 A Berlin 1985, p. PTh-10-190.

MAGNETORESISTANCE STUDIES OF AMORPHOUS FEZR ALLOYS

M. TRUDEAU, R.W. COCHRANE and J. DESTRY
Département de Physique, Université de Montréal, C.P. 6128
Succ. A, Montréal, Canada, H3C 3J7

1.ABSTRACT

We present results for the magnetoresistance in amorphous Fe_xZr_{100-x} alloys for $20 \leqq x \geqq 37$, between 1.5K and 30K and for $0 \leqq H \geqq 1$ Tesla. For $29 \leqq x \geqq 37$ the samples are paramagnetic and those with $x \leqq 29$ superconducting; all these samples exhibit a magnetoresistance that seems consistent with the theory of weak localization.

2.INTRODUCTION

The amorphous iron-zirconium alloys are of great interest for the study and the understanding of the transport processes, such as the Hall effect and the magnetoresistance, in amorphous transition metals alloys.

One of the main reason for this interest arises from an examination of the phase diagram for the Fe_xZr_{100-x} amorphous alloys(1). Depending upon the relative atomic concentration of the two constituents, a ferromagnetic ($x>37$), a paramagnetic or a superconducting regime ($x\leqq 29$) is observed at low temperature. Because of the possibility of producing sample continuously across this range of concentration, it has been possible to investigate the transport properties in the different intervals and follow their evolution from one regime to another.

3.EXPERIMENTAL DETAILS

All the alloys that have been studied here were made by melt spinning under helium atmosphere(2). The average thickness, typically 15 μm, was calculated from the measurements of the density, mass, length and width. All the contacts were made by ultrasonic bonding very fine aluminum wires between the sample and a copper contact pad. The samples were mounted on a copper substrate that was covered with a very fine sheet of paper to insure good thermal contact. This substrate was subsequently attached with varnish to a copper holder which was mounted into a liquid helium dewar for low temperature measurements.

To study the different transport parameters, resistivity, magnetoresistivity and Hall effect, it was necessary to develop a very sensitive AC ohmmeter that is capable of a resolution of the order of 2×10^{-8} Ω in a 1.0 ohm resistor, with a 10 mA current, at a frequency, typically, of 150 Hz(3).

4.RESULTS

We report here the measurements of the magnetoresistance as a function of the temperature ($1.5 \leqq T \geqq 30$ K) and field ($0 \leqq H \geqq 1$ Tesla) for alloys with

20 ≦ x ≧ 37, i.e. in both the superconducting and paramagnetic regimes. In these samples we find a positive magnetoresistance, leading to a negative magnetoconductivity ($\Delta\sigma = -\Delta\rho/\rho^2$), of the order of -40 to -400 $(\text{ohm m})^{-1}$ at 4.2K and 1 Tesla. The functional form of the variation with the field is H^n where n=2 at low field decreasing towards 0.5 at higher fields.

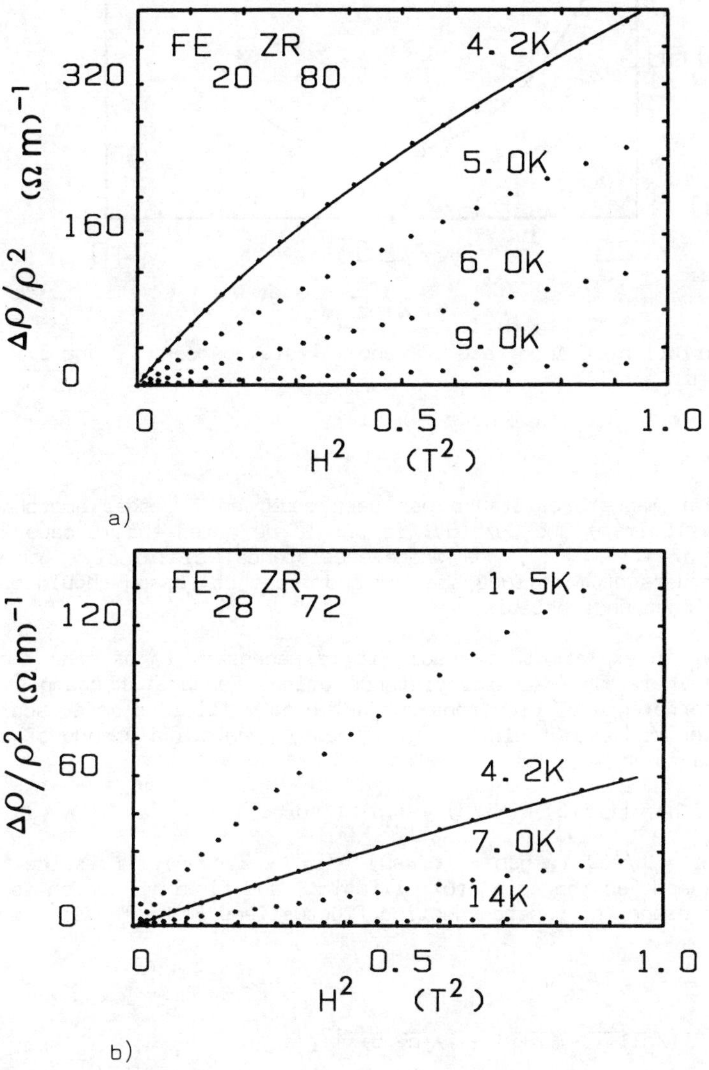

FIGURE 1. Magnetoresistance for a) $Fe_{20}Zr_{80}$ (Tc = 3.3K) and b) $Fe_{28}Zr_{72}$ (Tc = 0.6K), as a function of the temperature.

Also we observe a minimum in the magnetoresistance with the concentration at fixed temperature and field as illustrated in Fig.2.

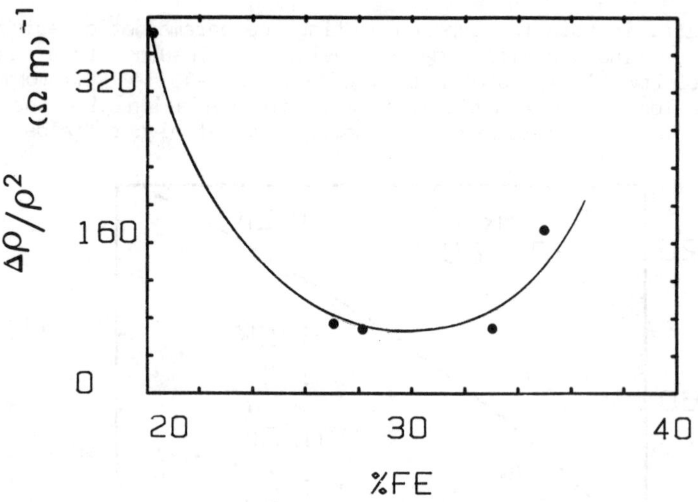

FIGURE 2. Variation of $\Delta\rho/\rho^2$ at 4.2K and 1 Tesla, between 20 and 37 atomic % of Iron.

5.DISCUSSION

A similar magnetoresistance has been observed in other amorphous alloys, such as CoZr(4) and CuZr(5). It should be noted that because of the small value of τ ($\simeq 10^{-16}$) the magnetoresistance, classically, should be at least 3 orders of magnitude smaller and for that reason should be experimentally almost undetectable.

In order to explain this result it is necessary to use the theory of localization where the magnetoresistance arises because of changes in the coherent interference of electrons subjected to multiple elastic scattering by the applied magnetic field(6). This theory predicts a change of conductivity of the form:

$$\Delta\rho/\rho^2 = (e^2/2\pi h)^{\frac{1}{2}}\ [(.5+\beta)\cdot F(Hi/H) - 1.5\cdot F(Hso/H)], \qquad (1)$$

where $Hi = h/4eD\tau i$, $Hso = (h/4eD)\ (\tau i^{-1} + 2\tau so^{-1})$, τi is the inelastic lifetime and τso the spin orbit lifetime, and finally β which is a term due to the presence of superconducting fluctuations (Maki-Thomson contribution).

Here

$$F(x) = \sum_{n=0}^{\infty} [2\ (\sqrt{n+1+x} - \sqrt{n+x}) - 1/\sqrt{n+.5+x}]\ , \qquad (2)$$

which we have found can be approximated to better than 1% by:

$$F(x) = 2\ (\sqrt{x+2} - \sqrt{x}) - (1/\sqrt{x+.5} + 1/\sqrt{x+1.5}) + (x+2)^{-3/2}/48\ . \qquad (3)$$

In Fig.1a and 1b we have included a least square fit (the solid lines) to the data at 4.2K using the localization expression. In these fittings, we have assumed a value of $\tau_{so} = 5.92 \times 10^{-13}$ s, as found in CuZr alloys at higher field(5); such a value renders the second term in eq(1) negligible for our range of T and H. For $Fe_{20}Zr_{80}$ the fit with τ_i and β variable gives at 4.2K:

$$\tau_i = 3.4 \times 10^{-11} \text{ s},$$
$$\beta = 8.3.$$

The value of β implies a superconducting transition temperature of 3.5K in reasonable agreement with the value of 3.3K found experimentally. As the temperature increases $\tau_i \simeq T^{-p}$ with p = 1.6.

On the other hand for x ≧ 28 the procedure discribed above leads to significant discrepencies between the fitted and measured values of Tc. These discrepencies result from a significant change in character of the magnetoresistance: its magnitude and near H^2 dependence are not consistent with the small value of β deduced from Tc. In order to analyse these data the value of β has been fixed using the Larkin(7) relation with Tc and a second term AH^2 has been added. The observation of spin fluctuations(8) in the alloys with 28≤x≥40 suggest a possible origin for this term. The value associated with the fit shomn in Fig.1b for $Fe_{28}Zr_{72}$ are:

$$A = 31.9 \ (\text{ohm-m-T})^{-1},$$
$$\tau_i = 2.5 \times 10^{-11} \text{ s}.$$

The additional term account for 50% of the mesured magnetoresistance at 1T. The inelastic scattering time is similar in magnitude to that found in the x = 20 sample.

6.CONCLUSION

In conclusion we have shown that for samples which are superconducting, the theory of weak localization seems to explain the experimental data. But as the Fe concentration increased towards the ferromagnetic transition, the discrepency between the theory and the data increases, suggesting that a new contribution plays a role in the magnetoresistance in a way not well understood at present. We believe that this additional term is tied directly to the appearance of spin fluctuations in these alloys.

REFERENCES

1. E. Batalla and al., Phys. Rev. B31, 577, (1985).
2. Z. Altounian and al., J. Appl. Phys. 53, 4755, (1982).
3. R.W. Cochrane and M. Trudeau, to be published.
4. M.Trudeau, R.W. Cochrane and J. Destry, to be published.
5. J.B. Bieri and al., Sol. State Comm. 49, 9, (1984).
6. G. Bergmann, Phys. Rep. 107, 1, 1, (1984).
7. Larkin, JETP lett. 31, 219, (1980).
8. Z. Altounian, J.O. Strom-Olsen, A.B. Kaiser and R.W. Cochrane, Phys. Rev. B 31, 9, 6116, (1985).

MAGNETOCONDUCTIVITY OF MgZn METALLIC GLASSES AT LOW TEMPERATURES

F. Küß, A. Schulte, P. Löbl, E. Lüscher, G. Fritsch*
Physik-Department, TU München, D-8046 Garching
*Phys.Inst.Fak.f.BAUV/JI, Universität der Bundeswehr München, D-8014 Neubiberg

ABSTRACT:

The conductivity σ of the "free electron" metallic glasses Mg_xZn_{100-x} (x=67, 72 and 77) has been measured as a function of the magnetic field ($0 \leq B \leq 6$ Tesla) in the temperature range between 1.5 and 80 K. The conductivity increases with B for low temperatures and small fields. At large B and T it decreases with B. Above 7 K $\sigma(B,T)-\sigma(B=0,T)$ is in quantitative agreement with perturbative corrections due to quantum interference in the presence of spin orbit scattering. We deduce spin orbit fields of 0.1, 0.08 and 0.06 Tesla for x=67, 72 and 77, respectively.

I. INTRODUCTION

The temperature dependence of the electrical resistivity ρ in simple metallic glasses ($\rho \lesssim 80\mu\Omega$cm, no d-electrons at the Fermi surface) has been described remarkably well within the nearly free electron model /1/. On the other hand, there is experimental and theoretical evidence /2/ - mainly from studies on 2-dimensional metals - that disorder drastically changes the motion of the conduction electrons at low temperatures. The phase coherence time of the electronic wave function becomes large when compared to the elastic scattering time. Hence, interference processes between a sequence of elastic scattering events can no longer be neglected. This mechanism is also sensitive to magnetic fields. Therefore, we discuss the magnetoconductivity which has been measured in 3-dimensional MgZn-glasses as a function of B, T and x with respect to these quantum corrections. These contributions are considered to be additive to the usual Boltzmann conductivity.

II. EXPERIMENTAL DETAILS

The amorphous MgZn alloys (starting materials Mg:4N and Zn:4N) were prepared by melt spinning in a protective Ar-atmosphere. The samples could be checked for amorphicity by X-ray diffraction. After production they were stored in liquid nitrogen, as the crystallization temperature is rather low. Conductivity measurements have been performed using the standard 4-point dc-technique described in more detail elsewhere /3/. Typical sample dimensions were 8 x 2 x 0.03 mm^3.

III. RESULTS AND DISCUSSION

The resistance of the samples as determined and normalized to its value at 4 K is shown in the inserts in Fig. 1-3. The temperature dependence observed compares favourably with literature data (for a discussion see /1/). In particular, we observe a maximum of ρ (T) at about 40-60 K and a minimum below 10 K. Hafner /1/ has shown by ab initio calculations using weak scattering theory that these pronounced features are the result of a rather delicate interplay between structural, electronic and

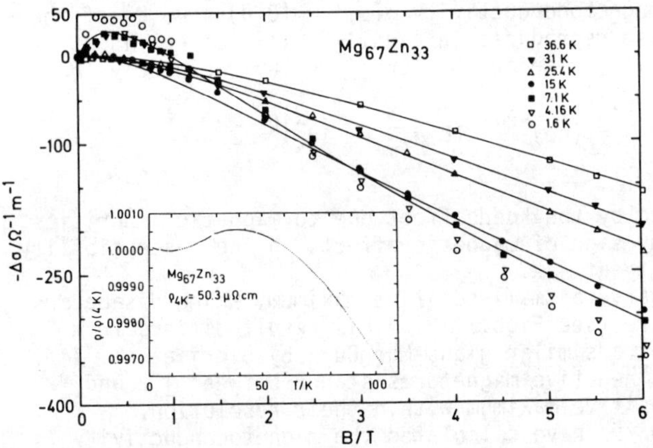

Fig.1
Change in the conductivity $\Delta\sigma$ versus magnetic field B for $Mg_{67}Zn_{33}$. Results for various temperatures are given. The full lines are theoretical fits to the data as explained in the text. The insert shows the resistance normalized to its value at 4 K.

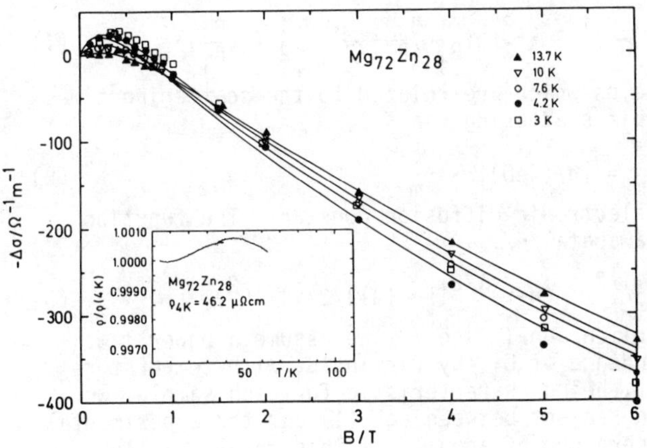

Fig.2
Same as Fig.1 for $Mg_{72}Zn_{28}$.

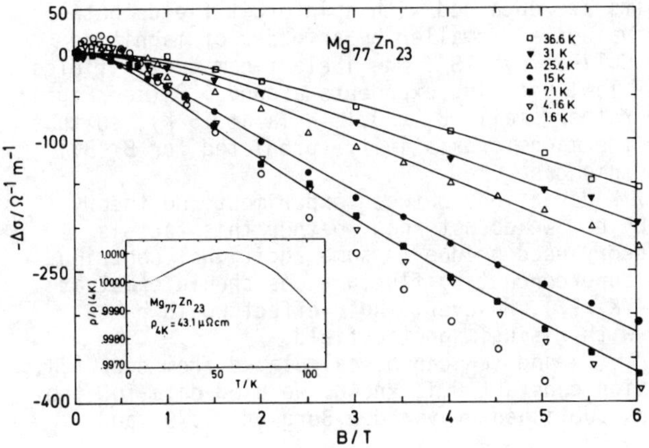

Fig.3
Same as Fig.1 for $Mg_{77}Zn_{23}$.

dynamical variables.

We have detected a magnetoconductivity $\sigma(B) - \sigma(B=0) = \Delta\sigma(B)$ of the order of 10^{-4} which is also reproduced in Figs.1-3. This effect is at least four orders of magnitude larger than predicted by the nearly free electron model:

$$|\Delta\sigma/\sigma| \approx (\omega_c\tau)^{-2} = (ne)^{-2} \cdot (B\cdot\sigma)^2 \approx 10^{-10}$$

for B = 1 Tesla.

It is also not caused by the Kondoeffect due to magnetic impurities, because we didn't find any sign of magnetic effects in the susceptibility and in the Hall resistivity at 4 K.

At low fields we observe a small positive maximum, which disappears with increasing temperature (see Figs. 1-3). This result differs from measurements, performed at a similar glass $Mg_{80}Cu_{20}$ by Bieri et al. /4/. These authors also find a negative magnetoresistance between 1.2 and 10 K, but no sign of a positive maximum within their resolution.

Fukuyama and Hoshino /5/ have calculated the magnetoconductivity $\Delta\sigma = \sigma(B,T) - \sigma(B=0,T)$ for the elastic scattering, described by the maximally crossed (Langer-Neal) diagram. In 3 dimensions, the result is /5/:

$$\Delta\sigma(B,B_i(T),B_{SO}) = \frac{e^2}{2\pi^2\hbar}\left(\frac{eB}{\hbar}\right)^{1/2} \cdot \left\{\frac{3}{2} f_3\left(\frac{B}{B_i+4B_{SO}/3}\right) - \frac{1}{2} f_3\left(\frac{B}{B_i}\right)\right\}. \quad (1)$$

Here B_i and B_{SO} denote fields which are related to the scattering times for inelastic and spin orbit scattering via

$$B_m = |\hbar/(4eD)| \cdot \tau_m^{-1} \quad (2)$$

with m = i, SO and D the electronic diffusion constant. The function $f_3(x)$ has been given by Kawabata /6/:

$$f_3(x) = \sum_{n=0}^{\infty} \{2[(n+1+x)^{1/2} - (n+x)^{1/2}] - (n+1/2+x)^{-1/2}\}. \quad (3)$$

In analogy to the 2-dimensional case /7/ we assume a power law: $B_i(T)=a\cdot T^p$ for the T-dependence of B_i. By varying B_i with temperature and keeping B_{SO} a constant, being characteristic for each sample, we were able to obtain good agreement between equ.(1) and the experimental data at least above a temperature of about 7 K. This result is also shown in Figs. 1-3.

The parameters necessary to fit the experimental data are summarized in TABLE 1. Best results are obtained with spin orbit fields between 0.06 and 0.1 Tesla. These values are smaller by an order of magnitude as the ones necessary for CuTi-alloys /8/. The inelastic magnetic fields $B_i(T)$ are close to a power law, yielding exponents around 1.8 (see Fig.4). Their absolute values are rather small ($B_i \lesssim 0.1$ Tesla at 15 K), so that a quadratic behaviour for the magnetoresistivity (predicted for $B \ll B_i$) can hardly be measured in this case.

At temperatures T < 7 K deviations between experiment and theory occur for higher fields. By now we do not know whether this fact is due to a failure of the theory used or due to some additional contribution to the conductivity. Superconducting fluctuations should exist at low temperatures ($T_c \lesssim 0.1$ K /9/), however, their effect would be to decrease the conductivity with rising magnetic field.

The scattering times $\tau_i(T)$ and τ_{SO} can be calculated from equ. (2), when the electronic diffusion constant D is known. We used data for the density of states $N(E_F)$, as published by van den Berg et al./9/ and

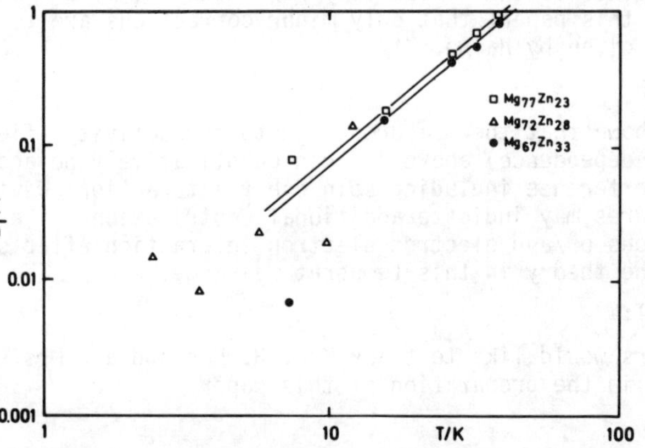

Fig. 4 Results for the inelastic magnetic fields $B_i(T)$ in a log-log representation. The lines shown refer to $B_i = 1.23 \cdot 10^{-3}\ T^{1.82}$ (Tesla) in case of $Mg_{77}Zn_{23}$ and to $B_i = 1.05 \cdot 10^{-3} \cdot T^{1.83}$ (Tesla) for $Mg_{67}Zn_{33}$. No fit was attempted for $Mg_{72}Zn_{28}$ because of the limited T-range available (large scatter at low T).

evaluated D from the resistivity

$$\rho^{-1} = e^2 N(E_F) \cdot D \tag{4}$$

as follows: $6.25 \cdot 10^{-4}$, $6.8 \cdot 10^{-4}$ and $7.3 \cdot 10^{-4}\ m^2 s^{-1}$ for x = 67, 72 and 77, respectively. The results are summarized in TABLE 1.

The quantum correction due to weak localization should also show up

TABLE 1: Results from a fit of equ.(1) to the data

$Mg_{67}Zn_{33}$ ρ(293K) = 50.3 μΩcm, $B_{SO} = 0.1$ T, $\tau_{SO} = 2.6 \cdot 10^{-12}$ s

T/K	1.6	4.16	7	15	25.4	31	36.6
B_i/Tesla	0.007	0.011	0.007	0.15	0.40	0.52	0.80
$\tau_i \cdot 10^{13}$/s	380	240	380	17.5	6.6	5.1	3.3

$Mg_{72}Zn_{28}$ ρ(293K) = 46.2 μΩcm, $B_{SO} = 0.08$ T, $\tau_{SO} = 3.0 \cdot 10^{-12}$ s

T/K	3	4.16	7.6	10	13.7
B_i/Tesla	0.013	0.008	0.018	0.017	0.11
$\tau_i \cdot 10^{13}$/s	190	300	130	140	22

$Mg_{77}Zn_{23}$ ρ(293K) = 43.1 μΩcm, $B_{SO} = 0.06$ T, $\tau_{SO} = 3.8 \cdot 10^{-12}$ s

T/K	1.6	4.16	7.1	15	25.4	31	36.6
B_i/Tesla	0.014	0.005	0.076	0.17	0.45	0.62	0.89
$\tau_i \cdot 10^{13}$/s	160	450	30	13.2	50	3.6	2.5

in the T-dependence of the resistivity /8/. However, it can be shown using the results of this paper, that only minor corrections are to be expected to the numbers given by Hafner /1/.

IV. CONCLUSION

We have shown that the measured magneto-conductivity (field- as well as temperature-dependence) above 7 K can quantitatively be accounted for by quantum interference including spin orbit interaction. Deviations at lower temperatures may indicate additional contributions like superconducting fluctuations or/and electron-electron interaction effects as well as a failure of the theory in this temperature range.

ACKNOWLEDGEMENT:

The authors would like to thank Mrs. H. Harland and Mrs. Schmitt for the assistance in the preparation of this paper.

LITERATURE:

/1/ Hafner,J.;J.Non-Cryst.Solids 69,325(1985)
/2/ Lee,P.A.,Ramakrishnan,T.V.;Rev.Mod.Phys.57,287(1985)
/3/ Küß,F.;Diplomawork TU München, in preparation
/4/ Bieri,J.B.,Fert,A.,Creuzet,G.,Ousset,J.C.;Solid State Comm.49,849 (1984)
/5/ Fukuyama,H.,Hoshino,K.;J.Phys.Soc.Japan 50,2131(1981)
/6/ Kawabata,A.;J.Phys.Soc.Japan 49,628(1980)
/7/ Bergmann,G.;Phys.Rep.107,1(1984)
/8/ Fritsch,G.,Schulte,A.,Lüscher,E.; this volume
/9/ Van den Berg,R.,Grondey,S.,Kästner,J.,von Löhneysen,H.;Solid State Comm.47,137(1983)

OPTICAL PROPERTIES OF Pt-Si GLASSES

E. Huber, U. Kambli, N. Baltzer and M. von Allmen

Institute of Applied Physics, University of Bern,
CH-3012 Bern, Switzerland

Abstract - Glassy Pt-Si alloys were produced by laser quenching in the composition range from 40-93 at.% Si. The dielectric functions were measured ellipsometrically from 1.4-5.5 eV and analyzed in terms of a classical Lorentz-Drude model. The optical conductivity at zero frequency compares favourably with the measured dc conductivity.

1. EXPERIMENTAL

Laser quenching is able to produce glassy alloys over a wide range of concentrations. Laser quenched glassy films of Pt-Si have first been prepared by von Allmen et al [1]. In this paper we report on the optical properties of glassy Pt-Si which we attempt to describe with a classical Lorentz-Drude type expression. Amorphous films of Pt-Si were prepared by vapor-deposition of alternating layers of Pt and Si onto sapphire substrates and subsequent irradiation with 50 ns laser pulses. The relevant cooling rates obtained with this procedure were of the order of 10^{10}K/s [2]. The irradiated films are amorphous in the concentration range from 40-93 at.% Si.

The dc conductivities of the films were measured with a four-point probe. The results are presented in Fig. 1 as a function of composition. There is an overall decrease with increasing Si-content but the rate of decrease is not constant. Hardly any change occurs between 50 and 80 at.% Si ("shoulder") whereas the conductivity drops sharply above 80 % Si. The optical conductivity was calculated from the optical data as described in section 2. The dielectric function ε between 1.4 and 5.5 eV were determined with a polarization-modulation ellipsometer [3]. Fig. 2 shows the real and imaginary parts for films of various compositions. For Pt-rich samples metallic

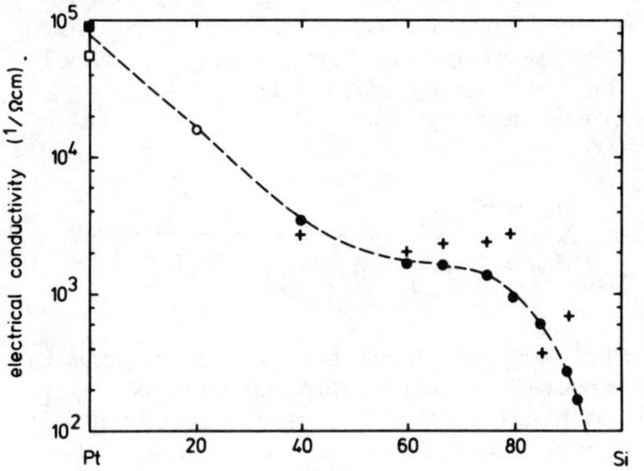

Fig. 1 Composition dependence of electrical conductivity of Pt-Si films. ● glassy; ○ partially crystalline; ■ crystalline Pt; □ liquid Pt extrapolated to room temperature. ✚ optical conductivity calculated from optical data as described in the text.

behavior is predominant, changing into non-metallic behavior for the Si-rich films.

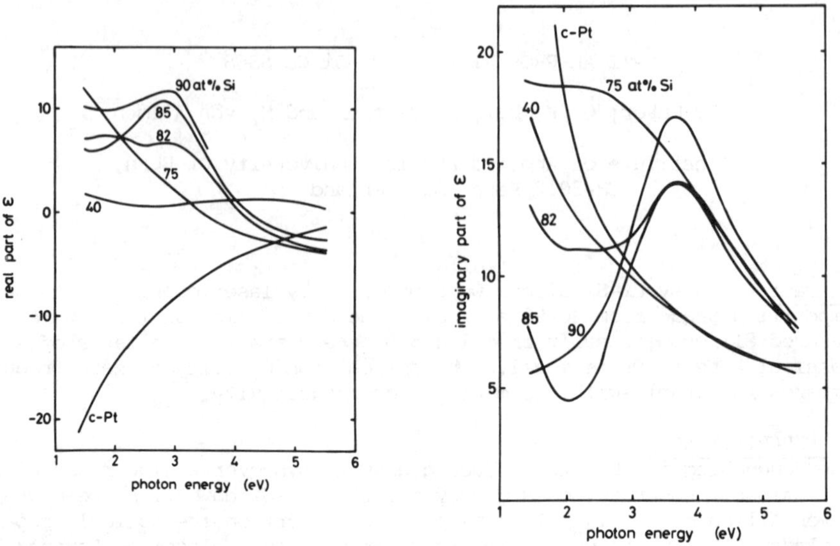

Fig. 2 Real and imaginary parts of measured dielectric function for glassy Pt-Si films of various compositions.

2. MODELLING OF THE DIELECTRIC FUNCTION

Two different types of transitions are assumed to be relevant for the electronic properties: intraband transitions of the free electrons and interband transitions of virtual bound states of Pt and Si electrons. A combined Lorentz-Drude expression was therefore used to model ε, where the Drude part accounts for the free electrons and the Lorentz part represents the various resonant states of the bound electrons. It is assumed that the interaction due to alloying does not affect these resonances, and their contributions are taken proportional to the corresponding elemental concentrations. For pure Pt Drude-like behaviour is found only for photon energies below 0.2 eV with a free electron density of about 0.4 electrons per atom [4]. At higher photon energies interband transitions strongly influence the optical properties. Experimental density-of-states data measured by photoelectron spectroscopy [5] was used to characterize the resonances. Pure amorphous Si can be described within the Lorentz model with a resonance at 3.4 eV [6]. The following combined Lorentz-Drude expression is used for the dielectric function:

$$\varepsilon(\omega) = 1 + \frac{4\pi e^2}{m_o}\left[-\frac{n_f}{\omega(\omega+\frac{i}{\tau})} + (1-x)\sum_{j=1}^{2}\frac{n_j}{\omega_j^2-\omega^2-i\Gamma_j} + x\,\frac{n_{Si}}{\omega_{Si}^2-\omega^2-i\Gamma_{Si}}\right], \tag{1}$$

where x is the molar fraction of Si. The effective free electron density n_f and the relaxation time τ were treated as adjustable parameters. All the other parameters were held at constant values for each concentration: n_1, n_2 and n_{Si} are the effective electron densities of the resonant states

ω_1, ω_2 and ω_{Si} of Pt and Si, respectively, and Γ_1, Γ_2 and Γ_{Si} are the corresponding damping constants (their numerical values were obtained by fitting equation (1) to the measured ε's of the pure elements). The effective free electron densities n_f and relaxation times τ were then obtained for the different glasses by fitting equation (1) to the experimental ε's. This model describes the experimental data satisfactorily over the measured concentration range. However, it has to be pointed out that the effective free electron density used in the Drude term also includes contributions from transitions with small excitation energies. Best-fit values for the effective free electron densities and relaxation times are presented in Fig. 3. The relaxation times are very short, of the order of

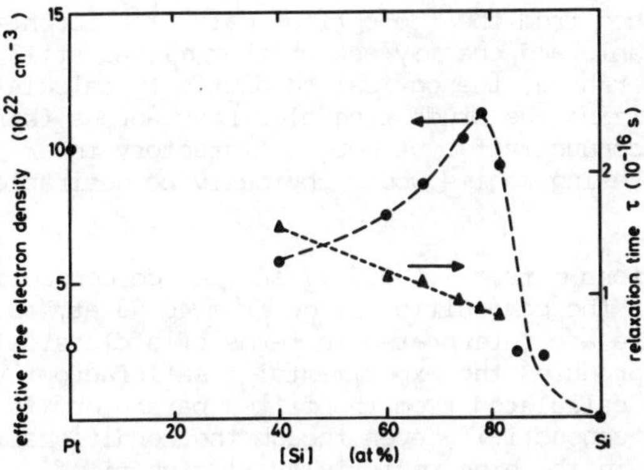

Fig. 3 Composition dependence of effective free electron densities and relaxation times obtained from fitting eq.1 to the experimental dielectric functions of glassy Pt-Si films. The open circle for pure Pt is the free electron density obtained from optical measurements in the far IR [4].

10^{-16} s. The concentration dependence of the free electron density in the Pt-Si glasses is striking. With decreasing Pt concentration it first increases markedly and considerably exceeds even the dc value of pure Pt; above 80 at.% Si it drops sharply. From the data of Fig. 3 the optical conductivity at zero frequency $\sigma_{opt}(0)$ included in Fig. 1 was calculated according to

$$\sigma_{opt}(0) = (e^2/m_o)n_f\tau . \qquad (2)$$

3. DISCUSSION

Both the electrical conductivity presented in Fig. 1 and the effective free electron densities obtained from the optical data of Fig. 3 show two distinct regions. Between 50 and 80 at.% Si the conductivity remains approximately constant at $2\cdot10^3$ $(\Omega cm)^{-1}$ and the free electron density even increases with decreasing Pt content, reaching a surprisingly high value of roughly twice the atomic density at 80 at.% Si. Beyond 80 at.% Si there is a sharp drop in both quantities, indicating a metal-nonmetal transition (MNT). The Drude free electron model has been applied to analyze the optical properties of metal-rich glassy alloys of Au-Si [7,8], Ag-Ge and Au-Ge [9] containing less than 50 at.% Si or Ge. For those metal-rich alloys short relaxation times as well as an increase of n_f with decreasing metal content have been reported. The relative increase of n_f was explained by the contribution of 4 free electrons per Si or Ge atom as Si and Ge are in a metallic state [7,10]. In Si-rich alloys, however, Si is not in a metallic state as evidenced by the presence of the characteristic resonance at 3.4 eV in the spectra. The high n_f below the MNT can therefore not be explained by the contribution of free electrons of Si.

The "shoulder" in the dc conductivity which is indicative of some saturation effect in the transport mechanism, and the relative increase in the effective free electron density below the MNT are puzzling features. In view of the relatively high resistivities of the Si-rich alloys and the short relaxation times obtained from the the optical data, the Lorentz-Drude model is not really applicable and one may ask whether n_f has still a physical meaning. On the other hand, the optical conductivity calculated from the parameters characterizing the Drude free electrons agrees fairly well with the experimental dc conductivity. A more satisfactory theory for systems in the strongly scattering regime would obviously be desirable.

4. SUMMARY

The dielectric functions ε from 1.4-5.5 eV and the dc conductivities of glassy Pt-Si films in the composition range from 40-93 at.% Si are presented. The optical data are interpreted in terms of a classical Lorentz-Drude model. The model reproduces the experimental ε satisfactorily, and the optical conductivity calculated from the fitted parameters is consistent with the measured dc conductivity even though the conditions for its applicability are not met in the high resistivity Si-rich films. A more detailed study will appear elsewhere.

ACKNOWLEDGEMENTS

We thank E. Krähenbühl for technical assistance. This work was supported in part by the Swiss Commission for the Encouragement of Scientific Research.

REFERENCES

[1] M.von Allmen,S.S.Lau,M.Mäenpää,and B.Y.Tsaur,Appl.Phys.Lett.37,84,(1980).
[2] M.von Allmen,this volume.
[3] E.Huber,N.Baltzer,and M.von Allmen,Rev.Sci.Instruments,in press.
[4] Landolt-Börnstein,New Series III,Crystal and Solid State Physics, Vol 15b,p.219,(Springer,Berlin 1985).
[5] Landolt-Börnstein,New Series III,Crystal and Solid State Physics, Vol.13a,p.420,(Springer,Berlin 1981).
[6] D.T.Pierce and W.A.Spicer,Phys.Rev.B 5,3017,(1972).
[7] E.Hauser,R.J.Zirke,J.Tauc,J.J.Hauser,and S.R.Nagel, Phys.Rev.B 19, 6331,(1979).
[8] E.Huber and M.von Allmen,Phys.Rev.B 28,2979,(1983).
[9] M.L.Theye,V.Nguyen Van,and S.Fisson,Philos.Mag.B 47,31,(1983).
[10] E.Huber and M.von Allmen,Phys.Rev.B 31,3338,(1985).

ELECTRICAL RESISTIVITY OF LIQUID RARE EARTH-TRANSITION ALLOYS OF NICKEL WITH LANTHANUM AND CERIUM

B. KEFIF, J.-G. GASSER
Laboratoire de Physique des Liquides Métalliques, Faculté des Sciences, 57045 Metz Cedex 1, France

1. INTRODUCTION

To our knowledge there exist no experimental measurements (in the liquid state) on the electrical resistivity of alloys of rare earth with transition metals. For the pure rare earth metals the experimental data are scarce and sometimes very inconsistent. We present below our measurements on the electrical resistivity of liquid lanthanum, cerium and lanthanum-nickel alloys (up to 50 at % nickel) and of cerium-nickel alloys (up to 55 at % nickel).

2. EXPERIMENT

The electrical resistivity has been measured in the quartz cell described by Gasser (1982). However the rare earth metals are very corrosive and react with silica at high temperatures. This has been put in evidence, indeed at constant temperature, the electrical resistivity grows slowly with time. For lanthanum alloys we have first determined that this phenomena disappears below 800 °C. For cerium alloys the drift could not be eliminated even at lower temperatures but has been estimated to 1.38 % per day. After the end of the experimental runs we observed the formation (on the internal surface of the capillary) of some solid film of undetermined composition which reduces the diameter of the capillary. This is consistent with the drift observed if the film is less conducting than the liquid alloy. The experimental raw measurements are plotted on the graphs. An asymetric error bar takes the drift into account. On some graphs we have plotted a "corrected measurement" by estimating the drift as a function of time.

2.1 LIQUID LANTHANUM

The electrical resistivity of liquid lanthanum is represented in Figure 1 together with the data of other authors. Our value at the melting point is very near that of Krieg but to our sense their results increase too fast with temperature. Gaibulaev's measurements are about 20 microohm·cm lower than ours but the temperature dependence seems more correct. Güntherodt's measurements published in different papers are near our values, sometimes they are lower, sometimes higher. As mentioned before a drift was superposed to the temperature dependence. It was estimated and corrected. The accepted values are 139.0 and 140.9 μohm·cm at 950 and 1050 °C, respectively. This gives us a corrected temperature coefficient of 19 nohm·cm/°C (Gaibulaev:22, Krieg:136). Our corrected curve is described by:

$$R_C = 0.01939 \cdot T_C + 120.5 .$$

It is consistent with Krieg's value at the melting point and with the last of Güntherodt's values at 1320 °C.

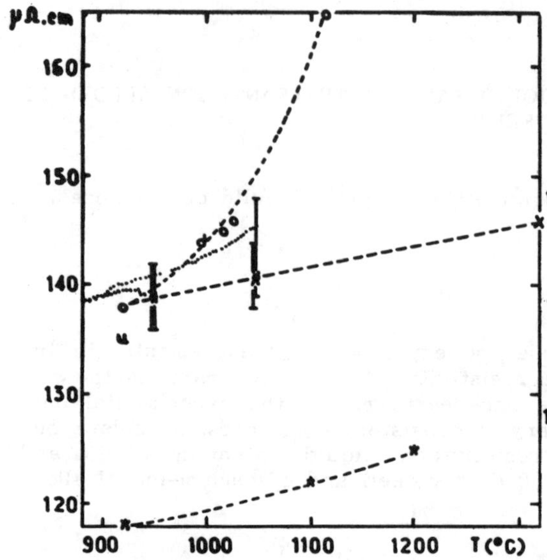

FIG.1: RESISTIVITY OF LIQUID LANTHANUM : *Gaibulaev (69) + Güntherodt (73a) o Krieg (69) u Güntherodt (73b) x Güntherodt (76) · Our work : raw measurements ⌘ Our work : drift corrected values .

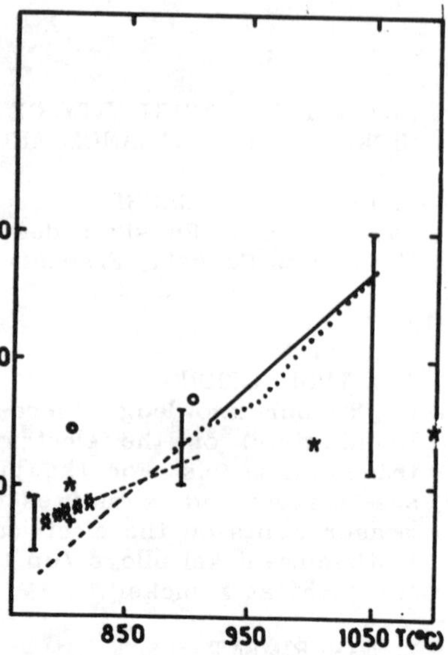

FIG.2: RESISTIVITY OF LIQUID CERIUM : * Gaibulaev (69) + Güntherodt(73) o Busch (70) · Our work : high temperature run ⌘ Our work : low temperature run .

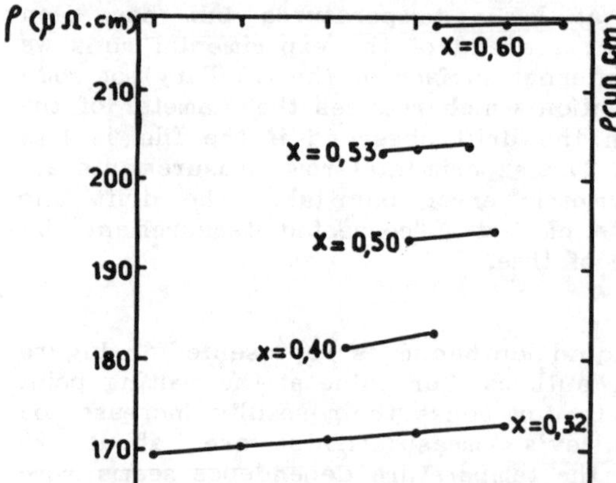

Fig. 3: Electrical resistivity of liquid La-Ni alloys as a function of temperature .

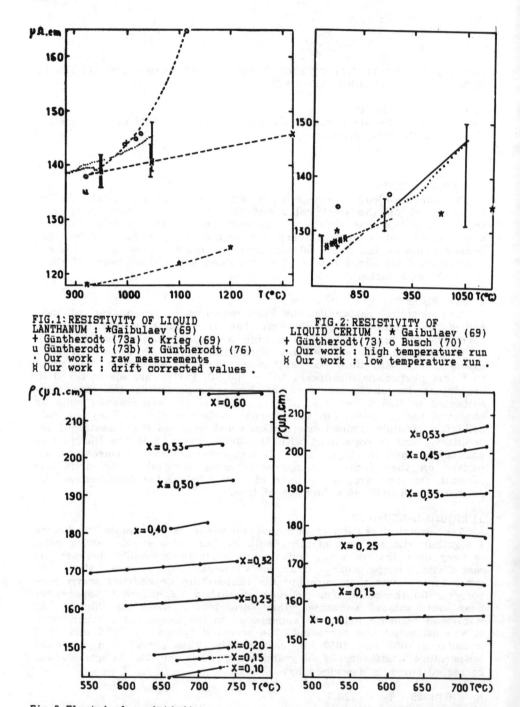

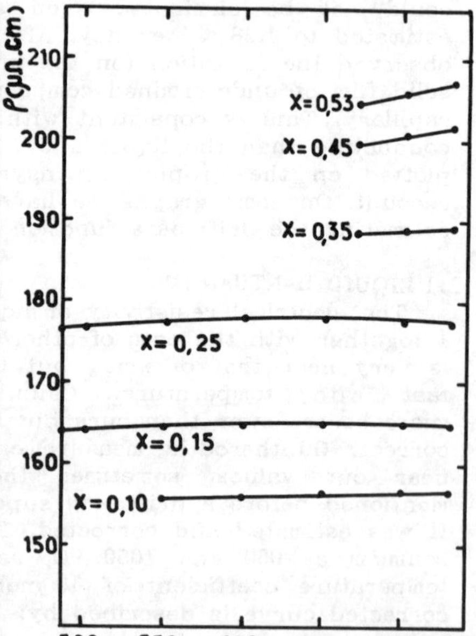

Fig.4: Electrical resistivity of liquid Ce-Ni alloys as a function of temperature .

2.2 Liquid Cerium

The electrical resistivity of cerium is presented in Figure 2 and compared to the values of other authors. Two experimental runs are reported, one at high, the other at low temperatures. The first run shows a drift with time. The slope is too large. The second run is more accurate. Our values lie within 1% of the data of Gaibulaev and Güntherodt and 6 % of those of Busch. This agreement may be considered as good with respect to the reactivity of these metals.

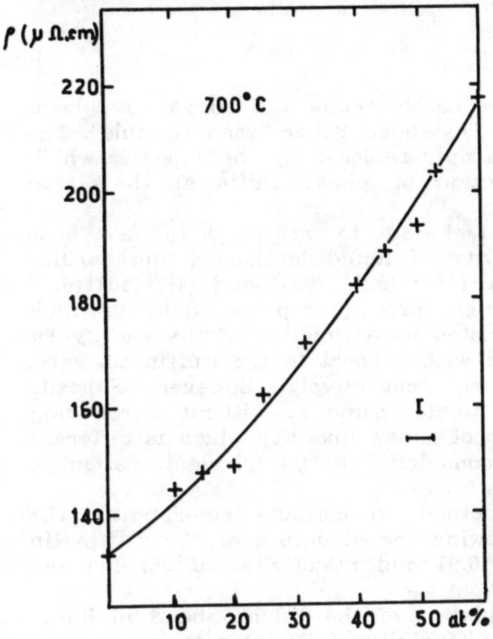

FIGURE 5

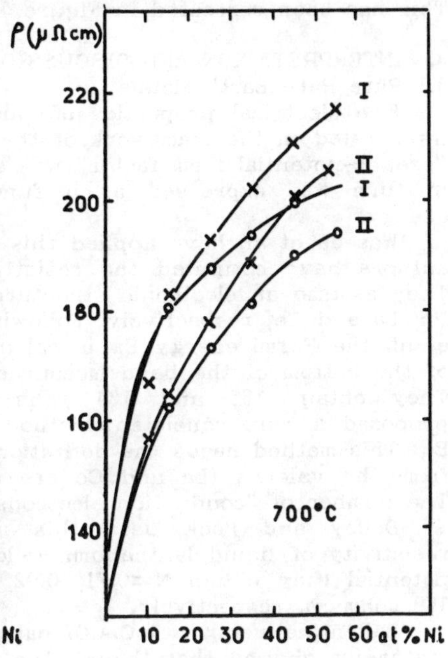

FIGURE 6

2.3 Liquid Lanthanum-Nickel Alloys

The electrical resistivity of 8 La-Ni alloys is shown as a function of temperature (Fig.3) and concentration (Fig.5) at 700 °C. The resistivity increases with nickel concentration. It reaches 219 μohms·cm. This alloy exhibits a very sharp maximum.

2.4 Liquid Cerium-Nickel Alloys

The electrical resistivity of 6 Ce-Ni alloys is plotted as a function of temperature (Fig.4) and composition at 700 °C (Fig.6). As for La-Ni the resistivity rises with nickel concentration. The maximum seems however to be less pronounced. Two series of measurements have been performed. The results are slightly different (about 5 %). This is due to the difficulty to perform such measurements because of:

- The reactivity of cerium with the container.
- The great sensitivity to air (rare earth metals must be handled in a glove box
- The high melting point of nickel (which must be fully dissolved in liquid cerium).

The second series of measurements is reproduced in Fig. 5. We observed that for 25 at% nickel the slope of the resistivity versus temperature curve becomes negative. This phenomena has also be observed in the first serie of measurements but not in the case of nickel-lanthanum.

Each series of measurements has been performed with the same quartz cell, by adding weighted quantities of metal in order to change the composition. Returning to the original composition several days later we have detected that the resistivity drift amounts to about 1.38% per day. This has been corrected in Figure 7.

3. INTERPRETATION AND DISCUSSION

3.1 Pure Rare Earth Metals

The electrical properties of rare earth metals and alloys have been interpreted in the framework of the "extended Faber-Ziman Formula". The "pseudo-potential form factor" was simply replaced by the T matrix which in turn was expressed as a function of phase shifts at the Fermi energy.

Waseda et al. have applied this technique to rare earth metals. These authors have computed the resistivity of liquid lanthanum and cerium. They assume an electronic structure $(4f)^o$ $(5d)^1$ $(6s)^2$ and $(4f)^1(5d)^1(6s)^2$ for La and Ce, respectively. Following a procedure proposed by Dreirach et al. the Fermi energy E_F is calculated as a function of the energy E_B of the bottom of the band measured with respect to the muffin tin zero. They obtain 165 and 134 μohm·cm, respectively. However, Esposito proposed a more consistent method to determine E_F without introducing E_B. This method needs the definition of a new quantity which is different from the valence (La and Ce are considered to be trivalent elements): The number of "conduction electrons N_c".

Delley and Beck used this method to compute consistently the resistivity of liquid lanthanum. Following the structure or the muffin tin potential they obtain N_c=0.71; 0.92; 0.91 and resistivites of 96; 133 and 158 μohm·cm, respectively.

Ballentine using a L.C.A.O. calculation of the valence band in liquid lanthanum showed that though the diffusivity of S-states is larger than the diffusivity of D-states, the "much larger density of D-states causes them to dominate the conductivity". In a hybridized S-D band model he obtains a resistivity of 151 μohm·cm. But in his conclusion he suggests that the "electrical conduction takes place via a kind of diffusion process for which the usual Boltzmann equation treatment is unjustifiable".

3.2 Alloys

The most important success of Ziman's formula has been the explanation of the negative temperature coefficient of the resistivity of liquid zinc in terms of the position of $2K_F$ compared to the position K_P of the main peak of the interference function. This has been extended to alloys for which $2K_F$ moves from the left (monovalent metals) to the right (polyvalent metals) of the main peaks of the partial interference functions when the mean valency increases with the polyvalent metal concentration. This effect explains the negative temperature coefficient of

- Noble-polyvalent alloys like Cu-Sn (Busch)
- Transition-polyvalent alloys like Mn-Sb (Gasser)
- Rare earth-polyvalent alloys like Ce-Sn (Güntherodt).

It implies that the transition or rare earth metals behave as monovalent ones (N_c=1).

The fact that Ce-Ni presents a slightly negative temperature coefficient at 25 AT % Ni while La-Ni does not, shows that the number N_C is probably very different when going from one rare earth to the other. Delley and Beck calculated values ranging from 0.71 (La) to 1.93 (Eu) but they did not compute it for cerium.

To our knowledge no quantitative theoretical result has been proposed for rare earth alloys.

4. CONCLUSION

Only a few data exist today in the domain of rare earth transition metal alloys. Experimental and theoretical difficulties may be the reason for this fact. We hope that we may be able to interpret quantitatively these results and to study other experimental properties in this kind of alloys.

REFERENCES

Ballentine L.E., Hammerberg J.E. Phys.Rev. B28, 1103 (1983)
Busch G. et al. Mat.Res.Bull. 5, 567 (1970)
Delley B., Beck H. J.Phys.F. 9, 517 (1979)
Dreirach O. et al. J.Phys.F 2, 709 (1972)
Esposito E. et al. Phys.Rev. B18, 3913 (1978)
Gasser J.G. Thése d'Etat, Metz 1982
Gaibulaev F. Sov.Phys.Sol.State 11, 819 (1969)
Güntherodt H.J., Künzi, H.U. Phys.Kond.Mat. 16, 117 (1973a)
Güntherodt H.J., Zimmermann A. Phys.Kond.Mat. 16, 327 (1973b)
Krieg G. et al. Inorg.Nucl.Chem.Lett. 5, 819 (1969)
Waseda Y. et al. J.Phys. F8, 125 (1978)

6. PRODUCTION OF AMORPHOUS MATERIALS

FORMATION OF NON-CRYSTALLINE METALLIC ALLOYS WITH AND WITHOUT RAPID QUENCHING

G. von MINNIGERODE, K. SAMWER, and A. REGENBRECHT

I. PHYSIKALISCHES INSTITUT DER UNIVERSITÄT GÖTTINGEN UND SONDERFORSCHUNGSBEREICH 126, BUNSENSTRASSE 9, D-3400 GÖTTINGEN, FRG.

1. INTRODUCTION

In a metallic glass (synonymously noncrystalline or amorphous metal) the microscopic atomic arrangement exhibits no periodicity or long range order, in contrast to the crystalline state, and the shear viscosity is sufficiently large that the material becomes rigid enough to be considered a solid, in contrast to fluids.

A metallic alloy in its amorphous phase is metastable, since its atoms exist in a higher configurational energy state than in its crystalline phase, but is unable to reach that state because of substantial energy barriers separating the two phases.

Generally the amorphous state can be achieved by four processes:

1) Supercooling or rapid cooling of the melt,
   and by circumvention of the molten state by
2) preparation from an aqueous solution (electrodeposition and chemical deposition),
3) preparation from the vapor phase (vacuum deposition, sputtering, gaseous decomposition),
4) preparation by solid state reaction (ion implantation, ion mixing, solid state diffusion reaction, mechanical alloying).

We focus on noncrystalline metals (omitting the nonmetallic glasses) and list the processes which are used for the production of these materials in three groups: In the first group the rapid quenching process is obvious. In the second group the quenching process is more hidden, but the absence of mobility of the atoms after a very short time is regarded to be responsible for the metastability of the noncrystalline state. Whereas in the third group the high mobility of the atoms for a long time might be even necessary for the formation of the glassy state. The last group will be the main topic of this lecture.

2. AMORPHIZATION METHODS AND MECHANISMS

2.1. Methods of rapid quenching

2.1.1. Vapor deposition onto a substrate at low temperatures (synonymously quench condensation or vapor quenching)

This method for the formation of amorphous metals and alloys in the shape of thin films was founded in the pioneering studies of W. Buckel and R. Hilsch [1] and recent reviews and references are given in [2] .

For some elements it only takes a trace of gaseous inclusions to stabilize the amorphous structure by quench rates of about 10^{+15} K/s. Others require in excess of 10 at% of a second constituent to form the glassy phase. However the metastability of amorphous structures in the almost pure metals is rather weak as indicated by the low temperatures of crystallization of these materials. The significance and amount of inclusions of residual gases for stabilizing the amorphous state of such films are still an open field of controversy in the literature. The inclusion of gases may be even more important in sputtered amorphous films. The mechanism of amorphization is the reduction of the surface mobility of the condensed atoms by the low temperature of the substrate and the other condensing atoms in a time which is short compared to the time of nucleation of a crystalline phase.

2.1.2. Cooling of liquid alloys at high rates (synonymously melt quenching). About 25 years ago several methods were developed to cool liquid alloys at very high rates of about 10^{+6} - 10^{+8} K/s and to avoid the crystallization of some metallic alloys under these extreme conditions. It is the merit of Pol Duwez [3] to open a fascinating field of research to the solid state physicists. The basic working principle of these methods has always been the same: Fast cooling of the melt is achieved by establishing sudden mechanical contact with a good thermal conductor. It is generally recognized that pure metals cannot be quenched from the melt into a glassy solid, at least with the presently achieved cooling rates. So all glassy metals are actually alloys containing two ore more elements. There is no theory capable of predicting the composition of metallic alloys for the formation of glasses. However, from the large number of experimental results empirical rules and guide lines have been formulated. The research on melt quenching is spreading too fast for a short review, therefore we refer to the last international conferences [4 - 9] .

2.1.3. Laser quenching (synonymously laser glazing): In this method heat is produced in a thin surface layer of a solid sample by absorption of very short laser pulses. The surface layer of a width of one micrometer or less is heated, melted and subsequently cooled by conduction into the bulk of the specimen. Cooling rates up to 10^{+10} K/s result in an extended range of materials that can be obtained in the glassy state [10]

2.2. Methods based on a missing atomic mobility

2.2.1. Ion implantation: An implanted ion loses its energy (50 - 200 keV) in a series of either elastic or inelastic collisions. Elastic collisions usually involve large angle deflections and the transfer of kinetic energy to target atoms. Such a recoilling lattice atom acts as a secondary projectile and can produce a cascade of displacements. Inelastic collisions involve energy transfer to the electronic systems of the target atoms and the deflections involved are often negligible. Both processes are independent but the former dominates at low energy implantation experiments. The depth below the surface at which the ion stops is in the range of about 30 nm for a specific example: 50 keV P^+ ions implanted into Ni |11| . In irradiation experiments with heavy ions, a recoiling target atom can carry a substantial fraction of the incident, primary ion energy. Therefore the recoiling ion can act as a secondary projectile and can produce a succession of further displacements in close proximity, so that atoms in a small local volume of a "spike" can be envisaged as moving like a hot "gas" or "liquid". Cooling rates of about 10^{+4} K : 10^{-10} s = 10^{+14} K/s are estimated for the spikes. But for light projectiles as in P^+ ion or more than ever in B^+ ion implantation experiments the average cascade density is so dilute, that the concept of a spike fails, and with it the quenching analogy. The transformation from the crystalline to the amorphous state during ion implantation may, perhaps, be best envisaged as a gradual accumulation of disorder, either in the form of simple defects (vacancies, interstitials) or of more complex faults [11]. The necessary condition for amorphization of metals by ion implantation is then the suppression of radiation enhanced diffusion in low temperature experiments in order to quench in the accumulated disorder.

In a recent paper of G. Linker [12] the amorphization of Nb films by B^+ ion implantation is described. Evaporated thin films of Nb have been homogeneously implanted with B^+ ions at 77 K up to B concentrations of 15 at% . The homogeneous distribution of the B impurities was obtained by multiple-energy implantation (e.g. 20, 30, 70, 160 keV ions for a film 250 nm thick). A boron concentration of up to 5 at% is incorporated into Nb interstitially, leading to small atomic displacements of the b.c.c. host lattice. Then in a narrow B concentration range (of about 3 at%) the bulk of the material is rendered amorphous and complete amorphization is achieved at 15 at% of boron. The amorphization process, which by ion implantation can be followed continuously, can be described likewise by stress accumulation and spontaneous transformation into the amorphous state at a critical impurity concentration. Therefore the lack of release of these stresses by the missing mobility of the implanted atoms is one necessary condition for the amorphization process.

2.2.2. <u>Ion mixing:</u> Ion mixing is complementary to ion implantation. In mixing experiments the interest lies in the redistribution of atoms in a solid brought about by its irradiation with energetic ions. This redistribution process is, in general, a complicated phenomenon that depends on many parameters. It includes collisional displacement mechanisms that are fast and athermal, and the migration of atoms and defects that are radiation-induced. The migration effects last much longer than the collisional effects, and are thermally activated. Following the paper of W.L. Johnson et.al. [13] we restrict ourself to the mixing of metal bilayers in the limit of heavy metals ($Z > 20$) and heavy energetic ions ($E > 100$ keV) and in the absence of delayed effects such as radiation enhanced thermal diffusion. The authors use the term "promt regime (PR)" for the time period from initial interaction of the ion up to the time required to reestablish a uniform ambient temperature throughout the solid. For the subsequent "delayed regime (DR)" the system may remain in a non-equilibrium state with respect to variables other than temperature (e.g. defect concentrations). In the experiments of mixing metal bilayers care was taken to eliminate delayed atomic rearrangements by the low temperature of the substrate. The PR of cascade evolution is divided into times of the order 10^{-13} s in which the incident ion and secondary ions lose most of their kinetic energy ("ballistic regime") and a time of the order 10^{-11} s in which further (particle-particle) interactions within the cascade volume lead to progressive thermalization of the kinetic energy distribution among the particles in the spike to a high effective temperature $T_{eff} \cong 10^{+4}$ K. ("thermalizing regime"). In the last period of PR the relaxation of the thermal spike from 10^{+4} K to sufficient low ambient temperature is completed without long range diffusion in a time of order 10^{-10} s ("relaxation regime"). The majority of atomic mixing takes place in the late stages of well developed cascades where particle energies are of order $k_B T_{eff} \cong 1$ eV. Thermochemical effects (negative enthalpy of mixing and binding enthalpy) are found to play an important role in biasing the random walk process of mixing. The resulting composition profile of this interdiffusion process is fixed at the high temperature $T_{eff} \cong 10^{+4}$ K . The final solid phases formed in the relaxation regime of order 10^{-10} s must evolve without long range diffusion or polymorphically from this concentration profil. A "polymorphic phase diagram" was constructed and shows the thermodynamic composition limits on crystalline solid solutions. Outside these limits crystalline phases are unstable. Growth of intermetallic compounds within polymorphic limits is subjected to extreme kinetic constraints by quench rates of 10^{+4} K : 10^{-10} s = 10^{+14} K/s . Therefore amorphization is expected beyond the polymorphic limits of the terminal solid solutions when radiation enhanced diffusion is suppressed by low temperature experiments.

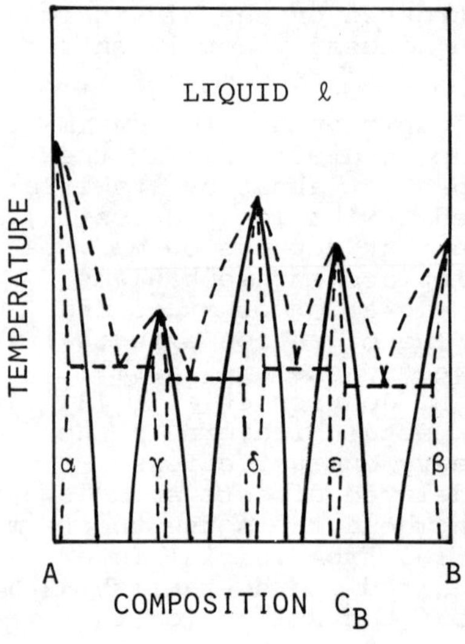

FIGURE 1. (from ref. [13]) Equilibrium phase diagram of an A - B alloy (dashed lines): Terminal solid solutions α,β and intermetallic compounds γ,δ,ε.

Polymorphic phase diagram (full lines): The two-phase regions of the equilibrium phase diagram are disallowed. The full lines represent the couples (T,C_B) where the free enthalpy of the crystalline solid phases without chemical segregation becomes smaller than the undercooled liquid (or amorphous phase) ℓ .

2.2.3 Electrodeposition and chemical deposition: Various intertransition metal and transition metal - metalloid amorphous alloys have been prepared by plating methods with and without the use of electrical current. The first report on an amorphous Ni - S alloy layer prepared by an electroplating method was made already 1930 [14] . Amorphous alloys prepared so far by plating techniques are listed in reference [15]. The advantage of these techniques is the potentiality of preparing bulky amorphous samples in every form as desired. But the precise composition of the product depends strongly on the deposition conditions and bath composition during sample formation, and hydrogen may be incorporated into the atomic structure of the sample. Typical bath compositions and deposition parameters have been described a) for electrodeposition [15 - 17] and b) for deposition by chemical reduction [15,18,19]. For electrodeposition the bath includes the metal salt and acids, an ammonium- or sodium-salt of an organic acid as buffer, and ammonium-hydroxide in order to adjust the pH value of the solution. Following the arguments of G. Dietz 1) the largely covering of the cathode with inhibitory ions is responsible for very small active spots, which fluctuate in time and position. Therefore the deposition rate is locally very high (the global current density is about 1000 $A/(m^2)$ - equivalent to a deposition rate of about 17 nm/s for Co - P). This fact together with the chemical binding between the deposited atoms reduces the surface mobility of the atoms so much, that the amorphous state can be quenched in.

The plating solution of Simpson [19] for chemical deposition was of the alkali hypophosphite type and had the following composition: 0.089 M cobalt sulphate ($CoSO_4 \cdot 7\ H_2O$), 0.318 M ammonium sulphate (NH_4SO_4), 0.306 M sodium citrate ($Na_3C_6H_5O_7 \cdot 2\ H_2O$) as buffer, and 0.3 M hypophosphorus acid (H_3PO_2). The bath temperature was 85 °C , and the pH value of the fresh solution was adjusted to 8.6 with ammonium hydroxide (NH_4OH). The amorphous Co-P alloy with 9 at% P was deposited with a rate of 1 nm/s on a catalytic substrate of nickel plated Al. A possible mechanism for currentless plating is that the presence of hydrogen on the catalytic metal (by the reaction: $H_2P^{1+}O_2^- + H_2O \rightarrow H_2P^{3+}O_3^- + H_2$) constitutes a galvanic cell, in which hydrogen is the anode and the metal the cathode. The process could then be considered as an electrolytic one 2).

2.3. Methods of amorphization without quenching

2.3.1. Formation of amorphous alloys by solid-state diffusion reactions ("SSR"):

2.3.1a. To the best of ours knowledge the first SSR studdies leading to an uniform amorphous semiconductor were done already 1973 [20,21]. Amorphous Ag-Te films were obtained by diffusing Ag atoms at 203 K from an already deposited crystalline Ag film into the condensing toplayer of tellurium, or

---

We wish to acknowledge gratefully discussions on these subjects with 1) Professor Dr. G. Dietz, II. Physikalisches Institut der Universität zu Köln, and 2) Professor Dr. K.G. Weil, Institut für Physikalische Chemie der TH Darmstadt.

by diffusing Ag atoms at 203 K into an amorphous Te film which was deposited first at 77 K. The temperature of 203 K is high enough to allow sufficient rapid Ag diffusion (0.2 nm/s through an 100 nm amorphous film), and low enough to result in an amorphous Te-Ag film. By measuring the rates of Ag consumption as a function of temperature the diffusion constants D and the activation energy ΔE for the Ag diffusion in the amorphous Ag_2Te film were determined: $\Delta E = 0.29$ eV, $D(203\ K) = 20\ (nm)^2/s$. Whenever Ag is deposited first amorphous Ag_2Te films are obtained which recrystallize close to room temperature. On the other hand, whenever Te is deposited first at 77 K, one obtains films with the composition AgTe which remain amorphous at room temperature. Analogous experiments of interfacial reaction in bilayers of amorphous Si and crystalline Rh thin films were carried out later [22]. Because of the increasing interest in the application of thin amorphous semiconducting films such SSR studies seems to us very important.

One may argue that in these experiments the amorphous state of the semiconductor is formed by vapor quenching (2.1.1.) and that this amorphous state is not destroyed or even stabilized by the penetrating metal atoms.

In this context an opposite experiment should be mentioned [23]. Metastable crystalline solid solutions of X-Ag (X = Ru, Rh, Os, Ir) were produced by vapor deposition at 77 K. The extraction of almost all Ag by a iodine treatment yield essentially pure amorphous X-films. These amorphous films are characterized by a low density (due to the fact that 70 % of the atoms are removed from the starting X-Ag film), by a high electrical resistivity of about $\rho(300\ K) = 10^{+3} - 10^{+4}$ µOhm·cm and by a strong negative temperature coefficient of resistivity. Strong percolation and localization effects can be studied in such amorphous structures.

2.3.1b. The first diffusion reaction with a transformation into an amorphous state was observed by hydriding a metastable crystalline intermetallic compound Zr_3Rh of the Cu_3Au structure [24]. For the full transformation of the melt quenched crystalline compound into the amorphous metallic hydride at 180 °C (453 K) and at 1 atm H_2 a time up to 6 days were necessary. Therefore quenching should be considered not at all. The diffusion of the metal atoms is nearly frozen out at the temperatures used here while the mobility of hydrogen is approximately ten orders in magnitude higher. The alternative to the disallowed chemical segregation into the well known stable crystalline hydrides is a polymorphic transformation to an amorphous hydride. Such a transformation does not require interdiffusion of metal atoms and yet provides a lowering of free enthalpy by the chemical reaction. Such a chemical "frustration" effect between the crystalline state on one side and the chemical reaction on the other side gives some understanding for the amorphization process by SSR: At least one of the constituents has to be motionless at the reaction temperature.

2.3.1c. The most puzzling aspect of amorphization by SSR arises in the recent studies of metal - metal systems which react in a multilayer sandwich configuration of polycristalline pure metal films. Up to now the formation of amorphous binary alloys have been reported in the following diffusion couples (the fast diffuser is marked by underlining): La-Au [25] and Y-Au [26] (films of 20 - 60 nm thickness were annealed for up to 5 h at 50°C - 80°C), Zr-Ni [27] (films of 30 - 100 nm thickness were annealed for 14 h at 300°C), Zr-Co [28] (films of 20 - 34 nm thickness were annealed for 24 h at 200°C), Hf-Ni [29] (films of 10 - 40 nm thickness were annealed for 12 h at 300°C - 340°C), Sn-Co [30] (bilayers of 6 nm thickness were annealed for up to 60 h at 17°C).

From these experiments conclusions have been drawn concerning favorable conditions for glass formation and growth by SSR [31,32]: (I) Glass formation and growth must be thermodynamically a downhill process. A large decrease of free enthalpy G should characterize the process (large negative value of ΔG of mixing). (II) Atomic mobility must be adequate to allow alloy formation and growth in practical times $t_p \cong 10^{+4} - 10^{+6}$ s. The characteristic time t_g required to form the initial glass layer and the time t_d required to transport the fast diffuser across the glass layer must be shorter than t_p. (III) The characteristic times t_g and t_d must be much shorter than the time t_x which is required to nucleate and growth a crystalline phase at the reaction temperature. (IV) The mutual terminal solubility of the components has to be small enough to avoid a regular solution of the system in the crystalline state and with it the possibility of a polymorphic transition from the glassy state into the crystalline state without chemical segregation.

Glass formation by SSR raises a number of unanswered questions: Why the glassy phase forms initially as opposed to more stable crystalline intermetallic compounds ? Once formed, the glass layer appears to be subject to similar crystallization kinetics as melt-quenched glasses, on the other hand, atomic mobility at least of the fast diffuser in the glass has to be adequate large (condition II) [33,34]. A imaginable answer to these questions may be the fact that the diffusion is driven by a large gradient in chemical potential (not only in the gradient of concentration) as long as the components have not yet reacted. The following observation may be a hint in this direction. Obviously the reaction temperature is very critical. A change in temperature of only 50°C can result in either no reaction in practical times or in the appearence of some first X-ray lines of an intermetallic compound. But if the amorphous phase has appeared with its broad maximum in the X-ray diffraction pattern we can increase the temperature by more than 5o°C in order to complete the reaction without crystallization. After the reaction is completely finished by the total consumption of one of the constituents the amorphous alloy is much more stable against crystallization which means a further increase in temperature more than 100°C up to the crystallization temperature is allowed. (The numbers given here are proper to the Zr-Co system with a reaction temperature of about 473 K and a crystallization temperature of about 650 K).

In this short review we restrict ourselves to some results of intertransition metal systems. The temperatures and times mentioned so far have been plotted in a schematic temperature-time-transition (TTT) diagram in figure 2.

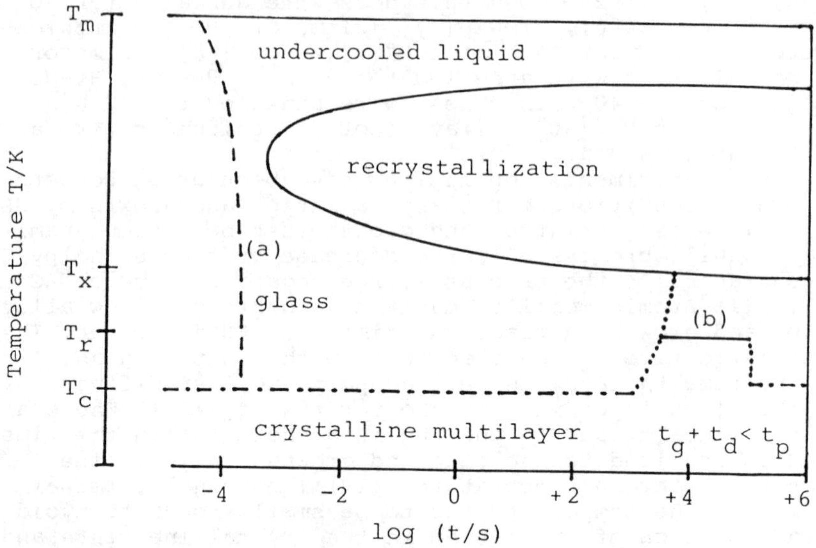

FIGURE 2. Schematic (TTT) diagram for formation of an amorphous alloy by a process of (a) melt quenching, (b) solid state reaction (SSR) of a crystalline multilayer condensed at T_c and annealed at $T_r << T_x$ (crystallization temperature). The time of the glass formation by SSR = (tg + t_d) has to be smaller than a practical time t_p of about 100 h.

In reference |28| we report on Zr-Co diffusion couples multilayers prepared in an UHV system with two rate-controlled electron beam evaporation sources. The multilayers were condensed onto different substrates held at temperatures below 273 K in order to avoid uncontrolled reactions. Each single layer had a thickness between 20 nm and 100 nm depending on the overall number of single layers and the total composition in the multilayer. The SSR annealing was carried out in the UHV system at temperatures of about 473 K (200°C) up to 24 h. The status of the multilayers was checked in different stages of reaction by X-ray analysis and by cross-sectional transmission electron microscopy (TEM) at sufficiently low temperatures outside of the UHV system.

Figure 3 shows X-ray diffraction patterns of the control multilayer (always kept at room temperature) and of three reacted multilayers. The X-ray diffraction data for the control multilayer shows the crystalline nature of the sample by exhibiting well defined (002) and (111) peaks of crystalline Zr and of crystalline Co respectively. The size of the crystals can be determined from the width of the peaks using the Scherrer formula. In the X-ray diffraction patterns of all three reacted samples a broad band characteristic of amorphous ZrCo has appeared. But a single amorphous phase is formed by SSR only from multilayers with a total composition centered around 55 at% Co.

The X-ray diffraction pattern of a reacted multilayer with 50 at% Co shows an additional (002) line of crystalline Zr, whereas the (111) line of crystalline Co is to be seen in the pattern of a reacted multilayer with 60 at% Co.

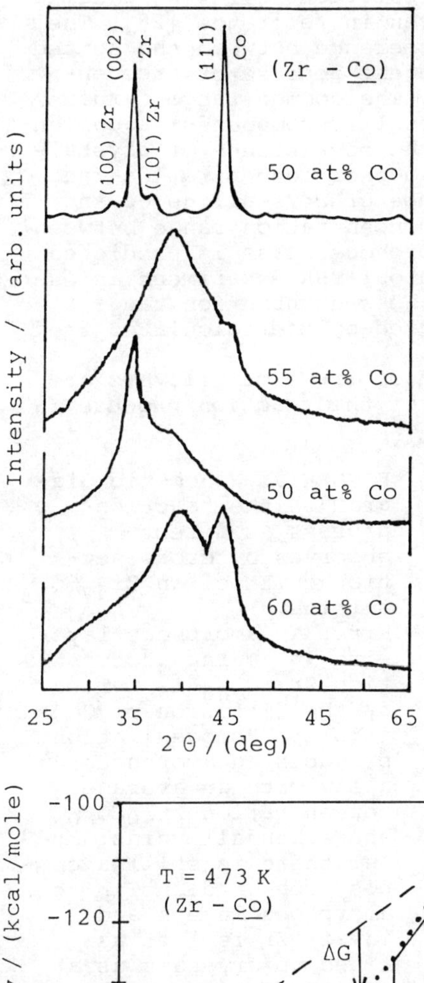

FIGURE 3. X-ray diffraction pattern (Cu-K_α radiation) vs. scattering angle for the crystalline control multilayer (always kept at room temperature) and three reacted samples (SSR annealing at 473 K for 24 h).

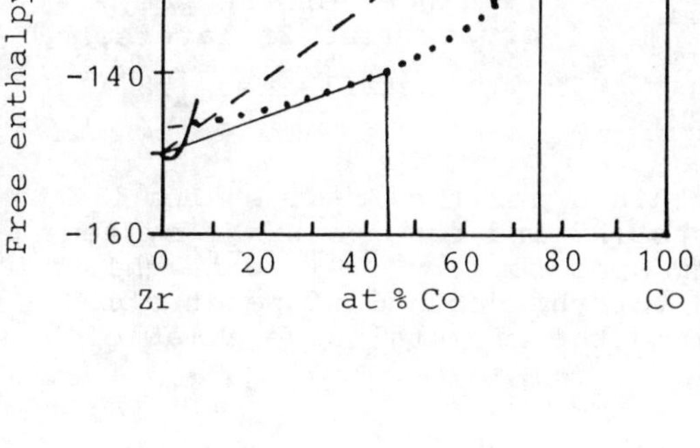

FIGURE 4. Free enthalpy vs. composition as calculated in reference |28| following Miedema's semiempirical approach. The strongest decrease ΔG of about 12 kcal/mole is predicted for an alloy of 55 at% Co. The dashed straight line ----- represents the free enthalpy G of an unreacted sample (as a mixture of the crystalline constituents). The dotted curve ····· represents G of an undercooled liquid (amorphous state of the reacted sample).

These experimental results may be compared with the predictions of a free enthalpy diagram which is plotted in figure 4. This diagram gives the calculated free enthalpy vs. composition at T = 473 K for a multilayer of Zr-Co before the reaction as a mixture of crystalline Zr and Co in the dashed straight line and after the reaction as an undercooled liquid (amorphous state) in the dotted curve. This calculation following a semi-empirical approach of Miedema is given in reference [28]. The driving force for the SSR is the difference between the dotted curve and the dashed line. The strongest decrase of free enthalpy G is predicted for 55 at% Co. The common tangent rule predicts (without additional intermetallic compounds) a concentration range of (0 - 45) at% Co for the coexistence of crystalline Zr and amorphous $Zr_{55}Co_{45}$ and another concentration range of (75 - 100) at% Co for the coexistence of crystalline Co and amorphous $Zr_{25}Co_{75}$. Whereas in the concentration range between 45 at% Co and 75 at% Co a single amorphous phase is predicted. The single amorphous phase formed in our SSR experiment is indeed centered around 55 at% Co, but the concentration range seems to be much smaller than predicted by the calculated free enthalpy diagram.

The results of cross-sectional TEM of Zr-Co multilayers are summarized in the schematic diagram of the reaction process in figure 5.

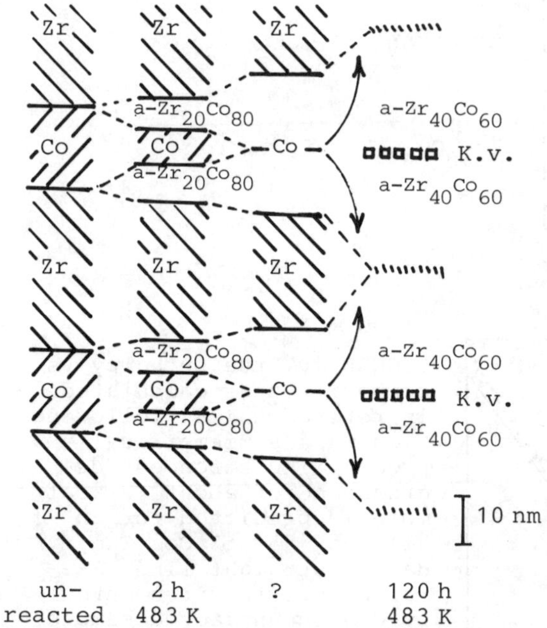

FIGURE 5. Schematic diagram of the reaction process as direct - observed by cross-sectional TEM of an $Zr_{47}Co_{53}$ multilayer.
First an amorphous layer with an average concentration $Zr_{20}Co_{80}$ is formed by the diffusion of Co into Zr. Homogenization produces an amorphous alloy with an average concentration $Zr_{40}Co_{60}$ an Kirkendall voids (K.v.). But there is still a composition gradient in the amorphous layers and 13 at% Zr is left as lined up crystals as a remainder of the polycrystalline Zr layers.
(For details see [28])

Similar results have been obtained for the Zr-Ni system by Auger depth profile measurements [27] and for the Hf-Ni system by 2 MeV He-ions backscattering spectrometry [29]. With this technique the authors observed for the depth d of penetration of the amorphous Hf-Ni layer into the Hf metal side a law of the form $d = \alpha \cdot t_o^{1/2} + \beta \cdot (D\,t)^{1/2}$ (where t is the reaction time, α and β are constants, D is the interdiffusion constant of Ni in the amorphous layer). The shift $t_o^{1/2}$ is temperature dependent and can be related to the interface kinetic mobility which limits growth of the layer in the limit of short times.

In spite of the stinging open questions we trust in SSR as a method for amorphization which has been now observed by different groups and for several systems. This new method is of interest as well for practical reasons as for a fundamental insight into the amorphous state of metals and in the following methods which may be even more important for applications in material science .

2.3.2. Vapor deposition onto hot substrates: The fast diffusion of one constituent is a necessary but not sufficient condition for amorphization in a solid state reaction process. Following this idea an amorphous alloy should also be formed in a different way of mixing the elements at the reaction temperature or even higher temperatures (but below the known crystallization temperature of the amorphous alloy formed by melt quenching ?). Therefore we produced a mixture of the elements by cocondensation of vapors onto hot substrates [35]. The temperatures of the substrate was kept constant within 10 K during the condensation. (The heat of condensation is the reason for these temperature fluctuations.) The films were prepared in the same UHV system as the multilayers for the SSR experiments [36].Immediately after the cocondensation the temperature of the substrate was decreased to room temperature. The films were produced with a thickness up to 1000 nm which is far above the thickness of quench condensed transition metal films limited by their internal stresses. The structures of the films were checked by X-ray analysis outside of the vacuum system. First results of these experiments are plotted in figure 6 and figure 7.

These figures are similar to a phase diagram of cocondensed Zr-Co and Zr-Cu films. The temperature of the substrate during cocondensation is shown vs. composition of the vapor. The results of X-ray analysis of the condensed films are plotted on this diagram with different symbols: Crosses indicate that the particular film was found to be crystalline (without regard to the type of crystalline structure), while solid points are used for completely amorphous films and a cross inside a circle is used to indicate partly crystalline films. The solid curve separates the region of amorphous films from that of crystalline ones. The shaded region gives the crystallization temperatures of melt quenched glassy alloys by different authors [36].

Those compositions result in amorphous films at the highest temperatures above 573 K of the substrate for which the decrease of free enthalpy ΔG is calculated to be an optimum (according to figure 4). An increase of the deposition rate from 0.4 nm/s to 4 nm/s has no influence on the concentration range in which amorphous films can be produced on hot substrates. At least at the highest temperatures one suspects a sufficient mobility of the fast diffuser atoms in the amorphous film and at the surface more than ever from the SSR experiments. Consequently the question raises again, what is the reason for the metastability of the amorphous state in these films at rather high temperatures ?

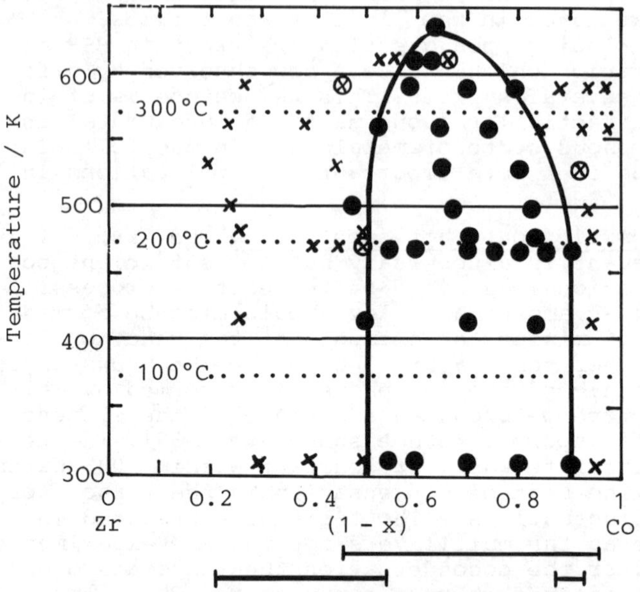

FIGURE 6. Zr_xCo_{1-x} composition range of glass formation vs. temperature of the substrate

x crystalline
● amorphous

substrate at 4 K
by melt quenching

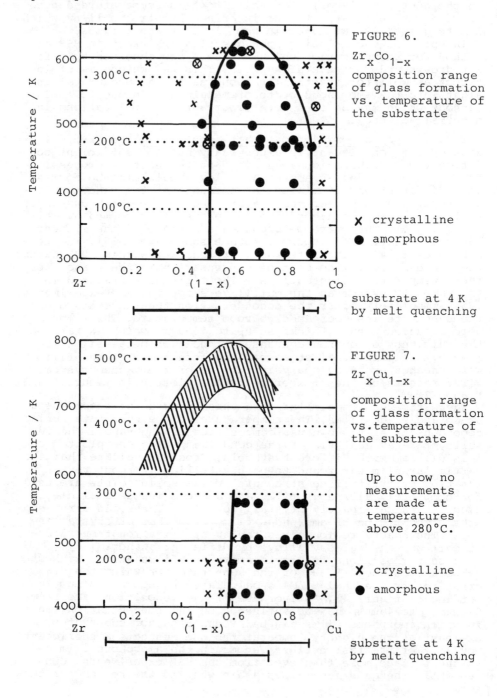

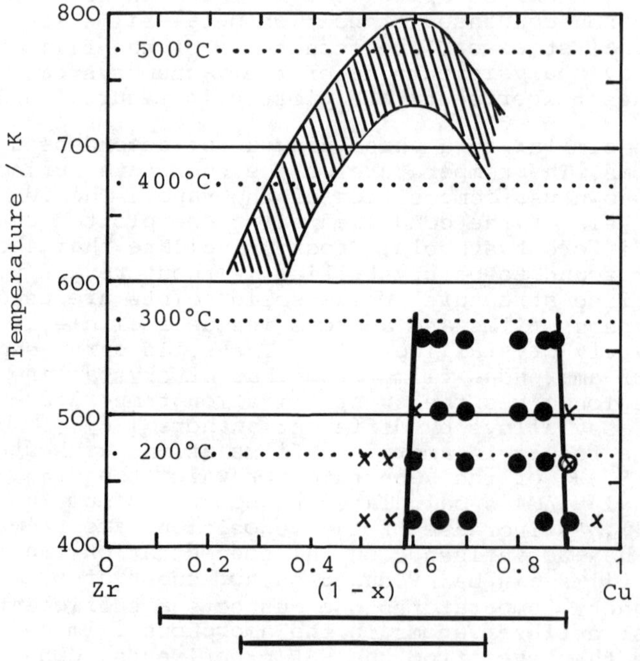

FIGURE 7. Zr_xCu_{1-x} composition range of glass formation vs. temperature of the substrate

Up to now no measurements are made at temperatures above 280°C.

x crystalline
● amorphous

substrate at 4 K
by melt quenching

We suppose the answer should be rather a barrier for nucleation than a lack of atomic mobility. But a lack of significant diffusion of Ni at 250°C (T = 523 K) through an annealed homgeneous $Zr_{50}Ni_{50}$ alloy (sandwiched at 383 K between pure Zr and pure Ni is reported in [34] ($D < 10^{-17}$ cm^2/s). But one has to consider the rather low temperature of 250°C in this experiment. A full set of data is given in [29] for the diffusion of Ni through amorphous Hf-Ni in a SSR experiment. At T = 613 K $D = 10^{-15}$ cm^2/s was measured. On the other hand one calculates with $D = D_o \exp - (Q/k_BT)$ and Q = 1.7 eV at T = 523 K (250°C) the low value $D = 4 \cdot 10^{-18}$ cm^2/s. So the diffusity is strongly dependent on temperature and more direct measurements of diffusivity in homogeneous amorphous films of interest in this context should be made.

2.3.3. Mechanical alloying: The SSR amorphization method has been extended to composites produced by cold-working of intercalated foils or elemental powders: Zr-Ni [37] (filaments of 120 nm thickness were produced and annealed at 300°C up to 153 h), Zr-Ni and Zr-Cu [38] (filaments of 100 nm thickness were produced and annealed at 250°C - 370°C for 1 - 32 h). In these experiments mechanical deformation serves to (1) consolidate the binary mixture, (2) increase the interfacial area between the two metals and (3) reduce at least one of the characteristic dimensions of the mixture. Moreover the defects produced by mechanical deformation may assist the diffusion process in the following annealing procedure by which most of the material becomes an amorphous alloy. Three dimensional amorphous samples produced by this method may be of interest for special applications. Differential scanning calorimetry was performed on these samples. For instance $Zr_{40}Cu_{60}$ yields an endothermic peak at 440°C and two sharp peaks at 475°C and at 500°C. The endothermic peak is characteristic of a glass transition and the first exothermic peak of the crystallization.

We have already earlier realized that cold-working produces alloying even at temperatures of 4 K [39]. Intense strength of deformation during repeated mechanical mixing, cold welding, fracturing and rewelding of ultrafine powders in a dry high energy ball mill produces mechanical alloys of even such elements which cannot be alloyed by other methods. Amorphous or partly amorphous alloys have been produced by this technique: Y-Co and Gd-Co [40], Nb-Ni [41], Ti-Ni [42], Nb-Ge and Nb-Ge-Al [43], Ti-Cu [44]. The mechanical alloying process (MA) seems to be not restricted to the formation only of the amorphous alloys mentioned above [44]. The MA process promises to be an unconventional new and powerful method of preparing amorphous materials. Moreover first succesful experiments of cold-pressing and sintering some of these MA powders were carried out. So we close this lecture with a promising outlook to those who are mainly interested in applications of amorphous metals.

3. SUMMARY

As far as the properties of the amorphous alloys are studied, no remarkable difference is found for samples produced by the different methods (with the exception of crystallization temperatures which seems to be size dependent). The amorphous state seems to be almost independent of the way in which it is reached. There are many streets to Roma !

REFERENCES

1. Buckel,W. and R. Hilsch: Z. Physik 138 (1954) 109.
2. Piecuch,M., Geny,J.F. and G. Marchal: In ref. [8], p.79.
3. Duwez: Progress in Solid State Chemistry, H. Riess (ed.), Pergamon Press, Oxford (1966), vol.3,p.377.
4. Rapidly Quenched Metals III (Proc.3rd.Int.Conf.),B.Cantor (ed.), Metals Society, London (1978).
5. Rapidly Quenched Metals IV (Proc.4th.Int.Conf.),T. Masumoto and K. Zuzuki (eds.) Sendai (1982).
6. Amorphous Metallic Alloys, F.E. Luborsky (ed.), Butterworth & Co. (1983).
7. Liquid and Amorphous Metals V (Proc.5th.Int.Conf.),C.N.J. Wagner and W.L.Johnson (eds.), Elsevier Science Publishers B.V. (1984), reprinted from J. Non-Cryst. Solids 61 & 62.
8. Amorphous Metals and Nonequilibrium Processing (Proc.Symp.) M. von Allmen (ed.), Les Editions de Physique (1984).
9. Rapidly Quenched Metals V (Proc.5th.Int.Conf.), S. Steeb and H. Warlimont (eds.), Elsevier Science Publ. B.V. (1985).
10. Glassy Metals II, H. Beck and H.J. Güntherodt (eds.), Springer Verlag (1983), 8. Laser Quenching: M.v. Allmen, p. 261 - 281.
11. Grant,W.A., Ali,A., Chadderton, L.T. and P.J. Grundy: In ref. [4], p.63.
12. Linker,G.:Materials Science and Engineering 69 (1984).
13. Johnson,W.L., Cheng,Y.T., Rossum,M.van and M.-A. Nicolet: Nucl. Instr. and Meth. B 7/8 (1985) 657.
A review is given by B.M. Paine and R.S. Averback in Nucl. Instr. and Meth. in Phys. Res. B 7/8 (1985) 666 - 675.
Rossum, M.van, Johnson,W.L. and M.-A. Nicolet: In ref. [8], p.99.
Rossum,M.van, Shreter,U., Johnson,W.L. and M.-A. Nicolet Mat. Res. Soc. (Proc.Symp.), Elsevier Science Publ. (1984), Vol. 27, p.127.
14. Bill,R.: Z. f. Krist. 75 (1930) 217.
15. Watanabe, T. and Y. Tanabe: In ref. [9] , Vol.I, p.127.
16. Brenner,A.: Electrodeposition of Alloys, Academic Press, New York - London (1963).
17. Brenner,A.,Couch,D.E. and E.K. Williams: J.Res.Nat.Bur. Standarda 44 (1950) 109.
18. Brenner,A. and G. Riddell: J.Res.Nat.Bur.Standards 39 (1974) 385.
19. Simpson, A.W. and D.R. Brambley: Phys. Stat. Solidi (b) 43 (1971) 291.
20. Bolotov,I.Ye. and A.V. Kozhin: Fiz. metal. metalloved. 35 (1973) 383 (Phys. of Metals and Metallog. 35 (1973) 146.)
21. Hauser,J.J.: J. Physique 42 (1981) C4 - 943.
22. Tu,K.N. and K.Y. Ahn: Appl. Phys. Lett. 42 (1983) 597.
23. Hauser,J.J.: Phys. Rev.B 28 (1983) 597.

24. Yeh,X.L., Samwer,K. and W.L. Johnson: Appl. Phys. Lett. 42 (1983) 242.
Samwer,K., Yeh,X.L. and W.L. Johnson: J. Non-Cryst. Solids 61 & 62 (1984) 631.
25. Schwarz,R.B. and W.L. Johnson: Phys. Rev. Lett. 51 (1983) 415.
26. Schwarz,R.B., Wong,K.L. and W.L. Johnson: J. Non-Cryst. Solids 61 & 62 (1984) 129.
27. Clemens,B.M., Johnson,W.L. and R.B. Schwarz: J. Non-Cryst. Solids 61 & 62 (1984) 817.
28. Samwer,K., Regenbrecht,A. and H. Schröder: In ref. [9], p. 1577.
Samwer,K.: In ref. [8], p. 123.
Schröder,H., Samwer,K. and U. Köster: Phys.Rev.Lett. 54 (1985) 197.
29. Rossum,M.van, Nicolet,M.-A. and W.L. Johnson: Phys.Rev.B 29 (1984) 5498.
Rossum,M.van, Shreter,U., Johnson,W.L. and M.-A. Nicolet: Mat. Res. Soc. (Proc.Symp.), Elsevier Science Publ. (1984) Vol. 27,p.127, and in ref. [9] , p. 1515.
30. Guilmin,P., Guyot,P. and G. Marchal: Phys. Lett. A 109 (1985) 174.
31. Johnson,W.L., Dolgin,B.P. and M. van Rossum: In Glass Current Issues, A.F. Wright and J. Dupuy (eds.), NATO ASI Series, M. Nijhoff Publ. (1984), p. 172.
32. Samwer,K.: In ref. [8], p. 123.
33. Rossum, M. van, Shreter,U., Johnson,W.L. and A.-M. Nicolet: Mat. Res. Soc. (Proc.Symp.), Elsevier Science Publ. (1984) Vol. 27, p. 127.
34. Barbour,J.C., Saris,F.W., Nastasi,M. and J.W. Mayer: Submitted to Phys.Rev. B.
35. Regenbrecht,A.: Unpublished results.
36. Minnigerode, G. von, Böttjer,H.G., Prüfer, G. and A. Regenbrecht: LT 17, (Proc. 17th. Int.Conf.), U. Eckern, A. Schmid, W. Weber and H. Wühl (eds.), Elsevier Science Publ. B.V. (1984) FI2.
37. Schulz,L.: In ref. [8], p. 135 and in ref. [9], p. 1585.
38. Atzmon,M., Verhoeven,J.D., Gibson,E.D. and W.L. Johnson: Appl. Phys. Lett. 45 (1984) 1052 and in ref. [9], p.1561.
39. Bergmann,G., Hilsch,R. and G. von Minnigerode: Z. Naturforschung 19 (1964) 580.
40. Ermakov,A.E., Barinov,V.A. and E.E. Yurchikov: Fiz. metal. metalloved. 52 (1981) 1184 and 54 (1982 935.
41. Mc Kamey,C.G. and J.O. Scarbrough: Appl. Phys. Lett. 43 (1983) 1017.
42. Schwarz,R.B.: Abstract submitted to ref. [9].
43. Politis,C.: Proc. Int. Conf. on the Materials and Mechanisms of Superconductivity, Ames, Iowa, May 1985.
44. Politis,C. and W.L. Johnson: We gratefully acknowledge private communications and a preprint.

MATERIALS AMORPHIZED BY LASER IRRADIATION

M. von Allmen
Institut für Angewandte Physik, Universität Bern,
CH-3012 Bern, Switzerland.

1. INTRODUCTION

That lasers can be used to melt and quench metallic alloys was realized some time ago [1], but active exploitation of the method has started only recently. Most of the present activities fall into one of two categories, namely work with scanned lasers [2] and work with pulsed lasers [3]. In the first group, a continous beam, typically from a high-power CO_2 laser (a few 100 W to several kW), is scanned across a bulk metal workpiece, sometimes covered with a thin coating. In this process, often referred to as "laser glazing", the coating and/or part of the bulk metal is molten by the beam and subsequently resolifies as the beam moves on. Sharply focussed beams must normally be used to overcome the high reflectivity of metals in the far infrared, but extended areas of a workpiece can be treated by applying partially overlapping line scans. Scanning velocities are usually limited by the absorbed power and range from tens of cm/s to a few m/s. Cooling rates in a scanned-laser heated surface region are comparable to those of mechanical quenching up to about 10^6 K/s - and enable glass formation only for easy glass formers [2]. For all its technological potential, no physically new phenomena have been observed with laser glazing.

The situation is quite different for the pulsed laser work. Irradiation of absorbing materials by intense laser pulses in the ns to ps regime causes extremely rapid heating and melting of a thin surface layer, followed by almost equally rapid quenching of the melt. Melt quenching at cooling rates around 10^{10} K/s can be had in thin (a few tenth of a micron thick) films with quite simple technology. As a result, metallic glasses and other metastable phases are obtained from a wide range of alloy systems [3,4]. In this lecture I shall review some experimental aspects of pulsed laser quenching and then discuss a few recent results.

2. EXPERIMENTAL

A proven method for preparing laser-quenched binary films calls for the following steps. First, a stack of 10 - 20 alternating elemental layers with the desired average composition is vapor-deposited onto a substrate of sapphire or refractory metal. Individual layers are made no more than 10-20 nm thick to ensure thorough intermixing of the elements during the short time a melt persists. Next, the film is irradiated with an unfocussed ns or ps pulse from a solid-state laser or a TEA gas laser, preferentially with a wavelength in the visible of ultraviolet. A homogenous spatial beam profile and a carefully selected fluence are necessary to avoid film damage by evaporation or peeling. A single pulse with an energy of a few tenths of a Joule is sufficient to quench a film area about 5 - 10 mm in diameter. More extended areas can be quenched with partially overlapping shots, although there can occasionally be problems within the overlap area, due to altered optical properties of the already irradiated material. A typical experimental setup is sketched in Figure 1. After irradiation, the structure of the quenched film is conveniently checked by Rutherford backscattering, x-ray diffraction and/or electron microscopy.

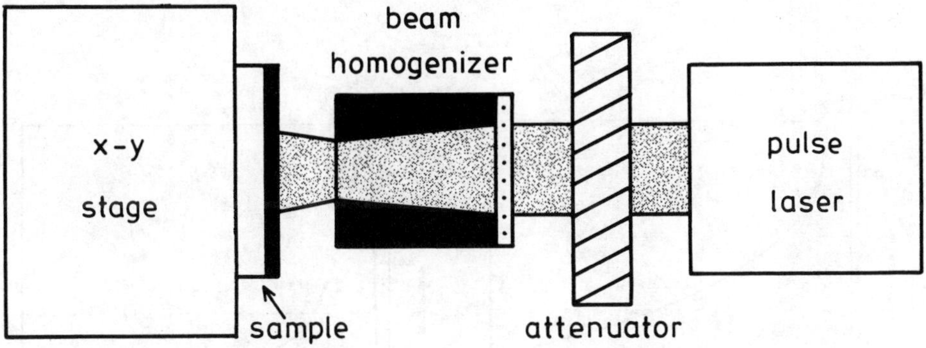

Figure 1. A typical setup for pulsed laser quenching of thin films.

The temperature-time history of the irradiated films cannot be measured at present, but it may be calculated by numerical simulation of the light absorption and heat flow processes. Since the molten area is several mm in diameter but only a fraction of a micron thick, heat flow in only one dimension needs to be considered. For the calculation, the sample (film plus substrate) is divided into a number of parallel slabs thin enough so their temperature can be taken as uniform. The transient temperature distribution is then obtained from repeated solution of the equation

$$T_n^{i+1} = T_n^i + [\Delta D_n^i - \Delta L_n^i + \Delta A_n^i]/C_n^i . \tag{1}$$

Here T_n^i and C_n^i are the temperature and the heat capacity, respectively, of slab n (n space increments Δz beneath the surface) at time i (i time steps Δt after the origin chosen). ΔD, ΔL and ΔA are energy increments due to heat flow, latent heat and light absorption, respectively:

$$\Delta D_n^i = (\Delta t/\Delta z^2)\{K_{n-1,n}^i (T_{n-1}^i - T_n^i) - K_{n,n+1}^i (T_n^i - T_{n+1}^i)\}, \tag{2}$$

$$\Delta L_n^i = \Delta t \; L_{sl} \; g(T,t), \tag{3}$$

$$\Delta A_n^i = \Delta t \; I_0 \; (1-R)\, \alpha\, e^{-\alpha z} f(t). \tag{4}$$

$K^i_{n,n+1}$ is the conductivity of the interface between slabs n and n+1, L_{sl} the latent heat, I_0 the mean laser irradiance, R the reflectance and α the absorption coefficient. The function g(T,t) describes uptake or emission of latent heat by the material, while f(t) represents the temporal pulse shape. All parameters can be made to depend on temperature, space or time as required by the experimental situation, and the resulting temperature curves are found to be quite reliable. Two examples are shown in Figure 2. Note that the cooling rate can be influenced, apart from the pulse duration, also by choice of the substrate material.

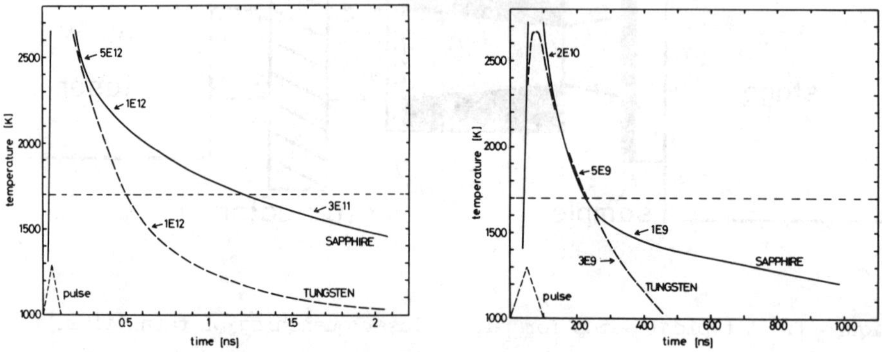

a) b)

Figure 2. Numerically calculated surface temperatures versus time in thin films (a: 60, b: 150 nm) of the alloy $Au_{50}Ti_{50}$ deposited on sapphire or tungsten substrates and irradiated by a: 50 ps and b: 50 ns (FWHM) laser pulses. The dashed lines indicate the melting point, exponential numbers give instantaneous cooling rates.

By going to shorter and shorter pulses (the shortest laser pulses produced today are a few tens of femtosec!) the cooling rate could be increased almost at will, however at a price: The usable material thickness (roughly given by $\sqrt{(\kappa\tau)}$ where κ is the thermal diffusivity and τ the pulse duration) decreases and becomes unpractically small for pulses much shorter than one ns. We have therefore conducted most or our experiments with 50 ns pulses, which yield quenched films up to a few hundred nm thick.

Given cooling rates like those of Figure 2, it is not surprising that metallic glass formation is not limited to the so-called "easy glass formers" known from mechanical melt quenching, but becomes quite common. Nevertheless, glass formation is still found to be easier in some systems than in others. What determines the glass forming ability at 10^{10} K/s?

3. GLASS FORMING ABILITY (GFA) [4,5,6]

Our studies have so far been concentrated on two kinds of binary alloys: (i) combinations of two transition metals (TM), and (ii) combinations of a transition metal with a group IV (GF) element. Figure 3 shows some of the phase diagrams we have investigated [4,5], with the ranges of glass formation indicated by dark areas (the upper boundaries of dark areas indicate the crystallization temperatures determined by slowly heating the glassy films). The observed glass forming ranges can be quantified in terms of simple kinetic parameters like the time the undercooled melt spends within a single-phase region during the quench, or the difference between the compositions of the melt and the "nearest" crystal [6]. There is, on the other hand, little correlation with traditional indicators of easy glass forming ability, such as melting point depression or differences in group number or atomic radius of the elements. The classical criterion for easy GFA, a deep eutectic, obviously doesn't correctly predict the dark areas either. It is true, however, that dark areas are almost always situated in two-phase regions. The available results support the prediction that ns laser quenching will in general fail to yield a glass from a particular alloy only if either

- single-phase crystallization (perhaps into a metastable crystal) is possible from the melt, or if
- the glass is unstable against crystallization at ambient temperature.

TM-TM systems in which we have obtained glassy phases by ns laser quenching include Au-Ti, Cu-Ti, Co-Ti, Cr-Ti, V-Ti and Ag-Cu; the only system where it positively failed was Zr-Ti, a continous solid-solution former. As for GF elements, they are well known to facilitate glass formation in alloys. Even pure Si can be melt-quenched into an amorphous phase (not a glass, the transition being first-order) by 10 ns pulses. Most combinations of transition or noble metals with Si, Ge or Sn seem to yield glasses over extended ranges of composition. One exception is the Ag-Si system which does so only near 80 at.% Si (i.e., far removed from the eutectic).

The contrasting behaviors of Ag-Si and Au-Si (see Figure 3) led us to undertake a quantitative investigation of GFA in the two systems [5]. Critical cooling rates were calculated essentially on the basis of the classical TTT approach [7], but with the important modification that the thermodynamic input data (free energy and heat of crystallization) were allowed to depend on composition and determined according to the common-tangent method for every crystal phase. Metastable crystalline phases can be treated the same way as stable ones. The procedure yields a curve of the critical cooling rate versus composition for every crystal phase included (Figure 4), from which the range of GFA can simply be read off for any cooling rate desired. E.g., at ~10^6 K/s, Figure 4 predicts GFA near the eutectic in Au-Si but none in Ag-Si, whereas near 10^{10} K/s GFA ranges close to those of Figure 3 result, in excellent agreement with experience. The differences between the two systems ultimately stem from differences in their heats of formation. More generally, these results illustrate the principle that it is the underlying thermodynamic parameters which govern the GFA of an alloy, rather than the topology of its phase diagram. The combined thermodynamic and kinetic treatment appears, in any case, suitable to treat GFA even around 10^{10} K/s, and perhaps beyond.

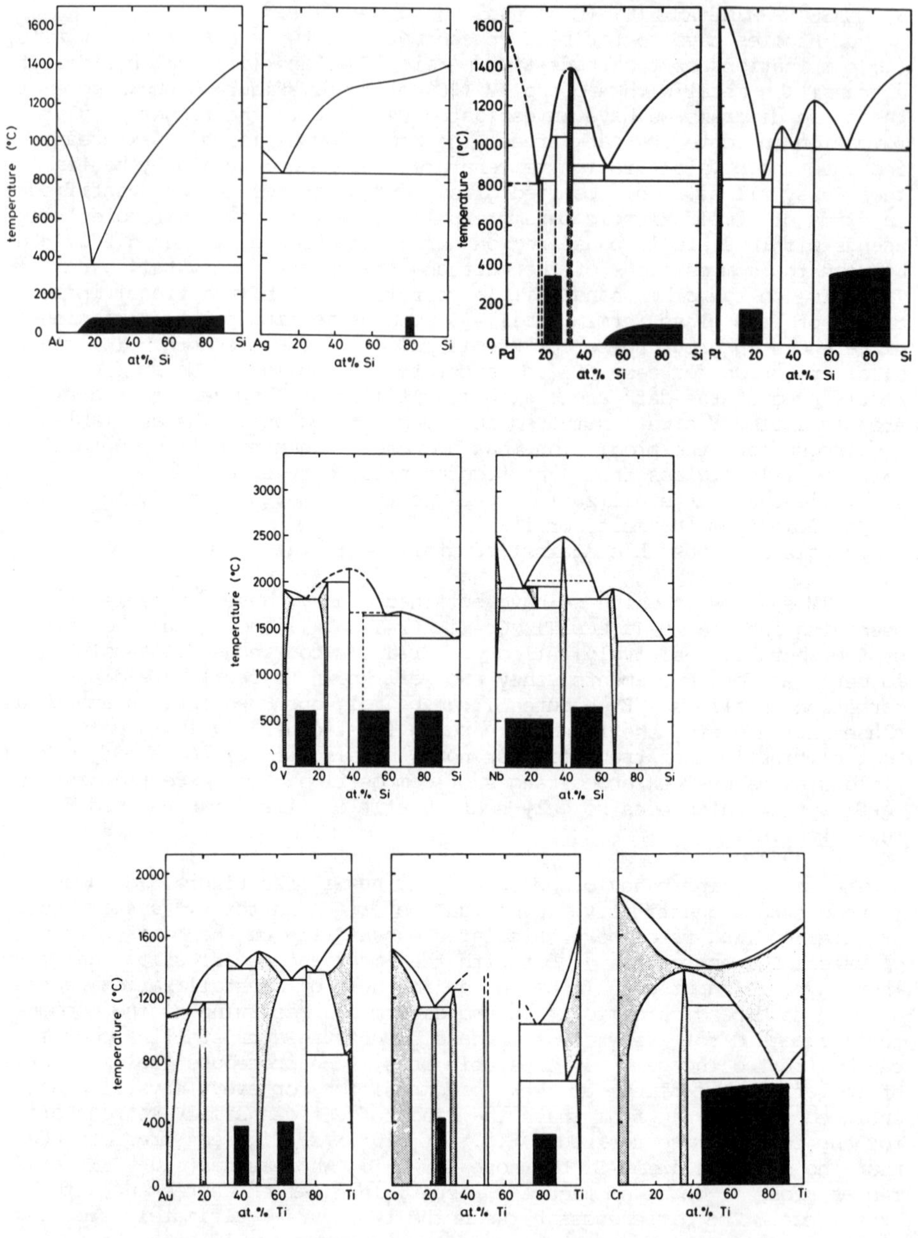

Figure 3. Phase diagrams of some binary systems studied with 50 ns laser quenching. Dark areas indicate compositional ranges of glass formation, with upper boundaries indicating crystallization temperatures (only selected compositions of the systems shown were tried).

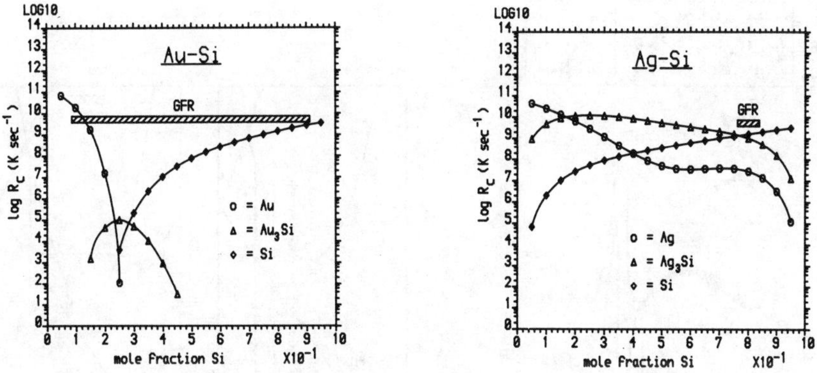

Figure 4. Calculated critical cooling rates for the suppression of three possible liquid-crystal transformations in the Au-Si and Ag-Si systems. Au_3Si and Ag_3Si are metastable hcp phases. From [5].

So much for the technique of pulsed laser quenching and the regime of glass formation it makes accessible. Let me now turn to some specific materials we have studied during the past few years.

4. REVERSIBLE AMORPHIZATION IN Cr-Ti [8]

Quenching a melt, even if at rates too slow to achieve glass formation, tends to leave behind a metastable structure. Some of the metastable crystalline phases show the remarkable property of turning amorphous upon subsequent slow furnace annealing. Examples include Au-Ti and Cr-Ti. Particularly interesting are films of the latter system, in which the solid-state amorphization process is diffusionless and, in some cases, even reversible.

The Cr-Ti system yields glasses on sapphire substrates between 45 and 60 at.% Ti; the use of W extends the glass forming range from 45 to at least 90 at.%. Crystallization temperatures for glasses of all compositions are near 650 °C (Figure 3, bottom right). As-quenched non-amorphous films contain the bcc high-temperature solid solution (β phase) that is metastable between room temperature and at least 670 °C. As-quenched films were annealed isothermally at 600 and 800 °C (usually for 30 min.) and then analyzed at room temperature. The results are schematized in Figure 5a. Films with the metastable β phase turned amorphous upon annealing at 600 °C. Subjected to a second annealing step at 800 °C, the freshly amorphous films crystallized, forming - depending on composition and substrate material - either the equilibrium

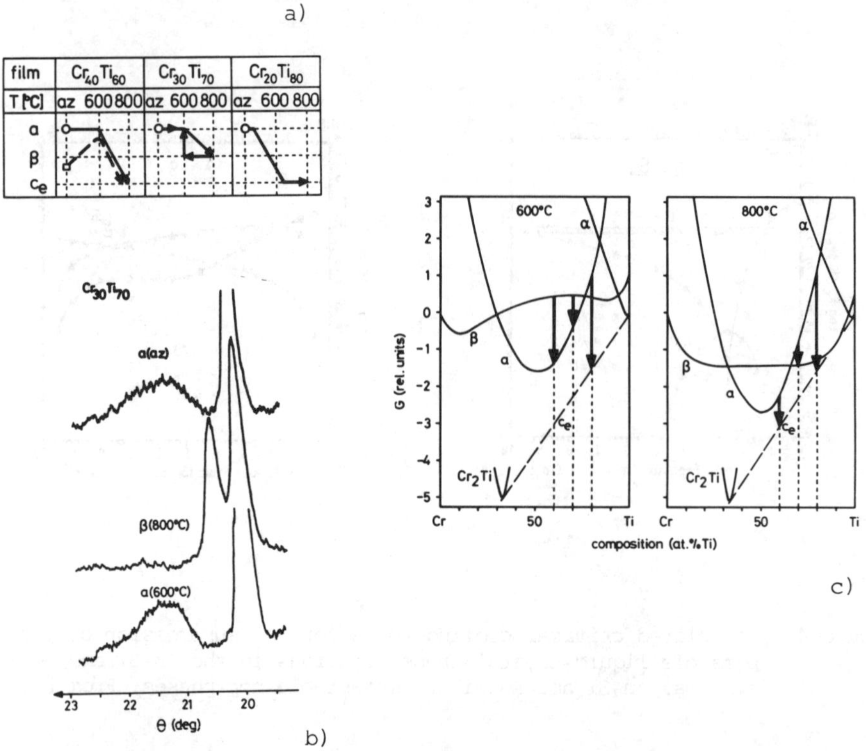

Figure 5. a) Schematics of annealing results of Cr-Ti films (a = amorphous, β = metastable bcc, c_e = equilibrium; o = tungsten, □ = sapphire substrate). The behavior on sapphire (dashed) is the same for all three compositions. b) Sections of diffractometer trace illustrating the reversible a→β→a transition in $Cr_{30}Ti_{70}$ films on tungsten. The diffraction line at 20 deg is from the substrate. c) Gibbs free energy diagram of Cr-Ti at 600 and 800 °C. Arrows indicate the transitions shown in Figure 5a. α stands for α-Ti (hcp). From [8].

configuration c_e (Cr_2Ti + α-Ti) or returning to the β phase. Films with 70 at.% Ti on W could be switched forth and back between the β and amorphous phases by applying alternating annealing steps at 600 and 800 °C. Figure 5b shows sections of diffractometer traces taken from the same film after irradiation (az), after annealing to 800 °C (β), and after annealing to 600 °C (a). The line at θ = 20 deg belongs to the W substrate, whereas that at 20.6 deg is the (110) line of the β phase. The β→a transition is evidenced by the disappearance of all sharp diffraction lines, and the concomitant appearance of the characteristic halo at 21.3 deg. Transitions β→a were also seen on sapphire substrates, but the amorphous phase formed was somewhat different from that on W and did not show the reversible amorphization feature.

Solid-state amorphization has been observed before in inhomogeneous (e.g., layered) materials composed of fast-diffusing elements [9]. Annealing of such structures at temperatures below the crystallization temperature leads to formation of a homogenous amorphous phase. However, the case of Cr-Ti is different, since the metastable β crystal has the same composition as the amorphous phase that grows from it. This is, to our knowledge, the first demonstration of diffusionless solid state amorphization. Moreover, the β→a transition is reversible in some cases. We believe that thermodynamic, in addition to kinetic, constraints are responsible for these unusual phase transitions.

Free energy curves at 600 and 800 °C consistent with the results of Figure 5a are shown in Figure 5c. The curves for crystalline phases are calculated from accepted thermodynamic data, whereas those of the amorphous phase is based on a regular solution model with an excess energy parameter chosen to fit our experiment. It takes the amorphous phase to be far more stable than the extrapolated melt, but is consistent with the prediction from Miedema's model as well as with a recent prediction of enthalpy differences between liquids and glasses by Saunders [10]. Based on these curves, amorphization of the β phase at 600 °C can be seen to be allowed for 60 as well as 70 at.% Ti. At 800 °C the relative positions of the "β" and "a" free energy curves are reversed near 70 at.%, allowing the observed a→β transition. The reason for the reversal of the free energy curves is that the amorphous phase is taken to have a constant negative heat of mixing, while the β phase is known to have a positive heat of mixing that decreases with temperature.

The significance of the demonstrated diffusionless solid-state amorphization process is that it enables the production of bulk amorphous metals (mm to cm thick) in a very simple way [11]. Similar behavior is expected to arise in a number of other systems which have high-temperature phases.

5. ELECTRONIC PROPERTIES OF TM-GF GLASSES [12]

Rapid quenching can be regarded as a trick to force "chemically incompatible" atoms into a homogenous solid compound where their mutual influences can conveniently be studied. Particularly interesting in this respect are metal-nonmetal glasses. The electronic properties of such glasses can be varied by means of the composition from metallic to nonmetallic. The chemical states of certain constituent atoms can vary dramatically according to the glass composition.

Consider the plot of resistivity versus composition of Au-Si and Au-Sn glasses in Figure 6. On the Au-rich end the resistivity varies linearly with composition, as expected for a mixture of two metals. Indeed, the Si atoms turns out to be in a metallic state (with two free electrons) in this range. The resistivity of the "dsm" (disordered solid metallic) Si can be estimated by comparing the slopes of the Au-Si and Au-Sn resistivity curves; it turns out to be higher than that of liquid Si (extrapolated to room temperature) but lower than that of the high-pressure metallic (β-tin) phase of Si.

Above 30 at.% Si the metallic character of Si is lost, while the alloy as a whole still behaves as a true disordered metal. This holds for compositions up to some 85 at.% Si, where it undergoes a metal-nonmetal transition. The critical composition corresponds to a mean distance between Au atoms of about 7 Å, which is compatible with a percolation type as well as a Mott (or Anderson) type of transition. All together, the resistivity of glassy Au-Si films can be varied via the composition over four orders of magnitude. Similar behavior is observed in other TM-Si systems such as Pt-Si and Pd-Si.

The underlying changes in the electronic structure also manifest themselves in the optical properties. Figure 7 shows, as an example, reflectance spectra of Au-Si glassy films of various compositions. Among the notable features are the screening of the Au d-band transition in Au-rich glassy films (only a few at.% of Si turn the gold into "silver"!) and marked reflectance minima of Si-rich ones. Both of these features visibly and abruptly disappear when the films crystallize into their equilibrium phases.

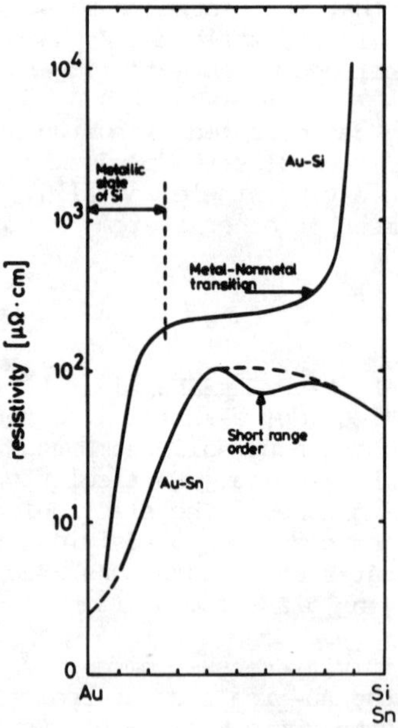

Figure 6. Resistivity as a function of composition for laser quenched Au-Si and Au-Sn glassy films. The dashed lines are for vapor-quenched films.

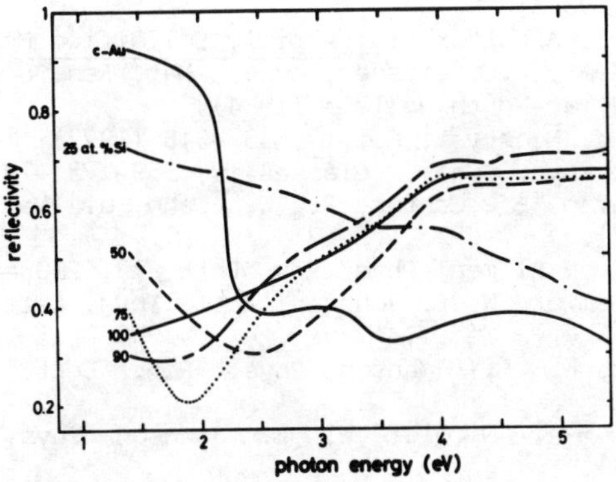

Figure 7. Reflectance spectra of pure crystalline Au and of Au-Si glasses of various compositions (expressed in at.% Si). From [12].

ACKNOWLEDGEMENT. Much of the work described here was supported, in part, by the Swiss Commission for the Encouragement of Scientific Research.

REFERENCES

1. W. A. Elliot, R. P. Gagliano, G. Krauss, Met. Trans. 4, 2031-2037 (1973).
2. Breinan E. M., Kear B. H., Banas C. M., Physics Today, November 1976, 44-50;
Becker R., Sepold G., Ryder P. L., Scripta Metall. 14, 1283-1285 (1980);
Bergmann H. W., Mordike B. L., J. Mat. Science 16, 863-869 (1981);
Yoshioka H., Asami K., Hashimoto K., Scripta Metall. 18, 1215-1218 (1984).
3. M. von Allmen, K. Affolter, M. Wittmer, in: Laser and Electron Beam - Solid Interactions and Material Processing (J. Gibbons, T. Sigmon, L. D. Hess, eds.), Mat. Res. Soc. Symp. Proc. 1, 559-566, Elsevier-North Holland (1981);
M. von Allmen, in: Glassy Metals II, (H. Beck, H. J. Güntherodt, eds.), Topics in Applied Physics 53, 261-281, Springer Verlag (1983);
M. von Allmen, E. Huber, A. Blatter, K. Affolter, Int. J. Rapid Solidif. 1, 15-25 (1984);
A. Blatter, M. von Allmen, in Amorphous Metals - Nonequilibrium Processing (M. von Allmen, ed.), MRS(E) Symp. Proc. 2, 73-78, Editions de Physique (1984).
4. M. von Allmen, S. S. Lau, M. Mäenpää, B. Y. Tsaur, Appl. Phys. Lett. 36, 205-207 (1980); ibid. 37, 84-86 (1980);
K. Affolter, M. von Allmen, H. P. Weber, M. Wittmer, J. Noncryst. Solids 55, 387-393 (1983);

K. Affolter, M. von Allmen, Appl. Phys. A 33, 93-96 (1984).
5. U. Kambli, M. von Allmen, N. Saunders, A. P. Miodownik, Appl. Phys. A 36, 189-192 (1985);
6. M. von Allmen, K. Affolter, in: Rapidly Solidified Metastable Materials, (B.H. Kear, B.C. Giessen, eds.), Mat. Res. Soc. Symp. Proc. 28, 81-86, Elsevier-North Holland (1984).
7. D. R. Uhlmann, J. Noncryst. Sol. 7, 337-348 (1972);
H. A. Davies, Phys. Chem. Glasses 17, 159-173 (1976);
P. Ramachandrarao, B. Cantor, R. W. Cahn, J. Mat. Science 12, 2488-2502 (1977).
8. A. Blatter, M. von Allmen, Phys. Rev. Lett. 13, 2103-2106 (1985).
9. X. L. Yeh, K. Samwer, W. L. Johnson, Appl. Phys. Lett. 42, 242-244 (1983);
R. B. Schwarz, W. L. Johnson, Phys. Rev. Lett. 51, 415-417 (1983);
M. von Rossum, M-A. Nicolet, W. L. Johnson, Phys. Rev. B, 29, 5498-5501 (1984).
10. A. R. Miedema, J. Less-Common Met. 41, 283-298 (1975); ibid. 46, 67-83 (1976);
N. Saunders, submitted to Int. J. Rapid. Solidif. (1985).
11. A. Blatter, M. von Allmen, to be published.
12. E. Huber, M. von Allmen, Phys. Rev. B 28, 2979-2984 (1983); ibid. 31, 3338-3342 (1985).

HOW TO PRODUCE WELL-DEFINED AMORPHOUS ALLOYS

HANS WARLIMONT

Vacuumschmelze GmbH, D-6450 Hanau,
Universität Stuttgart, D-7000 Stuttgart

Although amorphous structures are difficult to analyse and to define, some structural and microstructural features can be assessed as a function of alloy composition and specimen history. Variations in structural features with composition, irreversible and reversible changes with heat treatment, and homogeneous and heterogeneous microstructures can be distinguished. Experimental evidence of property variations as a function of composition and treatment is also becoming available. These findings suggest that structure - property relations of amorphous metals can be analysed quantitatively, if further progress in structural and microstructural analysis and definition is achieved.

Based on the application of principles derived from the behaviour of crystalline metals and of specific features of amorphous metals established recently, amorphous metallic materials can be developed by methods of controlled alloy design combined with a limited amount of intelligent empiricism. The development of some amorphous soft magnetic alloys on this basis is treated as an example.

This paper deals with the structures and structural defects of amorphous metals first; it goes on to dealing with structural changes and closes with a treatment of structure-property relations. In an appendix, schematic graphs and experimental data are presented which illustrate the basic features discussed and some properties and processes applied in the design of amorphous magnetic materials.

STRUCTURES AND PHASES

The occurrence of different structures and phases in amorphous metals [1] are outlined here first because they are an essential basis of what follows. The structural information on amorphous metals obtained to date is essentially due to three sources: measurements by scattering methods, conclusions drawn from the evaluation of measurements pertaining to local atomic environment such as nuclear magnetic resonance and Mößbauer spectroscopy, and results derived from theoretical models.

The results are conducive to an interpretation in terms of the occurrence of "structural units" in the amorphous structure resembling characteristic building blocks of related crystal structures. Such relations have, indeed, been found in a number of cases. I am stressing such findings as important and frequent features of amorphous structures and term such units "structure elements" in accordance with common usage for oxide glasses. Findings of this kind have been treated particularly extensively by Gaskell [2] who included oxide glasses and amorphous semiconductors along with amorphous alloys in his consideration.

From the structural information available it can be concluded that

- disordered solid solutions and

- ordered structures

may be distinguished in amorphous metals. Among the disordered solid solutions interstitial and substitutional solutions can be differentiated. The ordered structures may be ordered topologically only (being related to fcc, hcp, bcc... elemental structures) or chemically also (with structure elements related to crystal structures of intermetallic compounds).

Various structural investigations indicate the following, probably common features of amorphous alloys:

- the occurrence of solid solution and ordered structures in the form of identifiable phases, and

- the decomposition into several phases where pertinent mestable equilibria prevail. This particular point is considered more extensively in another paper [1]. It is appropriate to treat the amorphous state and its phase relations in terms of free enthalpy-composition-temperature plots.

As a final item regarding structures it should be pointed out that structural anisotropy as induced by magnetic field or stress annealing an unstable state of amorphous as well as crystalline short range ordered structures.

STRUCTURAL DEFECTS AND MICROSTRUCTURES

Free volume - which is of interest here - is the fractional specific volume exceeding that of the most densely packed, metastable amorphous structure the alloy considered could take at 0 K. We can, for the purpose of the present paper, distinguish between

- unstable and
- metastable

fractions of free volume. The unstable fraction, typically induced e.g. by any of the preparation techniques, is annealed out by irreversible structural relaxation processes. The metastable fraction is part of the metastable equilibrium glass structure and can, thus, vary with this structure as a function of composition and, reversibly, with heat treatment as discussed previously [3]; i.e. insofar free volume is a structure parameter.

Defects of high energy density such as vacancy-like, dislocationlike, and interface-like defects, e.g., do not appear to be stable in amorphous solids [4].

Localized disorder may be introduced in amorphous metals, e.g., by effects of the production process and of plastic deformation. By its very nature, localized disorder is unstable. Inhomogeneities of local disorder in the present context are ranging from several nm to μm in extent. The disorder induced by plastic deformation is localized in shear bands [5].

Internal stresses are predominantly unstable and are characteristically medium to long range volume and shear stresses. The sensitivity of magnetic domain structure and magnetic properties to internal stresses makes pertinent magnetic measurements particularly elucidating. Thus, domain structures have revealed tensile and compressive medium range stress centres as well as long range stress distributions to exist in amorphous metal ribbons [6]. Measurements of the coercive field strength as a function of annealing temperature indicate the removal of internal stresses by structural relaxation [7]. The determination of stress relaxation effects of externally stressed specimens by annealing is a well established technique. It shows that the underlying processes occur in rather close proximity to the crystallization temperature only.

Finally, **decomposition into two or more phases,** i.e. the existence of metastable heterogeneous phase equilibria in amorphous systems, needs to be considered. The formation of multiphase microstructures has been recognized for a long time to occur in oxide glasses [8] and is exploited to produce various special glasses. However, in metallic glasses phase separation has been treated as a curiosity up to now. Yet a number and variety of observations indicate that heterogenous states are frequent in amorphous systems [1] albeit they may have escaped observation in numerous capes so far because of the fine dispersion of phase separation, due to low diffusion rates up to the crystallization temperature, and because two-phase structures were not expected to be present.

MEANS TO AFFECT STRUCTURE AND ENSUING PROPERTY CHANGES

An obvious primary consideration in alloy design of amorphous materials is the selection of alloy components according to basic properties such as low or high melting point in relation to cohesive forces and, thus, strength, propensity to ferromagnetic or non-magnetic coupling, suitability for high surface activity or the formation of passivating films for corrosion protection etc., and one or more glass forming components. A more subtle and as yet unexplored aspect could be to choose the composition in such a way that not only desired basic properties and glass forming ability are obtained but that also a specific structure or microstructural configuration can be formed in the amorphous state either immediately during production or by subsequent treatments. Each as-produced amorphous alloy can undergo

- irreversible and
- reversible structural changes

which are most commonly induced by annealing treatments.

Thermally induced, irreversible structural relaxation from the (production induced) unstable towards a metastable state or to a constrained unstable state of lower energy involves a decrease in excess free volume and an increase in short range order. It can result in monophase or multiphase amorphous states, depending on the alloy system [1].

Structural reversibility in the sense that the degree (and perhaps the type) of metastable short range order is a function of temperature, is a frequent if not common property of metallic glasses. This behaviour has been observed with bond related properties such as Young's modulus E, specific volume V, enthalpy H, electrical properties such as resistivity ρ, and magnetic properties such as Curie temperature T_c, induced anisotropy K_u and magnetostriction λ_s. Structural reversibility is commonly revealed by annealing treatments and simultaneous or subsequent property measurements. If the annealing temperature is alternated, alternating property values are attained. They may either be fully reversible or an irreversible, monotonic property change may be superimposed. In the case of complete reversibility, metastable equlibrium of the structural state can be surmised to prevail. This can serve to determine an energy of formation of thermal disorder. It can be seen from these considerations of reversible structural changes that they can be utilized for property control where required.

A further means of deliberate structure and property control is to induce unstable states under specified conditions. Such treatments may involve, for example irradiation, magnetic or stress field annealing, leading to unstable states which are frozen in. Subsequent heat treatment in the absence of the factor of influence previously imposed reverts the structure back to its initial state of metastable equilibrium.

A most important aspect of deliberate structure control is the kinetics of all of the relevant processes and the effects of structural variables on the kinetics of structural variation. It is obvious that a multitude of interrelations exists between structural states and the ensuing driving force and atomic mobility. However, a treatment of this aspect is beyond the scope of the present paper. Recently, a very lucid and thorough treatment of the kinetics of structural changes in amorphous metals has been developed [24]. It is in keeping with an essential result of numerous experimental studies, namely the occurrence of a broad dispersion of structural relaxation times and a corresponding multitude of kinds and kinetics of thermally activated property changes. These phenomena are well accounted for by the high degree of disorder in the amorphous state.

BASIC STRUCTURE-PROPERTY RELATIONS

A number of fundamental effects of the amorphous structure on the properties of metals has been identified and interpreted by now. Table 1 shows a summary of most of the relations commonly considered. The individual properties will be briefly reviewed below.

The high residual electrical resistivity due to the disordered atomic array of amorphous metals has been discussed, e.g., in a review by Cote and Meisel [10]. They point out that the low or negative temperature coefficient is not a particular feature of the amorphous structure; it complies rather to Mooij's relation [11] which shows that for crystalline alloys an inverse relationship between the electrical resistivity and its temperature coefficient exists as well, with negative coefficients prevailing where $\rho > 150\ \mu\ \Omega cm$. The residual resistivity and the temperature coefficient at elevated temperatures vary with both irreversible and reversible variations of the amorphous structure. Balanzat [12] was first to prove the occurrence of reversible changes by succinct measurements of the comparatively small changes in resistivity. On the other hand phase separation in the Au-Si system was found to be accompanied by rather large changes, up to a factor of 4, in resistivity [13]. They can be attributed to the resulting mixture of a metallic and a semiconducting phase in this particular case.

The magnetic structure-property relations have been investigated most widely [14, 15]. The inherent non-periodic-

Table 1. BASIC STRUCTURE - PROPERTY RELATIONS OF AMORPHOUS METALS

Structural Feature	Property			
	Electrical	Magnetic	Mechanical	Chemical
non-periodic, amorphous structure	high residual resistivity ρ_o; low or negative coefficient $d\rho/dT$	no crystal anisotropy, $K_1 = 0$; low losses by hysteresis (p_h) and eddy current (p_w)	low elastic constants E,G,K; viscous flow	high surface activity; high rate of passivation; high solubility of H
reversible and irreversible changes in short range order	variations of ρ_o and of $d\rho/dT$	variations of K_u, λ_s and B(H)	variations of flow stress σ_o; variations of fracture toughness	
instability of extended defects		low coercive field H_c	high flow stress σ_o; low rate of work hardening $d\sigma/d\varepsilon$	high perfection of passivating layers
phase separation	variations of ρ_o		variations of fracture toughness	

ity of the amorphous structure leads to a corresponding absence of magnetic crystal anisotropy; in conjunction with the absence of strongly interacting structural defects this means that both domain wall motion and rotation of magnetic moments require small magnetic fields. Therefore, the coercive force H_c is small and amorphous metals are fundamentally soft magnetic. However, the magnetization process is affected, also, by induced magnetic and magnetoelastic anisotropies, the structural origins and magnitude of which require consideration. The induced magnetic anisotropy K_u can arise from magnetic field induced or stress induced structural anisotropy which can be caused by local atomic rearrangements at elevated temperature. The magnitude $10 \leq K_u = 10^3 J/m^3$, depending on alloy composition and heat treatment, and direction of K_u vs applied field H_a determine the shape of the hysteresis loop in wide limits. The magneto-elastic interactions which are proportional to the product of effecctive local stress α and magnetostriction λ_s can be reduced by stress relief annealing, i.e. high-temperature irreversible structural relaxation, and by choosing an alloy composition which provides a small magnitude of λ_s. It is interesting to note that λ_s also depends on structure and that it may vary with structural changes both irreversibly and even reversibly [3,21,27]. Finally, losses upon reversal of magnetization are low in amorphous metals due to the low magnitude of H_c, their high electrical resistivity ρ and the low strip thickness necessitated by the cooling conditions of the production process. That is to say that not only hysteresis losses but also eddy current losses are low due to effects of the amorphous structure.

The mechanical properties [18] are determined by two counteracting structural features: (i) the non-periodic packing of the constituent atoms and the inherent free volume cause the elastic moduli to be 20 to 30 % lower than those of comparable crystalline alloys; (ii) the absence of extended defects leads to a lack of nucleation sites for flow and, thus, to a high flow stress α_o as well as to a low or zero rate of work hardening $d\sigma/d\varepsilon$. Variations in flow stress can be caused by changes in short range order as induced by heat treatments. Two variants of plastic flow are observed which are dependent on temperature and applied stress: at low temperatures planar flow, restricted to narrow shear bands, prevails, whereas at high temperatures viscous flow, occurring throughout the stressed volume, dominates. It appears that both deformation processes are accompanied by structural changes: the shear bands of low temperature deformation appear to be zones of lower order indicated, e.g., by a higher rate of chemical attack [19]; the experiments on viscous flow of specimens at higher temperature clearly indicate a decrease in isothermal creep rate which is most likely a consequence of progressing structural relaxation, increasing order, and an ensuing decrease in diffusion coefficient rather than of work hardening. Finally, ductility and proneness to fracture are a function of composition and thermal history. Annealing leads to embrittlement rather frequently. Brittle behaviour may thus be associated either with certain types and degrees of short range ordered structure as such, or with phase separation.

Due to their unstable or metastable state and irregular atomic structure amorphous metals have an inherently higher chemical surface activity than their crystalline counterparts. This leads to high initial rates of dissolution but also to high rates of subsequent passivation if components forming passive layers such as Cr,Ti,Zr or P are present [20]. The passive layers provide a significantly higher corrosion resistance than for corresponding crystalline alloys. This is because of the absence of extended defects and of chemical inhomogeneities such as segregation effects or precipitates, and to an increased rate of dissolution of non-passivating alloy components.

EXAMPLES FOR THE EVIDENCE OF STRUCTURE-PROPERTY RELATIONSHIPS

The design and treatment of amorphous soft-magnetic materials makes most widespread use of structure-property relations implicitly. However, of the general relations outlined above, none has been proven directly by a set of correlated quantitative measurements of structure parameters and properties. Similarly, none of the other kinds of properties has been investigated in a quantitative relation to structure. The difficulties lie primarily in the lacking accessibility of quantitative structural data. Indirect evidence of structural variations with composition and changes with temperature is, therefore, an auxiliary tool to reveal the effects of structure. This will be shown in terms of two examples.

In the course of a systematic investigation of amorphous Fe-Ni-B alloys in a wide concentration range [21,22] the structural relaxation behaviour was studied by a combination of methods. Strong variations with composition were found, e.g., for the heat of structural relaxation and the rate of stress relaxation. Moreover, as fig. 1 shows, these two properties are strongly correlated: at compositions where the heat of structural relaxation is high, the degree of stress relaxation is also high, and vice versa. These observations suggest that in this particular example the variation in boron content leads to a number of different structural states. Specifically, an alloy at $X_B \cong 22$ at.% changes only weakly by annealing whereas an alloy at $X_B \cong$ 25 at.% shows a maximum variation in both properties, i.e. a peak in heat release as well as in degree of stress relaxation. Since we are combining thermodynamic with kinetic data here, a simple correlation with possible structural parameters cannot be established. However, in combination

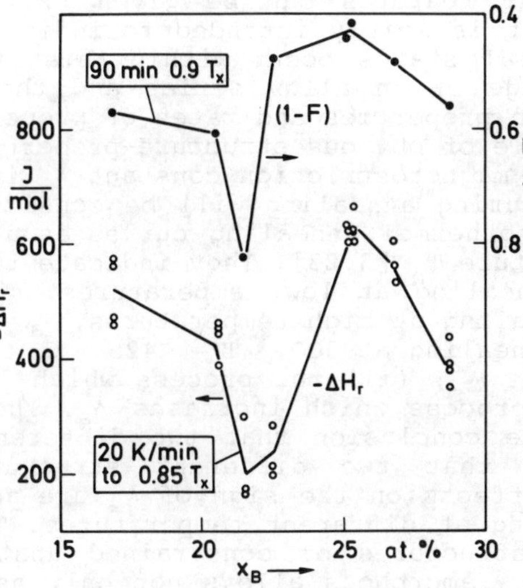

FIGURE 1. Index of residual stress (1-F) after stress relaxation, and relaxation enthalpy-ΔH_r after structural relaxation of amorphous $(Fe_{0.75}Ni_{0.25})_{100-x}B_x$ alloys [21,22].

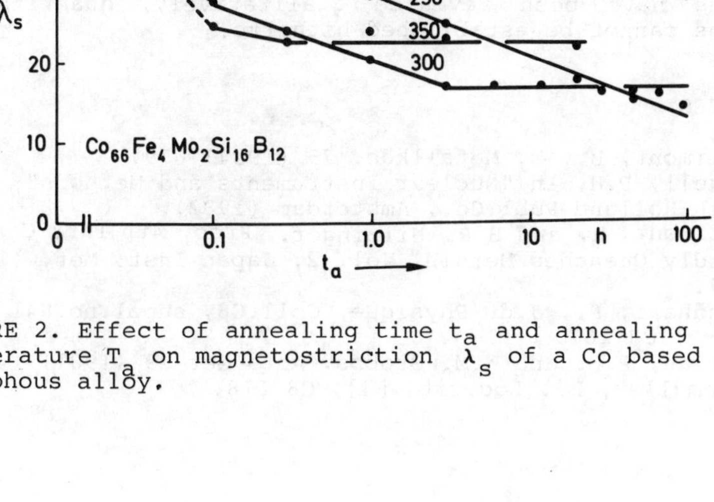

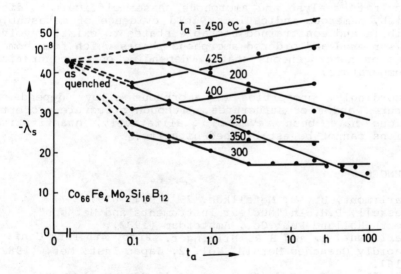

FIGURE 2. Effect of annealing time t_a and annealing temperature T_a on magnetostriction λ_s of a Co based amorphous alloy.

with other observations, an interpretation in terms of particular structural features can be given [21,22]. In the present context it is mainly intended to indicate that more than two structural states occur within this rather narrow concentration range of an alloy series and that they have drastic effects on properties and rates of property changes. As a second example of obvious structure-property relations, the change of the magnetostriction constant λ_s in a Co based amorphous alloy during annealing will be considered. Fig. 2 shows a set of isothermal annealing curves as a function of annealing temperature T_a [3,23]. They indicate two prominent features: (i) annealing at low temperatures, $T_a < 300$ °C, lowers λ_s, annealing at high temperatures, $T_a \geq 450$ °C, increases λ_s and annealing at $300 \leq T_a \leq 425$ °C first lowers and then increases λ_s; (ii) the process which lowers λ_s is faster than the process which increases λ_s. These observations lead to the conclusion that two different processes are involved and that two different structural states, having opposite effects on the sign of λ_s are approached on continued annealing at different temperatures. Thus, different metastable structures or constrained unstable structures are found in amorphous alloys not only as a function of composition but also as a function of temperature and time.

CONCLUSIONS

Amorphous alloy systems show close similarities to crystalline alloy systems. Although amorphous structures are difficult to analyse and amorphous phases difficult to distinguish, numerous indications yield evidence of metastable equilibria and constrained unstable states to exist, involving disordered and ordered amorphous phases which form homogeneous or heterogeneous states, depending on concentration and temperature.

Accordingly, properties of amorphous alloys depend on structures and microstructures. Although structure-property relations have been revealed qualitatively, quantitative relations cannot be established hitherto.

REFERENCES

1. Warlimont, H.. Z. Metallkde. 75 (1984) 679.
2. Gaskell, P.H. In "Nuclear Instruments and Methods", North-Holland Publ.Co., Amsterdam (1982).
3. Warlimont, H. and H.R. Hilzinger. Proc. 4th Int. Conf. Rapidly Quenched Metals, Vol. 2, Japan Inst. Met. (1982) 1167.
4. Chandhari, P.. J.de Physique, Coll.C8, suppl.no.841 (1980) C8-267.
5. Donovan, P.E. and W.M. Stobbs. Acta Met 29 (1981) 1419.
6. Kronmüller, H.. Loc.cit. [4], C8-618.

7. Takahashi, M., T. Miyazaki and A. Watanabe. J. Japan Inst. Met. 43 (1979) 339.
8. Vogel, W. Struktur und Kristallisation der Gläser, VEB Deutscher Verlag für Grundstoffindustrie, Leipzig (1971).
9. Hilzinger, H.R. Loc.cit. [3], 791
10. Cote, P.J. and L.V. Meisel. Glassy Metals I, Springer-Verlag, Berlin/Heidelberg/New York (1981) 141.
11. Mooij, J.H. Phys.stat.sol. A17 (1973) 521.
12. Balanzat, N. Scripta Met. 14 (1980) 173
13. Mangin, Ph., G. Marchal, C. Mourey and Chr. Janot. Phys. Rev. B 21 (1980) 3047.
14. Luborsky, F.E.. Amorphous Metallic Alloys, Butterworths and Co., London/Boston/Toronto (1983).
15. Hilzinger, H.R., A. Mager and H. Warlimont. J.Magn. Magnet.Mater. 9 (1978) 191.
16. Narita, K., J. Yamasaki and H.Fukunaga. J.Appl.Phys.50 (1979) 7591.
17. Kohmoto, O. N. Yamaguchi, K. Ohya, H. Fujishima, T. Ojima. IEEE Trans.magn. MAG 14 (1978) 949.
18. Kovács, I. and J. Lendvai. Rev. Deform. Behaviour Mater. 9 (1982) 5.
19. Pampillo, C.A. and H.S. Chen. Mat.Sci.Eng. 13 (1974) 181.
20. Hashimoto, K. and T. Masumoto. In: Glassy Metals, R. Hasegawa (ed.), CRC Press, Boca Raton (1983).
21. Gordelik, P. Dissertation, Universität Stuttgart (1983).
22. Warlimont, H., P. Gordelik and F. Sommer. To be published.
23. Hilzinger, H.R. Unpublished data.
24. Cunat, C. Thèe, Université de Nancy I (1985).

Appendix

Alloy Design of Amorphous Magnetic Materials

Alloy Design of Amorphous Magnetic Materials (I)

Components

- required intrinsic properties
  - magnetic moment (Fe, Co, Ni, RE ...)
  - other magnetic properties e. g. magnetostriction constant
  - high T_m, high T_x (Nb, Ta, Mo, W ...)
  - propensity to passivation (Cr, P ...)
  - others
- propensity to glass formation
  - ratio of atomic radii
  - bonding, degree of short range order
  - free enthalpy difference liquid/crystal
  - phase equilibria: deep eutectics
  - effects on crystallization kinetics: atomic radii, short range order, viscosity

Table 1

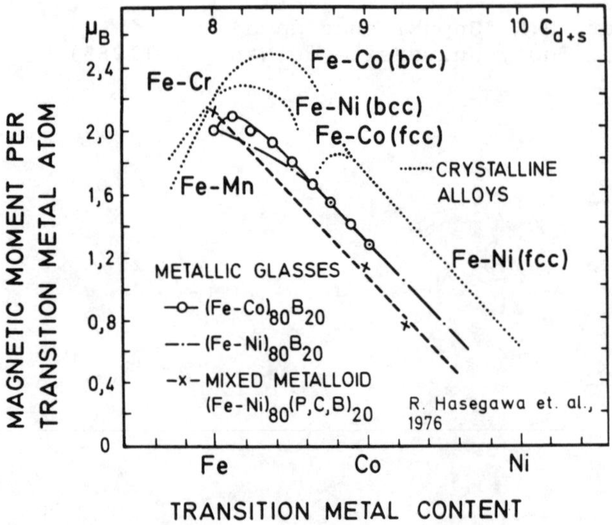

FIGURE 1. Bethe-Slater type relation of the magnetic moment of transition metals in amorphous alloys.

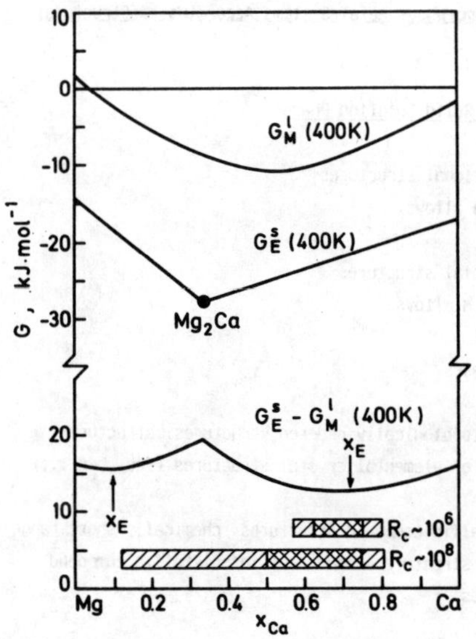

FIGURE 2. Mg-Ca system: free enthalpies of undercooled melt G_M^l and crystalline state G_E^S and their difference at 380 K. X_E - eutectic concentration; R_C-cooling rate K/s. (F. Sommer, Z. Metallkde. 72 (1981) 219).

Alloy Design of Amorphous Magnetic Materials (II)

Effects of Composition, Structures, Defects

- frozen-in state and metastable phase equilibria
  - amorphous phases, structures and phase equilibria
  - phase separation
  - composition dependence of thermodynamic and kinetic properties
- structure and property changes by structural relaxation
  - reversible and irreversible changes in structure
  - dispersion of relaxation times and underlying processes

Table 2

Phases In Amorphous Metal Systems According to Structural Features

Disordered Solid Solution Phases

- substitutional structures
  (e.g. B-B alloys)

- interstitial structures
  (e. g. T-M alloys)

Ordered Phases

- simple, topologically ordered structures; structure elements of elemental crystal structures (fcc, bcc ...)

- intermetallic compound structures, chemically short range ordered; structure elements of intermetallic compound structures

Table 3

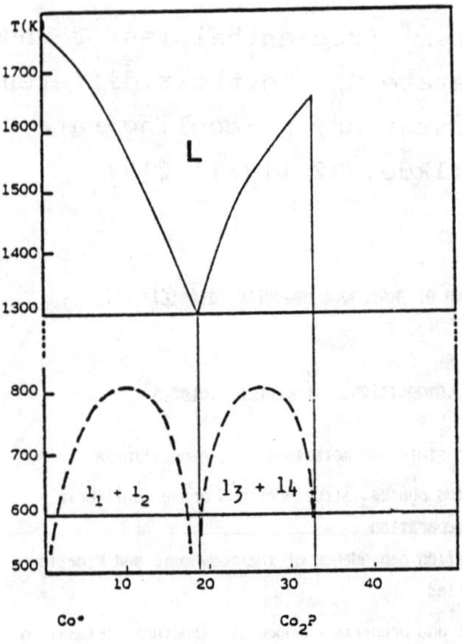

FIGURE 3. Co-P system: stable and hypothetic metastable undercooled liquid phase equilibria
(C. Cunat [24]).

$$(sro)_f = (sro)_f^o + (sro)_f^a + (sro)_f^t + (sro)_f^u + (sro)_f^x$$

anti-
structural
disorder

fully
relaxed   structural
free
volume

thermal uniaxial extra

$$V_f = V_f^o + V_f^s + V_f^t + V_f^x$$

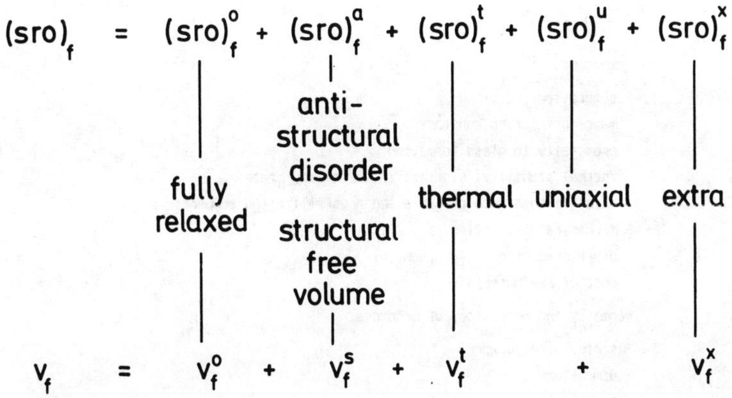

FIGURE 4. (H. Warlimont, H.R. Hilzinger [3]).

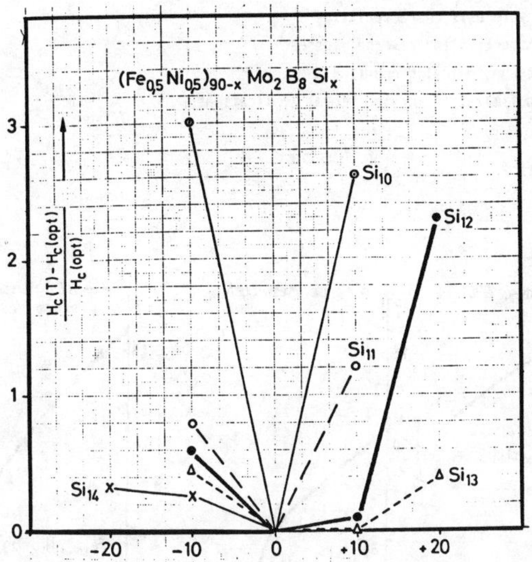

FIGURE 5. Examples of optimum annealing temperature T for minimum coercive force H_c in an alloy series with varying Si content.

Alloy Design of Amorphous Magnetic Materials (III)

Practical Design Aspects

- components
  - saturation induction
  - magnetostriction constant
  - propensity to glass formation
  - thermal stability, kinetics in the glassy state
  - relative magnitude of Curie and crystallization temperature
  - mechanical properties
  - other properties (e.g. chemical)
  - cost of raw materials
- geometry and morphology of products
  - strip, wire, powder
  - dimensions
  - tolerances, morphology
- heat treatments
  - stress relaxation
  - inducement of magnetic anisotropy
  - partial crystallization
- application related properties
  - hysteresis characteristics
  - magnetization reversal losses
  - technological properties
  - optimization for individual applications

Table 4

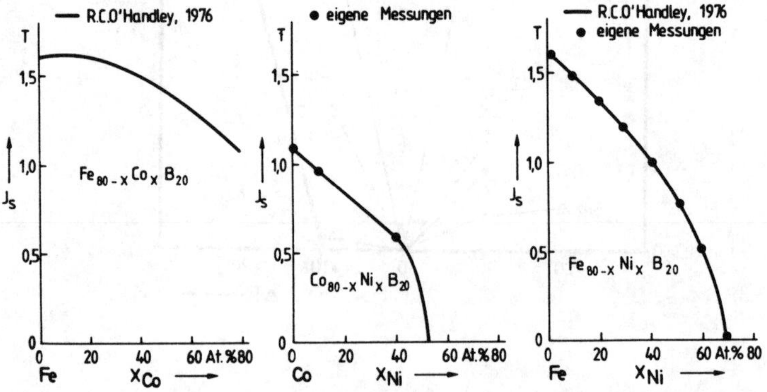

FIGURE 6. Saturation magnetization in $(Fe,Co,Ni)_{80}B_{20}$ alloy series.

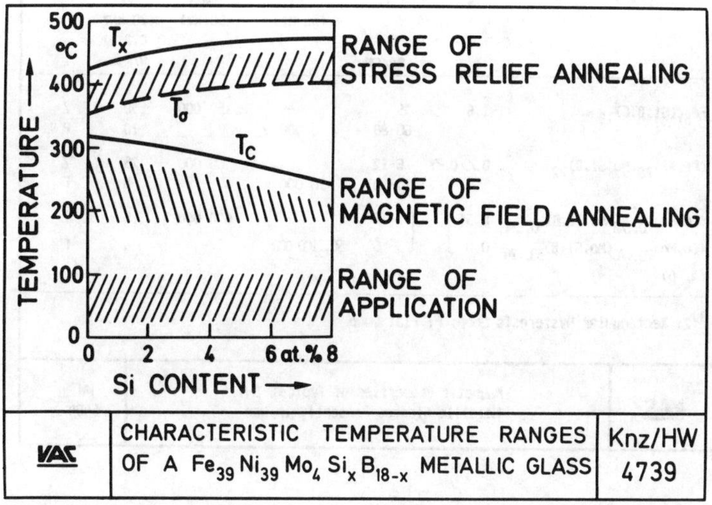

FIGURE 7

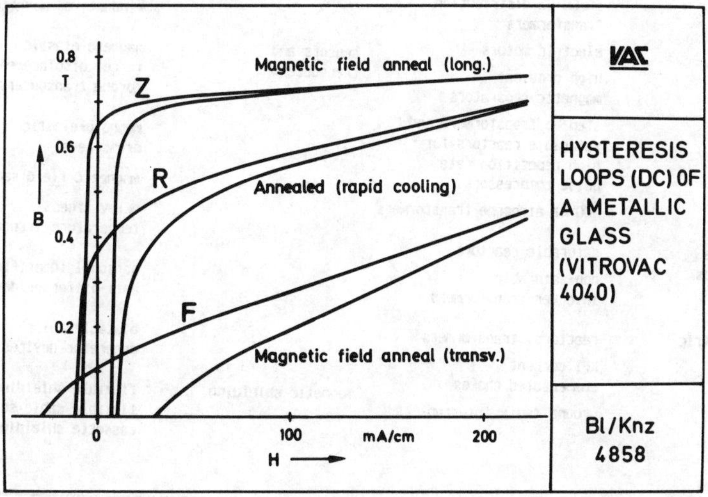

FIGURE 8

Alloy	J_s T	H_c mA/cm	μ_i (50 Hz)	μ_{max} (50 Hz)	p_{Fe} (20 kHz 0.2 T) W/kg	loop[1)]
$Fe_{81}(Si,B,C)_{19}$	1.6	35	-	150 000	30	Z
		60-80	8 000	-	10	F
$(Fe,Ni)_{78}(Mo,Si,B)_{22}$	0.75-0.85	6-12	-	300 000	25	Z
		10-40	10 000	-	8	F
$(Co,Fe)_{70..76}(Mo,Si,B)_{30..24}$	0.55	3- 5	-	600 000	6	Z
$(Co,Mn)_{70..76}(Mo,Si,B)_{30..24}$ (λ_s=0)	0.8	3- 5	50..100 000	-	4	F

1) Z: Rectangular Hysteresis Loop, F: Flat Loop

VAC — Magnetic Properties of Typical T-M Metallic Glasses Presently in Use — HW 5404

Table 5

Field of Application	Specific Application
power electric applications	50/60 Hz distribution transformers
	electric motors
	high gradient magnetic separators
	step-up transformers and saturable reactors for high repetition rate pulse compressors
	400 Hz airborne transformers
switched mode power supplies	saturable reactors
	non-saturable inverter-transformers
various electric circuits	reactors, transformers
	RFI current compensated chokes
	ground fault interrupters

VAC — HW 5405 — Magnetic Applications I

Table 6, I

Field of Application	Specific Application
magnetic heads	audio, video, data (inductive, magneto-resistive)
sensors and transducers	magneto-elastic force, displacement and torque transducers
	magneto-elastic press keys
	magnetic field sensors
	delay lines, temperature transducers, sensors
	personal identification and anti-pilferage devices
	bistable pulse generator devices
magnetic shielding	flexible shielding wrap, flexible cable shielding, cassette shielding spring

VAC — HW 5406 — Magnetic Applications II

Table 6, II

ION IMPLANTATION AND AMORPHISATION IN METALLIC SYSTEMS

R. BRENIER, J. DUPUY

Département de Physique des Matériaux (Laboratoire associé au C.N.R.S.)
Université Claude Bernard - Lyon I, 43 Boulevard du 11 Novembre 1918,
69622 Villeurbanne Cédex

1. INTRODUCTION

The applications of ion implantation and ion beam mixing to the improvement of surface properties of metals are now well established.

High dose ion implantation is a suitable technique for surface amorphisation or more generally surface treatment of metals. It consists in the insertion of atoms, having a kinetic energy, in a target. The doping process occurs to be out of thermodynamic equilibrium and is not limited in concentration (figure 1). The implantation of atoms at a definite depth in a target is always accompanied by an energy deposition which can result in defect creation along the particle-trajectories, and hence in amorphisation for high energy deposition. It has to be noticed that defects in metals are created by the elastic collisions of the particles with the atoms of the target, while the electronic excitations or ionizations cannot lead to defect creation. Useful applications of high dose implantation for the improvement of the surface properties of metals appear to exist in wear and corrosion.

However, high irradiation doses (of the order of 10^{22} ions $\cdot$ m^{-2}) are a limitation of the method with respect to the duration of the implantations and to the problem of surface sputtering of the target.

In order to remedy these disadvantages an original technique for surface treatment of metals using an ion beam has been developed recently. It is called ion beam mixing. An ion beam is mixed either with a thin film deposited on a substrate or with a multilayer of different compounds separated by interfaces (A-B-A-B ... for example).

The principal aim of this paper is to report on ion beam mixing in metallic compounds and includes:

1. An outline of the understanding of the basic processes of particle interactions with solids.
2. Selected experimental results and different models for ion beam mixing.
3. Empirical glass forming ability criteria associated with ion beam mixing.
4. New results and applications in high dose implantation and ion beam mixing.

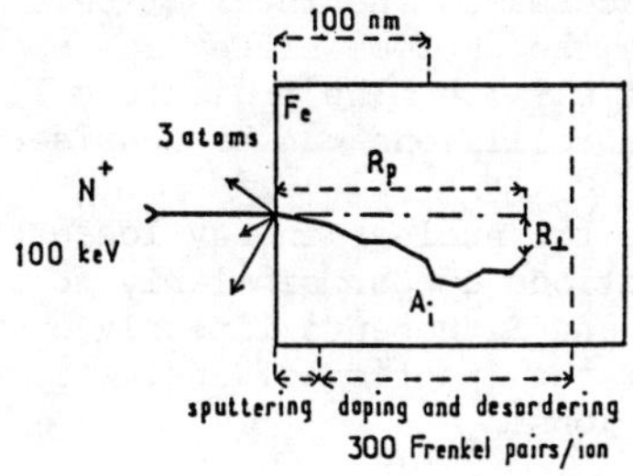

FIGURE 1.

2. BASIC PROCESSES IN THE PARTICLE MATTER-INTERACTIONS

2.1 The energy losses

An ion penetrating into a metal interacts with both the free and bound electrons, and with the nuclei. This leads to a modification of its trajectory, its electrical charge and to a decrease of its energy. Two contributions are mainly responsible for the ion energy losses along the path R (figure 1):

- the nuclei-ion interactions expressed by the nuclear stopping power $S_n = dE_n/dR$.
- the electrons-ion interactions expressed by the electronic stopping power $S_e = dE_e/dR$.

For an energy range lower than 200 keV/nucleon the calculation of these two quantities is based on the LSS theory (1,2,3). The nuclear interaction is described by a series of binary elastic collisions between the incident ion and the nuclei which are more or less screened by their electronic shell. The interaction potential is then a Coulomb potential screened by a Thomas-Fermi function. The result is shown in figure 2 in reduced units and in the case of Xe^+ ions moving in an iron target.

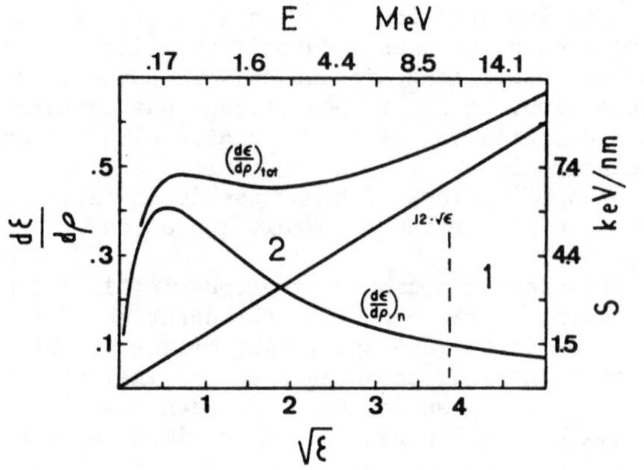

FIGURE 2.

In zone 1 the ion energy is high enough to allow small ion-nucleus approach distances, so that the potential is dominated by the Coulomb effect. In this case the weak energy transfers occur with a larger probability, hence the small angle deviations are the most likely. The ion trajectory is nearly a straight line and the nuclear stopping power is low but increases when the energy decreases. At lower ion energy (zone 2) the electronic screening effect is no longer weak. It decreases the interaction all the more because the ion energy is lower. Hence the Coulomb effect is compensated and the $S_n(E)$ curve shows a maximum at $\varepsilon \simeq 1/4$. The trajectory is also less straight. At the end of the ion path the collisions can be considered as hard sphere isotropic collisions.

In the energy range where the nuclear energy losses are important, the inelastic ion-electron interactions appear similarly to an ion slowing down in a viscous surrounding and S_e depends linearly on the ion speed (4). It is important to notice that the fraction of inelastic energy losses is less for heavy than for light ions.

2.2 The ion ranges

At the end of the slowing down processes the ion stops in the material in a random position. Different parameters can be defined on each individual path (figure 1): the total range $R = \Sigma A_i A_{i+1}$, the projected range R_p and the transverse range $R_\perp$. Because of the statistical nature of the processes the ions are implanted in the material according to a depth distribution which is experimentally nearly gaussian. Hence, we are interested in the mean values R, R_p and $R_\perp$, and in the corresponding range stragglings ΔR, ΔR_p and $\Delta R_\perp$. R_p and ΔR_p are easily measurable (figure 3).

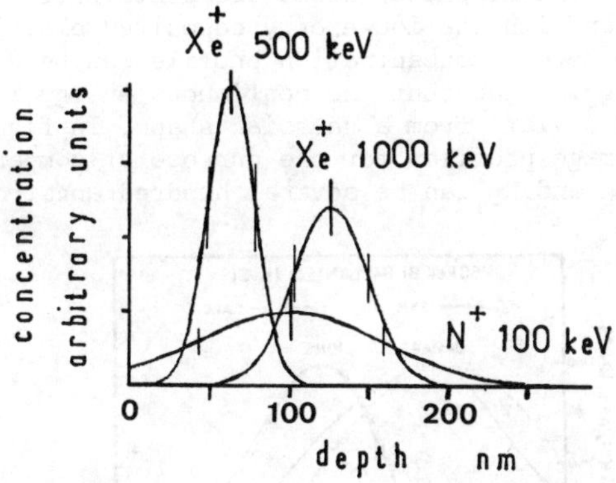

FIGURE 3.

Reference (4) gives two semi-empirical relations derived from the previously mentioned theory expressing R_p as function of the incident and the target features. As illustrated in figure 3 these relations show that an ion is penetrating all the more the more energetic and the lighter it is.For low energies ($\varepsilon < 0.1$), $\Delta R_p/R_p$ depends only on the mass ratio of the target and projectile $\mu = M_t/M_p$ (figure 4). Deeply implanted ions spread more than shallow ones.

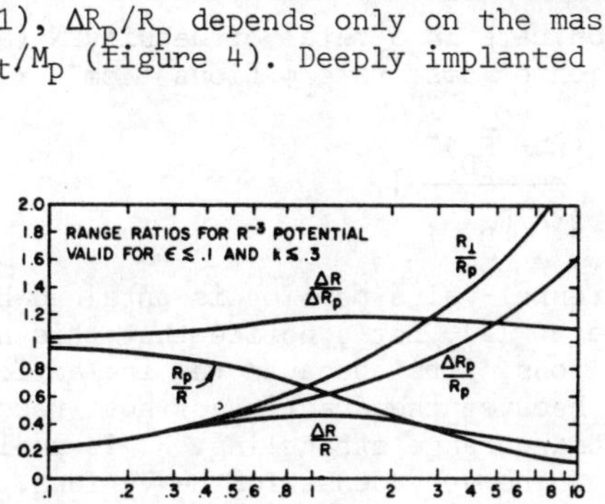

FIGURE 4.

2.3 The damages

Along the ion path the ion-atoms knockons can remove the atoms of the material from their equilibrium positions. In the Kinchin-Pease model (5) an atom is displaced if it receives a recoil energy bigger than a threshold value E_d. It can in turn remove another atom, and so on. As the displacement ranges are no more than a few interatomic distances, the defects occur in a limited zone around the ion path: the collisional cascade (≈ 100 Å). In metals the mean value for E_d is about 25 eV (6). In this purely collisional scheme, Kinchin and Pease evaluated the number of Frenkel-pairs per primary knockon which transfer the recoil energy E_r to be $\nu(E_r) = E_r/2E_d$ ($E_r > 2 \cdot E_d$). This calculation is based on hard sphere collisions. It has been improved by considering more realistic potentials by Robinson (7, 8). A better relation is the expression $\nu(E_r) = K\cdot(E_r/2E_d)$(1) where K is an efficiency factor lying between 0 and 1 (0 for a pure Coulomb potential). As the majority of the collisions occur at low energy K = 0.8 is a reasonable value (9). Hence the displaced atoms are distributed along the ion path in close relation with the ion energy deposited elastically.

In this way the defect concentration profile can be calculated from the LSS theory taking into account the continuous energy losses. These profiles (10,11) can deviate from a gaussian shape. In figure 5 we give the implanted and the damage profiles for the purpose of comparison (11). The difference between R_p and R_D can be several hundred angstroms.

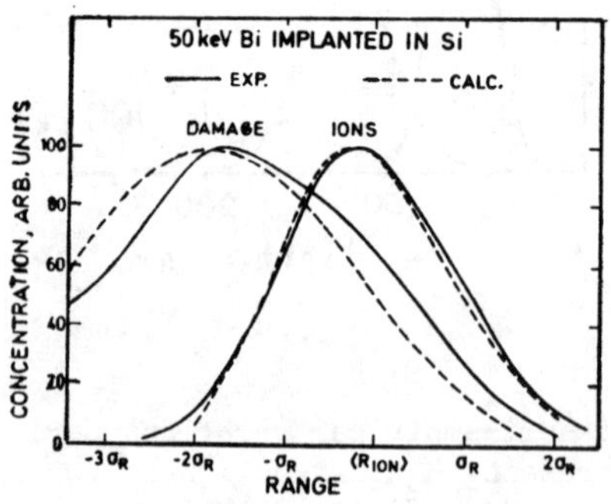

FIGURE 5.

The defect production rate in a metal of density N (atoms · cm^{-3}) produced by an ion irradiation of dose rate $\dot{\phi}$ (ions · cm^{-2} · s^{-1}) can be written as:

$$\dot{G}(x) = \frac{0.4\nu\dot{\phi}}{N\Delta R_D} \exp\left[-\frac{(x-\bar{R}_D)^2}{2\cdot\Delta R_D^2}\right]. \qquad (1)$$

The total number ν of Frenkel-pairs per ion is obtained by integrating equation (1) along the path (12). Let's notice that this number is larger for heavy than for light ions, first because the inelastic loss contribution is less, and second because the fraction of host recoils below $2\cdot E_d$ is smaller. The projected damage range straggling ΔR_p is smaller too, so that the collisional cascade can be very dense for heavy ions. This fact can lead to nonlinear and annealing effects and to discrepancies with this simple model. This has been shown by Averbach (13) who introduced an efficiency factor between the experimental and theoretical values of $\nu(\varepsilon_r = \nu^e/\nu^{th})$ lying between 1 and 0.3. For example in the modified Kinchin-Pease model Cu-ions of energy 500 keV produce about ν = 3400 Frenkel pairs/ion in a copper matrix (ε_r = 0.34). With a dose rate $\dot{\phi} = 10^{12} Cu^+ \cdot cm^{-2} \cdot s^{-1}$ the defect production rate is $\dot{G}(R_p) \simeq 10^{-2}$ dpa · s^{-1} (displacement per atom · s^{-1}) which leads to an accumulation of $G(R_p) = 10^2$ dpa for a dose of 10^{16} $Cu^+ \cdot cm^{-2}$. At such doses the defect concentration saturates due to the overlapping of the cascades. The profile curve shows a flat maximum (14). Near the surface the nuclear knockons are responsible for the sputtering of the target atoms. The sputtering yield can be calculated from equations given by reference (15).

3. ION BEAM MIXING

3.1 Definition (16,17,18,19)

Ion beam mixing consists of irradiating an inhomogeneous target with an ion beam of a rare gas (generally Ar^+, Kr^+ or Xe^+) which are free from chemical doping effects. The target multilayer configuration (figure 6a) composed of successive films of two elements A and B which are deposited on a substrate, is used to produce new phases. For this purpose the ion beam induces a spatial homogeneisation of the system where the overall composition A_xB_{1-x} is predetermined by the thickness of the monolayers (≈ 10 nm). The required irradiation doses are lower than the ones needed in ion implantation technique (a few 10^{16} ions · cm^{-2} at most), so ion beam mixing is free of high irradiation dose shortcomings (irradiation time and sputtering). To study the basic mechanisms involved in the interface A - B, the simpler bilayer configuration is more convenient (figure 6b). Choosing the A-atoms heavier than B, Rutherford Backscattering Spectrometry of α-particles (RBS) allows to get spatial information on the mixed interface. This interface can be characterized by a mixing parameter Q defined as the number of mixed B-atoms directly measured in the RBS-spectrum.

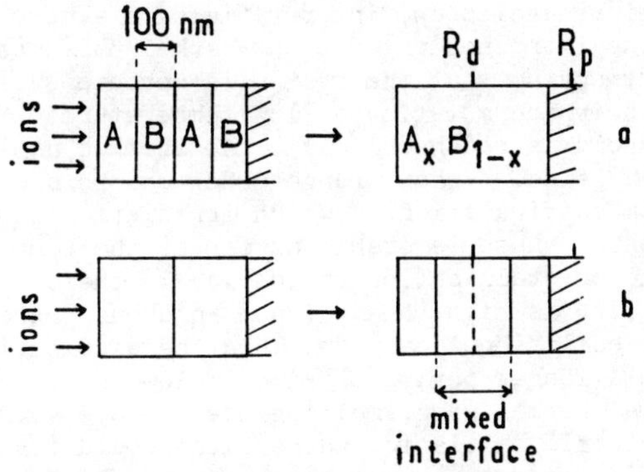

FIGURE 6a and 6b.

3.2 Classification of the models

The fundamental processes described in Section 2 exhibit three important features: they are independent of the temperature T, of the incident ion-flux ϕ (intracascade phenomena) and of the number ν of created defects. The value ν varies linearly with the deposited energy density F_D. Nevertheless the experimental agreement is good only if T and F_D are low (20-21) Let's give an experimental evidence (22) for such a discrepancy by considering Nb-Si bilayers, mixed at different temperatures. In such a target configuration the inhomogeneity of the abrupt interface Nb-Si is not expected to be an intrinsic and spontaneous reason for interdiffusion in the purely ballistic framework. Indeed, the results (Q-curve) (figure 7) reveal that the mixing parameter Q shows two temperature regimes: below a critical T_c, Q is nearly independent of temperature while above T_c, Q increases rapidly according to an Arrhenius-law.

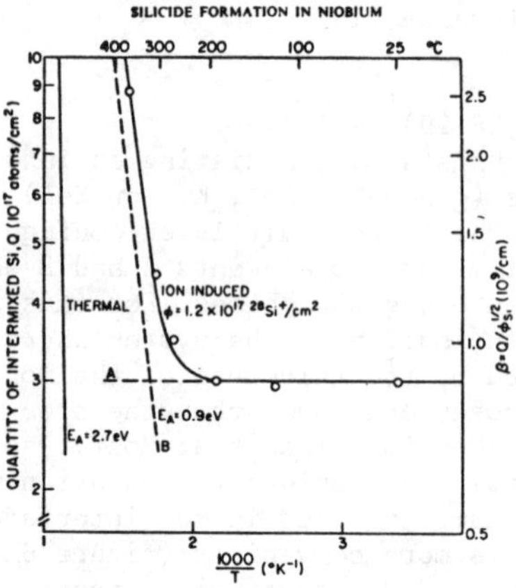

FIGURE 7.

Temperature dependent regimes clearly suggest that some thermally assisted mechanisms are involved in addition to atomic ballistic transport. The following rough scheme can be conceived: ion irradiation yields an equal quantity of vacancies and interstitials in a very short time ($\lesssim$ 1 ps) in a material. The introduction of each Frenkel-pair modifies the crystal-enthalpy and -entropy so that the system is moved away from thermodynamic equilibrium by their accumulation at low temperatures. Nevertheless, when a critical threshold is reached locally the thermodynamic forces slightly relax the system. Hence, a short range order can be built up instead of the completely random configuration to which irradiation should ineluctably lead. This amorphous phase is stable as far as the temperature is not increased above the crystallisation threshold. In the case where the deposited energy density is high (heavy ions) knockons between moving atoms can occur in the collisional cascade. Then the dissipation of the ion energy can last over a longer period of time ($\lesssim$ 100 ps): the displacement spike. An analogy with a thermal wave, melting the cascade volume has been proposed. This hypothetic ballistic undercooling could involve 10^{14} Ks^{-1} as a cooling rate. This leads to an amorphous phase. When the irradiation is performed at a temperature high enough to allow long range migration of the created defects, interstitials and vacancies intercascade fluxes occur for times ranging between 1 ps and a few hours. The defect transport greatly enhances their elimination by mutual recombination or at sinks. Very often these processes lead to the formation of a metastable crystalline phase.

In such a scheme it is convenient to classify the models of ion beam mixing according to the characteristic interaction time of the processes they involve (23). Table 1 gives this classification.

3.3 Ballistic models

On the basis of an analogy to diffusion in a gas (35), Andersen (28) has proposed a model in which the atomic motions under irradiation are macroscopically characterized by an effective diffusion coefficient $D \sim F_D \dot{\phi}/E_d$. The proportionality to the density of the deposited energy F_D and to the ion-flux $\dot{\phi}$ has been verified by performing Kr^+-irradiations on a Pt-marker embedded in a Si-matrix at low temperatures (36). Figures 8 and

9 offer evidence for the excellent experimental agreement.

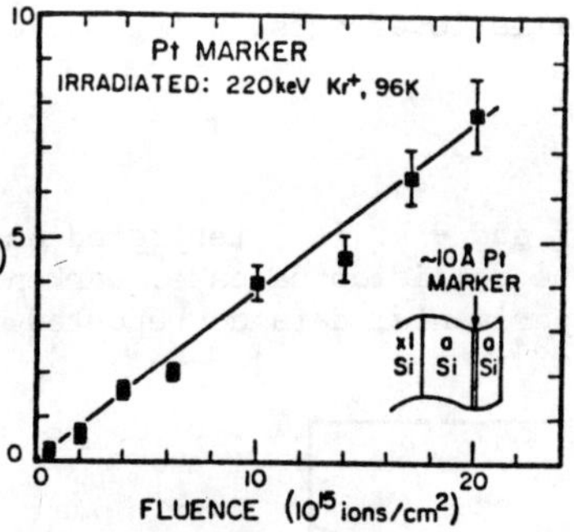

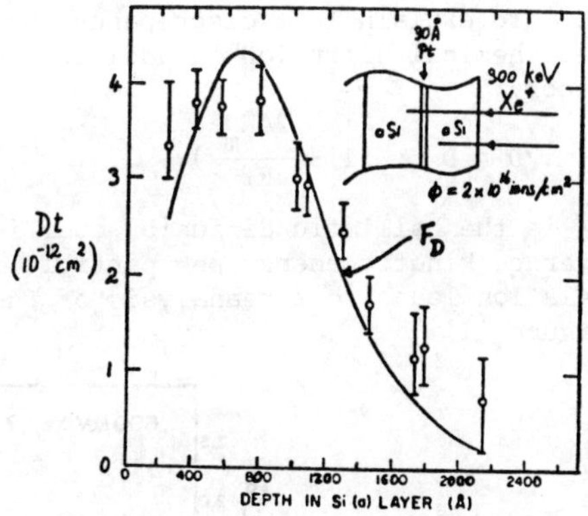

FIGURE 8 AND FIGURE 9.

Sigmund and Gras-Marti (27) described more microscopically the development of a cascade following the primary ion-atom knockons (isotropic cascade mixing) and the atomic recoils in head-on ion-atom collisions (recoil mixing). Nevertheless, the experiment shows quantitative discrepancies with these calculations (37).

3.4 High density cascades

In the case where the deposited energy density in the collisional cascade is high the previous simple ballistic picture becomes insufficient. Cheng et al. (32) performed a mixing of a Pt-M and Au-M interface with Xe^{++}-ions at L-N_2 temperature. M holds for transition metals (Ti,V,Cr,Mn, Co,Ni), elements chosen for their similarity in atomic numbers and masses, and for covering a wide range of heat of alloy formation ΔH_m at the nominal 1-1 composition. The spreading Dt of each bilayer interface as a function of irradiation dose ϕ is reported in figure 10. The linearity Dt = pϕ

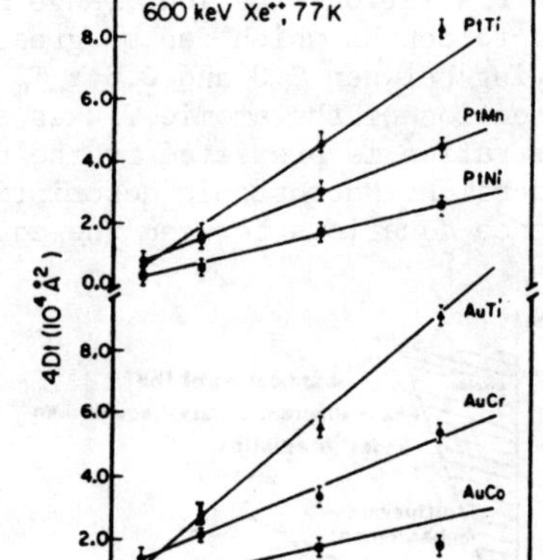

FIGURE 10.

assumed in the collisional picture is verified, but the different slopes p are not expected.

To explain this discrepancy the authors propose to take into account the chemical diffusion in addition to ballistic effects, using Darken's theory:

$$D = D_o \cdot \left(1 - \frac{2\Delta H_m}{RT}\right). \tag{2}$$

D_o is the ballistic diffusion coefficient and $\frac{3}{2}$ kT is interpreted as the average kinetic energy per particle in the collision cascade. Darken's relation leads to a reanalysis of the experimental data as reported in figure 11.

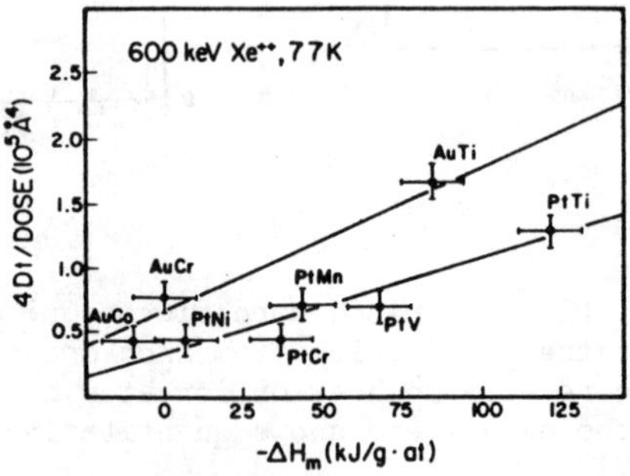

FIGURE 11.

The existence of two slopes is the evidence that the model is not sufficient. Nevertheless, the temperature deduced from the slopes is of the order of magnitude of 1 eV/at (10^4 K). Hence,this fact suggests that these chemical effects should play a role in the last stage of a dense cascade development.

3.5 The delayed processes

Since a long time, examples are known giving evidence of an enhancement of atomic diffusion (RED) (33), alloy ordering (38), or segregation in a solid solution (39) under irradiation. The long range redistribution of alloying components occurs to depths which can be greater than the ion range, for temperatures lying between 0.3 and 0.6 • T_m (T_m: absolute melting temperature). The direction of the atomic fluxes against a given gradient in the defect concentration is predicted by the solute size rule reported by Okomoto and Wiedersich. Macroscopic descriptions of these processes are founded on balance equations between the equilibrium defects,

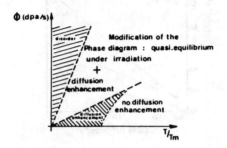

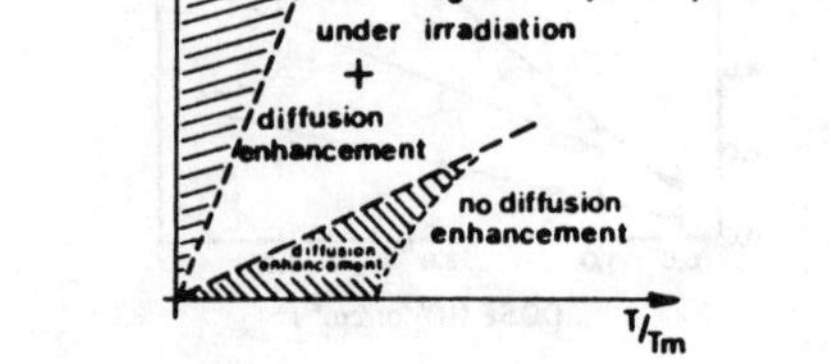

FIGURE 12.

the irradiation damage production enhancing the atomic mobility, and the defect elimination by mutual recombination or at sinks (33).

Following this scheme Adda et al. proposed a flux-temperature diagram containing all possible consequences introduced by irradiation (figure 12) (40).

Under high flux and low temperature the defect supersaturation is important since their production rate is high while their elimination is slow because of the weak atomic mobility. So this domain is priviliged for amorphous phase formation. At both low flux and temperature the irradiation effect is not important enough to profoundly modify the phase diagram. Atomic diffusion is only enhanced by irradiation (RED). Where both damage production and elimination are important, quasi-equilibrium conditions can exist leading to strong modifications in the phase diagrams (41). If the equilibrium defect concentration is high, the irradiation defect supersaturation does not exist any longer and no irradiation effect is to be expected.

A more sophisticated theoretical work dealing with the phase stability under irradiation is Martin's model (42) founded on the competition between ballistic mixing and irradiation enhanced diffusion driving the system back to low energy configurations. This leads to a steady state given by a "law of corresponding states": the equilibrium configuration of a solid solution under an irradiation at temperature T and flux $\dot{\phi}$ is the configuration the system would have outside irradiation ($\dot{\phi}$ = 0) at a temperature $T^* = T\cdot(1+\Delta)$. Δ is a function of T and $\dot{\phi}$.

The dependence of the effective temperature T* on T and $\dot{\phi}$ is shown in figure 13 in the case of regular solid-solutions.

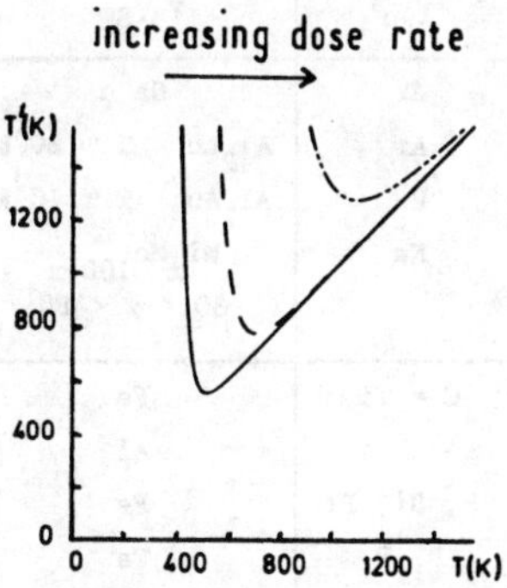

FIGURE 13.

We can notice that increasing the irradiation flux or decreasing the

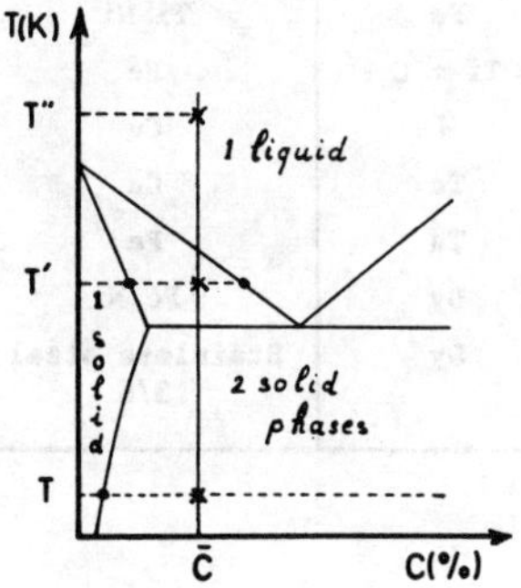

FIGURE 14.

Process lasting	Feature	References
< 1 ps prompt processes	- athermal - high energies - ballistic mixing	- cascades : ROUSH (24) ; WELLER (25) WALKER (26) - isotropic cascade mixing : SIGMUND (27) - recoil mixing : SIGMUND (27) - ballistic diffusion : ANDERSEN (28) ; MATTESON (29)
1 ps < t < 100 ps	- thermally assisted - intermediate energy	- energy spikes : SIGMUND (30) - thermal spikes : VINEYARD (31) - thermochemical effects : (32)
1 ps < t < 1 hour delayed processes	- thermally assisted - low energy	- radiation enhanced diffusion : (33) - radiation induced segregation : MARTIN (34)

TABLE 1.

	Ion	Target	Ref.
Inert gas	Ar	Ga α	62
	Ar	Al_2Au T ∿ 80°K	63
	He	Al_2Au T ∿ 10°K	
	Ne	Ni_xMo_{100-x} 30 < x < 80	65
Metalloïds	C = Ti	Fe	64
	P	Al	66
	B, Si, Pt	Fe	
	B, P	Ta	64
Metals	Ni	Ti Ni	67
	Si	Ti Ni	
	Fe	Ti Ni	
	Ti + C	Fe	68
	W	Cu	69
	Ta	Cu	
	Ta	Fe	
	Dy	Fe Ni	
	Dy	Stainless steel 3/6	

TABLE 2.

sample temperature have the same effect. The inefficiency of high irradiation temperatures is also assumed (T* = T). Figure 14 shows a possible application of this calculation to a binary phase diagram. Complete amorphisation occurs when T* lies in the liquid phase for a given composition.

4. EMPIRICAL GLASS FORMING ABILITY CRITERIA AND ION BEAM MIXING

By which criteria it can be judged whether a material can be amorphized by a given procedure is still an open question. Nevertheless empirical correlations or rules exist, mostly involving the physical and thermochemical parameters of the stability for the liquid and solid phases: atomic radii (43,44), ionicity (45), Miedema's coordinate (46,47) and electronic density at the Fermi level for metallic alloys (48,49,50). Sometimes these rules are only useful for a special technique or class of materials. Hence, ion implantation and ion beam mixing have their own ones. For example ion implantation in metal-non metal alloys are successfully sorted in a two coordinate-map by Rauschenbach and Hohmuth (51).

In this paragraph we will especially pay attention to Liu's structural rule (52), to the Alonso and Simozar criterion (53) and to correlations which can be established with the use of binary phase diagrams (54,55).

- Liu's structural rule

This empirical rule states that an amorphous phase can be obtained by ion beam mixing of multilayered systems if the two constituents have different crystalline structures, (frustration in the crystallisation process) whatever their atomic sizes and electronegativity differences are. These two last points are illustrated by Liu choosing six pairs of two elements A and B representing three possible combinations of structure: hcp/bcc, hcp/fcc, bcc/fcc, and covering a wide range in electronegativity and atomic size differences between A and B. In each case an amorphous phase is obtained. In order to apply this rule successfully, iron must be considered of having a fcc structure instead of bcc.

Nevertheless some exceptions seem still to exist: Co-Cu (hcp/fcc) and Co-Ag (hcp/fcc) have not been amorphised by ion mixing yet.

- Alonso and Simozar criterion

These authors propose to use as parameters, which determine the glass forming ability of metal-metal or metal-semiconductors pairs by ion mixing, the atomic radius ratio r_A/r_B and the heat of compound formation ΔH at the equiatomic composition. The latter has to be calculated according to Miedema's model (56). All the pairs which have been examined by ion mixing

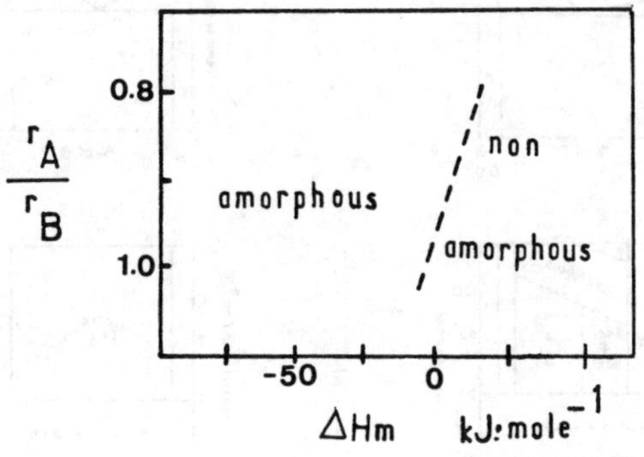

FIGURE 15.

are placed in a two coordinate map (figure 15). An empirical straight line can then be drawn between an amorphous and a non-amorphous zone.

In this work the glass forming ability appears to be strongly correlated with the existence of equilibrium compounds ($\Delta H < 0$). The binary phase diagrams of the seven non-amorphous alloys do not contain any ordered compound. So the amorphous phase and the ordered compounds seem to be competitors.

• Correlations with binary phase diagrams

Phase transformations under irradiation are phenomena far away from thermodynamic equilibrium. Nevertheless taking into account that thermochemical forces can be restoring forces under quasi-equilibrium conditions, correlations between the phases formed by ion mixing and the phase diagrams can be expected. In this picture, experimental data give evidence for the importance of the mixing temperature, the mutual solubility of the constituents, the existing intermediate phases and for their crystal structures. The latter ones can favor amorphisation for kinetic reasons if they are too complicated (57).

• Influence of a specific compound

In the case of transition metal-silicon bilayers the expected correlation is clear: ion mixing and thermal annealing lead to the same phases. The first phase formed is the highest congruently melting compound near the lowest eutectic composition (Walser and Bene rule). For example in the Ni-Si system it is Ni_2Si.

Nevertheless, this result cannot always be extended to metal-metal systems. As for the metal-metal multilayered systems the situation is complicated too. For example the case of Ni-Mo shows that the overall composition of the multilayer, corresponding to the compound defined by NiMo, is the best one for the formation of an amorphous phase: the required dose is lower than for other compositions (54). The presence of a well defined compound in the multilayer, formed first by preannealing, can prevent the system from amorphisation: the Fe-W system gives an amorphous phase except in the situation when the Fe_7W_6 compound is formed by preannealing (58a). But this is not a general rule: irradiation at room temperature to doses of $5 \cdot 10^{15}$ Xe^+/cm^2 induces amorphisation in the Ti-Au system, both in the preannealed samples containing TiAu and Ti_3Au compounds, and in the as-

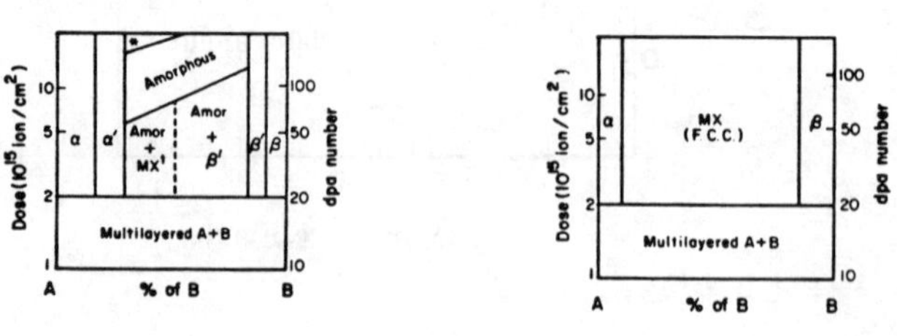

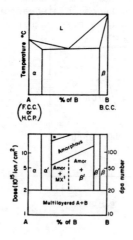

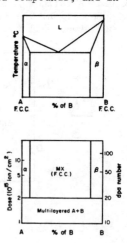

FIGURE 16.

deposited samples (58b). Thus, in some systems (FeW, PtAl (57)) the amorphous phase and well-defined compounds appear to be competitors.

• Influence of an eutectic

Many experimental investigations can be correlated with the existence of an eutectic. This is illustrated by Liu et al (55) using schematic simple eutectic diagrams (figure 16). If the two constituents have the same crystal structure, a metastable crystalline phase (Mx) showing the structure of the parent phases, is obtained after ion mixing in the central part of the diagram. If the two constituents have different crystal structures extended supersaturated solid solutions are obtained near the poles A and B. An amorphous phase is obtained in the vicinity of the eutectic if the dose is sufficiently high. This helpful influence of an eutectic is already known in rapid quenching techniques.

The zones near the poles remain crystalline: large stable solid solutions do not favor amorphization. This can be illustrated in the system Au-Ni showing high solubility (miscibility gap in the solid state only at low temperature) which cannot be amorphised by ion mixing at 77 K (58c). On the other hand very immiscible systems like Ag-Ni or Cu-W (59,60) mix only with difficulties under irradiation.

So, once a binary system is chosen, amorphization is favored if the overall composition of the multilayer is fixed away from regions with most stable compounds, in the vicinity of a deep eutectic and in a two phase region of the phase diagram. Nevertheless, phase diagrams do not allow to interpret the mixing behaviour of some systems: Si-Au and Si-Ag have similar phase diagrams but exhibit very different mixing characteristics (61).

5. NEW RESULTS AND APPLICATIONS

A non exhaustive list of new amorphous materials formed by ion implantation and ion beam mixing (IBM) are given in tables 2 and 3. The amorphization is produced under the general conditions given previously: low temperature (RT or LNT) and multilayer configuration which was proved more efficient for IBM.

Because IBM and ion implantation affect only small layers of material, specific properties which are characteristic of the surface, like resistance to chemical or electrochemical damages (corrosion and pitting) and resistance to mechanical damages (wear and friction) can be improved by these techniques. Moreover a very gradual interface between the bombarded layer and the bulk makes the adhesive property unquestionable.

Until recently it was not proved that the enhancement of the wear and corrosion resistance was related to an amorphization process. We have deliberately chosen illustrations where properties and a complete characterization have been discussed.

5.1 Ion implantation

5.1.1 Titanium ion implantation of AISI 52100 steel (72-75)

Several appropriate methods have been used to characterize the depth distribution of different elements: Auger analysis and nuclear reactions like (p,γ) for the titanium profile and (^{3}He, ^{4}He) for the carbon profile. This latter element is introduced during implantation performed in an uncleaned vacuum. Transmission electron microscopy (TEM) and channeling RBS allowed to show that the presence of slight carbon traces promote the formation of an amorphous Fe-TiC layer (figure 17) achieved under well defined conditions: the lower limit concentrations of carbon and titanium are respectively $4 \pm 2\%$ and 20% at..

An improvement in tribological behaviour as compared to steel has been obtained with high fluences ($5.10^{17} Ti^+/cm^2$). Friction and wear measurements with different lubrication tests are given in figures 18, 19 and 20.

Mixing of multilayers			Ref
Cu Ta, Ni Mo, Ni Nb, Al N_b, W Fe, Mo Co, Mo Ru, Co Cu, Co Au, Co Ag, Ti Au, Er Ni, Ti Ni, Co Ag.			64
Co Al Ni Al	75 K		70
$Zr_x Ru_{1-x}$	{ x: 75, 25	RT	71

TABLE 3.

		Critical current density ($\mu A/cm^2$)	Passivation current density ($\mu A/cm^2$) in 1M HNO_3 Sol
Mo Ni	Co-deposited film	8.10^{14}	10^3
	Amorphous	3.10^3	2.10^3
	Co-deposited film	3.10^3	2.10^2
Ti Ni	Amorphous	10^3	10
	Crystalline (annealed above Ti)	2.10^{15}	7.10^2

TABLE 4.

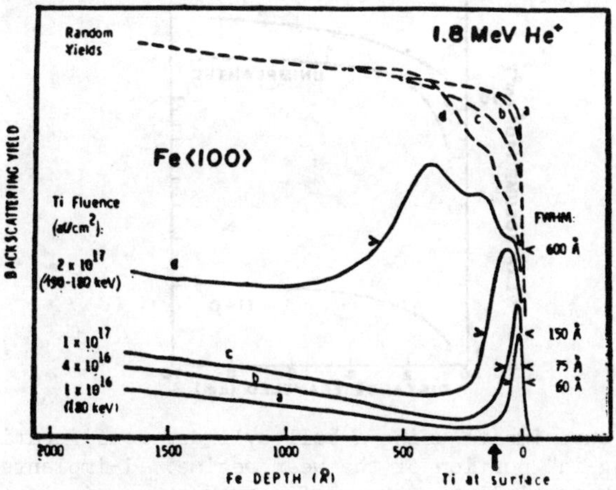

FIGURE 17. Helium-ion channeling spectra.

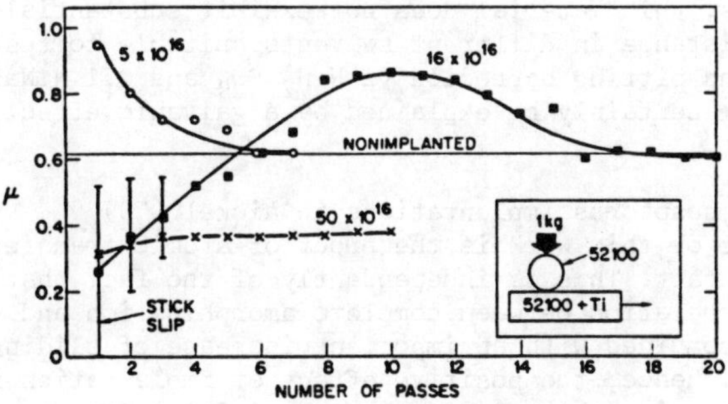

FIGURE 18. Coefficient of dry sliding friction versus number of passes for 52100 steels implanted with Ti.

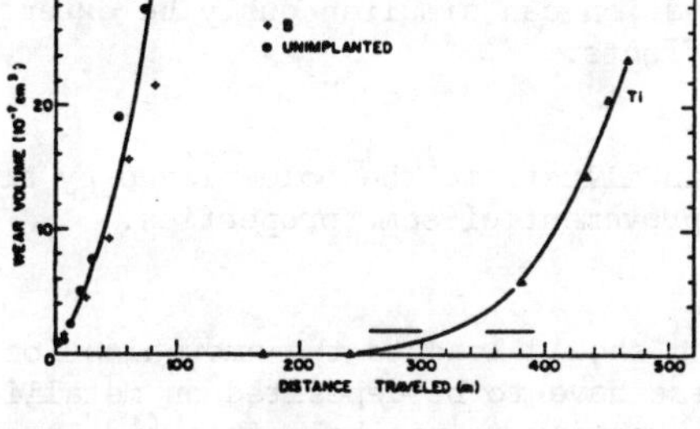

FIGURE 19. Wear volume for AISE-52100 ball on 52100 disk surfaces unimplanted, implanted with B, and implanted with 5 x 10^{17} Ti-ions/cm^2. The arrows beside the Ti curve indicate that the onset of severe wear is variable but begins at substantially larger distances than in the unimplanted specimen.

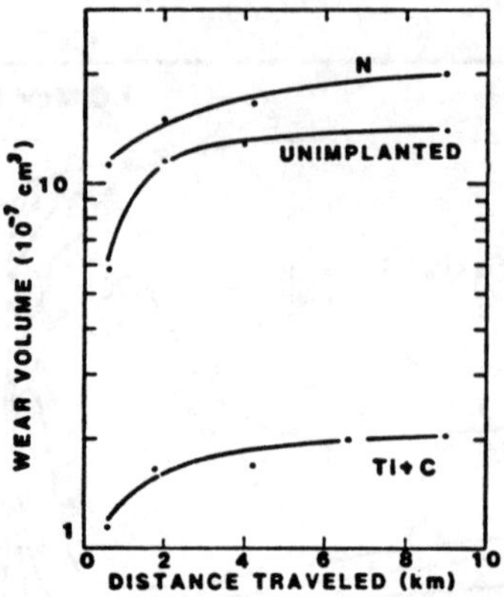

FIGURE 20. Wear experiments using a ball cylinder configuration to investigate the "running-in" portion of the wear regime. Ti-implants of 2 x 10^{17} ions/cm^2 show substantially better performance than unimplanted and nitrogen implanted specimens.

Nevertheless, this material does not exhibit substantial improvements in corrosion resistance in different solvents suitable to test active-passive behaviour and pitting corrosion (1 N H_2 SO_4 and 0.1 N NaCl). This negative result can certainly be explained by a galvanic effect between titanium and iron.

5.1.2 Boron and phosphorus implantations in nickel (76)

The interest of this work is the proof of atomic transfer and diffusion during contact. This is independently of the fact that evidence is given for the correlation between complete amorphization and a better friction resistance combined with an important increase of sliding distance between seizures. Hence, the positive effect of implantation is increased for a thickness considerably larger than the initial ion-range. Some crystallisation of the surface which is related to this diffusion effect, cannot be excluded.

The wear experiments of these two studies use the pin on disk configuration. It is not sure that the success registered for tribological properties by ion implantation can simultaneously be extended to mechanical and electrochemical effects.

5.2 Ion beam mixing

In this case we can illustrate the role played by structural and chemical aspects on the improvement of some properties.

5.2.1 Corrosion (77)

If ion beam mixing should lead to the amorphisation of thick MoNi and TiNi multi-layers, these have to be deposited on metallic substrates in order to ensure that the coating is pin-hole free and less sensitive to larger variations in thickness. Moreover, a certain thickness was dissolved by a corroding aqueous environment before a passivating layer could form.

In table 4 we report data for the critical current density and for the current density for passivation. TiNi amorphous layers exhibit much more resistance to corrosion attack than the polycrystalline multiphased alloys, and superior corrosion resistance properties as compared to NiMo. Such a behaviour can be due to a rapid formation of a titanium oxide passive film, stable up to higher potentials than the molybdenum oxides.

5.2.2 Superconductivity (78)

This property is not related to surface properties (see the Bennemann lecture), but the results obtained for a binary-system with little mutual solubility, Cu-Bi, are interesting for both the mixing process and the performance obtained. The bilayer CuBi which has been mixed at 10 K is amorphous with a superconducting transition below 5 K. This property persists with annealing up to temperatures as high as $\frac{2}{3}$ T_c.

REFERENCES

1. Linhard J, Vibeke Nielsen and Scharff M: Mat. Fys. Medd. Dan. Vid. Selsk. 36/10, 8 (1968).
2. Linhard J. Scharff M and Schiott H E: Mat. Fys. Medd. Dan. Vid. Selsk. 33/14, 1 (1963).
3. Linhard J, Nielsen K O, Scharff M and Thomsen P V: Mat. Fys. Medd. Dan. Vid. Selsk. 33/10, 1 (1963).
4. Schiott H E: Rad. Effects 6, 107 (1970).
5. Kinchin G H and Pease R S: Rep. Prog. Phys. 18, 1 (1955)
6. Lucasson P: in "Fundamental Aspects of Radiation Damage in Metals", Proc. Int. Conf., Gatlinburg, 1978, Edts: M.T. Robinson and F.W. Young.
7. Robinson M T: Phil. Mag. 12, 741 (1965).
8. Robinson M T: Phil. Mag. 17, 639 (1968).
9. Sigmund P: Rad. Eff. 1, 15 (1969).
10. Winterbon K B, Sigmund P and Sanders J B: Mat. Fys. Medd. Dan. Vid. Selsk. 37/14 (1970).
11. Bogh E, Hogild P and Stensgaard I: Rad. Eff. 7, 115 (1971).
12. Noriaki Matsunami: Atomic Data and Nuclear Tables 31, 1 (1984).
13. Sigmund P: Rev. Roum. Phys. 17/8, 969 (1972).
14. Robinson M T and Thorrens I M: Phys. Rev. B 3/12, (1974).
15. Averback R S, Bemdek R and Merkle K L: Phys. Rev. B 18/8,
16. Metastable materials formation by ion implantation, M.R.S. Symposium Proceeding, Vol. 7, S.T. Picraux and W.J. Choyke, Eds., North Holland, New York (1982) .
17. Matteson S and Nicolet M A: Ann. Rev. Mat. Sci. 13, 339 (1983).
18. Nicolet M A, Banwell T C and Paine B M: Mat. Res. Soc. Symp. Proc. 27, Elsevier Science Publishing Co., Inc.(1984).
19. Matteson S: IEEE Transactions on Nuclear Science NS-30/2, (1983).
20. Averback R S, Thompson L J Jr, Moyle J and Schalit: J. Appl. Phys. 53/3 (1982).
21. Andersen H H and Bay H L: Rad. Eff. 19, 139 (1973).
22. Matteson S, Roth J and Nicolet M A: Rad. Eff. 42, 139 (1979).
23. Proceedings of the Workshop and "Ion Mixing and Surface Layer Alloying", April 15 and 16, 1983, California Institute of Technology, Pasadena, California 9112 S, M.A. Nicolet, S.T. Picraux.
24. Roush M L, Goktede O F, Andradis T D and Davarya F: Nucl. Inst. Meth. 194, 64 (1982).
25. Weller R A and Rilurd Weller M: Rad. Eff. 60, 209 (1982).
26. Walker R S and Thompson D A: Rad. Eff. 37, 113 (1978).
27. Sigmund P and Gras-Marti A: Nucl. Instr. Meth. 168, 389 (1980) 182/183, 25 (1981).
28. Andersen H H: Appl. Phys. 18, 131 (1979).
29. Matteson S: Appl. Phys. Lett. 39, 288 (1981).

Matteson S, Paine B M and Nicolet M A: Nucl. Inst. Meth. 182/183, 53 (1981).
30. Sigmund P: Appl. Phys. Lett. 25 (1974).
Sigmund P and Clausen C: J. Appl. Phys. 52 (1981).
31. Vineyard G M: Rad. Eff. 29, 245 (1976).
32. Sigmund P: Appl. Phys. A 30, 43 (1983).
33. Adda Y, Beleyer and Brebec G: Thin Solid Film 25, 107 (1975).
34. Martin G, Cauvin R and Barbu A: In "Phase Transformations during Irradiation", edited by F.V. Nolfi Jr., Pergamon, London (1983).
Martin G and Barbu A: Nucl. Inst. Meth. 209/210, 203 (1983).
35. Haff P K and Switkowski Z E: J. Appl. Phys. 48, 18 (1977).
36. Paine B M: J. Appl. Phys. 53, 6828 (1982).
37. Sigmund P: Appl. Phys. A 30, 46 (1983).
38. Adam J and Dugdale R A: Nature 168, 581 (1951).
39. Rehn L E, Averback R S and Okamoto P R: Invited paper submitted to the International Conference on "Surface Modification of Metals by Ion Beams", University of Heidelberg, September 17-21, 1984.
Okomoto P R and Wiedersich H: J. Nucl. Mater. 53, 336 (1974).
40. Adda Y, Beleyer M and Brebec G: Thin Solid Films 25, 107 (1975).
41. Liu K Y and Wilkes P: J.M.N. 87, 317 (1979).
42. Martin F: Phys. Rev. B 30/3, 1424.
43. Mader S: Thin Solid Films 35, 195 (1976).
44. Egami T and Waseda Y: J. Non-Cryst. Sol. 64, 113 (1984).
45. Naguib H M and Kelly R: Rad. Eff. 25, 1 (1975).
46. Bangwei Z: Physica 121 B, 405 (1983).
47. Hohmuth K and Rauschenbach B: Nucl. Instr. Meth. 209/210, 249 (1983).
48. Nagel S R and Tauc J: Phys. Rev. Lett. 35, 380 (1975).
49. Mogro-Campero A and Walter J: Phys. Lett. 87 A/1-2, 49 (1981).
50. Moruzzi V L: Phys. Rev. B 27, 27 (1983).
51. Hohmuth K and Rauschenbach B: Phys. Stat. Sol. (a) 72, 667 (1982).
52. Liu B X, Johnson W L, Nicolet M A and Lau S S: Appl. Phys. Lett. 42, (1983).
53. Alonso J A and Simozar S: Sol. Stat. Com. 48, 765 (1983).
54. Liu B X, Johnson W L, Nicolet M A and Lau S S: Nucl. Instr. and Meth. 209/210, 229 (1983).
55. Lau S S, Liu B X, and Nicolet M A: Nucl. Instr. and Meth. 209/210, 97 (1983).
56. Miedema A R: Physica 100 B, 1 (1980).
57. Hung L S, Nastasi M, Guylai J and Mayer J M: Appl. Phys. Lett. 42, 672 (1983).
58a. Goltz G, Fernandez R, Nicolet M A and Sadana D K: in "Metastable Formation by Ion Implantation", Eds. S.T. Picraux and W.J. Choyke, M.R.S. Symposium Proc. Vol. 7, North Holland, Amsterdam (1982), p. 227.
58b. Tsaur B Y, Lau S S, Hung L S and Mayer J M: Nucl. Inst. Meth. 182/183, 67 (1981).
58c. Liu B X, Nicolet M A and Lau S S: Phys. Stat Sol. (a) 73, 183 (1982).
59. Tsaur B Y and Mayer J M: Appl. Phys. Lett. 37 (1980).
60. Westendorp H, Zhong Lie Whang and Saris F W: Nucl. Inst. Meth. 194, 453 (1982).
61. Lau S S, Tsaur B Y, Von Allmen M, Mayer J. M, Stritzkev B, Appleton B R and White C W: Nucl. Instr. Meth. 182/183, 97 (1981).
62. Holz M, Zieman P and Buckel W: Phys. Rev. Lett. 51, 1584 (1984).
63. Schmid A and Ziemann P: Nucl. Instr. and Meth. in Phys. Res. B 7/8, 581 (1985).
64. Delafond J: Ann. Chem. (Fr.) 9, 291 (1984).

65. Lawson R P W, Grant W A and Grundy P J: Nucl. Instr. and Meth. 209/210, 243 (1983).
66. Hohmuth K, Rauschenbach B, Kolitsch A and Richter E: Nucl. Instr. and Meth. 209/210, 249 (1983).
67. Moine P, Eymery J P, Gaboriaud R J and Delafond J: Nucl. Instr. and Meth. 209/210, 267 (1983).
68. Knapp J A, Follstaedt D M and Picraux S T: "Ion Implantation Metallurgy," Ed. C.M. Preece and J.K Hirvonen; Am. Inst. of Mining Metallurgy and Petroleum Engineers, Inc. N.Y.(1980), p. 152.
69. Soud D K: Rad. Eff. 63, 141 (1982).
70. Jaouen C, Riviere J P, Bellara A and Delafond J: Nucl. Instr. and Meth. in Phys. Res. B 7/8, 591 (1985).
71. Liu B X: Nucl. Instr. and Meth. in Phys. Res. B 7/8, 547 (1985).
72. Carosella C A, Singer I L, Bowers R C and Gossell C R: Nucl. Instr. and Meth. in Phys. Res. B 64, 103 (1985).
73. Singer I L, Carosella C A and Reed J R: Nucl. Instr. and Meth. 182/183, 923 (1981).
74. Knapp J A, Follstaed D M and Doyle B C: in ref. (68), p. 38.
75. Hubler G K, Trzaskoma P, McCafferty E and Singer I L: in "Ion Implantation for Material Processing", Ed. F.A. Smidt, Noyes Data Corporation (1983), p. 181.
76. Takadoum J, Pivin J P, Chaumont J and Roque Carmes C: J. of Material Sc. 20, 1480 (1985).
77. Bhottadaraya R S, Pronko P P, Rai A K, McCormich A W and Raffoul C: in Ref. (61), p. 694.
78. Averbach R S, Okomoto P R, Baily A C and Stritzker B: in Ref. (61), p. 556.

SURFACE AMORPHOUS ALLOYS OBTAINED BY ION-MIXING

L. Guzmàn, M. Elena, G. Giunta, F. Marchetti, P.M. Ossi,
G. Riontino and V. Zanini

1. INTRODUCTION

The favourable physical and technological properties of amorphous films account for the increasing research efforts towards an understanding of such systems (1).

In the specific case of amorphous alloys, in order to form compounds with a predetermined composition, many techniques are commonly used. Among these, an important role is played by ion-mixing (IM) (2).

In general it is possible to assert that by the IM technique amorphous alloys are formed over virtually unlimited composition ranges, in contrast to other "thermodynamic" techniques (rapid quenching from the liquid or the vapour phase) bounded by certain compositional limitations.

In the metglass family, the transition-metal/metalloid systems exhibiting high metal concentrations are interesting for the combination of unusual mechanical and magnetic properties (3), (4). The systems we consider are Fe-B and Ni-B, which, although apparently similar, have a somewhat different structure as evidenced by neutron scattering studies (5). The aim of the present work was to achieve by ion-mixing the formation of amorphous alloys and to investigate the role of the sample temperature during the ion-implantation with respect to the amorphization degree of the resulting films.

2. EXPERIMENTAL TECHNIQUES, RESULTS AND DISCUSSION

Thin multilayer films (Ni 285 Å, Fe 310 Å and B 50 Å, for a total thickness of about 1000 Å) with predetermined composition ($Fe_{80}B_{20}$) and $Ni_{80}B_{20}$) were obtained by radio frequency sputtering. The samples were then implanted with noble-gas ions of 160 and 180 keV with a dose of 2 10^{16} ions/cm^2 at different temperatures (from room temperature to 170 K).

2.1. Auger Electron Spectroscopy (AES) results.

AES depth profiles show (see Fig. 1 a,b) the distinct succession of the Ni and B (Fe and B) layers in the as evaporated samples. In the interface region, which is markedly sharper in Ni-B than in Fe-B, we find that the AES lineshape of boron - which gives information on its chemical state - indicates that this element remains unchanged in the case of Ni, while it is chemically bound in the case of Fe.

Fe-B MULTILAYER EVAPORATED Fig. 1a

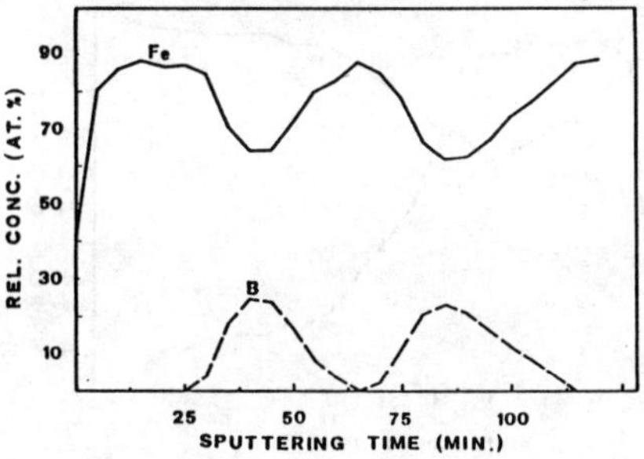

Ni-B MULTILAYER EVAPORATED Fig. 1b

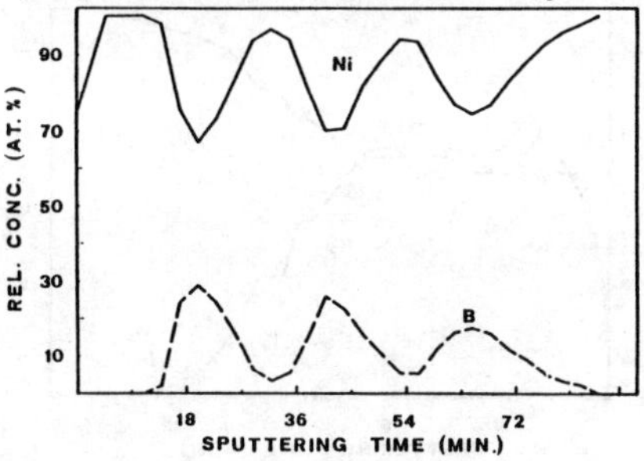

The periodicity of the as evaporated layers, although reduced, remained even after the Ar^+ implantation, indicating that, in this case, the metal-metalloid mixing is rather poor, while in the Kr^+ implanted samples (see Fig.2) we obtain a far better mixing of the whole multilayer. In this latter case, however, a small but significant difference is observed between the Fe-B and Ni-B systems.

Fe-B MULTILAYER Kr^+ IMPLANTED Fig. 2a

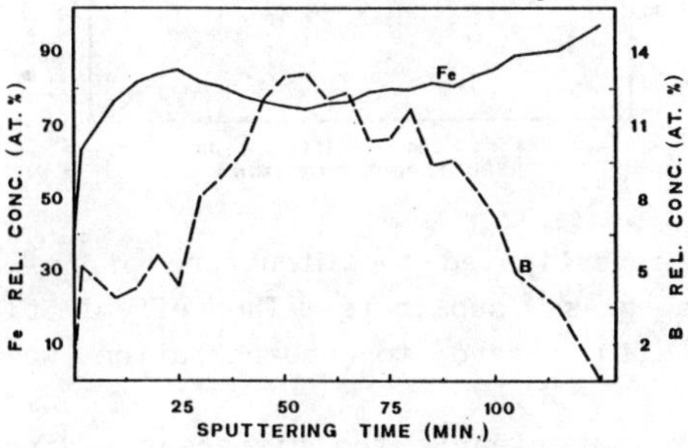

Ni-B MULTILAYER Kr^+ IMPLANTED Fig. 2b

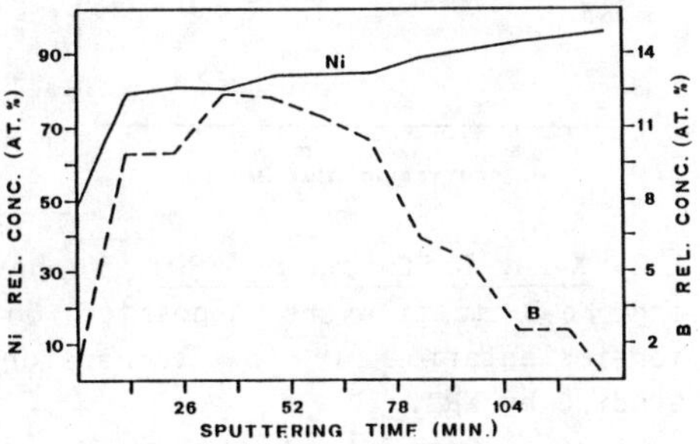

These facts could be connected to the different structure of the metal involved, which also helps explaining the results in Fig. 3, which shows the behaviour of simpler configurations (Fe-B-Fe and Ni-B-Ni) as evaporated, annealed in vacuum at 820 K, or ion-implanted at low temperature (about 170 K). One notices that the mixing (due to ion-implantation or to solid state diffusion) is more effective in the Ni-B system.

Fe-B-Fe EVAPORATED Fig. 3a

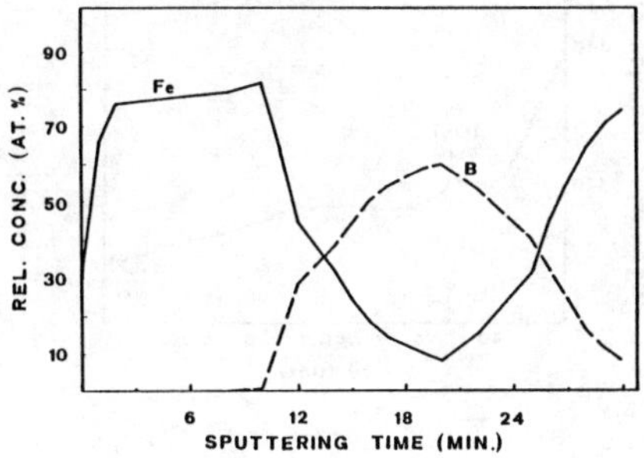

Ni-B-Ni EVAPORATED Fig. 3b

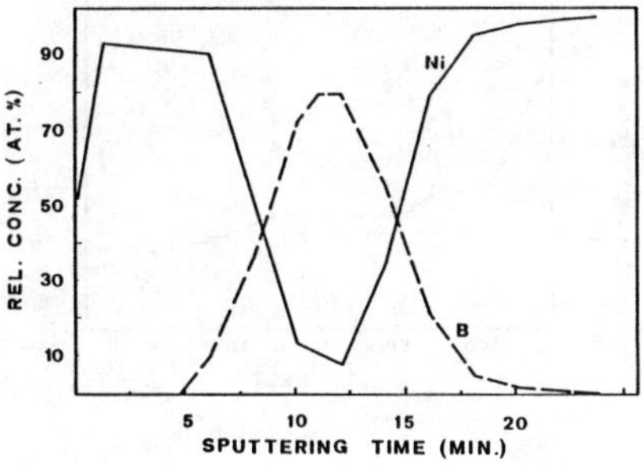

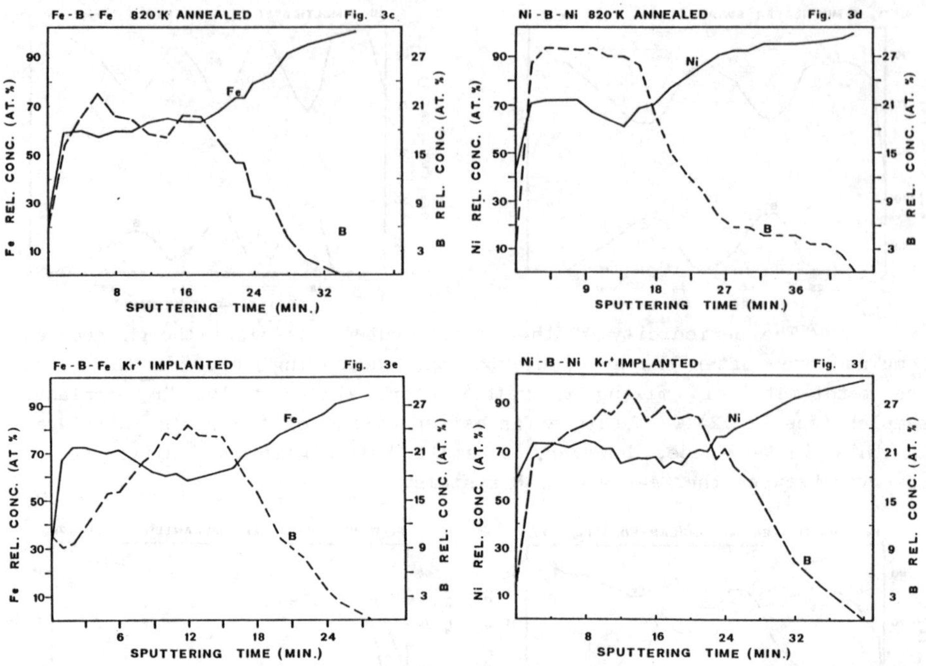

2.2. X-ray Diffraction (XRD). We have investigated the structure of Ni-B and Fe-B multilayers deposited on a glass substrate. The effect of ion-implantation at low temperature with regard to amorphization was studied by XRD.

Our data (see Fig. 4) point out clearly the different mixing behaviour of the two systems. In the case of Ni-B, the spectra show the presence of a Ni-B compound across the whole film thickness, and markedly lower intensity of the Ni (111) signal after the implant.

In the case of Fe-B we notice that there is no peak corresponding to the formation of a Fe-B compound and that the intensity of the Fe (110) signal does not change after ion-implantation.

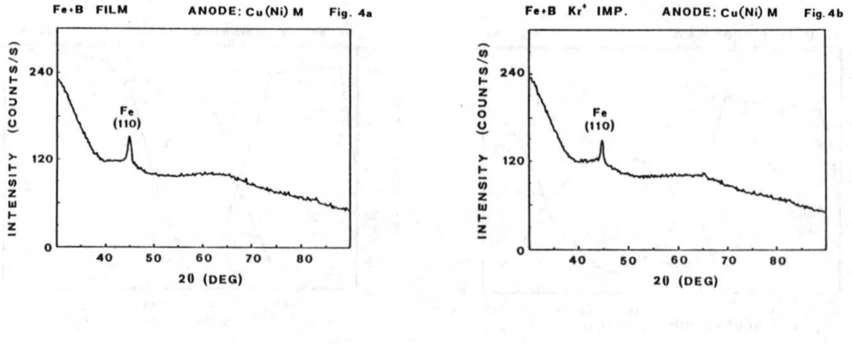

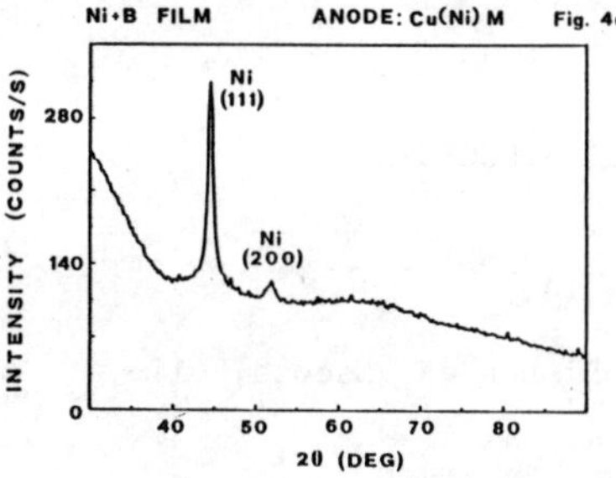

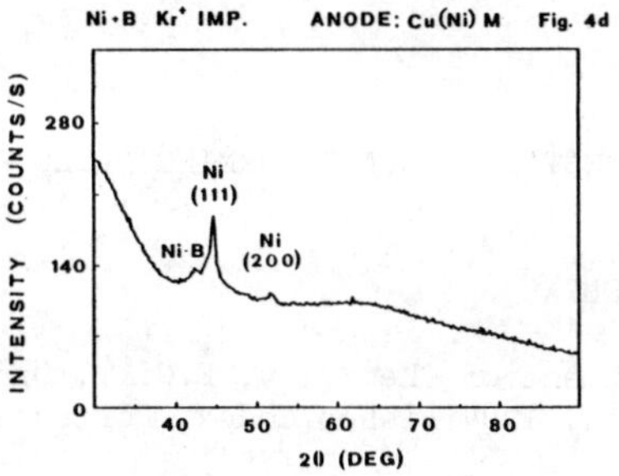

Due to the shallow film thickness (less than 100 nm) and the non-crystalline structure of the substrate, X-ray data do give partial evidence of amorphous phase formation.

3. CONCLUSIONS

Our results show that Fe–B and Ni–B alloys are obtained after ion-mixing. The Fe–B mixing appears to be far less efficient than the Ni–B one, under the same experimental conditions. We interpret this in terms of the different crystalline structure of Fe and Ni, influencing the interdiffusion and, consequently, the interface sharpness.

For the Ni–B system our data show the attainment of a partial amorphization after implantation at about 170 K. For the Fe–B system lower temperatures are needed to obtain the same result. Further research is necessary in order to give a quantitative determination of the resulting degree of amorphization.

REFERENCES.

1. Predel B.: Erzmetall 35 (1982) Nr7/8 350–357.
2. Alonso J.A., Simozar S.: Solid State Comm. 48 No. 9 (1983) 765–767.
3. Sperl W., Lamparter P., Steeb S.: Structure of Metallic Glassy Fe–B–, Co–B– and Ni–B–Alloys by means of X-ray Diffraction, Report No.MPI/84/Wi Max Planck Institut fuer Metallforschung, Stuttgart.
4. Steeb S., Falch S. and Lamparter P.: Zeitschrift fuer Metallkunde 75 (1984) H.8, 599–613.
5. Lamparter P., Sperl W., Steeb S.and Bletry J.: Z.Naturforsch. 37A (1982) 1223.

ORGANOMETALLIC GLASS-FORMING MATERIALS AND THEIR STRUCTURE

A.M. ELIAS

Department of Chemistry, F.C.L., University of Lisbon, R. Escola Politécnica, 1294 Lisbon Codex, Portugal

1. INTRODUCTION

Our knowledge of structure and properties of glasses has enormously advanced in recent years (1-2). However, little is known about the structure of organometallic glass-forming materials.

Associated with the vanishing of the liquid excess entropy occurs the vanishing of the ionic (or molecular) mobility, a tendency which is reflected in the temperature dependence of liquid relaxation rates W. These are described to good precision over several orders of magnitude by the rate law (3),

$$W = A_W \exp [- B_W/(T - T_{OW})]. \quad (1)$$

In addition, the experimental glass transition temperature T_g, permits to understand glass-formation structures in the glass.

In this paper a general approach to the glass-formation of cadmium (II) and 3d ions in organic salts is developed.

2. EXPERIMENTAL PART

α-Picolinium Chloride α-PicHCl, γ-Picolinium Chloride γ-PicHCl and α-Picolinium Fluoride α-PicHF were prepared by neutralizing a quantity of γ-, α-Picoline (Merck pro analysis) with conc. HCl (Mallinckrodt, ACS) and then by fractionally distilling, after which the crude distillation products were sublimed (4,5 and 6). All the transition metal halides were dried at 120°C and stored in dessicators before being used.

Transport properties, DTA and DSC measurements were performed under inert and dry atmospheres (4,5). In the case of fluoride salts a teflon cell and non corrosive materials were used (7).

3. RESULTS AND DISCUSSION

A discussion of 3d-ion spectra in oxide glasses detailing the ligand field theory account of the spectra is contained in an article by Bates (8), whereas a summary of current understanding of coordination numbers in glasses was given by Zarzycki (9). The general feature of the behaviour of these systems is that the particle mobility is temperature dependent, as is seen in Figure 1 and Figure 2. In fact, with the system α-PicHCl + γ-PicHCl (4:1) which is a glass-forming organic mixture, the addition of $CoCl_2$ turns out to be also glass-forming from 0 to 20 mol%. For a larger range of compositions the same behaviour was found in transport properties with systems like $FeCl_3$ and $CrCl_3$ in α-picHCl (4). Also a similar Arrhenius behaviour shows up with $CdCl_2$, $CoCl_2$ and $NiCl_2$ in α-picHCl (10). The change in T_o (ideal glass transition temperature) and T_g is a manifestation of the variation of the cohesive energy density of a liquid and hence reflects the effects of changing composition on the nature and strength of

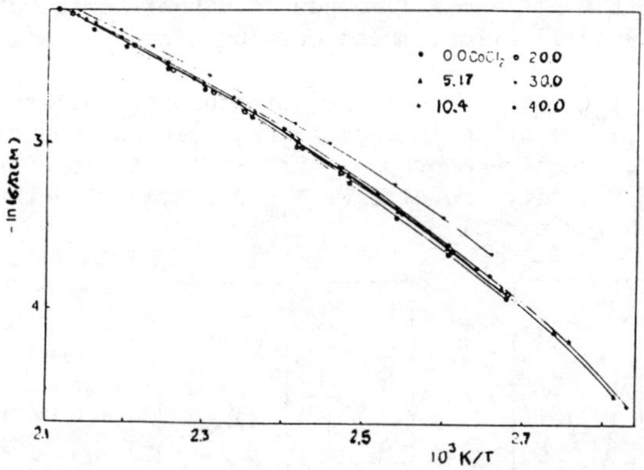

Fig. 1: Arrhenius plot of fluidity for α-picHCl + γ-picHCl + $CoCl_2$ system.

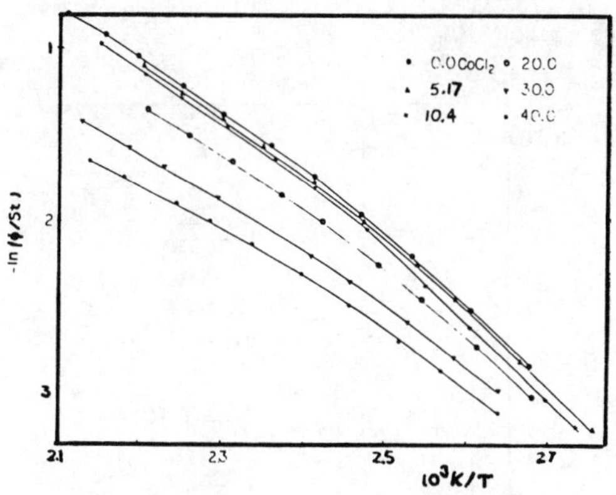

Fig. 2: Arrhenius plot of fluidity for α-picHCl + γ-picHCl + $CoCl_2$ system.

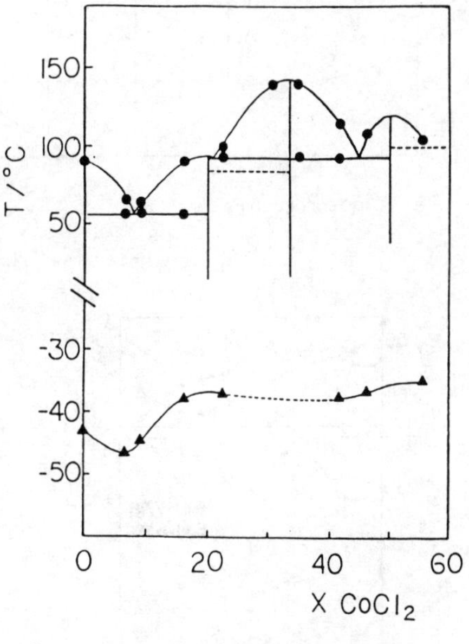

Fig. 3: Partial phase diagram and T_g values for the α-picHCl + $CoCl_4$ system.

interparticle interactions. Hence, details of the partial diagrams for $FeCl_3$ + α-picHCl, $CrCl_3$ + α-picHCl and $CoCl_2$ + α-picHCl are shown together with the plots of T_g. It is significant that in cases where the congruent melting compound corresponds to the 33.3 mol% $[CoCl_4]^{2-}$ and $[FeCl_4]^-$, the glass transition temperature values decrease or do not increase, see Figure 3 and Figure 4. The same is true for $[NiCl_4]^{2-}$ and $[CdCl_4]^{2-}$ (10). But, when the composition corresponding to the congruent melting compound is 25.0 mol% (which is the stoichiometry of the complex species $[CrCl_6]^{3-}$, Figure 4) the T_g values increase quite sharply. In the case of $[CoCl_4]^{2-}$

and $[FeCl_4]^-$, even the 1:2 and 1:1 compounds cannot be quenched to glasses without partial crystallization. These compounds have relatively low melting points.

The minima in $T_{o\sigma}$ and $T_{o\phi}$ near the tetrachlorocobaltate composition seem significant, Figure 5. We interpret these minima as a consequence of a decrease in Coulomb cohesive energy as the electrostatic charge density decreases. Maxima in fluidity are also well correlated with the decrease in T_g values for $FeCl_3$ + α-picHCl.

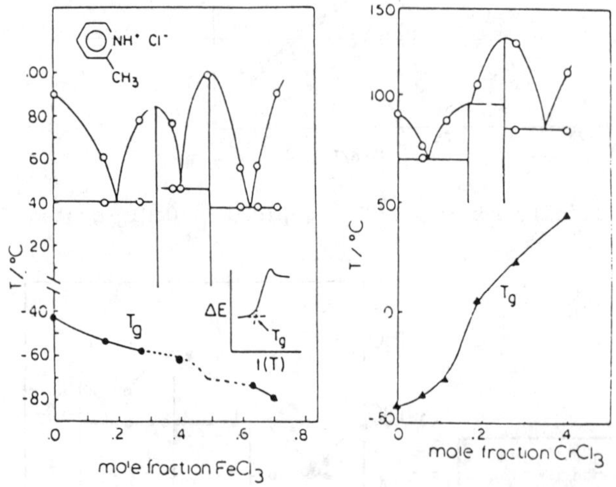

Fig. 4: Phase diagrams and glass transition temperatures for the binary systems α-picHCl+$FeCl_3$ and α-picHCl + $CrCl_3$. The insert shows the definition of T_g by DTA upscans. T_g is the starting point for the increase in heat capacity due to unlocking of "frozen" structures (4).

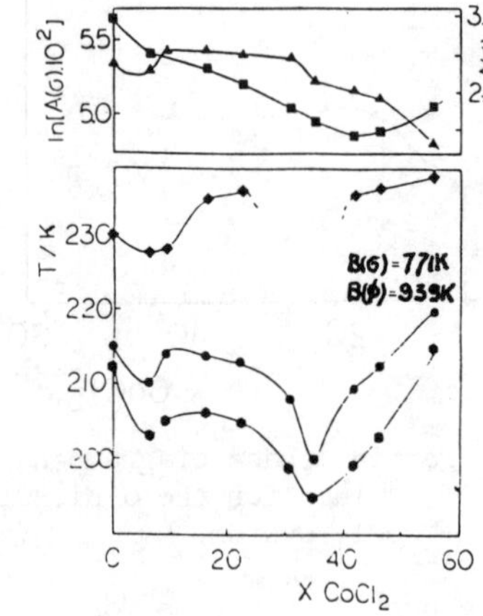

Fig. 5: Effect of composition on the parameters of equ. 1 and on T_g values for the system α-picHCl + $CoCl_2$. ◆:T_g, ●:$T_{o\sigma}$, ◆:$T_{o\phi}$, ■: $A\phi$, ▲:A_σ .

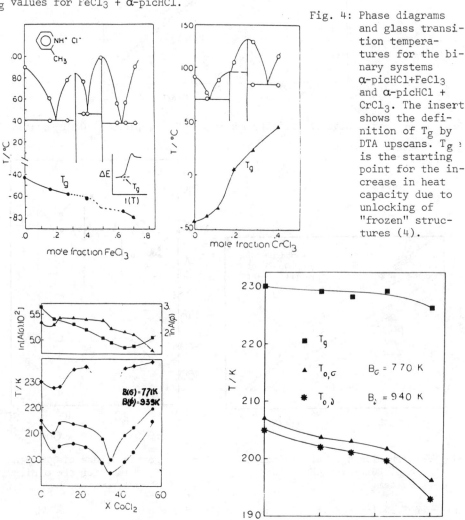

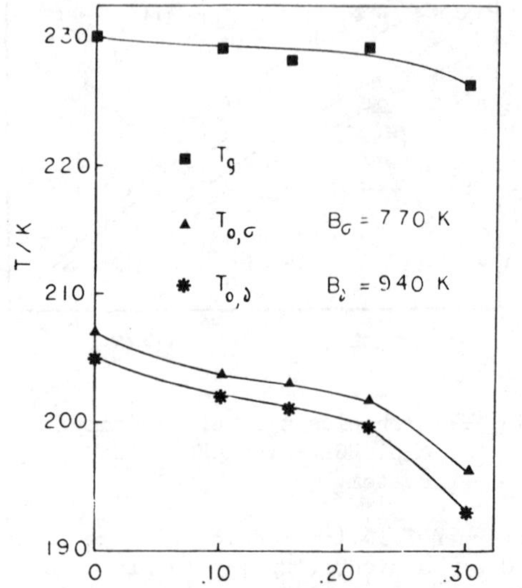

Fig. 6: Effect of composition on the parameters of equ. 1 and on the T_g values for the system α-picHCl + γ-picHCl + $NiCl_2$.

The behaviour of A_σ and A_ϕ for $CoCl_2$ + α-picHCl is shown in Figure 5. The addition of γ-picHCl (which is not glass-forming) to α-picHCl produces a glass with a T_g-value of 231 K, see Figures 6 and 7, for a composition ratio of 1:4. The effect of adding 4-methylpyridine-HCl to the organometallic glass-forming systems, $NiCl_2$ and $CoCl_2$ in α-picHCl is clear in those figures. In fact, the configurational entropy of both systems increases

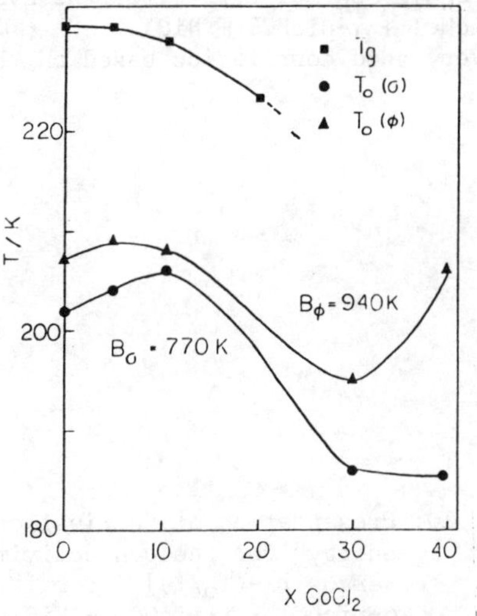

Fig. 7: T_g-, $T_{o,\sigma}$- and $T_{o,\phi}$-values for α-picHCl + γ-picHCl + $CoCl_2$ system.

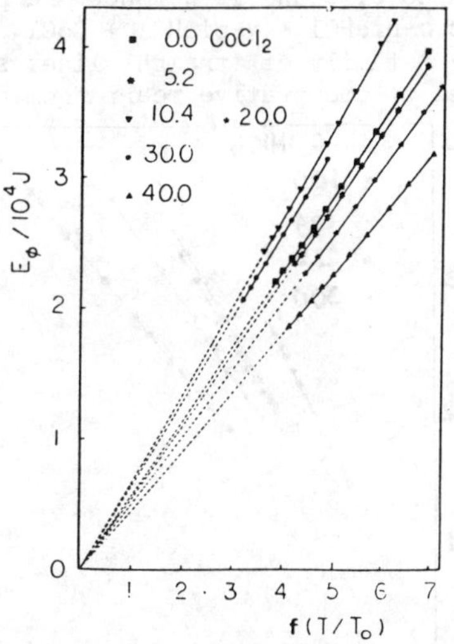

Fig. 8: Presentation of the "activation energy" for fluidities E_ϕ versus $[1-(T_o/T)]^{-2}$ for α-picHCl + γ-picHCl + $CoCl_2$.

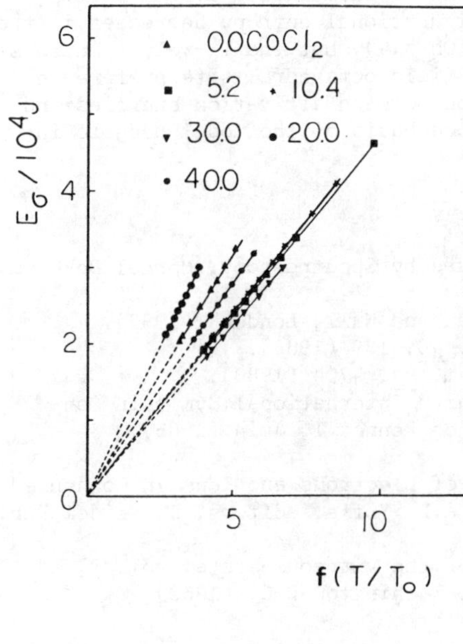

Fig. 9: Presentation of the "activation energy" for the conductivity versus $[1-(T_o/T)]^{-2}$ for α-picHCl + γ-picHCl + $CoCl_2$.

till near 33 mol%, after which for the Co (II) containing system the value of S_c decrease. And again the T_d-structure is found responsible for this behaviour.

Finally we found (11) a very striking behaviour of the "activation energy" for both mechanisms E_σ and E_ϕ as a function of $T \cdot T_o^{-1}$ (the inverse of ideal glass transition temperature times the actual temperature). In the Figures 8, 9 and 10 we show the plots of E_W versus $f(T \cdot T_o^{-1})$ for the systems α-picHCl + γ-picHCl + $CoCl_2$ and α-picHCl + γ-picHCl + $NiCl_2$.

A family of straight lines shows a very good correlation based on the model of cooperative rearrangements.

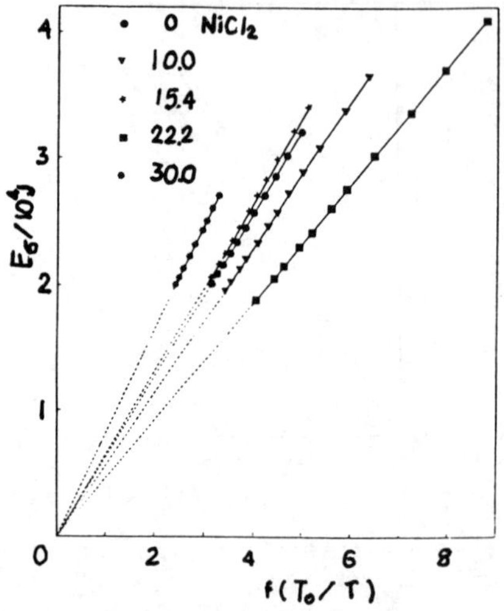

Fig. 10: Presentation of "activation energy" for the conductivity versus $[1-(T_o/T)]^{-2}$ for α-picHCl + γ-picHCl + $NiCl_2$.

4. CONCLUSIONS

We may say as a conclusion that in an organometallic system the T_g-values or, put in another way, the configurational entropy decreases if the prevailing species are structurally T_d (CN = 4), but the T_g-values increase if the structure has the highest ligand field octahedral site preference energies and CN = 6. A simple correlation between activation energies and (T/T_o) is always found confirming the good basis of the model used to interpret the transport data.

REFERENCES

1. Wong J and Angell C A: Glass Structure by Spectroscopy, Marcel Dekker, New York (1978).
2. Paul A: Chemistry of Glasses, Chapman and Hall, London (1982).
3. Adam G and Gibbs J H: J. Chem. Phys. 43, 139 (1965).
4. Angell C A and Elias A: J. Phys. Chem. 87, 4704 (1983).
5. Elias M E and Elias A M: Proc. of Third International Symposium on Halides Glasses, Part I, Université de Rennes I, Juin (1985).
6. Elias A M: to be published.
7. Elias A M: in Physics and Chemistry of Electrons and Ions in Condensed Matter, J.V. Acrivos, N.F. Mott and A.D. Yoffe, editors, D. Reidel Pub. Co., Boston, p. 435 (1984).
8. Bates T: pp. 195 in Modern Aspects of the vitreous state, vol. 2, ed. by J.D. Mackenzie, Butterworth Inc., Washington D.C. (1962).

9. Zarzycki J: pp. 525 in Physics of non-crystalline solids. Ed. by J.A. Prins. North Holland Publishing Co., Amsterdam (1965).
10. Elias A and Angell C A: to be published.
11. a) Elias A: Proc.9th Annual National Meeting of Port. Chem.Soc., C42, Braga (1985);
b) Elias A and Elias M E: in press J. Phys..

BULK AMORPHOUS METALS BY SOLID STATE REACTION

L. SCHULTZ
Siemens AG, Research Laboratories, D-8520 Erlangen, FRG

1. INTRODUCTION

Amorphous metals are usually produced by rapid quenching from the melt. In order to achieve the necessary high cooling rate, at least one dimension of the sample must be quite small In the case of melt spinning the ribbon thickness is rarely more than 50 micron. Expressed in thermodynamic terms rapid quenching from the melt is a fast change of a state variable. So it should also be possible to form the amorphous state by fast changing another state variable where the change of this variable must be fast enough to prevent the system from achieving the thermodynamical equilibrium. By changing the pressure quite fast, Mishima et al. /1/ succeeded in forming amorphous ice. Schwarz and Johnson /2/ showed that also by a "rapid" change of the alloy composition, i.e. by an interdiffusion reaction, the formation of an amorphous phase is possible. For these experiments a multilayer structure of two crystalline metals (Au and La) was produced by evaporation. A subsequent annealing treatment at temperatures well below the crystallization temperature led to the formation of the amorphous alloy. Since then a number of other binary transition metal alloys has been found which show the same reaction. (For a recent review see the paper by von Minnigerode et al. /3/ in this volume.)

Since this solid state reaction does not involve rapid quenching, it is possible by this process to produce bulk amorphous alloys of basically any shape and size if one can prepare a very fine composite of the two crystalline components by mechanical deformation /4-6/. In this paper we discuss different techniques to produce these composites, present a method how to check fast if a composite undergoes this amorphization reaction, and show some results on the reaction process.

2. PREPARATION TECHNIQUES

As starting materials for the preparation of fine composites one can either use thin foils, fine powders, thin wires or rods and tubes of the pure elements. Due to the kinetics of the diffusion reaction, at least one dimension D of the components of the composite should be in the range of 0.05 to 0.5 micron. The deformation ratio Q necessary to prepare the fine composite from starting material with the smallest dimension D_O must then be at least $Q = D_O/D$. In order to get along with reasonable deformation work, D_O should not exceed about 25 micron which would require a Q of 250 to achieve a 0.1 micron composite. From these considerations the rod and tube technique as known from the production of multifilamentary superconductors seems not to be practical, since the initial slab diameter is in the millimeter range. Also the production, handling and compaction of 25 micron wires seems not to be cost-effective. On the other hand, 25 micron foils and powders are commercially available. They are both easy to handle and to compact.

Thin foils have the additional advantage that they have a low amount of impurities, a homogeneous thickness and that their geometry results in a layered microstructure after compaction. They have been used to prepare wire-shaped /4,5/, sheet-shaped /4,6/ and tube-shaped composites. For the preparation of wire-shaped composites the 25 micron thick elemental foils are alternatively wound up to form a spiral structure. The compound is deformed in a steel jacket by swaging and wire drawing corresponding to a layer thickness reduction of about a factor of 25. After chemical etching of the steel jacket the wires are bundled 91 or 137 times and deformed again in a steel jacket which is finally removed. The final diameter of these wire-shaped composites is typically about 1 millimeter, but basically unlimited. By this technique we produced NiZr /4/, NiTi, FeZr, FeTi and NiNb composites. The resulting microstructure is rather homogeneous as can be seen in Fig. 2a for a NiZr composite where the individual layer thickness is about 0.15 micron. As discussed later, annealing of these composites results in complete amorphization for NiZr /4,5/, almost complete amorphization with beginning crystallization in NiTi and (so far) only partial amorphization in FeZr, FeTi and NiNb. These 1 mm composite wires are produced in length of several meters. They are easy to handle and can be used for electrical resistivity-, magnetic property- or DSC - measurements before or after annealing treatment.

In order to produce composite sheets /4/, the elemental foils are wound in a flat shape. The compound is deformed only by rolling in a steel jacket, which is later removed chemically. To achieve the necessary deformation ratio at a reasonable sample thickness the compound is piled up several times in a steel container and deformed again by rolling resulting in a rather inhomogeneously deformed composite. The layer thickness is in the range of 0.1 to even 1 micron. Independently from our experiments, Atzmon et al. /6/ used a similar technique. They wound two annealed foils of about 10 micron thickness in a spiral form and rolled them in a steel jacket. After removing the compound from the container it was processed by folding the compound so that its total thickness is doubled and by rolling it between stainless-steel plates. A very high deformation ratio is easily obtained by repeating this procedure 10 - 12 times. But the deformation without a steel jacket results in partly damaged sheets. In both experiments /4,6/ complete amorphization was neither achieved for NiZr /4,6/ nor CuZr /6/ sheets. Under the same conditions (and even on annealing together) where the wire composites did not show any crystalline intermetallic phases the composite sheets partly reacted to intermetallic phases (for example at 300 °C; see /4/). Atzmon and coworkers /6/ reduced the reaction temperature for NiZr to 260 °C resulting either in remaining elemental cystalline Ni and Zr or in new crystalline intermetallic phases. The same happened for CuZr /6/. We conclude that either the inhomogeneous deformation of the elemental layers, i.e. a size effect, or an unfavorable texture introduced by rolling is responsible.

For the preparation of tube-shaped amorphous metals the elemental foils are folded in a special self-supporting way. The compound is deformed by rolling it to a thin foil.Then it is wound on an inner steel tube. After covering it by an outer steel tube, it is deformed by tube swaging. This process again leads to rather homogeneous and very thin elemental layers which show a similar texture as the wires. This technique has been successfully applied for the preparation of amorphous NiZr tubes.

Instead of foils we also used elemental powders to prepare a fine composite of two elemental composites. The main problem in this technique results from the fact that by cold consolidation - unless they are specially treated - the powders are poorly bonded. On deformation they are, therefore, only shifted one besides the other and not deformed. So a much stronger deformation than for the foils is needed to get to a fine composite. Atzmon et al. /6/ also report on this technique but without detailed results. A much more promising technique is to use first a powder milling step which can produce a finely layered mixed powder /7/, which is later consolidated and further deformed to a fine composite. Details on this process will be reported elsewhere /8/. If powder milling is continued for longer times a true alloying occurs which also can result in the formation of amorphous powders as shown by Koch et al. /9/ for NiNb. Also a number of Zr-transition metal alloys show this reaction /7/.

3. INTERDIFFUSION REACTION

The measurement of the electrical resistivity versus temperature for constant heating rate is quite useful for the investigation of solid state amorphization in layered compounds. This is shown in fig. 1 (2.5 K/min) for a $Ni_{68}Zr_{32}$ composite wire with about 0.1 micron thick elemental layers. Due to the relative magnitudes of the specific resistivities of Ni and Zr the composite mainly shows the behavior of the Ni layers. At low temperatures the resistivity increases due to the temperature coefficient of Ni. At about 250 °C the diffusion of the Ni atoms into the Zr layers starts (Ni is a fast diffuser in Zr). Now, the resistivity increases sharply corresponding to the Ni layer reduction. This process is completed for this composite at about 420 °C. The resistivity remains constant up to about 510 °C where a sharp drop occurs. This drop indicates (or at least gives a strong hint for) a crystallization reaction. So it is quite obvious that during the interdiffusion reaction an amorphization by solid state reaction has occured. In the NiZr case a second crystallization step follows at about 590 °C. So

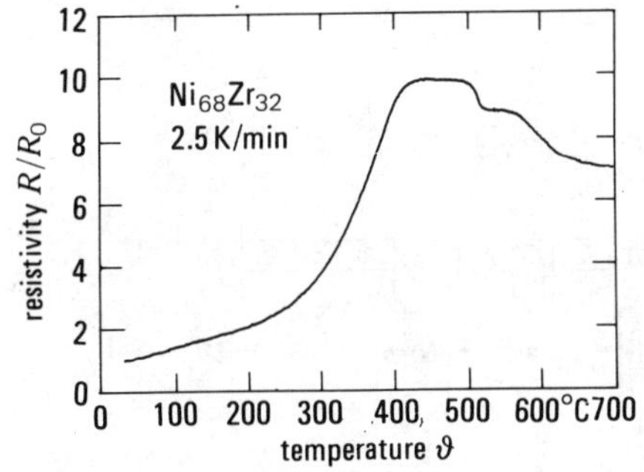

Fig. 1: Reduced resistivity versus temperature at a heating rate of 2.5 K/min for a $Ni_{68}Zr_{32}$ composite wire showing the diffusion and the crystallization reaction.

the measurement of electrical resistivity versus temperature at constant heating rate provides a quick check if a composite is a candidate for solid state amorphization. Of course, x-ray investigations should follow. Fig. 1 also emphazises that the temperature range between about 250 °C and 350 °C is important for the isothermal reaction of $Ni_{68}Zr_{32}$ composites. It should be used preferentially to produce bulk amorphous material.

The isothermal reaction is shown in detail for a $Ni_{68}Zr_{32}$ composite by SEM pictures (fig. 2) and x-ray diffraction patterns (fig. 3) for samples in the as-deformed state and after 1, 17 and 153 hours annealing at 300 °C. Fig. 4 gives a schematic description of this reaction when a surplus of Ni is

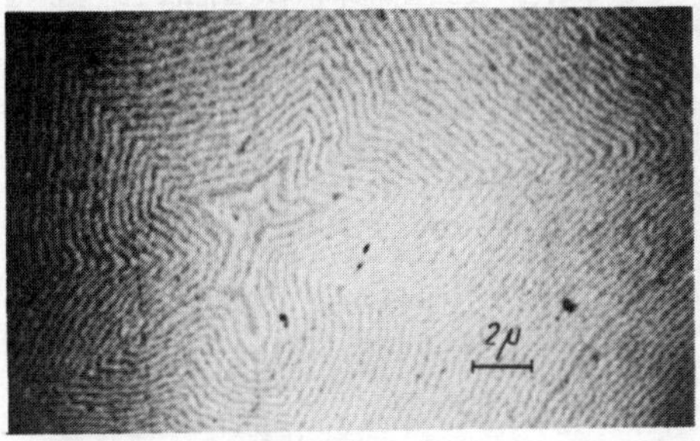

Fig. 2a: unreacted state

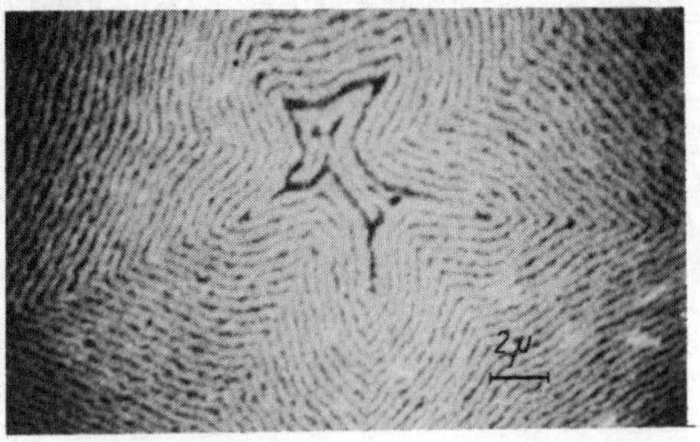

Fig. 2b: 1 h

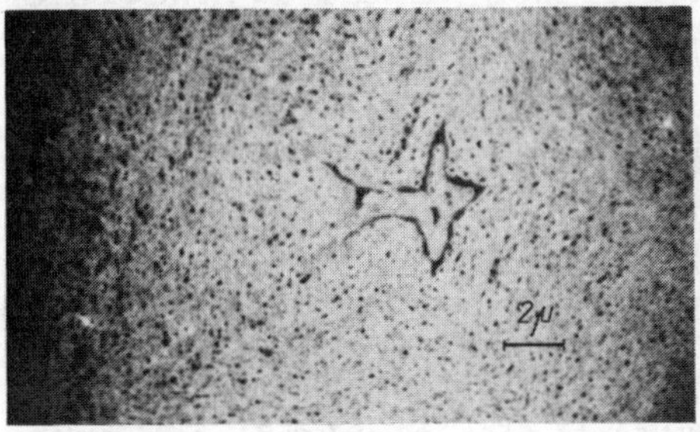

Fig. 2c: 17 h

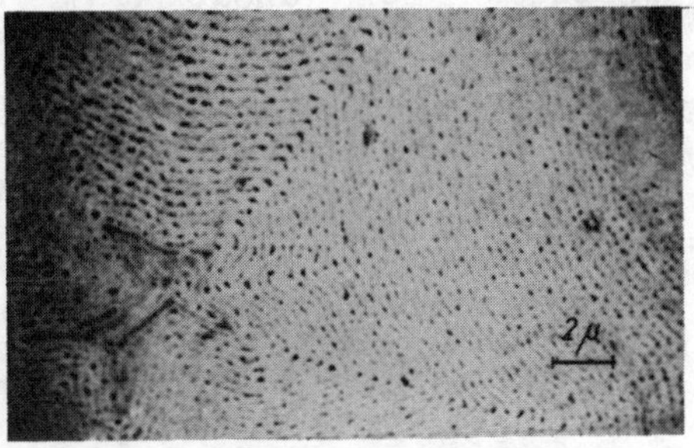

Fig. 2d: 153 h; amorphous sample with Kirkendall voids

Fig. 2: SEM pictures of a $Ni_{68}Zr_{32}$ composite wire in the unreacted state and after annealing at 300 °C.

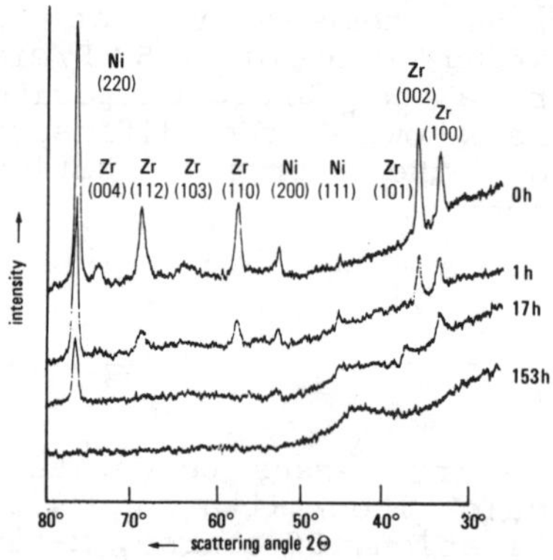

Fig. 3: X-ray diffraction patterns of unreacted and reacted (300 °C) $Ni_{68}Zr_{32}$ composite wires.

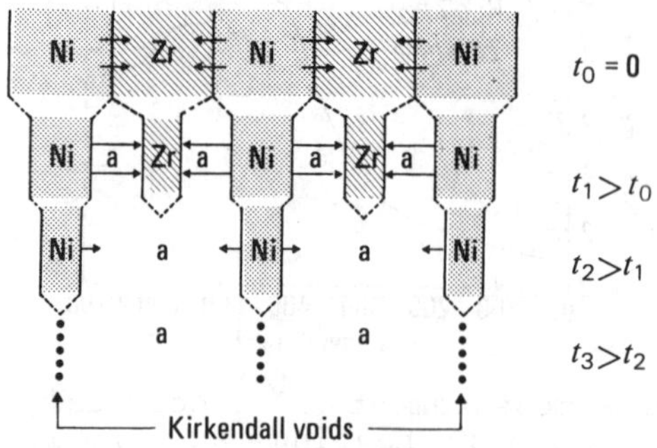

Fig. 4: Schematic description of the solid state reaction in NiZr composites for the case of a Ni surplus.

present. The diffraction pattern of the unannealed sample shows the Ni and the Zr intensity peaks due to the elemental crystalline layers (see fig. 2a). The deformation texture leads to the absence or to a weak intensity of some of the peaks, as, for example, the Zr-(101)- or the Ni-(111)-peaks. At the beginning of the reaction the Ni diffuses into the Zr layers forming amorphous NiZr at the interface. Later (fig. 4, time t_1) the Ni atoms have to cross the amorphous phase in order to react with the Zr at the interface between Zr and the amorphous phase. This has also been proved recently by Cheng et al. /10/ by marker experiments in evaporated NiZr sandwiches. At this temperature the Zr atoms are immobile. The amorphous phase grows into the Zr layers, whereas the Ni/amorphous phase interface is not changed but shifted by the disappearing Ni atoms as shown in fig. 4. The SEM picture shows a reduced width of the dark phase (Ni) after 1 hour annealing (fig. 2b). The bright phase comprises the crystalline Zr (about 1/5 of the layer thickness) and the amorphous NiZr (about 4/5). On further annealing the layer structure "Ni/amorphous/Zr/amorphous/Ni" changes to "Ni/amorphous/Ni" when all the Zr has been reacted (fig. 4, time t_2). Finally the residual Ni diffuses into the amorphous phase changing there the concentration and producing Kirkendall voids embedded in the amorphous material. The sequence of the void formation is shown in the SEM pictures. After 17 hours annealing (fig. 2c) the original microstructure is no more recognizable. In the amorphous NiZr there is some remaining Ni (grey spots). Some voids (dark spots) are also visible. After 153 h (fig. 2d) the Kirkendall voids are regularly arrange at the former positions of the Ni layers. The reaction is completed as is shown by the diffraction pattern (fig. 3). A diffuse intensity maximum of the amorphous phase is observed and crystalline peaks are absent.

4. CRYSTALLIZATION

As mentioned before, the existence of an amorphous phase can also be proved by the transition to the crystalline state. Fig. 5 shows the differential scanning calorimetric (DSC) trace of a $Ni_{68}Zr_{32}$ sample which has been amorphized by a 16 h heat treatment at 350 °C. Contrary to an unreacted composite sample /5/ there is no evidence for any diffusional reaction. At the chosen heating rate of 40 K/min the crystallization starts at 550 °C. The irregularity at 490 °C might be interpreted as the glass transition.

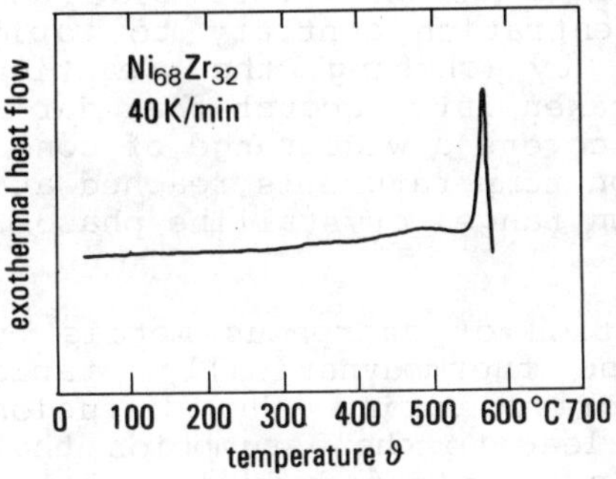

Fig. 5: DSC trace of an amorphous $Ni_{68}Zr_{32}$ wire.

5. BULK AMORPHIZATION IN OTHER ALLOYS

Besides NiZr we applied the foil winding technique also to produce NiTi, FeZr, FeTi and NiNb composite wires. In these samples complete amorphization has not been achieved so far which might be due to too thick elemental layers, to an annealing treatment not yet optimized or to thermodynanic reasons. In $Ni_{62}Ti_{38}$ with about 0.15 micron elemental layers a small amount of intermetallic, crystalline phase shows up on annealing at 325 °C besides two diffuse peaks which have to be attributed to phase separation in the amorphous state. Hence, two amorphous phases are in a metastable equilibrium to each other. Details will be published elsewhere. For $Fe_{67}Zr_{33}$ and $Fe_{60}Ti_{40}$ with about 0.2 micron elemental layers the resistivity versus temperature plot at constant heating rate (10 K/min) shows that the diffusion reaction starts at about 500 °C. The crystallization of the part amorphized before begins at 720 °C or at 700 °C, respectively. These surprisingly high crystallization temperatures in a concentration range which is not accessible by rapid quenching corresponds quite well (in the case of FeZr) to recent results of mechanically alloyed amorphous powders /8/. For FeZr, isothermal annealing at 600 °C leads both to the formation of amorphous and of intermetallic crystalline phases. The formation of crystalline Ni_3Nb besides the amorphous phase could not be avoided in $Ni_{60}Nb_{40}$ even on isothermal annealing at 400 °C. It must be concluded that the formation of the Ni_3Nb phase is more favourable when compared to the amorphous phases in the other cases. More detailed investigations have to be performed.

6. DISCUSSION

These results show that it is possible to produce bulk amorphous metals by solid state reactions in mechanically prepared composites. Although complete amorphization has been achieved so far only for NiZr wires there are strong hints that this techniques enables also the production of bulk amorphous samles in a large number of other alloy systems. This is also documented by the fact that by mechanical alloying, which seems to be a related process, amorphous powders of a large number of alloys have been produced /3,7,9/. The concentration range where the process works seems to be rather wide and independent of any eutectic concentration contrary to rapid quenching. On the other hand, by chosing the reaction temperature properly, it must be taken into account that during formation the amorphous phase covers a wide range of concentrations. If the crystallization temperature is reached at one point within this concentration range, crystalline phases will form.

The main question in the formation of amorphous metals by solid state reaction is, why the thermodynamically stable intermetallic phases are not formed during the diffusion reaction. The experiments so far lead to the assumption that their nucleation is hindered by kinetic effects which might be due to interfacial energies, stress effects or, as recent results on NiZr wires have shown, to a diffusion-related growth effect. Details will be discussed elsewhere.

ACKNOWLEDGEMENTS

The author is grateful to F. Gaube, Y. Uzel, P. Förster and B. Holzapfel for technical assistance and to K. Wohlleben and E. Hellstern for stimulating discussions.

REFERENCES

1. O. Mishima, L.D. Calvert and E. Whalley, Nature 310 (1984) 393.
2. R.B. Schwarz and W.L. Johnson, Phys. Rev. Letters. 51 (1983) 415.
3. G. von Minnigerode, K. Samwer and A. Regenbrecht, NATO ASI "Amorphous and Liquid Materials", Passo della Mendola, 1985 (this volume).
4. L. Schultz, Proc. MRS Europe Meeting on "Amorphous Metals and Non-Equilibrium Processing", Strasbourg 1984, ed. M.v. Allmen, p. 135.
5. L. Schultz, Proc. 5th Int. Conf. on Rapidly Quenched Metals, Würzburg 1984, p. 1585.
6. M. Atzmon, J.D. Verhoeven, E.D. Gibson and W.L. Johnson, Proc. 5th Int. Conf. on Rapidly Quenched Metals, Würzburg 1984, p. 1561
7. E. Hellstern and L. Schultz, Appl. Phys. Letters 48 (1986) (in print)
8. E. Hellstern and L. Schultz, to be published
9. C.C. Koch, O.B. Calvin, C.G. McKamey and J.D. Scarbrough, Appl. Phys. Letters 43 1983) 1017.
10. Y.T. Cheng, W.L. Johnson and M.A. Nicolet, Appl. Phys. Letters, 47 (1985) 800.

Subject Index